Lecture Notes in Computer Sc

Commenced Publication in 1973
Founding and Former Series Editors:
Gerhard Goos, Juris Hartmanis, and Jan van Leeuwen

Editorial Board

Jacky Akoka   Stephen W. Liddle
Il-Yeol Song   Michela Bertolotto
Isabelle Comyn-Wattiau   Samira Si-Saïd Cherfi
Willem-Jan van den Heuvel Bernhard Thalheim
Manuel Kolp   Paolo Bresciani   Juan Trujillo
Christian Kop   Heinrich C. Mayr (Eds.)

Perspectives in Conceptual Modeling

ER 2005 Workshops CAOIS, BP-UML
CoMoGIS, eCOMO, and QoIS
Klagenfurt, Austria, October 24-28, 2005
Proceedings

 Springer

Volume Editors

Jacky Akoka E-mail: Jacky.Akoka@int-evry.fr

Stephen W. Liddle E-mail: liddle@byu.edu

Il-Yeol Song E-mail: songiy@drexel.edu

Michela Bertolotto E-mail: michela.bertolotto@ucd.ie

Isabelle Comyn-Wattiau E-mail: isabelle.wattiau@cnam.fr

Samira Si-Saïd Cherfi E-mail: sisaid@cnam.fr

Willem-Jan van den Heuvel E-mail: W.J.A.M..vdnHeuvel@kub.nl

Bernhard Thalheim Thalheim@is.informatik.uni-kiel.de

Manuel Kolp E-mail: kolp@isys.ucl.ac.be

Paolo Bresciani E-mail: bresciani@itc.it

Juan Trujillo E-mail: jtrujillo@dlsi.ua.es

Christian Kop
Heinrich C. Mayr
E-mail: {christian.kop, heinrich.mayr}@ifit.uni-klu.ac.at

Library of Congress Control Number: 2005934274

CR Subject Classification (1998): H.2, H.4, H.3, F.4.1, D.2, C.2.4, I.2, J.1

ISSN   0302-9743
ISBN-10   3-540-29395-7 Springer Berlin Heidelberg New York
ISBN-13   978-3-540-29395-8 Springer Berlin Heidelberg New York

Springer is a part of Springer Science+Business Media

springeronline.com

© Springer-Verlag Berlin Heidelberg 2005
Printed in Germany

Typesetting: Camera-ready by author, data conversion by Scientific Publishing Services, Chennai, India
Printed on acid-free paper   SPIN: 11568346   06/3142   5 4 3 2 1 0

Preface

We are pleased to present the proceedings of the workshops held in conjunction with ER 2005, the 24th International Conference on Conceptual Modeling.

The objective of these workshops was to extend the spectrum of the main conference by giving participants an opportunity to present and discuss emerging hot topics related to conceptual modeling and to add new perspectives to this key mechanism for understanding and representing organizations, including the new "virtual" e-environments and the information systems that support them.

To meet this objective, we selected 5 workshops:

- AOIS 2005: 7th International Bi-conference Workshop on Agent-Oriented Information Systems
- BP-UML 2005: 1st International Workshop on Best Practices of UML
- CoMoGIS 2005: 2nd International Workshop on Conceptual Modeling for Geographic Information Systems
- eCOMO 2005: 6th International Workshop on Conceptual Modeling Approaches for E-business
- QoIS 2005: 1st International Workshop on Quality of Information Systems

These 5 workshops attracted 18, 27, 31, 9, and 17 papers, respectively. Following the ER workshop philosophy, program committees selected contributions on the basis of strong peer reviews in order to maintain a high standard for accepted papers. The committees accepted 8, 9, 12, 4, and 7 papers, for acceptance rates of 44%, 33%, 39%, 44%, and 41%, respectively. In total, 40 workshop papers were selected out of 102 submissions with a weighted average acceptance rate of 40%.

Together with three invited main-conference and two invited workshop keynote speeches, a demo and poster session, and a concluding panel discussion, ER 2005 featured 14 technical conference sessions, 15 technical workshop sessions and 7 up-to-date tutorials presented by outstanding experts in their fields. We were enthusiastic about the quality of this year's program in all its particulars.

These proceedings contain the selected workshop papers and the tutorial abstracts. Numerous people deserve appreciation and recognition for their contribution to making ER 2005 a success. First of all we have to thank the keynote speakers and the authors for their valuable contributions. Similarly, we thank the workshop chairs and those who organized the various tracks for their effectiveness, and particularly the members of the program committees and the additional reviewers, who spent much time assessing submitted papers and participating in the program discussions on acceptance or rejection. Special appreciation is due to Christian Kop and our student Peter Jelitsch, who assembled these proceedings and who had to adapt almost all the papers to conform to the LNCS layout rules. Likewise we are grateful for the engagement and enthusiasm of all members of the organization team, who gave their best to make ER 2005 an unforgettable event. Last but not least we thank our sponsors and supporters, in particular the University of Klagenfurt, the Governor of Carinthia, and the Mayor of Klagenfurt, who helped us to organize a high-level event at comparably low cost.

The workshop and tutorial co-chairs also express our deep gratitude and respect for the ER 2005 General Chair, Heinrich C. Mayr, whose leadership and organization were outstanding. The high-quality program is a reflection of the countless hours he spent working so hard for all of us.

October 2005   Jacky Akoka
                                                                 Stephen W. Liddle
                                                                      Il-Yeol Song

ER 2005 Conference Organization

Honorary Conference Chair

Peter P.S. Chen   Lousiana State University, USA

General Conference Chair

Heinrich C. Mayr   Alpen-Adria University of Klagenfurt, Austria

Scientific Program Co-chairs

Lois M.L. Delcambre   Portland State University, Portland, USA
John Mylopoulos   University of Toronto, Canada
Oscar Pastor López   Universitat Politècnica de València, Spain

Workshop and Tutorial Co-chairs

Jacky Akoka   CEDRIC – CNAM, France/Institut National des
                                     Télécommunications, Evry, France
Stephen W. Liddle   Brigham Young University, Provo, USA
Il-Yeol Song   Drexel University, Philadelphia, USA

Panel Chair

Wolfgang Hesse   Philipps-Universität Marburg, Germany

Demos and Posters Chair

Tatjana Welzer   University of Maribor, Slovenia

Industrial Program Chair

Andreas Schabus   Microsoft Austria, Vienna, Austria

ER Steering Committee Liaison Manager

Veda Storey   Georgia State University, USA

Joint Conferences Steering Committee Co-chairs

Ulrich Frank   University of Essen-Duisburg, Germany
Jörg Desel   Catholic University Eichstätt-Ingolstadt,
                               Germany

Organization and Local Arrangements

Markus Adam   Peter Jelitsch   Christine Seger
Stefan Ellersdorfer Christian Kop   Claudia Steinberger
Günther Fliedl   Heinrich C. Mayr
Robert Grascher   Alexander Salbrechter

ER 2005 Workshops Organization

ER 2005 Workshop and Tutorial Chairs

Jacky Akoka — CEDRIC – CNAM, France/Institut National des Télécommunications, Evry, France
Stephen W. Liddle — Brigham Young University, Provo, USA
Il-Yeol Song — Drexel University, Philadelphia, USA

BP-UML 2005 Program Chair

Juan Trujillo — Universidad de Alicante, Spain

AOIS 2005 Program Chairs

Manuel Kolp — Université Catholique de Louvain, Belgium
Paolo Bresciani — ITC-irst, Italy

CoMoGIS 2005 Program Chair

Michela Bertolotto — University College Dublin, Ireland

eCOMO 2005 Program Chairs

Willem-Jan van den Heuvel — Tilburg University, The Netherlands
Bernhard Thalheim — Christian-Albrechts-Universität Kiel, Germany

QoIS 2005 Program Chairs

Isabelle Comyn-Wattiau — CEDRIC – CNAM, France/ESSEC, France
Samira Si-Saïd Cherfi — CEDRIC – CNAM, France

BP-UML 2005 Program Committee

Doo-Hwan Bae	KAIST, South Korea
Michael Blaha	OMT Associates Inc., USA
Cristina Cachero	Universidad de Alicante, Spain
Eduardo Fernandez	Universidad de Castilla-La Mancha, Spain
Paolo Giorgini	Università di Trento, Italy
Jaime Gómez	Universidad de Alicante, Spain
Peter Green	University of Queensland, Australia
Manfred Jeusfeld	Tilburg University, The Netherlands
Ludwik Kuzniarz	Blekinge Institute of Technology, Sweden
Jens Lechtenbörger	University of Münster, Germany
Pericles Loucopoulos	University of Manchester, UK
Kalle Lyytinen	Case Western Reserve University, Cleveland, USA
Hui Ma	Massey University, New Zealand
Antoni Olive	Universitat Politècnica de Catalunya, Spain
Andreas L. Opdahl	University of Bergen, Norway
Jeff Parsons	Memorial University of Newfoundland, Canada
Oscar Pastor López	Universitat Politècnica de València, Spain
Witold Pedrycz	University of Alberta, Canada
Mario Piattini	Universidad de Castilla-La Mancha, Spain
Colette Rolland	Université Paris 1 Panthéon-Sorbonne, France
Matti Rossi	Helsinki School of Economics, Finland
Manuel Serrano	Universidad de Castilla-La Mancha, Spain
Keng Siau	University of Nebraska-Lincoln, USA
Il-Yeol Song	Drexel University, Philadelphia, USA
Bernhard Thalheim	Christian-Albrechts-Universität Kiel, Germany
Ming Tjoa	Vienna University of Technology, Austria
Ambrosio Toval	Universidad de Murcia, Spain
Panos Vassiliadis	University of Ioannina, Greece

BP-UML 2005 External Referees

James Goldman	Drexel University, Philadelphia, USA
Mary E. Jones	Drexel University, Philadelphia, USA
Francisco Lucas Martinez	Universidad de Murcia, Spain
Fernando Molina	Universidad de Murcia, Spain
Joaquin Nicolas	Universidad de Murcia, Spain
Choi Nam Youn	KAIST, South Korea
Faizal Riaz-ud-Din	Massey University, New Zealand

AOIS 2005 Steering Committee

Yves Lespérance	York University, Toronto, Canada
Gerd Wagner	Technische Universiteit Eindhoven, The Netherlands
Eric Yu	University of Toronto, Canada
Paolo Giorgini	Università di Trento, Italy

AOIS 2005 Program Committee

Carole Bernon	Université Paul Sabatier, Toulouse 3, France
Brian Blake	Georgetown University, Washington, DC, USA
Paolo Bresciani	ITC-irst, Italy
Jaelson Castro	Federal University of Pernambuco, Brazil
Luca Cernuzzi	Universidad Católica "Nuestra Señora de la Asunción", Paraguay
Massimo Cossentino	ICAR-CNR, Palermo, Italy
Luiz Cysneiros	York University, Toronto, Canada
John Debenham	University of Technology, Sydney, Australia
Scott DeLoach	Kansas State University, USA
Frank Dignum	University of Utrecht, The Netherlands
Paolo Donzelli	University of Maryland, College Park, USA
Bernard Espinasse	Domaine Universitaire de Saint-Jérôme, France
Stéphane Faulkner	University of Namur, Belgium
Behrouz Homayoun Far	University of Calgary, Canada
Innes A. Ferguson	B2B Machines, USA
Alessandro Garcia	PUC Rio de Janeiro, Brazil
Chiara Ghidini	ITC-irst, Italy
Aditya K. Ghose	University of Wollongong, Australia
Marie-Paule Gleizes	Université Paul Sabatier, Toulouse 3, France
Cesar Gonzalez-Perez	University of Technology, Sydney, Australia
Giancarlo Guizzardi	University of Twente, The Netherlands
Igor Hawryszkiewycz	University of Technology, Sydney, Australia
Brian Henderson-Sellers	University of Technology, Sydney, Australia
Carlos Iglesias	Technical University of Madrid, Spain
Manuel Kolp	Université Catholique de Louvain, Belgium
Daniel E. O'Leary	University of Southern California, USA
C. Li	University of Technology, Sydney, Australia
Graham Low	University of New South Wales, Australia
Carlos Lucena	PUC Rio de Janeiro, Brazil
Philippe Massonet	CETIC, Belgium

Haris Mouratidis	University of East London, UK
Jörg P. Müller	Siemens, Germany
Juan Pavón	Universidad Complutense Madrid, Spain
Omer F. Rana	Cardiff University, UK
Onn Shehory	IBM Haifa Labs, Israel
Nick Szirbik	Technische Universiteit Eindhoven, The Netherlands
Kuldar Taveter	VTT Information Technology, Finland/ University of Melbourne, Australia
Nhu Tran	University of New South Wales, Australia
Viviane Torres da Silva	PUC Rio de Janeiro, Brazil
Michael Winikoff	RMIT, Australia
Carson Woo	University of British Columbia, Vancouver, Canada
Bin Yu	North Carolina State University, USA
Amir Zeid	American University in Cairo, Egypt
Zili Zhang	Deakin University, Australia

AOIS 2005 External Referees

Alessandro Oltramari	LOA-CNR, Italy
Ghassan Beydoun	University of New South Wales, Australia
Renata S.S. Guizzardi	ITC-irst, Italy
Rodrigo de Barros Paes	PUC Rio de Janeiro, Brazil
Aneesh Krishna	University of Wollongong, Australia
Gustavo Robichez de Carvalho	PUC Rio de Janeiro, Brazil

CoMoGIS 2005 Program Committee

Natalia Andrienko	Fraunhofer Institut AIS, Germany
Masatoshi Arikawa	University of Tokyo, Japan
Patrice Boursier	Université de La Rochelle, France/
	Open University, Malaysia
Elena Camossi	IMATI-CNR, Genoa, Italy
James Carswell	Dublin Institute of Technology, Ireland
Christophe Claramunt	Naval Academy, France
Maria Luisa Damiani	University of Milan, Italy
Max Egenhofer	University of Maine, USA
Stewart Fotheringham	NCG, Maynooth, Ireland
Andrew Frank	Vienna University of Technology, Austria
Michael Gould	Universitat Jaume I, Spain
Jihong Guan	Wuhan University, China
Bo Huang	University of Calgary, Canada
Zhiyong Huang	National University of Singapore, Singapore
Christian S. Jensen	Aalborg University, Denmark
Menno-Jan Kraak	ITC, Enschede, The Netherlands
Ki-Joune Li	Pusan National University, South Korea
Dieter Pfoser	CTI, Greece
Philippe Rigaux	Université Paris-Sud 11, France
Andrea Rodriguez	University of Concepcion, Chile
Sylvie Servigne	INSA, France
Stefano Spaccapietra	EPFL, Switzerland
George Taylor	University of Glamorgan, UK
Nectaria Tryfona	Talent Information Systems/
	Hellenic Open University, Greece
Christelle Vangenot	EPFL, Switzerland
Agnes Voisard	Fraunhofer ISST, Germany/
	Freie Universität Berlin, Germany
Nancy Wiegand	University of Wisconsin-Madison, USA
Thomas Windholz	Idaho State University, USA
Stephan Winter	University of Melbourne, Australia
Ilya Zaslavsky	San Diego Supercomputer Center, USA
Shuigeng Zhou	Fudan University, China

eCOMO 2005 Program Committee

Fahim Akhter	Zayed University, United Arab Emirates
Boldur Barbat	Lucian Blaga University, Sibiu, Romania
Boualem Benatallah	University of New South Wales, Australia
Anthony Bloesch	Microsoft Corporation, USA
Antonio di Leva	University of Turin, Italy
Vadim A. Ermolayev	Zaporozhye State Univ., Ukraine
Marcela Genero	Universidad de Castilla-La Mancha, Spain
Martin Glinz	University of Zurich, Switzerland
József Györkös	University of Maribor, Slovenia
Bill Karakostas	UMIST Manchester, UK
Roland Kaschek	Massey University, New Zealand
Christian Kop	Alpen-Adria University of Klagenfurt, Austria
Jacques Kouloumdjian	INSA, Lyon, France
Stephen W. Liddle	Brigham Young University, Provo, USA
Zakaria Maamar	Zayed University, United Arab Emirates
Norbert Mikula	Intel Labs, Hillsboro, USA
Oscar Pastor López	Universitat Politècnica de València, Spain
Barbara Pernici	Politecnico di Milano, Italy
Matti Rossi	Helsinki School of Economics, Finland
Klaus-Dieter Schewe	Massey University, New Zealand
Joachim Schmidt	Technische Universität Hamburg-Harburg, Germany
Michael Schrefl	Johannes-Kepler-Universität Linz, Austria
Daniel Schwabe	PUC Rio de Janeiro, Brazil
Il-Yeol Song	Drexel University, Philadelphia, USA
Eberhard Stickel	Europa-Universität Viadrina, Frankfurt(Oder), Germany
Bernhard Thalheim	Christian-Albrechts-Universität Kiel, Germany
Jos van Hillegersberg	Erasmus University Rotterdam, The Netherlands
Ron Weber	University of Queensland, Australia
Carson Woo	University of British Columbia, Vancouver, Canada
Jian Yang	Tilburg University, The Netherlands

QoIS 2005 Program Committee

Jacky Akoka	CEDRIC – CNAM, France/Institut National des Télécommunications, Evry, France
Laure Berti	IRISA, France
Mokrane Bouzeghoub	Université de Versailles Saint-Quentin-en-Yvelines, France
Tiziana Catarci	Università di Roma "La Sapienza", Italy
Corinne Cauvet	Université Aix-Marseille III, France
Marcela Genero	Universidad de Castilla-La Mancha, Spain
Michael Gertz	University of California, Davis, USA
Markus Helfert	Dublin City University, Ireland
Paul Johannesson	Stockholm University, Sweden
Jacques Le Maitre	Université du Sud Toulon-Var, France
Oscar Pastor López	Universitat Politècnica de València, Spain
Geert Poels	Ghent University, Belgium
Nicolas Praat	ESSEC, France
Jolita Ralyté	University of Geneva, Switzerland
Il-Yeol Song	Drexel University, Philadelphia, USA
Bernhard Thalheim	Christian-Albrechts-Universität Kiel, Germany
Juan Trujillo	Universidad de Alicante, Spain
Dimitri Theodoratos	New Jersey Institute of Technology, USA

QoIS 2005 External Referees

Mary E. Jones	Drexel University, Philadelphia, USA
Ann Maes	Ghent University, Belgium
Luis Reynoso	Universidad Nacional del Comahue, Neuquén, Argentina
Manuel Serrano	Universidad de Castilla-La Mancha, Spain

Organized by

Institute of Business Informatics and Application Systems, Alpen-Adria University of Klagenfurt, Austria

Sponsored by

ER Institute
The Governor of Carinthia
The City Mayor of Klagenfurt

In Cooperation with

GI Gesellschaft für Informatik e.V.
Austrian Computer Society

Table of Contents

Seventh International Bi-conference Workshop on Agent-Oriented Information Systems (AOIS-2005)

Invited Talk

Positions in Engineering Agent Oriented Systems

Agent Oriented Methodologies and Conceptual Modeling

Agent Communication and Coordination

Second International Workshop on Conceptual Modeling for Geographic Information Systems (CoMoGIS 2005)

Invited Talk

Spatial and Spatio-temporal Data Representation

Spatial Relations

Spatial Queries, Analysis and Data Mining

Data Modeling and Visualisation

Sixth International Workshop on Conceptual Modeling Approaches for e-Business (eCOMO 2005)

First International Workshop on Quality of Information Systems (QoIS 2005)

Information System Models Quality

Quality Driven Processes

Tutorials

Preface to BP-UML 2005

Juan Trujillo

The Unified Modeling Language (UML) has been widely accepted as the standard object-oriented (OO) modeling language for modeling various aspects of software and information systems. The UML is an extensible language, in the sense that it provides mechanisms to introduce new elements for specific domains if necessary, such as web applications, database applications, business modeling, software development processes, data warehouses and so on. Furthermore, the latest approach of the Object Management Group (OMG) surrounding the UML even got bigger and more complicated with a more number of diagrams with some good reasons. Although providing different diagrams for modeling specific parts of a software system, not all of them need to be applied in most cases. Therefore, heuristics, design guidelines, lessons learned from experiences are extremely important for the effective use of UML and to avoid unnecessary complication.

BP-UML'05 (Best Practices of the UML) is the first edition of this International Workshop held with the 24th International Conference on Conceptual Modeling (ER 2005). This workshop will be an international forum for exchanging ideas on the best and new practices of the UML in modeling and system developments. The workshop will be a forum for users, researchers, analyzers, and designers who use the UML to develop systems and software. To keep the high quality of former workshops held in conjunction with ER, a strong International Program Committee was organized with extensive experience in the UML and also taking into consideration its relevant scientific production in the area.

The workshop attracted papers from 13 different countries distributed over all continents such as The Netherlands, France, Spain, Israel, Korea, USA, Canada and Australia. We received 27 submissions and only 9 papers were selected by the Program Committee, making an acceptance rate of 36%.

The accepted papers were organized in three different sessions. In the first one, two papers will present valuable experience reports and another one will describe how to apply UML for multimedia modeling. In the second one, one paper will be focused on evaluating the cardinality interpretation by users in a UML class diagram, and the other two papers will be focused on the Use case diagrams of the UML. Finally, in the third session, while one paper will present how to analyze the consistency of a UML diagram, the other two will be focused on the Model Driven Architecture (MDA) and metamodeling.

I would like to express my gratitude to the program committee members and the additional external referees for their hard work in reviewing papers, the authors for submitting their papers and the ER2005 organizing committee for all their support. This workshop was organized within the framework of the following projects: MESSENGER (PCC-03-003-2), METASIGN (TIN2004-00779) and DADASMECA (GV05/220). Thanks to the number of submissions of this first edition together with the high quality of the accepted papers, my intention is to organize the second edition of BP-UML again next year with ER2006.

J. Akoka et al. (Eds.): ER Workshops 2005, LNCS 3770, p. 1, 2005.
© Springer-Verlag Berlin Heidelberg 2005

cCurrent Practices in the Use of UML

Brian Dobing[1] and Jeffrey Parsons[2]

[1] Faculty of Management, University of Lethbridge,
4401 University Drive W., Lethbridge, AB, T1K 3M4, Canada
brian.dobing@uleth.ca
[2] Faculty of Business Administration, Memorial University of Newfoundland,
St. John's, NF, A1B 3X5, Canada
jeffreyp@mun.ca

Abstract. Despite widespread interest in the Unified Modeling Language (UML), there is little evidence about the extent and nature of UML use. This paper reports results of a survey of UML use by practitioners. Results indicate varying levels of use, and perceived usefulness, of different UML diagrams. In addition, we found significant involvement of non-IT professionals in the development of UML diagrams. An understanding of the range of current practices is an important foundation for determining "best practices."

1 Introduction

The UML has been widely accepted as the standard for object-oriented analysis and design (OOAD) [13]. A large number of practitioner articles and dozens of textbooks have been devoted to articulating various aspects of the language, including guidelines for using it. More recently, a substantial body of academic research on UML has developed, ranging from proposals for extending the language [16], [17] to ontological analysis of its modeling constructs [9], [10], to analysis of the language's complexity [19], [20] and experiments that evaluate various aspects of the effectiveness of UML models [4], [5].

Recently, there has been a growing interest in understanding the state of practice in software engineering, as evidenced by (for example) a special issue of *IEEE Software* on this topic (Vol. 20, No. 6, Nov-Dec 2003). However, despite the practitioner and academic interest in and attention to UML, to our knowledge there are no published empirical data on the extent to which, and ways in which, UML is used in practice. This research seeks to address this issue by surveying UML use in practice. Our findings give an overall picture of how UML is used. The study provides a useful point on which to anchor future discussions of 'best practices' of UML use and, thus, fits well with the theme of the BP-UML workshop.

2 Motivation

The UML combines and extends modeling notations previously proposed by its developers. The language was not developed based on any theoretical principles regarding the constructs required for an effective and usable modeling language for analysis and

J. Akoka et al. (Eds.): ER Workshops 2005, LNCS 3770, pp. 2–11, 2005.

design; instead, the UML itself arose from "best practices" in parts of the software engineering community [2], [3]. Furthermore, the language contains many modeling constructs, and has been criticized on the grounds that it is excessively complex [7], [14]. But at the same time, the language has also been criticized for lacking the flexibility to handle certain modeling requirements in specific domains [8].

The UML per se is a language, not a methodology. However, it is designed to be "Use Case driven" and iterative. In addition, many published books on UML contain prescriptions for applying the language in modeling (e.g., [15], [21]). These prescriptions sometimes differ. For example, some recommend using only Use Case Narratives (or, more simply, Use Cases) in verifying requirements with users [12], while others explicitly or implicitly indicate that other UML artifacts can be used for this purpose. For example, Activity Diagrams "can be safely shared with customers, even those unfamiliar with software engineering" [18, p.67]. Similarly, there are other wide variations in guidelines for using the language, in particular variations with respect to Use Cases [6].

In view of these issues, it would not be surprising to find a variety of practices followed by UML users. We believe understanding current practice is a prerequisite for conducting theoretical or applied research on UML. From a theoretical perspective, understanding how the language is used can support or challenge theoretical analyses of UML capabilities and deficiencies [9], [10]. Such an understanding can also support the development of theory to explain observed usage patterns. From a practical point of view, understanding how the language is used can help support the development of best practices. For example, if certain parts of the language are not widely used or seen as useful, further research is needed to understand why this is so, and may lead to evolution or elimination of those parts.

3   Research Methodology

To get a broad picture of UML use, a web survey was developed based on a literature review and preliminary interviews with about a dozen practitioners. To obtain a sample of analysts familiar with object-oriented techniques and the UML in particular, the Object Management Group (OMG) was contacted and they agreed to support the project. Their members were informed of the survey and the OMG endorsement. A link to the survey was also provided from the main OMG web page. OMG members were encouraged to share the link with others in the organization who were using the UML. Subsequently, an invitation to participate in the survey was posted to the comp.object Usenet newsgroup. No participation incentive was offered. The paper primarily reflects experiences with UML 1.x, and thus uses its terminology.

Surveying primarily OMG members and those who use its web site may produce biased responses. However, given that the goals of this research were to examine how UML users were using the language, rather than the extent to which it is being used in software development in general, the participation of the OMG seemed appropriate.

In addition, there are some obvious limitations with using a convenience sample. The number of people who received or read the invitation to participate is unknown because of the possibility of it being forwarded. It is also likely that some people found the survey through search engines, since the survey was, for some time, the top result of a Google search on "UML survey." Despite the lack of control over respondents, reviewing the comments and contact information suggests that the group as a whole does belong to the target population. Whether respondents are representative of the target population, analysts using UML, is difficult to determine.

4   Results

The survey received 299 usable responses, which either contained data on UML component usage (182) or reasons why the UML was not being used (117). Of the 182 analysts using UML components, most (171) were using the UML while 11 were using several UML components as part of another methodology. Respondents came from organizations of varying sizes and a broad range of industries. However, the limited responses in each category preclude a detailed analysis by either of these variables. Thus, the results presented here do not reflect possible differences in usage patterns in organizations of different sizes or from different industrial sectors.

Respondents had a wide range of experience in the IT field, reporting up to 45 years and 200 projects. UML experience was understandably less, with the median number of projects worked on by respondents being 4.0. The minimum Years of Experience in IT was 2 and the minimum number of IT Projects worked on was 3. While respondents report more project experience with UML than other object-oriented methodologies, it represents less than a quarter of their projects and about a third of their total years of experience.

Table 1 shows the relative usage of seven major UML analysis components. We restricted the survey to Use Case Narratives and UML diagrams covering system structure and behavior that are used to document system functionality, and therefore did not ask questions about model management diagrams such as deployment diagrams. Respondents were asked, "What proportion of the object-oriented/UML projects that you have been involved with have used the following UML components? The five-point Usage scale was: None, <1/3, 1/3 – 2/3, > 2/3 and All. Although UML is often presented as Use Case-driven, Class Diagrams were the most frequently used component among respondents to this survey with a 4.19 mean score and are used 2/3 or more of the time by 73% of respondents. Use Case Narratives were ranked fourth, behind Sequence Diagrams and Use Case Diagrams. The number of respondents to this question varied from 152 (Statechart) to 172 (Class Diagram). However, respondents were generally familiar with all the components, ranging from only 3% who have never used Class Diagrams to 25% who have never used Collaboration Diagrams. Use Case Narratives have never been used by 15%.

Since more experienced analysts would seem more likely to be following best practices, respondent experience measures were correlated with use of each UML component. The strongest relationships involved Statechart Diagrams and years of experience in object-oriented analysis and design (0.45, p<0.01) and years of experience with UML (0.35, p<0.01). Class Diagram usage also correlated significantly (p<0.01) with these two experience measures at 0.36 and 0.40 respectively, and with years of object-oriented programming (0.31). No other correlations between experience measures and component usage exceeded 0.30.

Table 1. UML Component Usage

UML Component	Usage[1]	>2/3 usage (%)	New Info	Some – All New Info (%)
Class Diagram	4.19**	73	3.51	86
Use Case Diagram	3.56**	51	2.42[††]	48
Sequence Diagram	3.51*	50	3.37	78
Use Case Narrative	3.25	44	NA	NA
Activity Diagram	2.87**	32	2.89[††]	63
Statechart Diagram	2.82**	29	3.38[†]	79
Collaboration Diag.	2.54**	22	2.98[††]	67

[1] Responses were on the scale: 1 - None; 2 - < 1/3; 3 – 1/3 to 2/3; 4 - > 2/3; 5 - All
*,** Significantly different from Use Case Narrative mean, ** $p<=0.01$, * $p<0.05$
[†],[††] Significantly different from Class Diagram mean, [††] $p<=0.01$, [†] $p<0.05$

There are a number of reasons for using multiple diagram types to describe system functionality. Perhaps the most important is that different diagrams convey different information. To investigate this, the survey asked which components provide new information beyond that contained in Use Case Narratives. The question used a five-point scale from "No New Info" to "All New Info," with "Some New Info" as the midpoint (3). This survey took a Use Case-driven perspective (which, given the results, may not have been most appropriate). Those who said they did not use Use Case Narratives were not asked this question so there were fewer respondents, from 89 (Collaboration Diagram) to 125 (Class Diagram). Also, some respondents had not used all of the diagrams.

Table 1 shows that the component of highest value for conveying new information not already contained in the Use Case Narratives was the Class Diagram, with a score of 3.51 on the five-point scale, and 85.6% of respondents believe it offers at least some new information (at least 3 on the 5-point scale). The Use Case Diagram was least useful in providing additional information, which is not surprising given its role is to present an overview of the project.

Best practices could also involve tradeoffs among diagrams. For example, given that Sequence Diagrams and Collaboration Diagrams are "isomorphic" [3, p.25], one might expect to find that analysts use either the Collaboration Diagram or the Sequence Diagram but not both. However, usage rates of the different UML components were all positively correlated, from an r^2 of 0.64 between Use Case Narratives and Use Case Diagrams to 0.16 between Use Case Narratives and Statechart Diagrams. Use of Collaboration and Sequence Diagrams correlated at 0.38 ($p < 0.01$). However, there was also a strong correlation (0.77) between the beliefs that Collaboration and Sequence Diagrams provide new information beyond Use Case Narratives, the highest correlation found among all pairs of components. This could be attributed to that isomorphic relationship.

Stronger relationships were expected between the belief that a UML component provides additional information beyond the Use Case Narrative and the usage level of that component. For Activity Diagrams, the correlation was 0.42 ($p<0.01$). However, no other correlation of this type exceeded 0.30. While this could indicate that Use

Case Narratives contain most of the needed information, many projects did not take a Use Case-driven approach so the results must be interpreted with caution.

Most respondents are making only partial use of the seven UML components studied (Table 2). Of the 135 respondents who reported their usage levels of all seven UML components studied, 51% were using five or more of them in at least a third of their projects while only 21% were using five or more in at least two-thirds of their projects.

Table 2. Number of UML Components Used

No. Of UML Components Used	>1/3 Projects (%)	>2/3 Projects (%)
0	6	13
1	4	14
2	8	13
3	10	23
4	21	16
5	16	10
6	19	3
7	16	8

Less experienced analysts might be expected to use only a few UML components in their initial projects, an approach recommended as part of Agile Modeling [1, p.46]. With more experience, analysts could make fuller use of all components. However, the data provide little support for this. There is a very low correlation (0.22, p<0.05) between the respondent's number of UML projects and the number of components used at least a third of the time. The correlation is even lower for the number of components used at least two-thirds of the time. Correlations are slightly higher when years of UML experience is used rather than the number of UML projects. However, there is still only a 0.32 correlation (p<0.01) with the number of components used at least a third of the time, dropping to 0.26 for the number used two-thirds of the time. Very similar correlations were found when using years of experience with object-oriented systems analysis and design. The remaining correlations among experience measures and number of components used one-third or two-thirds of the time were all less than 0.30.

There could be a similar expectation that larger projects would see wider use of UML components. However, there were only weak relationships between project size measures and component usage levels, in the range of -0.17 to 0.25.

Table 3 examines reasons for including each UML component in a project. Each respondent who reported using a particular component at least a third of the time was asked about four possible purposes. As expected, Use Case Narratives had the highest score for "Verifying and validating requirements with client representatives on the project team" at 4.00 (on a 5-point scale) and 87% of respondents rated them from "Moderately Useful" to "Essential." The use of other components for this purpose was higher than expected, based on our review of the literature. The least useful were Collaboration Diagrams, but still 51% reported them to be at least "Moderately

Useful" in verifying and validating requirements with clients. These high levels of client involvement are encouraging. The survey also included a single item that asked, "How successful has the UML been in facilitating communication with clients?" The items used a five-point scale from Not to Very Successful. The mean was 3.28 with about 24% choosing the lowest two levels. When defining best practices, the role of the client is critical and components that enhance analyst-client communication are thus particularly important.

Table 3. Role of Major UML Components

UML Component	Verifying and validating requirements with clients	Specifying system requirements for programmers	Documenting for future maintenance and other enhancements	Clarifying understanding of application among technical members
Use Case Narrative	4.00	$3.62^{\dagger}$	$3.15^{\dagger\dagger}$	$3.52^{\dagger\dagger}$
Activity Diagram	3.50**	$3.43^{\dagger\dagger}$	$3.35^{\dagger\dagger}$	$3.50^{\dagger\dagger}$
Use Case Diagram	3.36**	$3.06^{\dagger\dagger}$	$2.90^{\dagger\dagger}$	$3.17^{\dagger\dagger}$
Sequence Diagram	2.91**	$3.71^{\dagger}$	$3.76^{\dagger\dagger}$	$4.14^{\dagger}$
Class Diagram	2.90**	4.06	4.18	4.35
Statechart Diagram	2.63**	$3.51^{\dagger\dagger}$	$3.35^{\dagger\dagger}$	$3.74^{\dagger\dagger}$
Collab'n Diagram	2.62**	$3.25^{\dagger\dagger}$	$2.96^{\dagger\dagger}$	$3.40^{\dagger\dagger}$

** Significantly different from Use Case Narrative mean, ** $p<=0.01$
†,†† Significantly different from Class Diagram mean, $^{\dagger\dagger} p<=0.01$, $^{\dagger} p<0.05$

The other three purposes listed are more related to communication within the project team, among analysts, programmers and maintenance staff. In each case, the Class Diagram was considered most useful with the Use Case Diagram least useful. As noted earlier, the Use Case Diagram provides an overview of the project while programming tends to focus on implementing particular functionality. In Table 3, the usefulness levels reported for Sequence Diagrams are all significantly higher ($p<0.01$) on the three project team communication measures than those for the isomorphic Collaboration Diagram.

These results point to a potential disconnect in projects using the UML, where analysts rely on Use Case Narratives when communicating with clients and Class Diagrams when communicating with programmers. There is a risk that either the Class Diagram will convey additional information not contained in the Use Case Narratives that the client does not fully understand or that information in the Use Case Narratives could be incorrectly interpreted when creating the Class Diagram.

Those who reported using a particular component less than a third of the time (including not at all) were asked why they were not using it more often. There were fewer respondents for these questions, ranging from only 8 for Class Diagrams to 59 for Collaboration Diagrams. Table 4 shows the percentage of respondents who selected each possible reason. Respondents were encouraged to select all reasons that applied so row totals exceed 100%. A lack of understanding by analysts was the primary factor among the few not using Class Diagrams (50%). Similar concerns were expressed by 48% of respondents about Activity Diagrams. Leading concerns for the remaining components were over usefulness (Statechart), value (Sequence and Use Case Diagrams and Narratives) and redundancy (Collaboration).

Table 4. Reasons for not Using Some UML Components (% responses)

UML Component	Not well under-stood by analysts	Not useful for most projects	Insuffi-cient value to justify cost	Infor-mation cap-tured would be re-dundant	Not useful with clients	Not useful with pro-gram-mers
Class Diag.	50	13	13	25	25	25
Sequence	32	23	36	14	23	23
Use Case Narra-tive	29	26	37	29	11	26
Use Case Diag.	32	32	42	19	29	42
Statechart	35	42	28	12	28	33
Activity	48	23	35	35	14	25
Collaboration	27	32	24	49	29	24

Table 5. Client Involvement

UML Component	Develop (%)	Review (%)	Approve (%)	No. Resp.
Use Case Narrative	76	63	54	78
Use Case Diagram	57	69	46	77
Activity Diagram	47	60	19	57
Sequence Diagram	37	52	16	87
Class Diagram	33	53	20	103
Collaboration Diagram	38	48	13	48
Statechart Diagram	28	36	20	61

User participation has long been considered as crucial to the system development process. The survey also asked about the client's role in relation to each of the UML components being studied. Respondents were able to select more than one (e.g., they could report that clients helped to develop Use Case Narratives, reviewed some or all of them upon completion and had formal approval authority). The results are summarized in Table 5. For example, 76% of respondents who used Use Case Narratives reported that clients were involved in their development.

The results show that clients were most likely to be involved in developing, reviewing and approving Use Case Narratives and the Use Case Diagram. Of the remaining components, Activity Diagrams are probably the easiest for clients to understand and almost half the analysts report some involvement by clients in their development (consistent with the comment above). While clients were less likely to be involved in developing the Class Diagram, just over half were involved in reviewing this critical component.

Clients were least likely to be involved in developing or reviewing Statechart diagrams. The fact that about one quarter to a third were involved in these tasks may reflect the technical sophistication of some clients in the survey sample, since the composition of OMG membership includes many large companies in the computer industry.

Respondents were also asked about possible difficulties that occurred that "could be attributed to the UML." They could check any or all of the five categories listed. User interface concerns were checked most frequently (36%), followed by roles and responsibilities of particular users (23%), security (19%), data requirements (17%), and system capabilities and functionality (12%). We did not investigate these concerns in greater detail, but this clearly is an area meriting futher research. We also observed a wide range of client involvement practices in the survey results, not unexpected with a new approach.

5   Conclusion and Implications for "Best Practices"

This survey is the first we are aware of investigating how and why UML analysis components are used. Overall component use was similar to an earlier study which found highest usage levels for Use Case Diagrams and Class Diagrams and lowest for Collaboration Diagrams [22]. We found variations in the level of use, and perceived usefulness, of different UML component models. Such variations appear to be somewhat inconsistent with the notion of the UML as a "unified" language. Moreover, we found that use of only a subset of UML components on a project is widespread. The data also show a variety of reasons why certain UML components are not used.

In general, the wide variation in practice suggests that "best practices" of UML use have not permeated the survey respondents. Moreover, since most of the survey respondents were associated with the OMG, they might be expected to be on the leading edge of UML use. Best practices are more likely to be seen in the way the more experienced practitioners are using the UML. However, when comparing their usage to that of the least experienced UML practitioners in our survey, no major differences appear.

In addition, the degree to which UML components are developed, reviewed, and approved by clients is higher than might be expected based on the extant prescriptive literature on 'how to use' the language. This level of involvement suggests that UML, as it is used, is not exclusively a language for software professionals, and that a greater understanding of the usability of UML diagrams among non-experts is needed. In particular, research is needed to understand which components of UML can best facilitate communication between users and analysts. Such research is vital to developing an overall set of best practices for use of the language. In addition, a survey of clients involved in UML projects may yield interesting insights about which practices are most effective from their perspective.

The findings of this research contribute in other ways to understanding best practices. First, we found that Class Diagrams are the preferred method for communication among IS professionals on project teams. They would be expected to play a key role in any best practice. On the other hand, Collaboration Diagrams are used less often, deemed to be less useful, and appear to offer little additional value in relation to Sequence Diagrams. These findings suggest there may be little need for Collaboration Diagrams in a suite of best practices. Statechart Diagrams are also used less often than most other diagrams and seem to be less useful most of the time, but are rated highly for providing new information in some situations and have low redundancy. As one interview subject put it, "When they are useful, they are very useful." One possible use of this finding in the context of best practices is to use it as a basis for exploring and articulating the conditions in which Statechart Diagrams should be used.

Second, there is some support for taking a Use Case-driven approach as advocated in much of the UML literature. Survey results show Use Case Narratives are not only rated as most valuable for establishing requirements with clients but that clients are frequently (76%) involved in their development. The role of Use Cases must be considered in any UML Best Practices approach.

However, our findings also suggest that best practices in involving clients/users in the development process may extend to the development and review of artifacts beyond Use Case Narratives. We also found that the use of components other than Use Case Narratives among clients/users was higher than expected. While the UML practitioner literature generally seems to assume that many UML components are too complex or technical to be understood by clients, our results show they are frequently approved, reviewed, and even developed by them. This is not surprising given that many projects are not developed under a Use Case-driven approach. Since clients might be expected to be relatively unfamiliar with the UML, further work is needed to identify problems that might arise and find ways of preventing them, before best practices in user involvement can be articulated with supporting evidence. At the same time, there is limited empirical evidence to support the belief of our respondents that Use Case Narratives are a more effective way to communicate with clients than the other UML components.

In summary, we believe this survey is an important first step in identifying best practices for UML usage. By identifying current practices, we open the door to future research that can examine whether what is being done is providing the benefits expected, or whether there are costs associated with the relatively low level of use of some types of diagrams. However, much work remains. Grounded Theory [11] has been used in other fields, such as nursing and education, to provide a framework for Best Practices research and would seem appropriate here as well.

References

1. Ambler, S.W.: Agile Modeling: Effective Practices for eXtreme Programming and the Unified Process. Wiley Computer Publishing, New York (1992)
2. Booch, G.: UML in Action. Communications of the ACM, 42(10) (1999) 26-28
3. Booch, G., Rumbaugh, J., Jacobson, I.: The Unified Modeling Language User Guide. Addison-Wesley, Reading, MA (1999)
4. Burton-Jones, A., Meso, P.: How Good are these UML Diagrams? An Empirical Test of the Wand and Weber Good Decomposition Model. Proceedings of the Twenty-Third International Conference on Information Systems, Barcelona (2002) 101-114

5. Burton-Jones, A., Weber, R.: Properties do not have properties: Investigating a question-able conceptual modeling practice. In: Batra, D., Parsons, J., Ramesh, V. (eds.): Proceed-ings of the Second Annual Symposium on Research in Systems Analysis and Design (2003)
6. Dobing, B., Parsons, J.: Understanding the Role of Use Cases in UML: A Review and Re-search Agenda. Journal of Database Management. 11(4) (2000) 28-36
7. Dori, D.: Why Significant UML Change is Unlikely. Communications of the ACM. 45(11). (2002) 82-85
8. Duddy, K.: UML2 Must Enable A Family of Languages. Communications of the ACM. 45(11) (2002) 73-75
9. Evermann, J., Wand, Y.: Towards ontologically based semantics for UML constructs. In: Kunii, H., Jajodia, S., Solvberg, A. (eds.): Proceedings of the 20th International Confer-ence on Conceptual Modeling, Yokohama, Japan (2001) 354-367
10. Evermann, J., Wand, Y.: An Ontological Examination of Object Interaction in Conceptual Modeling. In: Parsons J., Sheng, O. (eds.): Proceedings of the Eleventh Workshop on In-formation Technologies and Systems, New Orleans (2001) 91-96
11. Glaser, B.G., Strauss, A.L.: The discovery of grounded theory: strategies for qualitative re-search. Aldine, Chicago (1967)
12. Jacobson, I., Ericsson, M., Jacobson, A.: The Object Advantage: Business Process Reen-gineering with Object Technology. Addison-Wesley, Reading, MA (1994)
13. Kobryn, C.: UML 2001: A Standardization Odyssey. Communications of the ACM. 42(10) (1999) 29-37
14. Kobryn, C.: Will UML 2.0 Be Agile or Awkward? Communications of the ACM. 45(1) (2002) 107-110
15. Larman, C.: Applying UML and Patterns: An Introduction to Object-Oriented Analysis and Design and the Unified Process. 2nd edn. Prentice Hall, Upper Saddle River, NJ (2002)
16. Moore, A.: Extending UML to Enable the Definition and Design of Real-Time Embedded Systems. Crosstalk: The Journal of Defense Software Engineering. 14(6) (2001) 4-9
17. Odell, J., Van Dyke, P., Bauer, B.: Extending UML for Agents. In: Wagner, G., Lesper-ance, Y., Yu, E. (eds.): Proceedings of the Agent-Oriented Information Systems Workshop at the 17th National Conference on Artificial Intelligence, Austin, TX (2000) 3-17
18. Schneider, G., Winters, J.: Applying Use Cases: A Practical Guide. 2nd edn. Addison-Wesley, Boston, MA (2001)
19. Siau, K., Cao, Q.: How Complex Is the Unified Modeling Language? Advanced Topics in Database Research. Vol. 1 (2002) 294-306
20. Siau, K., Erickson, J., Lee, L.: Theoretical vs. Practical Complexity: The Case of UML. Journal of Database Management. 16(3) (2005) (In press)
21. Stevens, P., Pooley, R.: Using UML: Software Engineering with Object and Components, Addison-Wesley, Reading, MA (2000)
22. Zeichick, A.: Modeling Usage Low; Developers Confused About UML 2.0, MDA. SD Times, July 15 (2002) (Available at http://www.sdtimes.com/news/058/story3.htm)

An Empirical Study of the Nesting Level of Composite States Within UML Statechart Diagrams

José A. Cruz-Lemus[1], Marcela Genero[1], Mario Piattini[1], and Ambrosio Toval[2]

[1] ALARCOS Research Group,
Department of Computer Science, University of Castilla – La Mancha,
Paseo de la Universidad, 4 – 13071 Ciudad Real, Spain
{JoseAntonio.Cruz, Marcela.Genero, Mario.Piattini}@uclm.es
[2] Software Engineering Research Group,
Department of Computer Science and Systems, University of Murcia,
Campus de Espinardo – 30071 Murcia, Spain
atoval@um.es

Abstract. As UML statechart diagrams are the core for modeling the dynamic aspects of software systems, we have been studying their understandability for the last three years. In previous researches, we have already studied the relationship between many of the constructs of the UML statechart diagrams and the effect that they have on the understandability of the diagrams themselves. We have also performed a family of experiments whose results indicated that the use of composite states make UML statechart diagrams easier to understand. This fact motivated us to go a step further and investigate if the Nesting Level of Composites States (NLCS) has an impact on the understanding of the diagrams through a controlled experiment and a replication. In this paper, we present the experimental process and the main findings of them. Unfortunately, the obtained results are not quite conclusive and we have not been able to find an optimal use of nesting within UML statechart diagrams and further empirical research is needed, considering more complex UML statechart diagrams.

1 Introduction

New approaches in software engineering like MDA (Model Driven Architecture) [17] and MDD (Model Driven Development) [1] are enabling a shift in focus from software to models of software. These approaches consider models as end-products rather than just mean to produce software.

In truly 'model-driven' software engineering, the quality of the models used is greatly important. For that reason, models like UML ones are gaining more relevance in the development of software, as the quality of the models used will later determine the quality of the software systems produced.

As UML statechart diagrams are the core for modeling the dynamic aspects of software systems [13], we have been studying their understandability for the last three years. Our main idea was that if diagrams are difficult to understand this will affect their maintainability. In previous researches, we have studied the relationship between many of the constructs of the UML statechart diagrams and the effect that they have

J. Akoka et al. (Eds.): ER Workshops 2005, LNCS 3770, pp. 12–22, 2005.
© Springer-Verlag Berlin Heidelberg 2005

on the understandability of the diagrams themselves. First, we defined and validated a set of metrics [11] for evaluating if the structural properties of UML statechart diagrams, such as size and complexity, influenced the understanding of UML statechart diagrams. In these researches we had found that the usage of composites states had apparently no influence on the understandability of UML statechart diagrams. This fact seemed to be a bit suspicious. For that reason, we decided to run another experiment, and a further replication, for specifically studying if the use of composite states facilitated or not the understanding of UML statechart diagrams [10]. The results of this empirical study indicated that the use of composite states improves the understandability efficiency of UML statechart diagrams, i.e. how accurately the different stakeholders understands the diagrams, if the subjects have a certain level of experience in working with this kind of UML diagrams. These findings motivated us to go a step further and define a new metric named *Nesting Level in Composite States* (NLCS) which indicates the maximum number of nested composite states in an UML statechart diagram. We based on the measure DIT (Depth of Inheritance Tree) defined in [9], as we think that there is a certain similarity between the NLCS within an UML statechart diagram and the depth of a within a generalization hierarchy in an UML class diagram.

In this paper, we will investigate in the NLCS affects the understanding of UML statechart diagram and try to find the optimal nesting level within a diagram through a controlled experiment and a replication of it. For designing the experiment we took several ideas from the different experimental experiences performed related to the DIT metric [3-8, 12, 14, 18, 19, 21].

Not only do we want this paper to be taken under a research point of view, but also to be useful for designers and software engineering teachers at universities.

We will begin defining our research question. The description of the experimental process, covering the design, tasks and performance of the experiment is explained in section 3. Section 4 describes the data analysis and the interpretation of the obtained results. Section 5 tackles all the features related to the replication of the experiment. Finally, conclusions and future work are presented in section 6.

2 Research Question

As we commented in section 1, our research question can be stated as:

> *Does the use of different nesting levels of composite states in UML statechart diagrams affect the understandability of the diagrams?*

In order to answer this question we have defined the previously presented metric NLCS. Based on the guidelines exposed in [18], we have formulated the following experimental hypotheses:

- H_{0-ij}: the understandability of UML statechart diagrams with i and j composite states nesting levels is not significantly different,
- H_{1-ij}: the understandability of UML statechart diagrams with i and j composite states nesting levels is significantly different,
  In both cases, i, j $\in$ {0, 1, 2} and i≠j.

This way, there are three distinct null hypotheses (H_{0-01}, H_{0-02}, H_{0-12}), taking account of symmetries ($H_{0-12}=H_{0-21}$).

3 Experimental Process

In this section, we describe the controlled experiment that we carried out at the University of Murcia (Spain) in May 2005 for testing the hypotheses stated in the previous section. All the experimental process is based on the guidelines outlined in [23].

3.1 Subjects

38 subjects from the University of Murcia participated in this experiment. 11 of them were on their 4th year of Computer Science whilst the rest had finished their Computer Science studies less than one year before.

The tasks to be performed did not require high levels of industrial experience, so experiments with students could be considered as appropriate [2, 15]. Moreover, students are the next generation of people entering this profession, so they are close to the population under study [16]. Besides, working with students implies a set of advantages [22], such as the fact that the prior knowledge of the students is rather homogeneous. The availability of a large number of subjects is another plus point.

All the subjects had received a short training session before the performance of the experiment, in which the main constructs of UML statechart diagrams were commented on and where some examples of the tasks to be performed by them were explained by the conductor of the experiment.

3.2 Experimental Design

The dependent variable was the understandability of UML statechart diagrams measured by:

- **Effectiveness:** number of correct answers vs. total number of asked questions.
- **Efficiency:** number of correct answers vs. time spent on answering the questions.

The independent variable was the nesting level of the different UML statechart diagrams, measured by the metric NLCS. We used three different diagrams with 0, 1 and 2 nesting levels respectively that modeled exactly the same system (an ATM) and were conceptually equivalent. An example of the experimental material is shown in the Appendix A, at the end of the document. Moreover, for the interested readers, the original experimental material is available at http://alarcos.inf-cr.uclm.es.

3.3 Experimental Task

Each subject received one diagram out of the three possibilities. The universe of discourse (UoD) of the diagrams was quite usual and not exceptional at all, so that there was no need for extra effort in understanding them.

Each diagram had a test which contained 9 questions which were exactly the same (questions and answers) for the three different diagrams. The questions inquired about what state would be reached after the triggering of some events, which state would be reached after a certain sequence of events and guard conditions or what sequence was the minimum possible for going from one given state to another, for instance. The subjects had to note down the answers to the questions and the times at which they started and finished answering the whole questionnaire.

3.4 Experimental Procedure

The experiment started with a twenty-five-minute introductory session in which the conductor briefly explained the main motivation for the experiment as well as the main elements of an UML statechart diagram. After that, the materials for the experiment were randomly distributed to the subjects.

In order to increase the motivation and interest of the subjects, the students were explained that the exercises that they were going to perform could be similar to those that would find in their exam at the end of the term.

At this point some examples in shortened version were performed by the conductor, who explained the correct answer to each question and the way of fulfilling the questionnaires properly.

Throughout this time, the subjects were allowed to ask the conductor about any doubt that they might have, and they could make any remarks they wished to.

4 Data Analysis and Interpretation

All the data analysis presented in this section was carried out by means of SPSS [20].

First, we carried out an analysis of the descriptive statistics of the data. The boxplots of the data shown in Figures 1 and 2 illustrate the statistics summarized in Table 1 and Table 2.

In order to check the hypotheses presented in section 2, we performed some t-Tests with $\alpha=0.05$. The obtained the results for the different dependent variables taking into account all the possible NLCS values are shown in Table 3.

Table 1. Summary statistics for effectiveness

NLCS	Mean	Median	Min.	Max.	St. Dev.	Skew.	Kurtos.
0 (N=13)	0.820513	0.778	0.667	1	0.096635	0.8663	0.3516
1 (N=13)	0.790598	0.778	0.611	0.944	0.096635	-0.1927	-0.5104
2 (N=12)	0.736111	0.750	0.444	1	0.185206	-0.4030	-1.0046

Table 2. Summary statistics for efficiency

NLCS	Mean	Median	Min.	Max.	St. Dev.	Skew.	Kurtos.
0 (N=13)	0.014647	0.014675	0.0090	0.0224	0.003844	0.3130	-0.3562
1 (N=13)	0.013717	0.013722	0.0062	0.0202	0.004191	-0.2557	-0.8077
2 (N=12)	0.012510	0.012262	0.0062	0.0182	0.004149	-0.1652	-1.1057

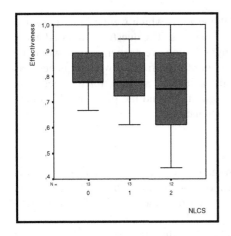

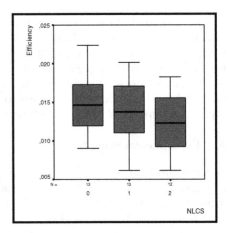

Fig. 1. Effectiveness box-plot

Fig. 2. Efficiency box-plot

Table 3. t-Tests results

Dependent variable	NLCS	df	t Stat.	Sig.
	0 vs 1	24	0.789	0.438
Effectiveness	0 vs 2	16.281	1.411	0.177
	1 vs 2	16.281	0.911	0.376
	0 vs 1	24	0.587	0.563
Efficiency	0 vs 2	23	1.334	0.195
	1 vs 2	23	0.723	0.477

Both for effectiveness and efficiency, the mean values for 0 and 1 nesting levels were quite close, while the mean values for 2 nesting levels were much lower. In our opinion, this means that a flat nesting level (0 or 1 levels) helped the subject to a better understanding of the diagrams than a bigger nesting level.

The results of the t-Tests performed did not allow us to reject any of the null hypotheses that we presented in section 2, as all the significance levels are above 0.05.

Anyway, these results were considered as preliminaries. That is why we performed the replication of the experiment that we present in the following section.

5 Replication

Most of the features of the replication are exactly the same that in the original experiment, so in this section we will only comment the main differences:

– The replication took place at the Universidad de Castilla – La Mancha (Spain) at the end of May 2005.

– It was performed by 64 undergraduate students. They all were on their 3rd year of Computer Science and had already received a nearly complete Software Engineering course in which they had been taught the main features of UML. Anyway, these subjects received the same training session by the same experiment supervisor and performed the same examples than in the first experiment.
– The experimental material was also exactly the same than in the original experiment and it was randomly given out to the subjects.

5.1 Data Analysis and Interpretation of the Replication

Again, our first step was carrying out an analysis of the descriptive statistics of the data. In this case, we can find that the box-plots of the data in Figures 3 and 4 illustrate the statistics summarized in Table 4 and Table 5.

Table 4. Summary statistics for effectiveness (replication)

NLCS	Mean	Median	Min.	Max.	St. Dev.	Skew.	Kurtos.
0 (N=21)	0.830689	0.889	0.667	1	0.083289	-0.304651	-0.07462
1 (N=21)	0.698413	0.722	0.389	0.889	0.113273	-1.072197	1.46562
2 (N=22)	0.739899	0.778	0.5	0.889	0.116769	-0.701001	-0.46467

Table 5. Summary statistics for efficiency (replication)

NLCS	Mean	Median	Min.	Max.	St. Dev.	Skew.	Kurtos.
0 (N=21)	0.015473	0.015801	0.0103	0.0216	0.003071	0.1515	-0.7263
1 (N=21)	0.012659	0.012302	0.0050	0.0190	0.003947	-0.0588	-0.6591
2 (N=22)	0.012927	0.011998	0.0081	0.0186	0.003439	0.2409	-1.0887

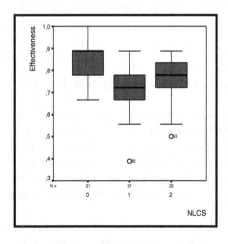

Fig. 3. Effectiveness box-plot (replication)

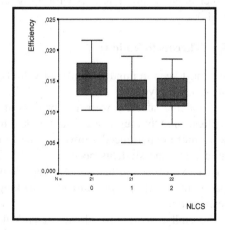

Fig. 4. Efficiency box-plot (replication)

Again, we found the best values for both effectiveness and efficiency when the value of NLCS is 0.

We also checked the hypotheses presented in section 2, by performing some t-Tests with $\alpha=0.05$. Table 6 presents the results for the different dependent variables in function of the NLCS values.

Table 6. t-Tests results (replication)

Dependent variable	NLCS	df	t Stat.	Sig.
Effectiveness	0 vs 1	39	4.300	**0.000**
	0 vs 2	40	2.693	**0.010**
	1 vs 2	39	-1.210	0.233
Efficiency	0 vs 1	40	2.578	**0.014**
	0 vs 2	41	2.556	**0.014**
	1 vs 2	41	-0.238	0.813

In this case there are some statistically significant values, when relating the effectiveness and efficiency obtained for values 0 vs. 1 and 0 vs. 2 of NLCS. This would allow us to reject the hypotheses H_{0-01} and H_{0-02}.

This would indicate that the optimal nesting level within a UML statechart diagram is 0, that is, not using composite states. An explanation to this finding could be the size of the UML statechart diagrams used in the experiment. In fact, the diagram with a value for NLCS of 0 has only 13 simple states, so it could be more effective and efficient to have no nesting level in the diagram as it can be understood quite immediately. It seems that introducing nesting levels unnecessarily overloads the designer without adding any positive contribution to the understandability of UML statechart diagrams. In order to check this explanation, some more experiments using diagrams with a bigger size and more complexity must be carried out.

5.2 Threats to Validity

We must keep in mind a number of validity issues that are typically related to experiments of this type.

First, the subjects were not professional modelers in the experiment or in the replication. Obviously, we would expect much more accurate results if the subjects were more experienced. However, the limited difficulty of the tasks and the UoD used make the students become suitable experimental subjects, as they are much easier to work with than some others. Nevertheless, further replications of the experiment using people already working in this profession would be really interesting.

Secondly, as we have already remarked, the diagrams that have been used represent a relatively simple model and it is possible that if real-projects, with more complex diagrams were used, we would obtain different results.

6 Conclusions and Future Work

Worried about how UML constructs impact on the understandability of UML statechart diagrams we carried out several empirical studies. The results obtained in a previous empirical research [10] had revealed that the use of composite states improved the understandability efficiency of UML statechart diagrams, i.e. how accurately the different stakeholders understands the diagrams, if the subjects have a certain level of experience in working with this kind of UML diagrams.

Going a step further, in this research we have investigated if the Nesting Level in Composite States (NLCS), which indicates the maximum number of nested composite states in an UML statechart diagram, affects the understanding of UML statechart diagrams.

The findings obtained through a controlled experiment and a replication of it have not been really conclusive. We have not been able to find an optimal use of nesting within UML statechart diagrams, and we can only partially conclude that a flat nesting level (0 or 1) within a relatively simple UML statechart diagram makes it more understandable.

As a future work, we must perform some new experiments with more complex diagrams in order to obtain more conclusive results which would allow us to establish some useful guidelines for designers. Moreover, we think that our research can have educational repercussions. The findings obtained until now give justify as special emphasis on the use of composite states when teaching UML statechart diagrams in software engineering courses at universities.

Acknowledgements

This research is part of the MESSENGER project (PCC-03-003-1) financed by 'Consejería de Ciencia y Tecnología de la Junta de Comunidades de Castilla-La Mancha (Spain)', the DYNAMICA/PRESSURE and the DYNAMICA/CALIPO projects supported by 'Dirección General de Investigación del Ministerio de Ciencia y Tecnología (Spain)' (TIC2003-07804-C05-05) and the ENIGMAS project (PBI-05-058) financed by 'Consejería de Educación y Ciencia de la Junta de Comunidades de Castilla-La Mancha (Spain)'.

The authors would like to thank Professor José Manuel García Carrasco, Professor Crescencio Bravo and Professor Félix Óscar García for allowing us to perform our experiments with their students.

And finally, we would like to thank all our subjects from the Universities of Murcia and Castilla – La Mancha for their patience and commitment with our studies.

References

1. Atkinson, C. and Kühne, T.: Model Driven Development: a Metamodeling Foundation. IEEE Transactions on Software Engineering. 20 (2003) 36-41
2. Basili, V., Shull, F. and Lanubile, F.: Building Knowledge through Families of Experiments. IEEE Transactions on Software Engineering. 25 (1999) 456-473
3. Briand, L., Bunse, C. and Daly, J.: A Controlled Experiment for Evaluating Quality Guidelines on the Maintainability of Object-Oriented Designs. IEEE Transactions on Software Engineering. 27 (6) (2001) 513-530
4. Briand, L., Morasca, S. and Basili, V.: Property-Based Software Engineering Measurement. IEEE Transactions on Software Engineering. 22 (1) (1996) 68-86

5. Briand, L., Wüst, J., Daly, J. and Porter, V.: Exploring the Relationships between Design Measures and Software Quality in Object-Oriented Systems. The Journal of Systems and Software. 51 (2000) 245-273
6. Briand, L., Wüst, J. and Lounis, H.: Investigating Quality Factors in Object-oriented Designs: An Industrial Case Study. Technical Report ISERN 98-29, version 2 (1998)
7. Cartwright, M.: An Empirical View of Inheritance. Information and Software Technology. 40 (4) (1998) 795-799
8. Chidamber, S., Darcy, D. and Kemerer, C.: Managerial Use of Metrics for Object-Oriented Software: An Exploratory Analysis. IEEE Transactions on Software Engineering. 24 (8) (1998) 629-639
9. Chidamber, S. and Kemerer, C.: A Metrics Suite for Object-Oriented Design. IEEE Transactions on Software Engineering. 20 (1994) 476-493
10. Cruz-Lemus, J. A., Genero, M., Manso, M. E. and Piattini, M.: Evaluating the Effect of Composite States on the Understandability of UML Statechart Diagrams. Proc. of ACM/IEEE 8th International Conference on Model Driven Engineering Languages and Systems (MODELS / UML 2005). Montego Bay, Jamaica (2005)
11. Cruz-Lemus, J. A., Genero, M. and Piattini, M.: Metrics for UML Statechart Diagrams. Chapter 7 in Metrics for Software Conceptual Models. Genero, M., Piattini, M. and Calero, C. (eds.) Imperial College Press, United Kingdom (2005)
12. Daly, J., Brooks, A., Miller, J., Roper, M. and Wood, M.: An Empirical Study Evaluating Depth of Inheritance on Maintainability of Object-Oriented Software. Empirical Software Engineering. 1 (2) (1996) 109-132
13. Denger, C. and Ciolkowski, M.: High Quality Statecharts through Tailored, Perspective-Based Inspections. Proc. of 29th EUROMICRO Conference "New Waves in System Architecture". Belek, Turkey (2003) 316-325
14. Harrison, R., Counsell, S. and Nithi, R.: Experimental Assessment of the Effect of Inheritance on the Maintainability of Object-Oriented Systems. The Journal of Systems and Software. 52 (2000) 173-179
15. Höst, M., Regnell, B. and Wohlin, C.: Using Students as Subjects - a Comparative Study of Students & Proffesionals in Lead-Time Impact Assessment. Proc. of 4th Conference on Empirical Assessment & Evaluation in Software Engineering (EASE 2000). Keele, UK (2000) 201-214
16. Kitchenham, B., Pfleeger, S., Pickard, L., Jones, P., Hoaglin, D., El-Emam, K. and Rosenberg, J.: Preliminary Guidelines for Empirical Research in Software Engineering. IEEE Transactions on Software Engineering. 28 (8) (2002) 721-734
17. OMG: MDA - The OMG Model Driven Architecture. Object Management Group (2002)
18. Poels, G. and Dedene, G.: Evaluating the Effect of Inheritance on the Modifiability of Object-Oriented Business Domain Models. Proc. of 5th European Conference on Software Maintenance and Reengineering (CSMR 2001). Lisbon, Portugal (2001) 20-29
19. Prechelt, L., Unger, B., Philippsen, M. and Tichy, W.: A Controlled Experiment on Inheritance Depth as a Cost Factor for Code Maintenance. The Journal of Systems and Software. 65 (2003) 115-126
20. SPSS: SPSS 11.5, Syntax Reference Guide. SPSS Inc. Chicago, USA (2002)
21. Unger, B. and Prechelt, L.: The Impact of Inheritance Depth on Maintenance Tasks - Detailed Description and Evaluation of Two Experimental Replications. Technical Report, Karlsruhe University (1998)
22. Verelst, J.: The Influence of the Level of Abstraction on the Evovability of Conceptual Models of Information Systems. Proc. of 3rd International Syposium on Empirical Software Engineering (ISESE 2004). Redondo Beach, USA (2004) 17-26
23. Wohlin, C., Runeson, P., Hast, M., Ohlsson, M. C., Regnell, B. and Wesslen, A.: Experimentation in Software Engineering: an Introduction. Kluwer Academic Publisher (2000)

Appendix A: Experimental Material

In this appendix we present an example of the original (Spanish) experimental material handed out to the subjects in the experiment and a translated version of the questionnaire attached to the diagrams.

The complete experimental material can be found at http://alarcos.inf-cr.uclm.es

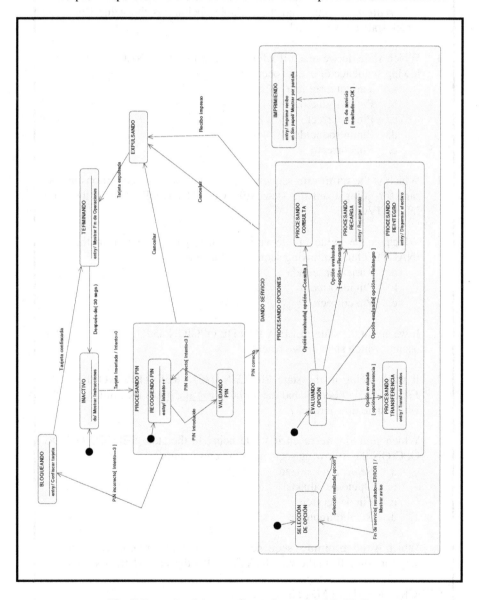

Fig. 5. Example of the experimental material (NLCS=2)

CHECK TIME (HH : MM : SS) __ : __ : __

1. If we are in the state IMPRIMIENDO and the event *Recibo impreso* occurs, which state do we reach?

2. If while being in the state SELECCIÓN DE OPCIÓN the event *Selección realizada* occurs and the variable opción has the velue Consulta, which state do we reach?

3. Which state do we reach if while being in the state INACTIVO the following sequence of events occurs?
    a. Tarjeta insertada
    b. Pin introducido
    c. Pin incorrecto
    d. Pin introducido
    e. Pin correcto

4. Which is the **minimum** sequence of events and guard conditions necessary for going from the state SELECCIÓN DE OPCIÓN to the state INACTIVO?

5. Which is the value of the variable Intento if starting from the state INACTIVO the following sequence of events occurs?
    a. Tarjeta insertada
    b. Pin incorrecto
    c. Pin correcto

6. If we are in the state SELECCIÓN DE OPCIÓN and the event *Cancelar* occurs, which state do we reach?

7. If while being in the state PROCESANDO REINTERGRO the event *Fin de servicio* occurs and the variable resultado has the value ERROR, which state do we reach?

8. Which state do we reach if while being in the state SELECCIÓN DE OPCIÓN the following sequence of events occurs?
    a. Selección realizada
    b. Opción evaluada
    c. Fin de servicio
    d. Recibo impreso

9. Which is the **minimum** sequence of events and guard conditions necessary for going from the state INACTIVO to the state TERMINANDO?

CHECK TIME (HH : MM : SS) __ : __ : __

Utilizing a Multimedia UML Framework for an Image Database Application

Temenushka Ignatova and Ilvio Bruder

Database Research Group, Computer Science Department,
University of Rostock, Germany
{temi, ilr}@informatik.uni-rostock.de

Abstract. To support the design of data models for multimedia applications, we employ the concept of a framework introduced in object-oriented design. We define a UML framework, which can be used for deriving application-specific multimedia database models. With the UML framework, we define the core elements of a multimedia database model, such as mediatype- and application-independent structure, content, relationships and operations. Thereby, the advantages of using UML for representing multimedia data as well as shortcomings of this approach are discussed. Furthermore, we describe the utilization of the UML framework for the instantiation of a model for an image database of scanned handwritten music scores.

1 Introduction

Multimedia databases have to provide support not only for representing the data itself, but also for the application-specific behavior and characteristics of this data. Therefore, the latter have to be considered in the design of conceptual data models for building multimedia database applications.

There are a number of proposals for multimedia data models originating from a database, an information retrieval, as well as from a media-specific points of view [1]. Inside [2], Subrahmanian sets the theoretical fundamentals of multimedia database systems as well as the requirements towards the underlying data models. A model for multimedia documents, which implements the information retrieval aspect is proposed by Chiaramella, Mulhem, and Fourel in [3]. A logical model for multimedia documents based on the description logic ACL has been enhanced with fuzzy logic based functionality by Meghini, Sebastiani, and Straccia [4]. In chapter 2 of [5], Wu, Kankanhalli, Lim, and Hong have represented content-based definitions for multimedia documents and systems from the computer vision point of view.

These models propose concepts with a strong theoretical background. However, they concentrate more on the multimedia data itselfs rather than on the design of multimedia applications. Therefore, we study the possibilities to build application-specific conceptual models for multimedia databases. The MPEG-7 standard, for example, offers an extensive multimedia description interface, for media-specific descriptors, as well as media type independent description schemes. Thus, making use of MPEG-7 should be considered in designing

J. Akoka et al. (Eds.): ER Workshops 2005, LNCS 3770, pp. 23–32, 2005.

database models. However, the Data Definition Language offered by the standard XML Schema does not provide the right structure for managing the data efficiently in a database, as mentioned by Kosch in [6]. A possible solution to the problem was offered in the doctoral thesis of Westermann [7], in which the author represents a Typed Data Object Model as a generic data model for XML documents.

Furthermore, the complexity of the data and the structural and semantic dependencies, as well as the need of integrating functionality set challenges for the design of multimedia database applications. The abstraction of this information considering the possibilities for a later transformation and implementation within an object-relational database environment is not trivial.

In this paper, we propose a conceptual framework for modeling multimedia data using UML and describe its practical adaptation for a specific image retrieval application. In Section 2, the structure and elements of the framework are represented. Section 3 explains how the framework can be used for modeling an image database application. And finally, in Section 4 conclusions and future work are represented.

2   UML Framework for Multimedia Database Applications

A Framework in the context of object-oriented programming languages [8] and UML [9] refers to a customizable, extendable skeleton of a software architecture, which can be used for sub-classing domain specific applications. In this paper, we use the term framework to represent a set of UML classes and relationships between them for deriving application-specific multimedia data models. The proposed framework is based on a three-level structure: the media abstraction level, the document level, and the collection level.

A media abstraction is defined similarly to the definition found in [2]: it describes the content, properties and the structure of an object of a specific media type. A media abstraction has an identifier, an object type – text, audio, video, image –, the raw data as large objects, features as attribute-value-pairs, which describe extracted properties, relationships within the object as a mapping between features and data and mappings from features onto knowledge concepts, and finally, interpreted content and properties of the media object as attribute-value-pairs.

In the document level, we define the logical structure of a document through components (structural elements of a document), which contain a set of media abstractions and other components (sub-components) building together a hierarchy. Components, which are not sub-components are root elements of a document. Components have also an identifier and a type (document, component, or media object component), which can be determined using the position within the hierarchy. They have a parent property and other metadata as attribute-value-pairs. There exist relationships between components and mappings onto knowledge concepts.

A collection is an aggregation of documents and sub-collections. These structures should explain overall properties and relationships of a set of documents. Documents have sometimes different roles within a collection. A collection has an identifier and properties, such as a parent identifier and collection metadata. There are also possible relationships between documents and sub-collections.

2.1 Multimedia Structures

An overview of the UML framework is shown in the package diagram on Figure 1.

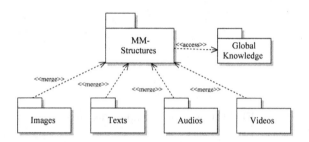

Fig. 1. Package Diagram of the Framework

The package MM-Structures defines the media types and provides references to global knowledge. The document structure for the storage of a multimedia document is modeled with the MM-Structures package. Components of a document, which contains exactly one media type are modeled by the respective media package. The knowledge is used for structure analysis, domain specific interpretations and media transparent analysis. It is accessed by the corresponding methods. However, in our framework we do not specify further the structure and content of the knowledge, because it is application-dependent.

To begin with, we introduce components (documents) and collections and their metadata. These represent structures, which are independent from a specific media type.

The UML diagram shown on Figure 2 represents a Collection class as a higher-level concept for managing multimedia documents. A collection has an identifier and a parent determined by the aggregation relationship to build a hierarchy of collections and sub-collections. Components are organized in a similar structure. They have either an aggregation relationship to collections or a child/parent aggregation to another component (component – subcomponent). Components and collections can have metadata, which is represented by the abstract class Metadata in the diagram. Two instances of the Metadata class are predefined in the framework: Dublin Core as a general metadata set and Physical metadata. More specific or other metadata can be defined for an application using these metadata classes.

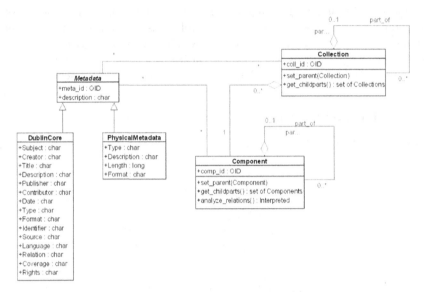

Fig. 2. Components and Collections

Figure 3 shows components and possible relationships between them. The relationship concept allows to model relationships and interaction between the media types to represent data analysis and query processing possibilities. We distinguish four low-level relationships: spatial, temporal, logical, and interactional relationships in the framework. Furthermore, we define high-level relationships, which have to be analyzed and interpreted from other relationships. Spatial and temporal relationships are used for spatial and temporal arrangements of components. Interactional relationships are relationships, which define possible user activities. Logical relationships describe relationships such as a book has chapters and chapters have sections. The logical arrangement is represented by an aggregation in UML. All other relationships are represented by associations. In real-world applications it is often necessary to restrict the association to an aggregation or composition or to alter the multiplicity. UML does not provide such "association overriding" concepts, therefore such changes have to be made by the application developer. Interpreted relationships can be build by combining any other relationships. The Interpreted Relationship class is represented by an n-ary UML association and is related with only one other relationship of any kind, but has two relations with the original data in Components. In reality there are also cases of this relationship with more than one related classes, but according to the UML standard it is not possible to model an abstract relationship with unknown number of related classes (e.g., binary and 3-ary). The interpreted relationship class is an abstract class. The concrete analysis function for the interpretation has to be defined for a specific application by the framework user. The classes and relationships we designed so far can be used, extended, and specialized by sub-models of a specific media type and for a concrete application.

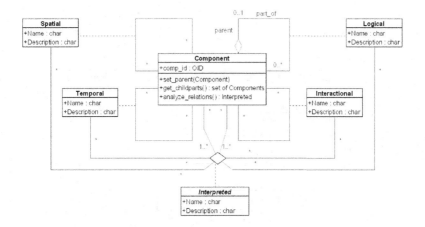

Fig. 3. Components and their Relationships

2.2 Image Component

We consider digital raster images as an example of a multimedia component, in order to meet the requirements of the example application described in Section 3. Analogously, other types of components, e.g., text and vector image components can be specified. In Figure 4, we have extended the multimedia document framework with classes for describing images and their content.

The class Image Component allows the representation of the raw image data as a binary sequence or a reference to a file, through a path description. Digital Images have attributes, which are content independent, which means that they are not in direct relationship with the visually expressed information in the image. Such attributes can be image size, image format etc. These data are represented by the class Image Technical Metadata, which is derived from the multimedia class Metadata in MM-Structures. Furthermore, metadata related to the application context can be assigned to digital images. However, we leave these kind of data out of our Framework, since it is application-dependent. Such metadata can be added by defining new attributes or new classes and appropriate relationships to the Image Component class.

The Image Component represents information about the content of the image based on spatial abstractions, which are derived from the segmentation of the image. Thereby, the content of an image is interpreted as a set of regions. Each image contains on one side other images and on the other regions or segments of an image. These containment possibilities are modeled as an aggregation relationship. The relationships allow building a hierarchy of regions of an image. Application-specific regions can be defined by the developer by implementing the abstract class Region.

A feature can be assigned to each region of an image, where the whole image can also be described as a region. Various features can be defined to describe the content of images, by inheriting from the abstract class Feature. These can be

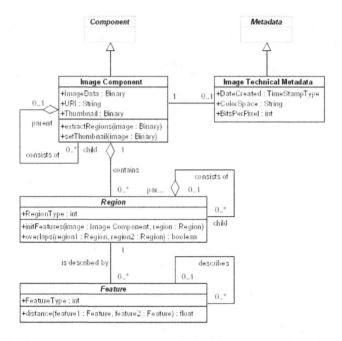

Fig. 4. Image Component

low-level features such as dominant color or color histogram, as well as high level features, such as names of objects or concepts. Therefore, relationships between features have been defined with an association. These relationships are used to link low-level features with high-level features.

2.3 Methods/Operations

The possibility to design operations within the data model is one of the reasons for choosing UML as a modeling approach for our framework. Besides the data management, operations are also important for building an application. We consider the following cases for making use of operations:

- Constructors and destructors to build the base structure of multimedia data.
- Feature Extraction: encapsulation of operations for the automatic extraction of features. We consider the combination of defining feature extraction methods as image class methods as well as defining them as methods specific only to a feature, or as global methods for the application.
- Operations for interpreting features, such as distance functions, transformations etc. should be defined as methods of the corresponding feature.
- Operations to support derived attributes and relationships

Methods are represented by their signatures within the class definition in a UML class diagram. However, defining the semantic of an operation with

UML sets certain challenges. State charts, communication, and activity diagrams, among others, can be used to specify further information on methods in UML. Nevertheless, there are problems to combine all these diagrams with the class diagram to receive a complete specification of an operation. Some constraints within a method can be defined by OCL. However, OCL is only usable in a few restrictive cases. Using comments is also not an applicable, general possibility. The best way for later model transformation is the pre-implementation of general methods. The disadvantage is that we have to implement the method in different programming languages for a general use of the model. But, we have to implement the method only once for many applications.

It only makes sense to pre-implement methods of general use. Therefore, in our framework we provide constructor and destructor methods (implemented in Java) for building the structure of documents. Application-dependent methods have to be implemented by the application developer.

2.4   Mapping onto an ORDBMS Schema

A mapping mechanism for the UML Model can be defined for a specific implementation platform. We consider mapping onto the SQL99 standard for building object-relational database schemas. Currently, we provide a simple transformation, similar to the one introduced in [10, 11]. Classes are translated into structured types and corresponding typed tables. Complex attribute data types, which are not supported by SQL99 are mapped onto structured types and a reference to a structured type respectively. Derived attributes are represented by methods and triggers. Associations and aggregations are translated into references to structured types. Composition is represented as an association with a cascading constraint. Association classes are treated like ordinary classes. Inheritance is expressed using the UNDER clause for building structured types hierarchies.

Due to the general character of the framework, it is not possible to create an optimized, well-formed database schema, because we do not know the application and its possible data and queries. Therefore, we have to consider making more often use of the standardised multimedia options of SQL99 and SQL/MM or concrete platform extensions such as the IBM DB2 Extenders. The UML framework described here has been implemented in *IBM Rational Rose Data Modeler*[1] as a Rational Rose framework, which can be used for building application-specific models with this modeling tool.

3   How to Use This Framework - Example

To demonstrate the usage of the framework we designed a database application for storing images of music scores and their handwriting characteristics, used for the automatic identification of their scribes. The application is implemented for a digital archive of historical music scores in the project eNoteHistory [12].

---

[1] http://www-306.ibm.com/software/awdtools/developer/datamodeler/

Fig. 5. Automatic Object Recognition in Music Scores

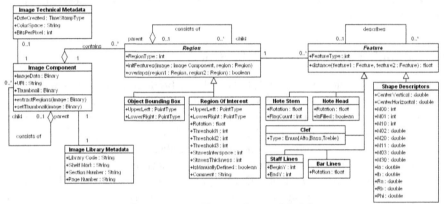

Fig. 6. eNoteHistory Image Model

The application steps involved in the automatic handwriting analysis and content-based retrieval are carried out as follows. At first, for all digital scores in the database, for which information about the scribe (e.g., name of scribe) exists, image processing algorithms are applied to extract the visual features of the images, representing the handwriting characteristics. Figure 5 shows the recognized objects in the manuscript. For each recognized object: note heads, note stems, bar lines a set of geometrical features is extracted, such as: height and width of the bounding box, radius of the bounding ellipse, x, y coordinates of the centroids, orientation etc. The handwriting characteristics of scores with unknown scribes can subsequently be compared with the set of extracted features in the database and using distance metrics for calculating the similarity between features a query result of the type: a list of k-most similar scores with associated scribes can be generated.

For this database application we have derived the model shown on Figure 6 from the UML Framework. The Image Component class represents the scanned page image of the music score. In addition to the Technical Metadata a class Image Library Metadata has been defined, which was derived from the abstract class Metadata. The classes Regions and Features and the associations between them were imported without extensions or changes from the UML framework. Application-specific implementations of these classes have been derived to represent the possible types of regions: a Region of Interest is a selected region of the image, which participates in the extraction of the features; an Object Bounding Box represents the bounding box of an image segment, which is a salient object. These two types of regions can be organized in a hierarchy through the aggregation association of the parent class. The hierarchy, which is meaningful for this application has a Region of Interest as a root node and many object bounding boxes under the root node, which can furthermore contain other bounding boxes. The Region of Interest has attributes, which in this case do not need to be represented as features, because they are only a limited number and do not need special operations for their comparison. For the Object Bounding Box however multiple features can be associated. The ones needed for the application are derived from the abstract class Feature. We have defined five high-level features: Note Stem, Note Head, Clef, Staff Lines and Bar Lines, as well as a class of low-level features: Shape Descriptors. Through the association modeled for the Feature class it is possible to associate the shape descriptors with the high-level features, which are described with these descriptors such as the shape of a note head or of a note stem.

The operations, defined in the model are intended to be used on one side to generate the data, which has to be derived from the image by image processing algorithms and on the other side to support similarity queries on the images, by providing a distance function for the features representing the content of the images. The operation extractRegions(Image) has to generate the instances of regions for a specific image. setThumbnail(Image) is responsible for the creation of a thumbnail for the image and setting as a value for the attribute Thumbnail. The operation initFeatures(Image Component, Region) should extract the high-level and low-level features corresponding to a region and create the instances of objects to be stored in the database. The operation overlaps(Region, Region) implements a function, which analyzes the position of two regions and returns a true value if the two regions have overlapping points. This function can be used for deriving spatial relationships between the objects of an image or other spatial image analysis. And finally the function distance(Feature, Feature) implements a distance function, which compares two instances of a Feature and returns a scalar value representing their degree of similarity. This function can be used for processing similarity search queries on the images. The implementation of these operations is currently provided as a Java application. For the transformation of the model onto an ORDBMS we have implemented an Add-In for the *IBM Rational Rose Data Modeler*, which translates the model classes, attributes and relationships into a DDL script for creating the database schema. The integration of the operations however is currently left to the developer.

4   Conclusions and Future Work

This paper introduces an UML framework for multimedia data to support the development of multimedia database applications. We focus on the description of the image component and its usage for the design of an application for the identification of scribes of handwritten music scores. The experience with this application proved the advantages of the framework, such as the facilitated design and maintenance of the application, and the seamless integration with other applications. However, some problems remain still open. One of the main challenges in the design of the framework is the definition of methods to be used in concrete applications. Therefore, research is still going on to improve this general modeling approach. We are also working on combining more media types from a multimedia document using the relationships of the general framework.

References

1. Ignatova, T., Bruder, I.: Utilizing relations in multimedia document models for multimedia information retrieval. In: Proc. of the Int. Conf. - Information, Communication Technologies, and Programming -, Varna. (2003)
2. Subrahmanian, V.S.: Principles of multimedia database systems. Morgan Kaufmann Publishers Inc., San Francisco, CA, USA (1998)
3. Chiaramella, Y., Mulhem, P., Fourel, F.: A Model for Multimedia Information Retrieval. (1996)
4. Meghini, C., Sebastiani, F., Straccia, U.: A Model of Multimedia Information Retrieval. (1998)
5. Wu, J., Kankanhalli, M., Lim, J.H., Hong, D.: Perspectives on Content-based Multimedia Systems. Kluwer Academic Publishers (2000)
6. Kosch, H.: MPEG-7 and Multimedia Database Systems. SIGMOD Rec. **31** (2002) 34–39
7. Westermann, G.U.: A Persistent Typed Document Object Model for the Management of MPEG-7 Media Descriptions. Diss. (2004) Techn. Universität Wien.
8. Gamma, E., Vlissides, J., Johnson, R., Helm, R.: Design Patterns CD: Elements of Reusable Object-Oriented Software. Addison Wesley, Boston, USA (1998)
9. Booch, G., Rumbaugh, J., Jacobson, I.: The Unified Modeling Language user guide. Addison Wesley, Redwood City, USA (1999)
10. Marcos, E., Vela, B., Cavero, J.M.: A Methodological Approach for Object-Relational Database Design Using UML. Inform., Forsch. & Entwickl. **18** (2004) 152–164
11. Muller, R.J.: Database Design for Smarties : Using UML for Data Modeling. Morgan Kaufmann (1999)
12. Bruder, I., Finger, A., Heuer, A., Ignatova, T.: Towards a Digital Document Archive for Historical Handwritten Music Scores. In: 6th International Conference of Asian Digital Libraries ICADL, Kuala Lampur, Malaysia (2003)

Object Class or Association Class? Testing the User Effect on Cardinality Interpretation

Geert Poels, Frederik Gailly, Ann Maes, and Roland Paemeleire

Management Informatics Research Unit,
Faculty of Economics and Business Administration,
Ghent University – Ugent,
Hoveniersberg 24, B-9000 Gent, Belgium
{geert.poels, frederik.gailly, a.maes,
roland.paemeleire}@UGent.be

Abstract. In UML class diagrams, a many-to-many relationship with attributes can be represented by an association class or by a connecting object class. It is unclear which modeling construct is preferred in particular modeling scenarios. Because of lack of theory, this paper investigates the issue empirically. An experiment was conducted that tested the effect of representational form chosen on the performance of model users at cardinality interpretation tasks. It was shown that, controlling for cardinality knowledge, business users can better interpret the information that a UML class diagram conveys about a many-to-many relationship with attributes if this relationship is represented as an association class. The implication for 'best practices' in UML modeling is that modelers should refrain from objectifying such relationships if the goal is an effective communication of domain semantics to users that are not modeling experts.

1 Motivation

In conceptual modeling practice, it is accepted that semantic relationships between objects can have attributes of their own [24]. The UML construct to specify such relationships is the association class, which is both an association and a class [14]. Although the representation by an association class seems a logical choice, widely-read literature on conceptual modeling (e.g. [6], [12]) recommends that also objects be used to represent semantic relationships. Relationship objectification means that the association class between two classes X and Y (Figure 1) is replaced by an object class, which is not an association, but has associations with X and Y (Figure 2).

Comparing both figures, it can be seen that objectification introduces a level of indirection in the relationship between X and Y. Another consequence is the changed positioning of the multiplicities. These multiplicities specify the allowable cardinalities of the modeled relationships (e.g. an object of class X is related to at least c and at most d objects of class Y).

In a UML class diagram representing a structural model of the domain, the modeled cardinality constraints pertain to laws that hold for the semantic relationships between objects. A major function of such models is to communicate domain semantics [10]. It is therefore imperative that relationships and their laws be rightfully

J. Akoka et al. (Eds.): ER Workshops 2005, LNCS 3770, pp. 33–42, 2005.

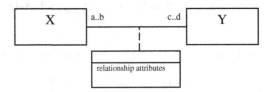

Fig. 1. Association class represents a semantic relationship with attributes

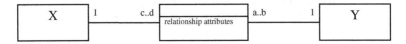

Fig. 2. Objectifying a semantic relationship with attributes

interpreted by model users, including business users that might have limited modeling experience and training in UML.

The question addressed in this paper is whether the representation chosen for a relationship with attributes affects the ability of model users to understand the information conveyed by a UML class diagram. It might be that the indirection introduced through objectification makes it more difficult to interpret the modeled cardinalities as intended by the modeler. It might as well be that the pattern of multiplicity specifications is simpler after objectification. For class diagrams, there are no theoretically-grounded prescriptions to distinguish situations in which object classes should be used from situations in which association classes should be used [16]. Nevertheless, modeling practitioners expect clear guidance as to which modeling constructs should be used under particular circumstances [25]. With communication towards business users in mind, what is the best way to represent a relationship with attributes: object class or association class? This paper aims at finding an answer by investigating the issue empirically.

Section 2 lists some advantages and disadvantages of objectification found in the literature, that are used to further refine the research question. Section 3 presents our research framework. The design, operation and data analysis of an exploratory experiment are presented in section 4. Section 5 concludes the paper.

2 Objectifying Relationships: Advantages and Disadvantages

The practice of relationship objectification is motivated by a number of empirical studies indicating that database designers find relationships more difficult to model than objects [24]. But also for other applications than database design, objectification is recommended. For instance, [7] presents a formally defined enterprise modeling method that objectifies all many-to-many relationships between objects into contract objects that are existence dependent on the objects participating in the relationships.

Objectification reduces complexity by making the relationships between objects simple and easy to understand [20]. In particular when the maximum cardinality constraint is 'many' on both sides of the relationship (i.e. when the *b* and *d* in Figures

1 and 2 are '*'), objectification simplifies the many-to-many object connectivity (Figure 1) by replacing it with one-to-many links to a connecting object (Figure 2).

Other researchers have argued against objectification. In [4], [24] it is shown that this practice violates an important rule of the Bunge-Wand-Weber (BWW) representation model, which is an ontological theory used to evaluate conceptual modeling grammars. According to the BWW model, objects should be used only to represent things. Furthermore, things can be linked only by mutual properties. A semantic relationship is a mutual property of two or more things, but is not a thing. Therefore, an object should not be used to represent a semantic relationship.

Lack of ontological clarity leads to semantic ambiguities in models, which in turn affects the user understanding of the models; a prediction supported by theories of cognition [3]. On the other hand, objectification circumvents the direct representation of many-to-many relationships, which are more difficult to conceptualize than one-to-many relationships. It is not uncommon that the application of the BWW model contradicts widely used practices in conceptual modeling [19]. Therefore, ontology-based predictions should be tested empirically [10]. Based on the observations made in this section, the study presented next focuses on many-to-many relationships.

3   Research Framework

Our research framework (Figure 3) is derived from Cognitive Fit theory [23]. The task performance of model users is affected by cognitive fit, which is the match between the model representation and the task that the user has to perform using/on the model. The more suitable the representation is for the particular (type of) task executed by the particular (type of) user, the better the task performance is.

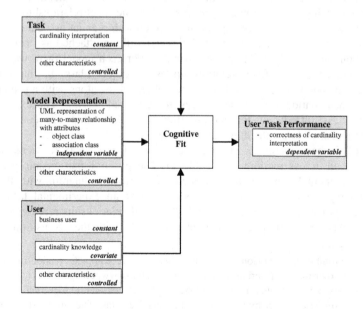

Fig. 3. Research framework

The research question addressed is whether the representation of a many-to-many relationship with attributes in a UML class diagram has an impact on the correctness of cardinality interpretation by business users. The representation chosen is the main factor under investigation and is varied at two levels: object class or association class. Given the exploratory nature of the research, the direction of the effect is not stated.

Consistent with the research question, the levels of the user and task factors are fixed at respectively business users and cardinality interpretation tasks. To investigate the research question in an experiment, other model representation, task, and user characteristics that may affect cognitive fit must be controlled to assure internal validity. Such characteristics include diagram layout [18], size [2], structural complexity [11], task complexity [1], domain knowledge [4], personal characteristics [9], modeling experience [15], and modeling language knowledge [13].

One user characteristic needs special attention. In [8] a significant impact of cardinality knowledge on user performance was found for tasks that involved cardinality interpretation and error identification. Low cardinality knowledge may affect someone's ability to correctly interpret relationship cardinalities, regardless of the representation chosen. To assess the impact of such a confounding effect, cardinality knowledge is used as a covariate in the study.

4   An Experiment

4.1   Design

The experiment employed two pairs of class diagrams representing two structural models from the auctioneering domain: a model for the sales/collection process and another model for the acquisition/payment process.[1] Each model included one many-to-many relationship with attributes (hereafter called the focal relationship). For each model there was a diagram showing the focal relationship as an association symbol (i.e. a solid path that connects the related classes) with an attached association class used as a container for the relationship attributes. For each model there was another diagram representing the focal relationship as an object class. In that case the relationship attributes are contained within the object class, along with a primary key that is the concatenation of the primary keys of the related classes.[2] These classes and the object class introduced for representing the relationship are connected by association paths, without attaching names to them.

Order of process modeled and representation of the focal relationship were counterbalanced across participants (Table 1). The group 1 participants performed first an experimental task on the sales/collection process diagram with the focal relationship represented as an object class and next an experimental task on the acquisition/payment process diagram with the focal relationship represented as an association class. The group 2 participants dealt first with the acquisition/payment

---

[1] Page limits prohibit the inclusion of the diagrams. A more complete version of the paper, including all experimental materials, can be found at http://www.feb.ugent.be.

[2] In [5] primary key attribute(s) of an object class are explicitly identified in the list compartment of the classifier by means of a {Primary Key} tag. Though not standard UML, this convention was followed in the course from which the study participants were drawn.

process diagram with the focal relationship represented as an object class and next with the sales/collection process diagram with the focal relationship represented as an association class. Counterbalancing reversed these orders for group 3 (i.e. reverse order of group 1) and group 4 (i.e. reverse order of group 2).

Table 1. Experimental design

Representational form	Object class				Association class			
Business process modeled	Sales/ Collection		Acquisition/ Payment		Sales/ Collection		Acquisition/ Payment	
Order	first	second	first	second	first	second	first	second
Group	1	3	2	4	4	2	3	1

This within-subjects experimental design was chosen to control differences in user characteristics. It necessitated, however, the use of two models. The models are conceptually very similar. The focal relationship fulfills in both models the role of relating a transfer of goods with a transfer of money. Only the perspective is different: an outwards transfer of goods and inwards transfer of money for the sales/collection process versus an inwards transfer of goods and outwards transfer of money for the acquisition/payment process.

Apart from the representation of the focal relationship, the other diagram elements are identical within each pair of diagrams. Also, there are no remarkable differences in size, structural complexity or aesthetics across the four diagrams.[3]

The minimum and maximum cardinality constraints specified for the focal relationship express important business policies that govern the business process modeled. The experimental task was directed towards the interpretation of these cardinalities. For each diagram there were six "yes or no" questions requiring participants to interpret the information about the focal relationship conveyed by the diagram and check this interpretation against a given scenario (described in the question). Discrepancy checking is a key aspect of model validation [13]. Furthermore this task is representative for the way that business users interact with conceptual models.

The same questions were used for the two diagrams representing a same model (Tables 2 and 3). Also across the two models, the questions used were conceptually similar. Participants were told that they had to answer these questions using the diagrams they received, even if they believed the diagrams were incomplete or invalid. The number of questions correctly answered was used as a measure of a participant's correctness of cardinality interpretation.

---

[3] The diagrams with the object class representation are slightly bigger in the sense that they show an additional diagram node. However, the questions used for the comprehension task (cf. infra) focused the attention on the semantic relationship between the transfer of goods and the transfer of money. The total size of the diagrams is not relevant for the type of comprehension tested.

Table 2. Interpretation task for class diagrams of Sales/Collection process

Questions for Sales/Collection process
1. Must, according to the diagram, a buyer pay the full amount of the bid price offered immediately upon acceptance of his/her bid ?
2. Is the diagram correct if a buyer is not allowed to pay for more than one accepted bid at a time ?
3. Can, according to the diagram, a buyer who's bid has been accepted, pay the bid price offered in parts ?
4. Is the diagram correct if some days after acceptance of their bid, buyers must pay the full amount of the bid price offered ?
5. Can, according to the diagram, payment be received from a buyer, even before his/her bid is accepted ?
6. Is the diagram correct if a buyer must pay part of the bid price offered, immediately upon acceptance ?

Table 3. Interpretation task for class diagrams of Acquisition/Payment process

Questions for Acquisition/Payment process
1. Is the diagram correct if an owner is paid only after some of his/her goods have been sold at the auction ?
2. Must, according to the diagram, every payment made to an owner, relate to at least one accepted offer of goods ?
3. Must, according to the diagram, some amount be paid to an owner, immediately upon acceptance of his/her offer of goods to auction ?
4. Is the diagram correct if more than one payment made to an owner, relates to the same accepted offer of goods?
5. Is the diagram correct if payments cannot be made unless they relate to an offer of goods that is accepted for the auction?
6. Must, according to the diagram, every payment made to an owner, relate to at most one accepted offer of goods ?

Forty-three graduate students took part in the experiment. As business students majoring in accounting, they were intimately familiar with transaction-oriented business processes and accounting policies. In the weeks before the experiment a thirty hours course on Accounting Information Systems was given. About half the time was spent on conceptual modeling of business processes. During classes the students were forced into the roles of (future) system users and system auditors that had to read business process documentation (such as structural models) in order to check processes as modeled against reality (as for instance described in textual scenarios). The course exercises on conceptual modeling focused especially on the interpretation of modeled business policies, considering cardinality constraints on relationships as a kind of internal accounting controls. Students also learned how to interpret business policies that cross-over relationships.

The notation used for structural models was that of the UML class diagram. The teaching approach taken was to consider UML as a communication vehicle for domain knowledge, meaning that, as future business professionals, the students must be able to comprehend conceptual models in UML notation (at least class diagrams and activity diagrams). A questionnaire administered in the first course session indicated that none of the students had taken prior UML training and only a couple of students were familiar with entity-relationship concepts, though most of them acquired elementary relational database (MS Access) knowledge. As the participating students possessed up to four years of working experience, they approximate a representative sample of the target population of business users.

The students had studied sales/collection and acquisition/payment processes before, but not for an auctioneer. Therefore participants' domain knowledge was assessed as low. In [17] it is shown that when participants are domain experts, they tend to interpret conceptual models using their domain knowledge, instead of using the information conveyed by the models.

The cardinality knowledge of the students was assessed using a procedure similar to [8]. At the beginning of the experiment each participant had to answer four "true or false" questions that tested the ability to identify the correct narrative corresponding to a cardinality notation. Following [17], diagram fragments were shown with symbolic labels (e.g. "A", "B", ...) attached to the diagram elements, instead of meaningful names. Hence any potential impact of domain knowledge was avoided. The number of questions correctly answered was used as a measure of a participant's cardinality knowledge.

4.2 Operation

When the students entered the room where the experiment took place, they were randomly assigned to the four experimental groups. Participants started with the cardinality knowledge test. Next they were given a first diagram and the interpretation questions corresponding to the process modeled in the diagram. After finishing the first interpretation task, participants were given their second diagram along with the corresponding questions.

No time limit was put upon the experiment; students could take whatever time they thought was necessary. The students were motivated to perform well in the experiment as it was part of their final exam. However, the experiment made up only one quarter of the exam and on the totality of the exam there was a time constraint.

4.3 Data Analysis

Thirty-seven participants managed to give correct answers to all questions for testing cardinality knowledge. Three students made one mistake; three others made two mistakes. Given the limited variability in the scores, cardinality knowledge could not be included as a covariate in the study, as originally planned. Instead, cardinality knowledge was controlled by excluding from the analysis the data of participants with a less than perfect score.

Descriptive statistics are shown in Table 4. For each representational form the mean number of questions correctly answered (i.e. a real value in the interval [0..6]) and its standard deviation are shown, per group as well as for all groups together. The mean score for correctness of interpretation was always higher for the association class representation.

The paired differences in correctness of cardinality interpretation scores between the treatments of the within-subjects factor (i.e. representational form) were normally distributed. Therefore a paired samples t-test was applied to test whether the representation of the focal relationship in the diagrams had an impact on the correctness of cardinality interpretation by the participants. The test results indicate that the mean correctness of cardinality interpretation score (for all groups, n = 37) was significantly higher for the association class representation ($t = -5.18$, $p < .001$).

Table 4. Experimental data – descriptive statistics & results of hypothesis test (controlled for cardinality knowledge)

Experimental groups (n = group size after controlling for cardinality knowledge)	Correctness of interpretation Mean score (Standard deviation)	
	Object class representation	Association class representation
Group 1 (n = 11) First: Sales/Collection, object class representation Second: Acquisition/Payment, association class representation	3.82 (0.87)	4.09 (0.83)
Group 2 (n = 9) First: Acquisition/Payment, object class representation Second: Sales/Collection, association class representation	3.56 (1.24)	5.11 (0.78)
Group 3 (n = 9) First: Acquisition/Payment, association class representation Second: Sales/Collection, object class representation	3.33 (0.50)	4.56 (0.88)
Group 4 (n = 8) First: Sales/Collection, association class representation Second: Acquisition/Payment, object class representation	3.63 (0.92)	4.63 (0.74)
All groups (n = 37)	3.59[a] (0.90)	4.57[a] (0.87)
Results of hypothesis test: [a] significantly different, $p < .001$ (two-tailed)		

Also tested were confounding effects that might be caused by the order in which the diagrams were shown to the participants or the business process modeled. As the normality requirement was satisfied, paired samples t-tests were used. The results show that there was no order effect ($t = 0.218$, $p = .829$). Neither was the correctness of cardinality interpretation different for the two business processes modeled ($t = 0.880$, $p = .384$). The latter posttest also confirms that characteristics related to the interpretation task (e.g. task complexity) were effectively controlled in the experiment, i.e. the sales/collection process questions and acquisition/payment questions were equally difficult.

5 Discussion

The results of the experiment indicate that, controlling for cardinality knowledge, business users can better interpret the information that a UML class diagram conveys about a many-to-many relationship with attributes if this relationship is represented as an association class instead of an object class. The implication for establishing 'best practices' in UML modeling is that modelers should refrain from objectifying such relationships if the goal is an effective communication of domain semantics to business users, who are not UML or modeling experts.

A weakness of the study is the use of only a limited number of class diagrams. In replication studies the use of more diagrams from different business domains should be considered to increase the validity of the study results.

The main limitation of the study concerns its external validity, which needs to be further investigated. Without further research it is difficult to assess whether the findings of this study can be generalized outside its specific task and user setting. The experiment was conducted in a very homogeneous environment, primarily defined by the course from which the participants were drawn. Future studies should consider more heterogeneous environments to address the possible impact of environmental setup on the results. Perhaps model users with more UML training and modeling experience react differently to the task and treatments presented? Perhaps objectification of many-to-many relationships results in a conceptual data model that is easier to implement in a relational database?[4] Notwithstanding these questions, the use of models for communication between system analysts and developers has been investigated much more than their use for communication between analysts and users [22]. This study contributes towards a better understanding of UML as a communication instrument towards business users of conceptual models.

In this exploratory study, user and task factors were controlled to increase internal validity. In future research, the interaction between the different factors that determine cognitive fit must be further investigated. In our own future work we will focus on the impact of representation familiarity on the task performance of conceptual model users. It is plausible that users experience less interpretation problems with a representational form simply because it is more familiar to them (e.g. because of learning). Therefore, it is better to explicitly consider the potentially confounding effect of representation familiarity by including this variable in the research model.

References

[1] Batra, D., Wishart, N.A., 2004. Comparing a rule-based approach with a pattern-based approach at different levels of complexity of conceptual data modelling tasks. International Journal of Human-Computer Studies 61, 397-419.

[2] Bowen, P.L., O'Farrel, R.A., Rohde, F.H., 2004. How Does Your Model Grow? An Empirical Investigation of the Effects of Ontological Clarity and Application Domain Size on Query Performance. In: Proceedings of the 25th International Conference on Information Systems. Washington DC, USA. 77-90.

[3] Burton-Jones, A., Meso, P., 2002. How good are these UML diagrams. An empirical test of the Wand and Weber good decomposition model. In: 23rd International Conference on Information Systems. Barcelona, Spain. 101-114.

[4] Burton-Jones, A., Weber, R., 1999. Understanding relationships with attributes in entity-relationship diagrams. In: Proceedings of the 20th International Conference on Information Systems. Charlotte, NC, USA. 214-228.

[5] Connolly, T., Begg, C., 2002. Database Systems: A Practical Approach to Design, Implementation, and Management, 3rd Ed. Addison-Wesley.

[6] Date, C.J., 1995. An Introduction to Database Systems, 6th Ed. Addison-Wesley.

---

[4] In [24] it is noted that the design and implementation perspective of relationships as links between objects is different from the conceptual modeling view of a domain (i.e. relationships as mutual properties of things). This argument is further supported in [21] specifically for binary UML associations.

[7] Dedene, G., Snoeck, M., 1995. Formal deadlock elimination in an object-oriented conceptual schema. Data & Knowledge Engineering 15(1), 1-30.

[8] Dunn, C.L., Gerard, G.J., Grabski, S.V., 2003. Visual Attention Overload: Representation Effects on Cardinality Error Identification. In: Proceedings of the 24th International Conference on Information Systems. Seattle, WA, USA. 47-58.

[9] Dunn, C.L., Grabski, S.V., 2001. An Investigation of Localization as an Element of Cognitive Fit in Accounting Model Representations. Decision Sciences 32(1), 55-94.

[10] Gemino, A., Wand, Y., 2003. Evaluating modeling techniques based on models of learning. Communications of the ACM 46(10), 79-84.

[11] Genero, M., Poels, G., Piattini, M., 2002. Defining and Validating Measures for Conceptual Data Model Quality. Lecture Notes in Computer Science 2548, 147-153.

[12] Halpin, T., 1996. Conceptual Schema and Relational Database Design: A Fact Oriented Approach. 2nd Ed. Prentice-Hall.

[13] Kim, Y.-G., March, S.T., 1995. Comparing Data Modelling Formalisms. Communications of the ACM 38(6), 103-115.

[14] OMG, 2004. UML 2.0 Superstructure Specification, Revised Final Adopted Specification (October 8, 2004). Object Management Group. http://www.omg.org.

[15] Parsons, J., 2003. Effects of Local Versus Global Schema Diagrams on Verification and Communication in Conceptual Data Modelling. Journal of Management Information Systems 19(3), 155-183.

[16] Parsons, J., Cole, L., 2004. An Experimental Examination of Property Precedence in Conceptual Modelling. In: Proceedings of the 1st Asia-Pacific Conference on Conceptual Modeling. Dunedin, New Zealand.

[17] Parsons, J., Cole, L., 2005. What Do the Pictures Mean? Guidelines for Experimental Evaluation of Representation Fidelity in Diagrammatic Conceptual Modeling Techniques. Data & Knowledge Engineering, *accepted for publication*.

[18] Schütte, R., Rotthowe, T., 1998. The Guidelines of Modeling – An Approach to Enhance the Quality in Information Models. Lecture Notes in Computer Science 1507, 240-254.

[19] Shanks, G., Tansley, E., Weber, R., 2003. Using ontology to validate conceptual models. Communications of the ACM 46(10), 85-89.

[20] Snoeck, M., Dedene, G., 1998. Existence Dependency: The key to semantic integrity between structural and behavioural aspects of object types. IEEE Transactions on Software Engineering 24(4), 231-253.

[21] Stevens, P., 2002. On the interpretation of binary associations in the Unified Modeling Language. Software and Systems Modeling 1(1), 68-79.

[22] Topi, H., Ramesh, V., 2002. Human Factors Research on Data Modeling: A Review of Prior Research, An Extended Framework and Future Research Directions. Journal of Database Management 13(2), 3-19.

[23] Vessey, I., 1991. Cognitive fit: A theory-based analysis of the graph versus tables literature. Decision Sciences 22(2), 219-240.

[24] Wand, Y., Storey, V.C., Weber, R., 1999. An Ontological Analysis of the Relationship Construct in Conceptual Modeling. ACM Transactions on Database Systems 24(4), 494-528.

[25] Wand, Y., Weber, R., 2002. Information Systems and Conceptual Modeling – A Research Agenda. Information Systems Research 13(4), 363-376.

Organizing and Managing Use Cases

James L. Goldman and Il-Yeol Song

College of Information Science and Technology,
Drexel University, Philadelphia, PA 19104, USA
{jimg, song}@drexel.edu

Abstract. The UML recommends that software system functionality and interactions be documented through use case narrative descriptions and use case
diagrams. The UML, however, provides no structure or framework for organizing a large number of use cases that may be required for complex systems. In
this paper, we present various taxonomies of existing use case classification
schemes and one additional scheme for classifying and organizing use cases.
We then discuss how we can effectively understand categorized use cases in
terms of project priority and personnel skills to achieve the best possible allocation of project resources to use case-driven development efforts. The proposed
method uses simple sequential questions to determine use case categories to aid
analyzers in real-world projects. Our method is moderately simple to understand and implement.

1 Introduction

The use case approach is widely used in capturing the functional requirements of a
system. A use case is a complete, external behavior of the system from its start to the
end, providing a value to the actor by achieving a goal of the actor. Thus, a set of use
cases collectively defines the functional requirements of the system.

UML includes use case diagrams and use case narrative structures to capture functional requirements of systems. UML, however, provides little structure or framework
for organizing the large number of use cases typically required for complex systems.
Such a situation contradicts an important motivation behind UML: clarity of exposition and complexity control.

While the concept of the use case is easy to understand in isolation and in textbook
examples, it becomes much less clear in large complex system applications. Knowing
how to write use cases does not, by itself, inform us about how to best use the information in the use cases. In this paper we are concerned with classification of use cases
and their use in assisting analysts to use the use case approach effectively. In a use
case-driven project, use case models will be reviewed frequently, both informally and
formally [1]. A structured organization of the use cases will facilitate any such review, thus enhancing the value of the use cases.

The use case driven approach is one of the most frequently cited best practices for
modern software development [4], [11]. Use case driven development implies that use
cases are carried over from requirement modeling through analysis, design, implementation, and testing in a system life cycle. However, even best practices use case-
driven software development methodologies may not adequately consider assignment

J. Akoka et al. (Eds.): ER Workshops 2005, LNCS 3770, pp. 43–52, 2005.
© Springer-Verlag Berlin Heidelberg 2005

of staff responsibilities according to skills and experience. In an examination of the Rational Unified Process, John, Bass, and Adams [10] (p.4) found that "RUP makes no such checks on staff or methods, assuming, perhaps, that all development teams are qualified to perform all necessary activities."

Therefore, an organizational framework of use cases benefits all the stakeholders throughout the use case driven project. A framework helps not only modelers but also project managers, designers, and developers who use the requirements captured in the use cases.

In this paper, we address two issues of use cases: organizing use cases based on a taxonomy and assigning use cases to development teams. We first present various taxonomies of existing use case classification schemes and one additional scheme for classifying use cases. We then discuss how we can effectively understand categorized use cases in terms of project priority and personnel skills to achieve the best possible allocation of project resources to use case-driven development efforts.

The rest of this paper is organized as follows: Section 2 discusses problems in organizing use cases. Section 3 discusses various existing taxonomies of use case classification schemes. Section 4 presents an overview of the proposed application method. Section 5 concludes our paper.

2   Difficulty of Defining Use Cases Hinders Organization

In this section, we argue that consistently organizing use cases is difficult, because their content boundary, length, and level of detail are variable.

In defining the use case concept, Jacobson was careful to avoid too much formalization so that his concept could be flexibly applied [5]. The "include," "extend," and "generalization" relationships between use cases that have been traditionally depicted in use case models are also quite flexible. The structure Jacobson proposed was simply the one that followed from object-oriented software design.

Use case narratives are frequently written at multiple levels of abstraction. For example, Jacobson et al. [9] describe how a use case description can be stated at three different levels – at the business level, at the class diagram level, and at the user interface or technology-dependent level.

Use cases, even if well organized, are often incomplete and imprecise: "People rarely have time to make the use cases formal, complete, and pretty. They usually only have time to make them 'sufficient,' which is all that is necessary" [6] (p. 5). Cockburn further points out that use cases may not fully capture all requirements.

Collins-Cope has proposed an approach to use case analysis that separates non-system-dependent, or "essential" use cases, from use cases that describe system functionality [7]. The approach uses a set of linked RSI (requirements/service/interface) use cases to reflect each aspect of the desired functionality.

According to Rosenberg and Scott [12], "a use case describes a unit of behavior; requirements describe the laws that govern that behavior." (p. 123) Although there is not a one-to-one mapping between requirements and use cases, the authors point out that "the result of use case modeling should be that all required system functionality is described in the use cases." According to Armour and Miller [2], "A system may be defined as the sum of its use cases" (p. 25). These statements imply that additional information describing the structure and organization of the use cases is necessary for fully understanding the model.

A full understanding of the use case model in use case-driven development is essential. Without it, requirements will be misinterpreted or overlooked, leading to errors that are costly to correct.

3 Taxonomy of Use Cases

In this section, we present a taxonomy that includes six different use case classification schemes. Figure 1 depicts the six schemes in a class diagram.

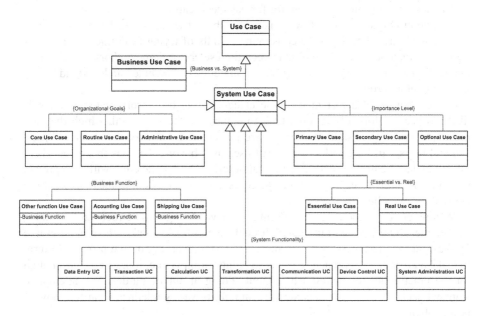

Fig. 1. A class diagram representation of taxonomy of use case classification schemes

3.1 Motivation

Complex systems require complex use case models. Unlike textbooks and learning exercises with single digit numbers of use cases, real systems may require dozens of use cases. An unorganized list of a hundred use case descriptions may not be the best solution.

Classifying use cases may offer some or all of the following benefits:

- Helps software designers and programmers understand a large project's needs more clearly.
- Assists management in assignment of project personnel
- Provides additional attributes to guide systematic or automatic translation of use cases from descriptive text into formal specifications, class models, or code.
- Aids understanding relationships, similarities, and differences among use cases
- Makes large corpus of use cases more approachable by end users

A classification effort will yield project management benefits, helping managers match use cases with the most appropriate personnel. Classifying use cases according to their business function (see Section 3.6) and system functionality (see Section 3.7) presents information about each use case that identifies the closest matching skills and experience.

Classification is expected to assist attempts at systematic or automatic translation of use cases into other documentation formats, into software designs, or even into code. In a review of UML-based formal systems analysis approaches, Whittle [15] notes the need for more formal semantics; classification is a step in conversion from textual descriptions into a language for formal specification.

Taxonomic organization allows a large number of use cases to be understood in context while avoiding the time-consuming pitfalls of trying to define relationships (typically "include" and "extend") between the use cases. Some authors [3], [12] recommend that use case modelers avoid the subtleties of "include" and "extend" early in the modeling process.

In the absence of a formal use case categorization framework, categorization is likely to occur informally. People tend to emphasize the tasks with which they are most familiar. Project staff will emphasize use cases in terms of their own abilities, and end users will lack understanding of use cases beyond the ones that affect them directly. Classification provides additional descriptive attributes that will help project staff and end users to understand the use cases that concern them in the context of the entire use case collection.

Writing effective use cases is difficult. Their value ought to be recovered as much as possible. Organization of use cases through categorization makes them more usable by the analysts, designers, programmers, and testers working on the project. In turn, to make a taxonomic approach to use case organization viable, it must be easy enough for the practitioner to accomplish without adding an undue burden. We attempt to augment each classification scheme with straightforward guidelines regarding how to best apply it.

3.2 Business Use Cases vs. System Use Cases

Business use cases differ from system use cases. Business level use cases may include both manual and system operations. Most business use cases include a mix of manual and system processes. Pure system use cases, in contrast, only describe direct interaction with a system. According to Cockburn, "Business process people write business use cases to describe operations of their business, while a hardware or software development team writes system use cases for their requirements" [6] (p. 7).

Business level use cases are more relevant to business process reengineering than to software system analysis and design. System use cases are derived from information collected in the business use cases, but they describe the functionality that the system delivers to actors in the business use case. Distinguishing business level use cases from system level use cases is helpful in establishing the system boundary. We limit our classification schemes to system use cases.

3.3   Essential Use Cases vs. Real Use Cases

An essential use case expresses a high level view while remaining free of implementation details. A real use case "concretely describes the process in terms of its real current design, committed to specific input and output technologies, and so on" [11] (p. 58). Essential use cases are closer to the "what" than the "how." "Use cases are requirements analysis and modeling tools that should describe what a system does (or should do), rather than how the system works (or should work)." [8] (p. 30)

Larman acknowledges that there is not a clear line between essential and real use cases. There are advantages to using both essential and real use cases. Essential use cases are reusable with successive implementations of the same system on newer technological platforms, whereas real use cases are tied to platform-specific designs. On the other hand, real use cases can inform programmers about what is needed with a higher level of detail.

Assessing where a use case falls on the essential vs. real scale can be accomplished by answering this question: Will this use case still be accurate and useful if the technological environment and constraints change significantly?

3.4   Organizational Goals: Core vs. Administrative vs. Routine Use Cases

Song has proposed that use cases may also be classified according to whether they are core use cases, administrative use cases, or routine use cases [14], documenting the extent to which each use case supports organizational goals.

Core use cases describe the new or distinctive functionality that is being specified for a system. Each core use case provides tangible functionality to the user of the system. Core use cases answer the question "What does the system do for us?" Core use cases address the main purpose for the system, and are not incidental to it.

Administrative use cases describe operations that are necessary for the integrity of the overall system's operation. Typical administrative use cases include "Back up the system," "Shut down the system," "Synchronize the remote databases," etc.

Routine use cases describe operations that users must do, usually repetitively, in order to realize the functionality provided by the system. Routine use cases do not themselves provide the functionality that is the purpose of the system. Examples of routine use cases would include "Log in," "Print usage log report," "Change password," etc. Routine use cases differ from administrative use cases in that routine use cases are most often performed by users in the normal course of using the system; administrative use cases are most often performed by administrators as distinct maintenance or system operational procedures.

A simple heuristic that can be used to classify use cases as either core, administrative, or routine, is to answer the following questions in order:

Was the system created in order to provide the functionality described in this use case? If the answer is yes, then the use case is a core use case.

Does this use case describe an operation without which the entire system would not operate properly over time? If the answer is yes, then this use case is required for the proper operation of the system, and is therefore an administrative use case.

Does this use case describe functionality without which user would not be able to properly utilize the system's features and interfaces? If the answer is yes, then this use case is a routine use case.

If the answer to all three questions is no, then perhaps the use case is not necessary.

3.5   Importance Level: Primary vs. Secondary vs. Optional Use Cases

Use cases may be categorized by their importance as primary, secondary, or optional use cases. The difference between primary and secondary use cases is that primary use cases provide essential functionality, and are performed frequently. Secondary use cases are of less importance and are performed less frequently. Optional use cases are those whose functionality is desirable but not essential.

The method to determine if a use case is primary, secondary, or optional is to answer two questions: *1. Can system operation begin if this use case has not been implemented?* If the answer is no, then this is a primary use case. *2. Can the system substantially deliver its intended value to the users if this use case has not been implemented?* If the answer is no, then this is a secondary use case. Otherwise, this is an optional use case.

3.6   Organizing Use Cases by Business Function

Several use case authors recommend that system use cases be organized by business function [2], [12], [13] such as accounting, order processing, manufacturing, etc. Business functions are familiar to people who will use an intended system. Organization of personnel by business function within a firm is a common practice. Departmental/business unit structures within the organization, or existing accounting cost centers, for example, may indicate how use cases relate along business function lines.

Software developers and managers will benefit from a business function classification of use cases because it will help them identify both developers and end users who will have the most appropriate expertise for writing the detailed use case and designing the corresponding software.

3.7   Organizing Use Cases by System Functionality

In this section, we present a use case classification scheme based on typical system functionality. The function types are (1) data entry/maintenance, (2) Transaction processing, (3) Complex Calculation, (4) Transformation, (5) Communication, (6) Device control, and (7) System administration.

When we adopt this classification, we caution modelers to check each use case against certain criteria that denote a primary use case. The properties we recommend to check against are [4], [9], [11], [14]:

- It is a goal of the actor of the use case
- It includes a complete process from start to finish
- It provides a value to the actor

A classification based on system functionality attempts to represent the nature of the experience, expertise, and skills that are most appropriate for the design and implementation of the underlying software required for a use case. Each use case is assessed on how its implementation is expected to make use of the system resources, combined with a judgment about what business knowledge is most relevant.

Data entry / maintenance. The primary purpose of these use cases is to manipulate the data in a database. Adding a new customer record, change customer data, or updating an inventory are typical examples of this category of use case.

Transaction processing. In a transaction, an event that occurs at a point in time is processed and recorded. Often, transactions can be rolled back. Much of the functionality of business information systems involves processing and recording business transactions. Examples include sales, refunds, hiring, registrations, and reservations.

Complex Calculation. Calculation use cases are often "include" use cases in that they are subsidiary to base use cases, but depending on the system, calculation use cases may be primary. For example: a payroll system may have a use case for computing various withholding tax amounts. Tax computations are common examples of calculation use cases.

Transformation. A transformation, or conversion, operation may result in a large change across many tables in a database or files in a file system. It can be based on business rules, a calculation, or a change in the external environment. Examples might include a fiscal period closing process or a merger-related change to information structure and content. Transformation use cases would frequently denote functions that prepare data for compatibility with external systems.

Communication. Use cases whose primary purpose is to move information to/from outside the system boundary fall into this category. Electronic mail is the obvious example, but data transfers to/from other systems also fit here. For example, preparing and sending eCommerce order information messages within a supply chain would be a communication use case. Reports are also included under the heading of communications. Reports are one-way communications from the system to users.

Device control. These are use cases that are intended primarily to specify control of physical devices like robot arms, parts conveyors, etc.

System administration. System administration use cases describe functions necessary for the good continuing operation of the system, but that do not themselves contribute directly to the business purpose for which the system is being created. Typical system administration functions include starting the system, backing up data, logging in, and generating usage reports.

3.8 Classification Summary

Five classification schemes for system use cases have been presented:

1. Essential vs. Real use cases
2. Organizational goals: core, routine, and administrative use cases
3. Importance level: Primary, secondary, and optional use cases
4. Business function classification
5. System function classification

Each system use case description should be assigned its designated place within each of the five schemes, with some allowance being made for grey areas and multiple proportional assignments.

4 Application Method

In order to unlock the value that the use cases represent, it is necessary to systematically apply a method of classifying and organizing them so that the project team can best understand how the use cases relate to each other and to the entire project. The advantages of systematic organization will grow with the complexity of the project and the number of use cases. The process of categorizing a set of system use cases on multiple schemes can be guided by applying the following method (see Figure 2). The goal of the suggested method is to immediately match the highest priority use cases with the appropriate project resources. Note that the suggested method is integral with writing the use cases, not applied after the fact.

First, identify each use case and write its title and brief description. As each use case is identified, determine whether it supports organizational goals as a core, routine, or administrative use case according to the questions in Section 3.4. Also determine whether it is a primary, secondary, or optional use case according to the questions in Section 3.5. The organizational goal and primary/secondary/optional designations may be made in parallel, but they explain different use case attributes. Use cases which are both core and primary are the most important. They should be prioritized ahead of the others.

Next, elaborate the use case descriptions for the core primary use cases. The level of technology specificity evidenced by the detailed descriptions and scenarios will help assess the degree to which each use case is an essential, as opposed to real use case (Section 3.3).

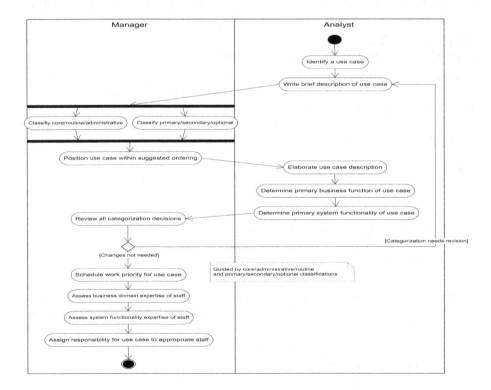

Fig. 2. Suggested use case categorization method

Using the now detailed descriptions, these use cases may next be categorized according to the business function (Section 3.6) and system functionality (Section 3.7) that each represents. After the core primary use cases have been handled (or once progress is well under way), the remaining use cases should then be elaborated with detailed scenarios generally in the following order: first primary, then secondary, and finally optional use cases. Within each of those priorities, elaborate first the core, then the routine, and finally the administrative use cases.

The last step is the assignment of responsibility. Using the priority and organizational goals classifications to guide the sequence, compare each use case's business functions and system functionality and match each with the matching skills and experience evidenced in the project staff. Once responsibility for each use case is assigned, project team members may begin the transformation of use cases into software and database designs.

5 Conclusion

This paper has examined a total of six different schemes for categorizing use cases. In addition to five schemes from existing literature, we have introduced an additional scheme based on system function types. We have also proposed a straightforward methodology to guide use case classification. The method rests on sequentially answering some simple questions. The resulting organization classifies use cases using five schemes. It is not complex and therefore easy to apply.

The proposed classification schemes organize use cases for the efficient allocation of project resources. Two schemes consider prioritization so that resources are allocated to high priority use cases first. The other two schemes consider the characteristics of the use cases so that proper resources are allocated to each. A business function dimension facilitates the assignment of domain expertise. A system function dimension facilitates the assignment of appropriate system development expertise.

In use case-driven methodologies, use cases are used as the basis for software design, implementation, and test plans. It is vital that project teams are able to fully understand the use cases and the relationships among them. Large projects may include many use cases which may be assigned to many project team members, making that task more difficult. Therefore, classifying use cases enhances informational value by identifying additional attributes. Categorization helps project managers schedule and allocate their team members according to project priority needs by putting use cases in order. Categorization also helps match use cases with the most appropriate personnel according to their skills and experience so that team members work on use cases for which they are best suited.

The proposed organizational structures and suggested classification methodology are not difficult to implement. The end result is a collection of use cases with enhanced informational value and a better allocation of resources. The benefits obtained will outweigh the costs.

References

1. Anda, B. and Sjøberg, D. I. K.: Towards an inspection technique for use case models. In: SEKE '02: Proceedings of the 14th international conference on Software engineering and knowledge engineering. ACM Press (2002) 127-134.
2. Armour, F. and Miller, G.: Advanced Use Case Modeling: Software Systems. Addison-Wesley (2001).
3. Bittner, K. and Spence, I.: Use Case Modeling. Addison-Wesley, Boston (2003).
4. Booch, G., Rumbaugh, J., and Jacobson, I.: The Unified Modeling Language User Guide. Addison-Wesley, Boston (1999).
5. Cockburn, A. and Fowler, M.: Question time! about use cases. In: Proceedings of the 13th ACM SIGPLAN conference on Object-oriented programming, systems, languages, and applications. ACM Press (1998) 226-229.
6. Cockburn, A.: Writing Effective Use Cases. Addison-Wesley, Boston (2001).
7. Collins-Cope, M.: The requirements/service/interface (RSI) approach to use case analysis. In: Proceedings of Technology of Object-Oriented Languages and Systems (TOOLS). IEEE Computer Society (1999) 172-183.
8. Dobing, B. and Parsons, J.: Understanding the Role of Use Cases in UML: A Review and Research Agenda. Journal of Database Management. Vol. 11, No. 4 (2000) 28-36.
9. Jacobson, I., Christerson, M., Jonsson, P., and Overgaard, G.: Object-Oriented Software Engineering: A Use Case Driven Approach. Addison-Wesley, Reading, MA (1992).
10. John, B. E., Bass, L., and Adams, R. J.: Communications across the HCI/SE divide: ISO 13407 and the Rational Unified Process. In: Proceedings of HCI International 2003. Lawrence Erlbaum Associates Mahwah, New Jersey (2003).
11. Larman, C.: Applying UML and patterns: an introduction to object-oriented analysis and design, third edition. Prentice Hall PTR, Upper Saddle River, NJ (2005).
12. Rosenberg, D. and Scott, K.: Use Case Driven Object Modeling with UML: A Practical Approach. Addison-Wesley, Boston (1999).
13. Schneider, G. and Winters, J. P.: Applying Use Cases, Second Edition: A Practical Guide. Addison-Wesley, Boston (2001).
14. Song, I.-Y.: Object-Oriented Analysis and Design Using UML: A Practical Guide. Pearson Publishing, Boston, MA (2004).
15. Whittle, J.: Formal approaches to systems analysis using UML: An overview. Journal of Database Management. Vol. 11, No. 4 (2000) 4-13.

A Comparative Analysis of Use Case Relationships

Margaret Hilsbos, Il-Yeol Song, and Yoo Myung Choi

College of Information Science and Technology,
Drexel University,
Philadelphia, PA 19104
{mhilsbos, song, ymc22}@drexel.edu

Abstract. Use case relationships are used to manage the complexity of use cases. The UML defines the three types of use case relationships: include, extend, and generalization. The appropriate use of the use case relationships, however, is one of the most contentious areas. We found that the suggestions of various authors overlap but conflict, leaving room for dissension. In this paper, we present a comparative analysis of the use case relationships discussed in eleven literatures, including the UML 2.0 specification. For a coherent approach for applying use case relationships, we present three rules derived from the review of the literatures and our own experience and illustrates the rules with examples. Our rules are based on the analysis of preconditions, postconditions of use cases, and characteristics of the behaviors being separated.

1 Introduction

Use Cases are a fundamental starting point of object oriented analysis and design. The Unified Modeling Language (UML) through release 1.5 has not defined all aspects of use case modeling explicitly, and does not provide a method for determining the correct modeling techniques for a given situation. Several authors have discussed various approaches; some different authors' approaches conflict with each other. The release of UML 2.0 clarifies some of the rules, but still leaves gaps.

Ambiguity and misuse of use cases and use case relationships have been cautioned against by many authors [9, 10, 11, 14, 16, 21, 22]. In this paper, we analyze the application of use case relationships. Use case relationships are used to manage the complexity of use cases. The UML defines the three types of use case relationships: *include*, *extend*, and *generalization* [20]. The appropriate use of the use case relationships, however, is one of the most contentious areas in a use case modeling. Metz et al. [17] review various meanings of 'alternative courses" discussed by several authors and summarize three different meanings of them – *alternative history*, *use case exceptions*, and *alternative part*. To our knowledge, however, there has been no comparative study on use case relationships by different views of various experts and literature. In this paper, we present a comparative analysis of the use case relationships discussed in eleven literatures, including the UML 2.0 specification. We present the agreed usages and different view points of the use case relationships. Several points of contention are identified and a logical resolution is argued for each. Finally, we propose three rules for applying use case relationships and illustrate them with examples.

J. Akoka et al. (Eds.): ER Workshops 2005, LNCS 3770, pp. 53–62, 2005.

The rest of this paper is organized as follows: Section 2 provides an extensive literature review in term of usages of use case relationships. Section 3 presents three recommended rules of using the use case relationship for a coherent approach. Section 4 concludes our paper.

2   Comparative Review

There are three types of relationships used in use case models: *include*, *extend*, and *generalization*. The "include" and "extend" relationships are represented as a stereotype dependency and are enclosed in guillemets as: «extend», «include».

Use case relationships enter the modeling process after an initial set of use cases have been determined and at least high-level descriptions have been written. The following general purposes have been suggested for applying relationships to the use case model:

1.   factor out reused behavior to remove redundancy [1, 175; 3, 169; 2; 8, 161; 13, 111; 15, 388]
2.   factor out requirements that can be implemented at a later time [3, 165; 5, 260; 6, 237; 8, 169]
3.   separate functionality to reduce the scope of an initial use case to a more manageable size [1, 206; 6, 110; 15, 388; 19]

2.1   Include

The following rules are generally agreed upon by the authors surveyed:

- The base use case "calls" the included use case, like a subroutine. After execution of the included use case, control returns to the base use case at the point just after the inclusion was called.
- An included use case must contain only one insertion segment. The entire included use case is executed when it is called.
- An included use case may be called by multiple other base use cases; that is the original intention of «include».

Using «include» for Behavior Common to More Than One Use Case. The UML 2 specification states that "the include relationship is intended to be used when there are common parts of the behavior of two or more use cases" [18, 518].   Most of the surveyed authors noted common behavior as a major reason for using include [1; 3; 4; 5; 8; 12; 13; 15; 20; 24].

Chonoles and Schardt suggest particularly looking for an opportunity to factor out common behavior when several use cases interact with the same external system as a secondary actor. It is very probable in this situation that the interaction with the external system *should* be common; factoring it out as an included use case helps prevent inconsistency [8, 163].

Using «include» for Optional or Alternative Paths. The UML 2.0 specification states "Note that the included use case is not optional, and is always required for the including use case to execute correctly" [18, 518].   But several authors propose that ‹‹include›› may be used for conditional behavior. Armour and Miller note that "no conditional guard is associated with the included use case at its inclusion point" but "this does not preclude the including use case from containing conditional logic that might result in the included behaviors not executing during a particular instance of execution" [3, 169]. Bittner and Spence do not discuss the point, but their example for «include» is conditional: "If the customer is a new customer, include use case Add Customer Information..." [5, 257]. Cockburn implies support for using ‹‹include›› for conditional behavior, by restricting the use of ‹‹extend›› to cases where the base use case is locked, or there are very many asynchronous extensions. He argues that ‹‹include›› is much easier for most people to understand and use than ‹‹extend›› or generalization, and should therefore be preferred [6, 116, 207]. Larman references Cockburn's statement and uses examples of ‹‹include›› which are conditional [15].

Using ‹‹include›› to Handle Asynchronous Actions. Larman suggests using «include» to handle asynchronous events, instead of using «extend» as others suggest [15, 387]. An asynchronous action is one that can be called at any, or almost any, point in the base use case.

Using ‹‹include›› to Decompose Overly Complex Use Cases. The UML 2.0 specification notes "The Include relationship allows hierarchical composition of use cases as well as reuse of use cases" [18, 518]. Larman also suggests using ‹‹include›› to "decompose an overwhelmingly long use case into subunits to improve comprehension" [15, 387]. Kulak and Guiney show examples of ‹‹include›› used to decompose business processes, but their examples are not functional decomposition of use cases [13, 172, 256]. Other authors also use ‹‹include› to decompose a complex use case [24, 215; 20, 80]. The logic of an Include use case, however, should be complex enough to deserve a separate use case documentation [23].

2.2 Extend

Regarding the extend relationship, the following rules are generally agreed by the authors surveyed:

- The base use case must be able to stand alone – it can execute completely and successfully without executing the extending use case.
- An extending use case cannot stand alone – it is never directly initiated by an actor, and depends on the base use case (i.e., it is always abstract).
- An extending use case may contain multiple insertion segments; an extension point must be defined for each segment.
- For each insertion segment, control must exit from and return to the base use case at the same point, the extension point.
- An extending use case may be re-used; it may extend more than one base use case. In this case, the base use cases establish the pre-conditions for the extending use case. If two use cases share an extension, then at the extension point for each base use case, the system conditions must satisfy the same precondition.

In the following subsections, we review reasons for using ‹‹extend›› proposed by various authors.

Using «extend» for a Complex Alternative Course of Action. Adolph and Bramble suggest using «extend» "when an alternative course of action interrupts a number of steps in a scenario." They use the example "Book Flight for Frequent Flier" as an extension of the use case "Book Flight" [1, 183-187]. See Figure 1 for the depiction of this example in the use case diagram.

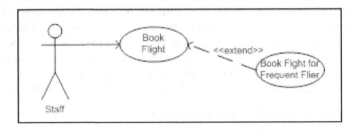

Fig. 1. An example of a complex alternative course of action

The extending use case "Book Flight for Frequent Flier" is said to have multiple insertion segments because the base use case is interrupted several times within one alternative scenario. The use of one extending use case to handle multiple insertion segments which occur at different points in the base use case is supported by several other authors [3; 4; 20]. UML 2.0 supports using «extend» when there are multiple insertion segments (behavior fragments) in the extending use case [18, 515].

Using «extend» for Optional Behavior or Exceptional Behavior. Using «extend» for optional behavior or exceptional behavior is supported by most of the reviewed authors   [1, 195; 3, 153; 4, 87; 5, 259; 7; 12, 130; 13, 44; 24, 53]. Armour and Miller caution that the extension must always return control to the base use case at the extension point; therefore it is not appropriate to use «extend» for exception or error handling that does not return control to the base use case, or does not return control at the step immediately following the extension point.

Using «extend» for Asynchronous Options. Constantine and Lockwood [7] and Cockburn [6, 237] discuss the use of «extend» to indicate the availability of asynchronous options. Cockburn uses the example of spell-check in a word processor. Constantine and Lockwood unfortunately give the example of resetting a test, which seems inappropriate for an «extend» because it seems to violate the rule that control must return to the base use case at the step immediately following the point of interruption by the extension.

Using «extend» to Add Behavior After the Base Use Case Is Locked. A few authors emphasize the value of «extend» for adding behavior to a base use case when its documentation is locked [4; 5; 6; 8; 15]. The argument for this is that the base use case doesn't need to know about the extension so therefore the extension can be added without updating the base use case documentation. Chonoles and Schardt note that in practice, this can result in confusing documentation, particularly if several releases have resulted in multiple nested extensions to the same base use case [8, 170].

Using «extend» to Separate Behavior to be Deferred in Development. Armour and Miller suggest using «extend» to separate behaviors that can be developed later from a base use case. They suggest this separation enables setting a lower priority for the extending use case than the base use case [3, 165]. Cockburn implies a similar approach [6, 237].

Extension Points. Some authors suggest that it can be desirable to depict extension points on the use case diagram [1, 188; 3, 163; 5, 265; 24, 55]. It can be anticipated that adding extension point information could clutter the diagram, as Cockburn suggests [6, 238]. Extension point information displayed in the use case diagram is removed from the context of the use case steps, so it is unclear what added value there is in showing it on the diagram.

2.3 Generalization

Use case generalization is fairly straightforward and there is general agreement among authors as to its usage. The reviewed authors suggest using use case generalization in the following situations:

- variations on a similar goal [3, 175; 4, 82]
- when different technologies are used to achieve a goal [5, 267; 8, 166; 24, 58; 19; 20, 675]
- when different actors achieve similar goals, or the same goal by different means [6, 241; 8, 166]

However, we note that use case generalizations can be restated using either extend or include relationships. Some authors recommend that generalization not be used as they can create confusion. [6, 12, 13, 15,]. Cockburn also discusses the hazards of using use case generalization. [6, 240]

3   Synthesis: A Coherent Approach to Use Case Relationships

The above review could be characterized as a list of differences. The lack of agreement among experts on so many points regarding use case relationships seems to ensure that there dwill be costly disagreements in the field. To avoid such disagreements, it would be helpful to have a coherent and dependable model for assigning use case relationships. The characteristics desired of such a model are that:

- The rules are unambiguous.
- The rules are easy to understand.
- The rules are simple to use.

These rules are difficult to achieve from a reading of the reviewed authors. Generally the applications proposed by the reviewed authors are understandable, and sometimes simple to apply. The problem is that when taken together, the suggestions of various authors overlap but conflict, leaving room for dissension. From our review we suggest to limit application of the reviewed authors' suggestions by the following guiding principles:

1. Minimize interleaving – [19]
2. Conform to the UML specification to the extent possible. Standards are created for the purpose of establishing a common language. Creating individual

"dialects" reduces portability and increases the learning curve for new team members.

3. The use case diagram is not a process flow. Do not use relationships for behavior that can stand alone with the proper specification of preconditions and postconditions (Example: *Log In* is not included, it establishes the precondition for other use cases)
4. Do not create a too small fragment of operations as an inclusion use case [23].

To further improve on this situation, we propose the three sets of rules for applying use case relationships. The first looks at using pre- and post-conditions to segregate terminal and initial behavior, respectively. The second determines the correct relationship based on general features of the behaviors being separated. The third evaluates relationships based on the assignment of postconditions. The rule sets operate from different perspectives and may be applied in tandem.

3.1 Rule Set 1: Avoid Relationship Overuse by Understanding Pre- and Post-conditions

An included use case should probably not be called at the very beginning or very end of a use case flow. "Encapsulatable" behavior at the beginning of a use case should be separated as a stand-alone use case. The postcondition of this use case becomes a precondition for the original use case. An example of this is the "Log In" use case for many applications. If "Log In" is treated as an included use case, the number of «include» arrows on the diagram could overwhelm and obscure other content.

Likewise, if behavior at the end of a use case seems to be a candidate for «include», consider whether the behavior can be treated as a stand-alone use case, with its precondition being the adjusted postconditions of the original use case (the postconditions of the original are adjusted to eliminate any set by the new use case). For example, one might be inclined to «include» "Process Payment" with use cases "Process Sale" and "Process Rental". Instead, "Process Payment" should, in most cases, be factored into a standalone use case. It is at the end of the sale or rental process, and usually contains sufficient complexity to warrant a separate use case. In the case that there are other processes after payment before the customer can leave the store, or exit the web transaction as the case may be, those processes similarly stand alone. Examples might be: "Disable Security Tag"; "Send Confirmation Email".

The precondition for "Process Payment" and the postcondition for "Record Order" become, "an order is prepared for payment". Note the matching of the postcondition of one use case to the precondition of another.

3.2 Rule Set 2: Characteristics of the Behaviors

Use include when the behavior fragment
- is required for at least one alternative described by the base use case, to fulfill the postconditions of the base use case AND
- is referred to more than once (multiple use cases, multiple times within the same use case, and/or is stand-alone as well as referred to from a use case).

Example: In some applications, there is a need to process coupons or discounts. Some of these (such as a manufacturer's coupon) are validated and rung as part of the order, before arriving at the total for payment. Then (for some applications) there are special discounts that apply only when, for example, a certain payment method is used. Particularly in the payment-specific case, it is difficult to factor out the processing of the discount as a standalone use case, because the discount must be returned and a new total calculated before "Process Payment" can complete. So in this circumstance, the use of ‹‹include›› may make the most sense. It should be noted that, absent the requirement for applying promotions during payment processing, "Process Promotion" *could* possibly be factored out as standalone, by adjusting the postcondition of "Record Order" to match the precondition of "Process Promotion": "order is complete and ready to apply discounts." The use case is shown in Figure 2.

Use extend when the behavior fragment
- is never required for successful completion of the base use case, and has its own postconditions AND
- needs to be separated from the base use case to improve clarity AND
- cannot be modeled as a stand-alone use case without overly fragmenting the base use case

Example: In a sales application where a customer may order several different items (e.g., an online bookstore), an item may not be in stock. If the order is to be processed anyway, the backordered item entails some additional processing before completing the order. This situation may be best modeled using an extending use case, as shown in Figure 2.

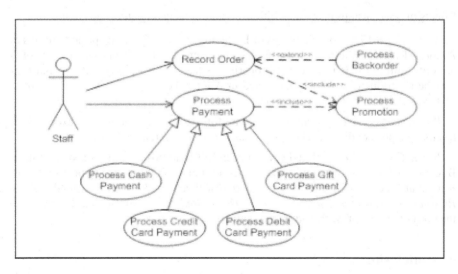

Fig. 2. Example of Rule Set 2

Use generalization when the behavior fragment

- represents an alternative, similar use case. [differences may include initiating Actor and/or postconditions] AND

- the common behavior of the similar use cases cannot be simply factored out as an *include* OR the project team prefers to see the commonality of the use cases indicated (i.e. the relative similarity of the use cases determines whether you just include common behavior, or show the use cases as specializations of the same general process).

For example, in the "Process Payment" use case (Figure 2), there are several types of payment methods. It would not be simple or straightforward to use ‹‹include›› for the common behavior, and it seems quite natural to use generalization in this case – the phrase "different means of achieving the same goal" is a tip that generalization probably applies.

3.3 Rule Set 3: The Postcondition Perspective

The following rules may be used to establish the appropriate segregation of behavior into use cases.

<u>Postconditions of alternatives – deciding whether to use relationships or unrelated use cases.</u>

If an alternative scenario of a use case results in different postconditions, a separate use case is required, which may be unrelated, or related through extension or generalization (not inclusion).

If an alternative is just another path to get to the same set of postconditions, a separate use case is not required and probably should not be used.

<u>Effects of relationships on postconditions:-deciding which relationships to use</u>

A includes B. The postconditions of B are necessary to fulfill the postconditions of A, for at least one alternative. For at least one scenario of A, the postconditions of A are only fulfilled if B is executed properly. B has its own postconditions, but they may be overridden by actions performed in A after the inclusion. *The postconditions of A include those of B.*

B extends A. The postconditions of A are fulfilled whether or not B is executed. B has its own postconditions. Therefore, *B and A have different postconditions.*

B and C specialize A. B and C must each fulfill all postconditions specified in A. B and C each may specify additional postconditions, and must specify postconditions if none are specified by A. *The set of postconditions which must be fulfilled is Ap + Bp when B is executed, Ap + Cp when C is executed.* Here, Ap, Bp, and Cp represent the postconditions of A, B, and C, respectively.

4 Conclusion

We have reviewed and summarized use case literatures regarding the application of use case relationships. We have noted areas of agreement and difference. We then discussed arguments made for resolution of differences. As a coherent approach for correctly applying to use case relationships, we have also proposed three rules derived from the review of the literatures and our own experience. Our rules are based on the analysis of preconditions, postconditions of use cases, and general features of the behaviors being separated.

From this analysis we conclude that practitioners should be aware of the nuances of appropriate application of each use case relationship, apply the relationships sparingly, and, when in doubt, develop several alternative models for complex problems. There may be more than one "correct" model to a given problem; the best solution is any solution that "works" and can be agreed to by the project team and stakeholders.

References

1. Adolph, S. and Bramble, P. (2003). *Patterns for Effective Use Cases*. Addison-Wesley.
2. Ambler, S. W. (2002). "Reuse in Use-Case Models: <<extend>>, <<include>>, and Inheritance" (Chapter 6 of the book *The Object Primer 2/e*.) Retrieved 9/2/2003 from http://www.agilemodeling.com/essays/useCaseReuse.htm#Figure1:UC%20Diagram
3. Armour, F. and Miller, G. (2001). *Advanced Use Case Modeling: Software Systems*. Addison-Wesley.
4. Arlow, J. and Neustadt, I. (2002). *UML and the Unified Process: Practical Object-Oriented Analysis and Design*. Addison-Wesley.
5. Bittner, K. and Spence, I. (2003). *Use Case Modeling*. Addison-Wesley.
6. Cockburn, A. (2001). *Writing Effective Use Cases*. Addison-Wesley.
7. Constantine, L., and Lockwood, L. (2001). "Structure and Style in Use Cases for User Interface Design," in van Harmelen, M., ed. *Object Modeling and User Interface Design*. Addison-Wesley.
8. Chonoles, M. J.; Schardt, J.A. (2003). *UML 2 for Dummies*. Wiley.
9. Firesmtih, D. (2005). *Open Process Framework*. http://www.donald-firesmith.com/ index.html?Components/WorkProducts/ModelSet/UseCaseModel/UseCaseModelingGuidelines.html~Contents.
10. Genova, G., Llorens, J. and Metz, P. (2005). "Open Issues in Industrial Use Case Modeling," *UML 2004 Satellite Activities*, LNCS 3297, 52-61.
11. Genova, G., Llorens, J. and Quintana, V. (2002). "Digging into Use Case Relationships," *Proc. of UML 2002*, 115-127.
12. Gomaa, H. (2000). *Designing Concurrent, Distributed, and Real-Time Applications with UML*. Addison-Wesley.
13. Kulak, D. and Guiney, E. (2000). *Use Cases: Requirements in Context*. Addison-Wesley.
14. Korson, T. (1998). "The Misuse of Use Cases," Object Magazine, 8(3), SIGS Publications, 1998, 18-20
15. Larman, C. (2002) *Applyng UML and Patterns: An Introduction to Object-Oriented Analysis and Design and the Unified Process*. Prentice Hall PTR.
16. Lilly, S. "Use Case Pitfalls: Top 10 Problems from Real Projects," in the *Proceedings of TOOLS*, USA'1999.
17. Metz, P., O'Brien, J., and Weber, W. (2003). "Specifying Use Case Interaction: Types of Alternative Courses", *Journal of Object Technology*, Vol. 2, No.2, 2003, 111-131.
18. Object Management Group. (August 2003). "UML 2.0 Superstructure Final Adopted specification." Retrieved 9/12/2003, http://www.omg.org/cgi-bin/doc?ptc/2003-08-02.
19. Rawsthorne, D. "CapturedAbstraction – A Pattern for Applying UML Generalization", in Adolph, S. et al. (2003). *Patterns for Effective Use Cases*, 198-200. Addison-Wesley.
20. Rumbaugh, J., Jacobson, I., and Booch, G. (2005). *The Unified Modeling Language Reference Manual*. Addison Wesley.

21. Simons, A J H. and Graham, I. "30 Things that go wrong in object modelling with UML 1.3", chapter 17 in: *Behavioral Specifications of Businesses and Systems* eds. H Kilov, B Rumpe, I Simmonds (Kluwer Academic Publishers, 1999), 237-257.
22. Simons, A. J H. "Use cases considered harmful", *Proc. 29th Conf. Tech. Obj.-Oriented Prog. Lang. and Sys., (IEEE TOOLS-29 Europe)*, 194-203.
23. Song, I.-Y. (2004). *Object-Oriented Analysis & Design Using UML: A Practical Approach*. Pearson Custom Publishing.
24. Schneider, G. and Winters, J. P. (2001). *Applying Use Cases*. 2nd ed. Addison-Wesley.

Applying Transformations to Model Driven Development of Web Applications*

Santiago Meliá and Jaime Gómez

Universidad de Alicante, Spain
{santi, jgomez}@dlsi.ua.es

Abstract. Nowadays, the maturity reached by the Web engineering research community can be assessed by the myriad of web design methods that have proven successful for the specification of the functional and navigational requirements posed by Web information systems. However, these proposals often fail to address architectural features, which results in Web specifications with rigid architectures, with no regard for their actual circumstances of use. To overcome this limitation, we propose a generic approach called WebSA. WebSA is based on the MDA (Model-driven Architecture) paradigm. It proposes a Model Driven Development made up of a set of UML architectural models and QVT transformations as mechanisms to integrate the functional aspects of the current methodologies with the architectural aspects. In order to illustrate our approach, in this paper we combine WebSA with the OO-H method to tackle the design of the well known J2EE Petstore specification.

1 Introduction

The rapid evolution of Internet has promoted in recent years intensive research in the field of functional modeling of Web applications. This fact has induced a new research trend within Software Engineering known as Web Engineering. In this context, different methods, languages, tools and design patterns for Web modeling have been proposed. These methods are centered mainly in the definition of functional aspects relative to the semantic of models to capture relevant properties of Web applications. However, few are the proposals that have tried to integrate in their methods the explicit consideration of architectural modeling features. Some authors have proposed the use of well known techniques in the Software Architecture discipline [1] in order to identify and formalize which subsystems, components and connectors (software or hardware) should make up the Web application.

These architectural features are especially important in methodologies that provide a code generation environment, such as WebML [5], OO-H [9], UWE [12], etc. The addition of an architectural view would cover the gap that nowadays exists between the Web design models and the code architecture. For this purpose, we propose the WebSA (Web Software Architecture) approach [13, 14], based on the standard MDA (Model Driven Architecture) [15]. Basically, WebSA provides the designer with a set of architectural models and transformation models to specify a Web application.

---

* This research has been partially sponsored by t the Spanish METASIGN (TIN2004-00779).

J. Akoka et al. (Eds.): ER Workshops 2005, LNCS 3770, pp. 63–73, 2005.
© Springer-Verlag Berlin Heidelberg 2005

Starting from these models the designer can integrate the Web functional models (domain, navigation and presentation) applying a set of model transformations following the Request for Proposals Query/View/Transformations (QVT) [16]. The result is an Integration model that captures the functional and architectural aspects of a Web application. Applying successive QVT transformations over the Integration model the Web specification can be converted to different implementation environments like J2EE or .NET.

This paper describes the first step, that is, how to achieve to the Integration model. We introduce the relevant concepts that are needed to understand our approach using the well-known Petstore running example [18].

The paper is organized as follows: Sect. 2 gives an overview of the WebSA development process and the modeling notation. Sect. 3 describes the functional viewpoint of by means of the Domain model of the Petstore blueprint. Next, the architecture of Petstore is specified by means of the WebSA Configuration model in sect. 4. Sect. 5 explains the QVT transformations showing how traditional Web functional models and the Configuration model can be merged into an Integration model. Sect. 6 gives an overview of the Integration model. Finally, sect. 7 and 8 outline the related work, conclusions and further lines of research.

2   The WebSA Approach: An Overview

WebSA is a proposal whose main objective is to cover all phases of Web application development focusing on software architecture. It contributes to fill the gap currently existing between traditional Web design models and the final implementation. In order to achieve this, WebSA defines a set of architectural models to specify the architectural viewpoint which complements current Web engineering methodologies [9, 12]. Furthermore, WebSA also establishes an instance of the MDA development process [11], which allows for the integration of the different viewpoints of a Web application by means of transformations between models.

The WebSA development process is based on the MDA development process in which the output artefacts of each phase must be models, which represent the different abstraction levels in the system specification. In the analysis phase the Web application specification is vertically divided into two viewpoints, as shown in the diagram flow of Fig. 1. On the one side, the functional-perspective is given by the Web functional models provided by Web methods like OO-H [9], UWE [12], etc. On the other side, the Subsystem Model (SM) and the Configuration Model (CM) define the software architecture of the Web Application. The SM and CM architectural models, defined by WebSA, use two different architectural styles to specify a Web application, namely, a subsystem (or layer style) and a component style respectively.

The first PIM-to-PIM transformation (from now on T1, see Fig. 1) turns the analysis models into platform independent design models. It integrates the information about functionality and architecture (see sect. 4) in a single Integration Model (IM). In the same way, the Integration Model, is the basis on which several PIM-to-PSM transformations, one for each target platform (from now on T2, see e.g. T2, T2' and T2'' in Fig. 1), can be defined. The output of these transformations is the specification of the Web application for a given platform (e.g. J2EE, .NET, etc).

In order to show the usefulness of the WebSA approach, we have chosen the J2EE Petstore example. This application constitutes a blueprint that uses best practices and design guidelines for a distributed component e-commerce Web application. Next section illustrates the design the Web Functional Viewpoint in the WebSA approach by means of the Domain model. The definition of this viewpoint with other notations would also be possible.

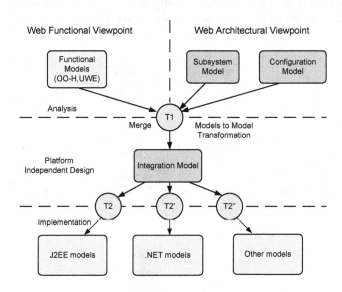

Fig. 1. WebSA Development Process

3   The Web Functional Viewpoint: Domain Model

The Web functional viewpoint of WebSA is made up by the Domain, Navigational and Process models defined by the traditional Web methodologies. However, for lack of space, in this paper we only focus on the Domain model in order to show how its information is merged with the architectural models. The Domain model represents the domain entities of the Web application and it is free from any technical or implementation details and represents an ideal class model.

Like a typical e-commerce site, the Petstore presents the customer with a catalog of products. The customer selects products of interest and places them in a shopping cart.

Fig. 2 depicts a class diagram which contains the most important domain entities of the Petstore. The *customer* class contains a set of different attributes that represent the personal data (userid, email, name, address, phone, etc.). Also, the application contains the customer preferences in the *Profile* class like favorite category, language, etc. In order to ease the accessibility to the different products, each product is classified in a Category (p.e a parrot corresponds to the bird category). Thus, the customer selects a particular product in the category list. At this moment, the application displays detailed information about the selected product. The product class contains the description and image. When there are several variants of the same product, each variant is shown as a separate *item*. For example, when showing details about an African parakeet, the items could be large male African parakeet or a small female African parakeet which has a different unit cost, supplier, stock, etc. When the customer decides to purchase a particular item and clicks a button to *addToCart* the item to the shopping *cart*. The customer may continue shopping, adding more items to the cart while there are items in stock. Finally, the customer can choose to order the items in the shopping cart at any time. The operation checkout from the Cart class is invoked.

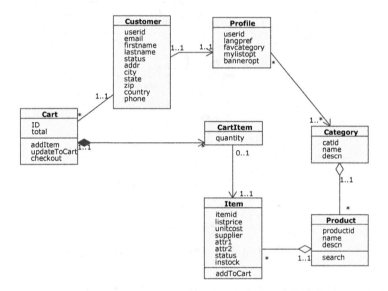

Fig. 2. Domain Model of Petstore

The next step in the WebSA analysis phase is to specify the Web architectural viewpoint. For the purposes of this paper, only the configuration model needs to be specified.

4 The Web Architectural Viewpoint: Configuration Model

The CM defines a component architectural style based on the structural view of the Web application. It defines a set of Web components and their connectors, where each component represents the role or the task performed by one or more common components identified in the family of Web applications. In this way, CM uses a topology of components defined in the Web application domain, and this allows us to specify the architectural configuration without knowing anything about the problem domain. At this level, we can also define architectural patterns for the Web application as a reuse mechanism. A Configuration model is built by means of a UML 2.0 Profile of the new composite structure model, which is well-suited to specify the software architecture of applications. The main modeling elements of the CM are WebComponent, WebConnector, WebPart and WebPattern. A description of their notation and semantics can be found in [14].

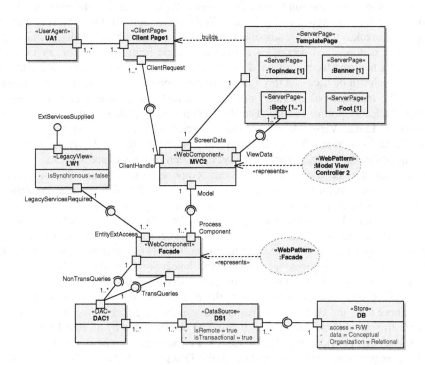

Fig. 3. Configuration Model of the Petstore

In order to represent its architectural style, the CM has been defined as an extension of the UML Composite Structure model, and includes Web components and properties of the Web application domain. The CM model also provides the necessary information for the T1 transformation defined in the WebSA development process (see Fig. 1) for integrating the functionality with the architecture in the IM model.

Fig. 3 shows a general view of the CM representing the Petstore architecture. In the front-end part of the model we find the component UserAgent (e.g. a browser) which receives the user's requests and renders the ClientPage set. Each ClientPage component contains the interface and functionality information and is responsible for sending messages to the MVC2 WebPattern component (MVC2 is detailed in [14]). The MVC2 Webpattern receives the requests through the WebPort ClientHandler and establishes the interface reaction through the WebPort ScreenData, which defines the ServerPage components.

In Petstore the ServerPages are specified following the pattern Master Template defined by Conallen [6]. Following this pattern, in Fig. 3 we have defined a TemplatePage that builds the client pages by instantiation of the WebParts TopIndex, Banner, Foot and Body.

Each instance of a Body ServerPage needs an interface to access the required data objects. Such interface is provided by the WebPort ViewData of the MVC2 Web Pattern. Looking at the MVC2, we can observe that this component needs information from the components that implement the business logic, which is obtained through the BLogic interface offered by the Façade WebPattern defined by [8]. This Façade component invokes the DAC component (Data Access Component), which contains the data access methods and decouples the business logic from the data. In our example DAC offers two interfaces, one for the non transactional queries, i.e. the data retrieval queries which can be accessed through the WebPort process component of Façade, and one for the transactional queries (insert, update and delete) which can be accessed through the Entity port of Façade. The WebComponent Façade is in turn related to the component LegacyView, which offers a series of services coming from the EntityExtAccess port to other applications and converts the received asynchronous calls into requests to the business logic. Finally, the specified remote and transactional data sources allow for the connection to a Store component that contains the information modelled in the domain model of the functional view of the Web application Petstore, and specifies a read/write access, as well as a relational organization.

Fig. 4 depicts the components of the Façade WebPattern which includes a set of stateless ProcessComponents (e.g., a Session Stateless EJB), which receives the requests through the *BLogic* WebInterface from the *MVC2*, and resends them to the Entity through the interfaces *createEntity* and *invokeServices*. This pattern requires an interface to DAC through the *nonTransactionalQueries* interface. Also, it has a set of EntityWeb components that represent the elements of the domain in the business logic. These have the tagged value isShare=true indicating that they can be shared by multiple transactions and users (e.g., it could be implemented by an EJB Entity). Note how this entity provides the *ExtEntityServices* interface for the View Legacy and sends the requests to the data layer through the *TransactionalQueries* WebInterface.

Once the WebSA analysis has been completely specified, the next step is to describe the transformation process to obtain the integration model.

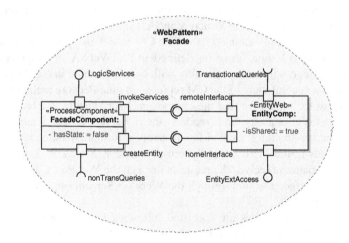

Fig. 4. Façade WebPattern

5 The WebSA Transformation Process

The WebSA transformation policy is driven by the architectural viewpoint, i.e. it is defined by a set of transformations in which first class citizens are the classes of the architectural view. The WebSA development process consists of two types of transformations: T1 and T2. T1 merges the elements of the architectural models of WebSA with those of the functional models, and translates them into a platform independent design model called Integration Model. T2 turns the Integration Model into a platform specific implementation model (e.g. J2EE or .NET). Both transformations are complex, i.e. they are made up of a set of smaller transformations, which are executed in a deterministic way in order to complete the transformation.

In MDA there are different alternatives for getting the information necessary for transforming one model into another (e.g. using a profile, using metamodels, patterns and markings, etc). WebSA has opted for a metamodel mapping approach to specify the transformations, because in this way it is possible to obtain the information of the different Web approaches just knowing their MOF metamodel.

As an example, let's take a closer look at the merging process of WebSA with the OO-H models (T1 in Fig. 1). In order to perform this merging, WebSA extends the MDA model transformation pattern of Bezivin [2]. In this pattern, MOF-based metamodels are the source of the transformation models that carry out the transformation to the target metamodel elements. The transformation models are defined in the QVT 2.0 language [16] which, as the reader might know, is an MDA standard also based on the MOF language.

The QVT specification has a hybrid declarative/imperative nature. The declarative part is split into a user-friendly part based on transformations which comprises a rich graphical and textual notation, and a core part which provides a more verbose and formal definition of the transformations. The declarative notation is used to define the transformations that indicate the relationships between the source and target models, but without specifying how a transformation is actually executed. QVT also defines operational mappings that extend the metamodel of the declarative approach with additional concepts. This allows for a definition of transformations that use a complete imperative approach. Next, we present an example of a T1 transformation using the graphical notation of QVT.

Fig. 5 uses the QVT graphical notation to define the *FacadeDomain2Integration* transformation. This transformation checks ('c' arrow) whether there is a class in the *Domain* model that contains a set of operations (see *NOperation o1set* in Fig. 5). Also, it checks in the *Configuration* model whether there is a *WebPattern* called *Façade* that contains both a *ProcessComponent* and an *EntityWeb* instances. If both patterns are found, the transformation enforces ('e' arrow) that both one stateless *Process Component* (that is, with its WebAttribute hasState=false) and an *EntityWeb* are created in the *Integration* model. In addition, the *o1set* from the class *c1* is transformed into a set of WebServices associated to the Process component *(s1set)* and a set of WebServices associated to the EntityWeb component *(s2set)*. Also, the *NAttributes* from the Domain model *(a1set)* are converted into a set of *WebAttributes* in the Integration model *(a2set)*. The links among the n-ary elements (depicted by two superimposed rectangles) in the T1 transformation are defined in the *Operations2WebServices* and *Attributes2WebAttributes* subtransformations included in its *Where* clause. On one hand, *Operations2WebServices* generates for all *NOperations* of each Class element a *WebService* in a Component. On the other hand, *Attributes2WebAttributes* generates for all *NAttributes* of each Class element a *WebAttribute* in an *EntityWeb*.

FacadeDomain2Integration

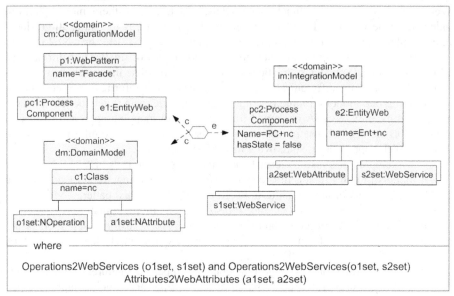

Fig. 5. Example of T1: FacadeDomain2ToIntegration

6   The Integration Model

IM defines a complete structural design of our application in a platform independent way. It integrates the SM and the CM with the functional viewpoint models. This model plays and outstanding role in WebSA, due to the fact that certain application characteristics are only identifiable when we consider together functional and architectural aspects. For instance, in order to determine the granularity of the business logic components, it is necessary to know both the architectural structure (e.g. whether this logic is likely to be distributed) and the business logic functionality itself (the tasks to be performed).

As previously stated (see fig. 1), the IM can be obtained by means of a PIM-to-PIM transformation applied on the SM and the CM together with the functional view, and reduces the modeling effort. Also, this automated mapping causes the IM to inherit the architecture patterns defined in the CM, which will be now reflected in the concrete application. The resulting model is the basis on which the designer may perform further refinements in order to fine-tune the architecture to the system needs.

Fig. 6 shows a portion of the Petstore IM that represents a simplified WebModule BusinessLogic. This module contains a set of WebComponents and their relationships obtained by the T1 transformation. On the left, the module has the IBLogic interface that gathers the requests from the client components. This interface grants access to the different ProcessComponents in charge of obtaining all the requests from the client and launching the transactions in the business logic. As stated in the T1 transformation, each ProcessComponent is obtained from one or more domain classes (Customer, Profile, Product, etc., see Fig. 2). When a ProcessComponent begins the transaction, it creates an EntityWeb by means of a Home Interface (e.g IHomeProduct), and subsequently invokes such EntityWeb to access their WebServices through the Remote interface. In our example, the EntityWeb stores the state of the class instances of the Domain model by means of a DAC component (which must be also generated by T1 for each domain class). The DAC component provides a mapping between the Object data and the relational store. In Fig. 6 we can observe how in our architecture all of DAC components require the IdataInterface in order to store the data in the persistent layer.

It is also important to stress that this model still centers on design aspects (WebComponents, their WebPorts and WebParts, WebInterfaces, WebModules and WebConnectors), and does not say anything about implementation. In this way, the model is still independent from the target platform. From this model, it is possible to define different T2 transformations (see Fig. 1) to specific platforms such as J2EE, .NET, PHP, etc.

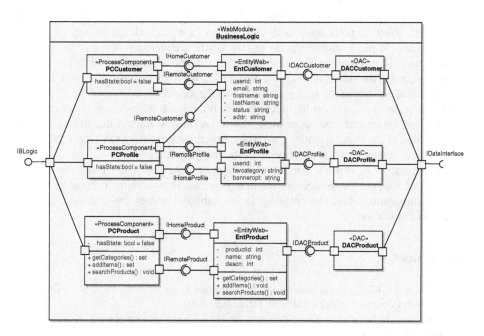

Fig. 6. Simplified BusinessLogic WebModule of the Petstore Integration Model

7 Related Work

This section compares our work with related research in the area of Web Engineering where MDA is applied to the development of Web applications.

In Tai et al [17] different kinds of artifacts are provided in a consistent and cohesive way by means of a metamodel. Our approach improves this idea with (1) the integration of proven successful models from the Web engineering field and (2) the formalization of the code generation phase by means of transformation rules.

Another model-driven methodology for Web Information System development is MIDAS [4]. This methodology uses XML and object-relational technologies for the specification of the PSMs. Unlike the WebSA approach, it does not establish the transformation mapping following the standard QVT, and it does not provide any Web application architectural aspect. In contrast, architectural aspects are present in other Web engineering proposals. OOHDM-Java2 [10] proposes a product line architecture in J2EE for simplifying the systematic construction of different families of applications, only useful for J2EE platforms. Similarly, WebML [5] proposes a static architecture based on the J2EE struts framework. The rigidity of these approaches in terms of architecture is a drawback that has been overcome in the WebSA approach with the definition of a set of flexible architectural models.

8 Conclusions and Further Work

WebSA is an approach that complements the currently existing methodologies for the design of Web applications with techniques for the development of Web architectures. WebSA comprises a set of UML architectural models and QVT transformations, a modeling language and a development process. The development process includes the description of the integration of these architectural models with the functional models of the different Web design approaches.

In this paper we focus on the development process of WebSA and describe how models are integrated and generated based on model transformations. For the specification of the transformations we have chosen the promising QVT approach that allows for visual and textual description of the mapping rules.

We are currently working on a tool to represent the set of QVT transformation models that support the WebSA refinement process. This work will allow to define the transformations while guaranteeing the traceability between those models and the final implementation.

References

1. L. Bass, M. Klein, F. Bachmann. Quality Attribute Design Primitives, CMU/SEI-2000-TN-017, Carnegie Mellon, Pittsburgh, December 2000.
2. J. Bézivin. In Search of a Basic Principle for Model   Driven Engineering, Novática n°1, June 2004, 21-24
3. F. Buschmann, R. Meunier, H. Rohnert, P. Sommerlad, M. Stal. Pattern-Oriented Software Architecture – A System of Patterns, John Wiley & Sons Ltd. Chichester, England, 1996
4. P. Cáceres, E. Marcos, B. Vela. A MDA-Based Approach for Web Information System, Workshop in Software Model Engineering, WisME 2004.
5. S. Ceri, P. Fraternali, M. Matera. Conceptual Modeling of Data-Intensive Web Applications, IEEE Internet Computing 6, No. 4, 20–30, July/August 2002
6. J. Conallen. Building Web applications with UML Second Edition. Adisón Wesley

7. Longman. September 2002.
8. E. Gamma, R. Helm, R. Johnson & J. Vlissides. Design patterns: elements of reusable object-oriented software. Reading Mass: Addison-Wesley, 1995.
9. J. Gómez, C. Cachero, O. Pastor. Conceptual Modeling of Device-Independent Web Applications. IEEE Multimedia, 8(2), 26–39, 2001
10. M. D. Jacyntho, D. Schwabe, G. Rossi. A Software Architecture for Structuring Complex Web Applications. Journal of Web Engineering, 1(1): 37-60, 2002
11. A. Kleppe, J. Warmer, W. Bast. MDA Explained: The Model Driven Architecture, Practice and Promise, Addison-Wesley, 2003
12. N. Koch, A. Kraus. The Expressive Power of UML-based Web Engineering, In Proc. of the 2nd. Int. Workshop on Web-Oriented Software Technology, CYTED, Málaga, Spain, 105-119, June 2002
13. S. Meliá, C. Cachero. An MDA Approach for the Development of Web Applications, In Proc. of 4th International Conference on Web Engineering (ICWE'04), LNCS 3140, 300-305, July 2004
14. S. Meliá, J. Gomez, N. Koch. Improving Web Design Methods with Architecture Modeling. 6th International Conference on Electronic Commerce and Web Technologies (EC-Web 2005), August 2005.
15. OMG. Model Driven Architecture, OMG doc. ormsc/2001-07-01
16. OMG. 2nd Revised submision: MOF 2.0 Query / Views /Transformations RFP, OMG doc. ad/05-03-02
17. H. Tai, K. Mitsui, T. Nerome, M. Abe, K. Ono. Model-Driven Development of Large scale Web Applications, IBM J. Res. & Dev. Vol. 48 No. 5/6, Sep/November 2004
18. TM J2EE Blueprint. Java Petstore 1.1.2, http://developer.java.sun.com/ developer/releases/ petstore/petstore1_1_2.html, November 2004

A Precise Approach for the Analysis of the UML Models Consistency*

Francisco Javier Lucas Martínez and Ambrosio Toval Álvarez

Software Engineering Research Group,
Department of Informatics and Systems,
University of Murcia (Spain)
{fjlucas, atoval}@um.es

Abstract. The UML notation is a well-know standard notation to describe OO systems. But the UML specification has certain imprecisions and ambiguities that, along with possible errors made by the modellers, may cause inconsistency problems in the models of the system. This paper presents a rigorous approach to improve the consistency analysis between UML diagrams.

This proposal is based on a previous formalization of the UML metamodel diagrams, [1–4], in *Maude*. The framework given by the specifications created helps to guarantee the consistency of models because all the specifications are integrated within the same formalism. This work focuses on the analysis of the inter-diagram consistency. Several examples of properties are shown that help to guarantee the consistency between UML Communication and Class Diagrams.

1 Introduction

UML [5] is a modelling language which was created as union of varied notations, and promoted by OMG. But UML specification has certain imprecisions and ambiguities that, along with possible errors made by the modellers, cause inconsistency problems in the models of the system. Within the UML-based development process, the main sources of inconsistency are, [6]:

1. The existence of multiple software artifacts or diagrams to describe the same system, which can cause inconsistencies in the information that appears in these diagrams.
2. The imprecise semantics of the UML, which means that a UML model may have multiple interpretations.

This paper presents a rigorous approach to analyze and improve the consistency between UML diagrams. This proposal is based on a previous formalization of the UML metamodel diagrams, [1–4], in *Maude* [7]. The framework given by the specifications created helps to guarantee the consistency of models, because

---

* Financed by the Spanish Ministry of Science and Technology, project DYNAMICA/PRESSURE TIC 2003-07804-C05-05.

J. Akoka et al. (Eds.): ER Workshops 2005, LNCS 3770, pp. 74–84, 2005.

all the specifications are integrated within the same formal technique (algebraic specifications). Furthermore, the semantic of each one of these specifications has a precise and no ambiguous interpretation due to its formalization in a formal language.

The language chosen for the realization of the formalization is *Maude*. This is a formal specification language that is based on equational logic and rewriting logic. Furthermore, Maude is a language that allows the execution of the specifications created, which allows one to animate models and create system prototypes to check the behavior of the system.

This work is based on the formalization carried out in previous work, in which the formalization of the following UML metamodel diagrams are treated: Class Diagram [1], Collaboration Diagram [2] (named Communication Diagram in UML 2.0), Statechart Diagram [3], and Sequence Diagram [4]. All theses formalizations have been updated to UML 2.0.

Thus, the integration of this formalization and the work produced that have been performed about them, such as: animating models, making transformation between models and verifying of properties can be used to improve the quality of a system.

Furthermore, all the applications of this formalization can be used in MDA, since UML language is usually used as the modelling language in MDA. We can use it to guarantee the consistency of the PIM (Platform Independent Model) models, before transforming them to PSM (Platform Specific Model) models. This formalization can also give support to the transformations that are made within the MDA (PIM→PSM, PSM→PSM, PIM→PIM).

This work focuses on the analysis of the inter-diagram consistency, and several properties are shown that help to guarantee the consistency between UML Class and Communication diagrams. This algebraic approach can be applied to any diagram which is formalized, see section 2.

After this introduction, in section 2, a general description of the algebraic formalizations of the UML diagrams used in this work is given. Section 3 shows the analysis of the consistency made for the UML Class and Communication Diagram. Section 4 identifies some related work. Finally, in section 5 conclusions and further work are given.

2   Algebraic Formalization of the UML Diagrams

As a previous step to the rigorous analysis of the consistency between the different UML diagrams of a system, it is necessary to have a rigorous representation of these models. We decided to make an algebraic formalization of part of the UML metamodel. Figure 1 presents the necessary algebraic modules to carry out an analysis of the consistency. The *Integration* module uses the available specifications of the UML diagrams and implements the equations that check the inter-diagrams consistency.

This paper focuses on the Class and Communication Diagrams. This method is generalizable to any combination of two or more diagrams, because we have

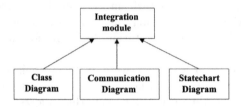

Fig. 1. Algebraic modules used in the analysis of the consistency

the corresponding formalization and integration with the rest of the diagrams, for example the UML Statechart Diagram [3].

The next sections give a description of the formalization of each diagram needed to understand the rest of the paper. For the sake of brevity, the description of the specification offers a very simplified view of the algebraic formalization. For more details, see [1, 2].

2.1   UML Class Diagram

The first diagram that will be commented is the UML Class Diagram. This diagram describes the static structure of a system and is made up of a set of elements such as classes, interfaces, and others; and relationships among these elements, such as associations and aggregations. The module that contains the formalization of the diagram is shown in Figure 2. For the sake of brevity, this is a very reduced part of the formalization (see [1] for more details).

```
(fmod CLASSDIAGRAM is sort ClassDiagram .
  ...
  op classDiagram :  ClassList ObjectList
                        AssocList LinkList -> ClassDiagram .
  op getCDClasses : ClassDiagram -> ClassList .
  ...
  var CLASSES : ClassList . var OBJECTS : ObjectList .
  var ASSOCIATIONS : AssocList . var LINKS : LinkList .
  eq getCDClasses(
       classDiagram(CLASSES, OBJECTS, ASSOCIATIONS, LINKS)) = CLASSES .
  ...
endfm)
```

Fig. 2. Module that formalizes the UML Class Diagram

This specification along with the one shown in the next section will be used to show the application of the formalization of the metamodel of the UML diagrams to guarantee the consistency between models.

In Figure 3, we can see an example of a Class Diagram despicted with a CASE tool. This diagram represents a reservation system and will be the example used in the paper to check the inter-diagrams consistency (section 3).

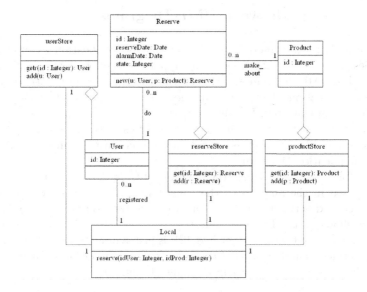

Fig. 3. Example of a UML Class Diagram

2.2   UML Communication Diagram

The next specification need is the one corresponding to the UML Communication Diagram. This type of diagram shows an interaction between objects. One of the most important aspects that is shown in this diagram is the context of the interaction. In Figure 4 appears part of this diagram formalization, the complete formalization used in this diagram is shown with more detail in [2].

```
(fmod COMMUNICATIONDIAG is sort CommunicationDiag .
   ...
   *** Constructor
   op communicationDiag : LifeLineList MessageList -> CommunicationDiag .
   ...

   op getCDLifeLines : CommunicationDiag -> LifeLineList .
   op getCDMessages : CommunicationDiag -> MessageList .

   var LIFELINES : LifeLineList .     var MESSAGES : MessageList .
   eq getCDLifeLines(communicationDiag(LIFELINES, MESSAGES)) = LIFELINES .
   eq getCDMessages(communicationDiag(LIFELINES, MESSAGES)) = MESSAGES .
   ...
endfm)
```

Fig. 4. Module that formalizes the UML Communication Diagram

3   Consistency Between UML Diagrams

Guaranteeing the inter-models consistency or the verification of inter-models properties is one of the most interesting applications of the formalization of the metamodel of UML diagram and this is what we shall see in this section.

As an application of the formalization performed, several properties have been already implemented. In this work, two of them are shown to illustrate the process of consistency verification. Other properties and operations implemented can be found in [8] such as class consistency, correct order of method invocation in a Communication Diagram,...

In the following subsections, the example of Class Diagram shown in Figure 3 will be used in the analysis of the consistency. This consistency verification has been realized between this Class Diagram and the Communication Diagrams which are shown in the Figures 6 and 9.

Although the first property verified, 3.1, is just syntactic, the second property, 3.2, also verifies richer features of the consistency between both diagrams, such as type checks of the parameter in the calls of methods.

3.1   Verification of the Consistency Regarding Associations

In this section we verify the following property:

"For each association that is defined in the Communication Diagram there must exist at least one association in the Class Diagram that allows the sender to send messages to the receiver"

This property guarantees that, for each association defined in the Communication Diagram, there exist at least one association in the Class Diagram that connects the classes that take part in the association of the interaction. In this verification we do not take into account derived associations.

We have two different alternatives to identify an association of the Communication Diagram in the Class Diagram. These are:

1. To compare the name of the association of the Class Diagram, AssocName, with the name that the association (*Connector*) of the communication diagram has.
2. To search associations in the Class Diagram which connect classes of which ClassName is the same ClassName which appears in the *Connectable Elements* (*Connector Ends*) of the association of the communication diagram.

It does not seem practical to require the same, exact, name in the association label of the Communication Diagram and in the association of the Class Diagram. Moreover, frequently this name does not appear, therefore we choose the second alternative as our criterion in the search of associations.

To implement this property, we only search that an association exists in the Class Diagram which connects the classes that are connectable elements in the association (connector) of the communication diagram. Al least one must exist in order to keep the property. Since we do not take into account the association

```
(fmod PROPERTY1 is ...
   op testProp1 : ClassDiagram CommunicationDiag -> String .
   op testProp1 : AssocList ConnectorList -> String .

   var CommunD : CommunicationDiag .
   var CLASSL : ClassList .              var OL : ObjectList .
   var AL : AssocList .                  var LL : LinkList .
   var CL : ConnectorList .              var C : Connector .

   eq testProp1(classDiagram (CLASSL, OL, AL, LL), CommunD) =
         testProp1(AL, getCDConnectorList(CommunD)) .
   eq testProp1(AL, empty) = "" .
   *** C CL = ConnectorList. C = head, CL = tail.
   eq testProp1(AL, C CL) =
     if assocExists(AL, getClassName(getCE1(C)),
           getClassName(getCE2(C))) == nullAssoc then
       "ERROR. An association in the Class Diagram "
       + "between the classes " + string(getClassName(getCE1(C)))
       + " and " + string(getClassName(getCE2(C)))
       + ", which is necessary for the association "
       + string(getLabel(C)) + ", does not exist. " + testProp1(AL, CL)
     else testProp1(AL, CL) fi .
endfm)
```

Fig. 5. Formalization of the property regarding Associations (section 3.1)

name, an association in the Communication Diagram can be "made" by several associations in the Class Diagram.

Figure 5 shows the specification that formalizes the verification of this property. Note that other semantics for the fulfilment of this property could be specified too.

In Figure 6 we see an example of Communication Diagram that does not fulfil this property, and in Figure 7 we can see the property reduction that

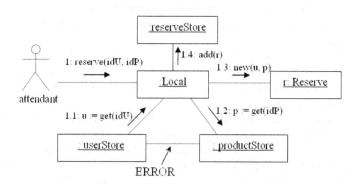

Fig. 6. Example of a UML Communication Diagram that breaks the property verified in 3.1

```
reduce in RESERVE :
  testProp1(classDiagEj,CDreserve)
result String :
  "ERROR. An association in the Class Diagram between
  the classes userStore and reserveStore, which is  necessary for
  the association userSt_resSt_label, does not exist. "
```

Fig. 7. *Maude* reduction of the example of the Figure 6

Maude produces. As we have already indicated, the Class Diagram used as the example is shown in Figure 3.

Another property implemented using this formalization, which is not shown here due to its similarity with the property of this section, is the verification of the consistency of the classes used, in other words, it guarantees that the classes used in the Communication Diagram are present in the Class Diagram to which it belongs.

3.2   Verification of the Consistency Regarding Methods

In this section, we verify that the use of the methods in the Communication Diagram is consistent with the information that appears in the Class Diagram. The property that we want to check is the following:

"The methods used in a communication diagram must be declared in the class diagram and their declaration, with regard to parameters and types, must be correct."

As we have already said, we will verify that the methods that are executed in a *ConnectableElement* (*Connector End*) exist in the class corresponding to their *ClassName*. In the case that this method exists, the property also verifies that the method has the same number of parameters, and the same types, as in the class. Furthermore, the overload of methods has been taken into account in the implementation of the property. The module that verifies this property is shown in Figure 8.

As we can see, we check that each method used by a *ConnectableElement* exists in the class to which it belongs. To do this, first we look for the class and then look for methods with the same *OpName* as in the message of the Communication Diagram.

If no method is found, an error is given as output. If one is found, we check that the number of parameters with the method invoked in the communication diagram is the same as in its definition and that these parameters have the same types. If the method has the same number of parameters but the types are different, the reduction of the property also informs us. The reasons for this error might be that the information of the types has not been included in the communication diagram or that this information has been included incorrectly.

```
(fmod PROPERTY2 is...
  op testProp2 : ClassDiagram CommunicationDiag -> String .
  op testProp2 : ClassList MessageList -> String .
  *** Params: operation list of a class, receiver and its message list.
  op testProp2 : OpList ConnectableElement Message -> String .
  *** Verify that the messsage is among the OpNames from OpList.
  op testProp2 : Message OpList ConnectableElement Int -> String .
  ...
  eq testProp2(classDiagram(CL,OL,AL, LL),
      communicationDiag(LifeLineL, ML)) = testPropInter2 (CL, ML) .

  eq testProp2(CL, empty) = "" .
  eq testProp2(CL, M ML) =
      testProp2(getOperations(getClassbyName(getClassName(
        getReceiver(M)), CL)), getReceiver(M), M)
      + testProp2(CL, ML) .

  eq testProp2(OpL, CE, M) =
    if findOps (OpL, M) =/= nullOp then
      testProp2(M, findOps(OpL, M),CE,0)
    else
      "ERROR. The message (" + string(getMsgNumber(M))
      + ". " + string(getMsgLabel(M)) + ") "
      +" doesn't exist in the class "+ string(getClassName(CE)) + ". "
    fi .
  ...
  eq testProp2(M, nullOp, CE, I) =
    if I == 0 then "ERROR. The signature of the message ("
      + string(getMsgNumber(M)) + ". " + string(getMsgLabel(M)) + ") "
      +"doesn't concur with the method of the class " +
      +string(getClassName(CE))+" neither in number of parameters nor "
      + "type of them. "
    else "ERROR. The signature of the message ("+string(getMsgNumber(M))
      + ". " + string(getMsgLabel(M)) + ") "
      +" doesn't concur with the method of the class "
      + string(getClassName(CE))
      + " because the types of the parameters are not correct. "
    fi .
  eq testProp2(M, Op OpL, CER, I) =
    if length(getOpParamList(Op)) == length(getMsgParamList(M)) then
      if getTExpr(getOpParamList(Op)) == getTExpr(getMsgParamList(M))
            then ""
      else testProp2(M, OpL, CE, I + 1) fi
    else testProp2(M, OpL, CE, I)
    fi .
endfm)
```

Fig. 8. Formalization of the property regarding Methods (section 3.2)

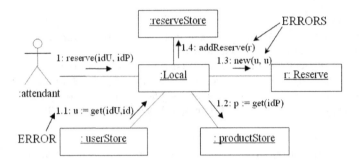

Fig. 9. Example of a UML Communication Diagram that breaks the property verified in section 3.2

```
reduce in RESERVE :
  testProp2(classDiagEj,CDreserve)
result String :
  "ERROR. The signature of the message (\001\001. get) doesn't concur
  with the method of the class userStore neither in number of
  parameters nor type of them.
  ERROR. The signature of the message (\001\003. new)  doesn't concur
  with the method of the class Reserve because the types of the
  parameters are not correct.
  ERROR. The message (\001\004. addReserve)  doesn't exist in the class
  reserveStore. "
```

Fig. 10. Reduction in *Maude* of the property verified in section 3.2 on the example of the Figure 9

Finally, in the verification of this property, we cannot take into account the syntactic identity of the name of the parameters, since the identifiers that appear in the method definition, which are called *formal parameters*, cannot be the same as the identifiers of the parameters in the invocation, called *actual parameters*, and which will replace the formal parameters in the body of the method.

Figure 9 shows an example that contains the three possible errors that this property detects. The first error is the message *1.4 addReserve*, which does not exist in the class *reserveStore*. The second error is found in the message *1.1 get(idU,id)*, this method exists in the class *userStore*, but does not have the same numbers of parameters. The last error is produced in the message *1.3 new(u,u)* (the method is declared in the class *Reserve*) also has two parameters, but the type of the second parameter is *User* instead of *Product*. The property reduction are shown in the Figure 10.

4   Related Work

In [6], a general and updated view of the consistency problems within the UML based development process are given. In this work, the use of techniques to avoid problems of consistency is justified, because UML is considered as a standard in the development of systems.

The approach of formalizing the UML metamodel to guarantee the correct development of models has been dealt with in many papers, although the formalization of the UML Communication Diagram has not been deeply studied. Most of the approaches formalize other UML diagrams such as UML Class or Statechart diagrams.

In [9], a proposal to verify UML models using B abstract machines for UML Class Diagram is presented. Another paper [10] tackles the formalization of UML models and discusses the integrity consistency check between different models. In this approach, the formal language Object-Z is used, which allows the authors to implement each UML element as a class. None of these approaches offers the possibility of making a automatic translation from the models to the formal specification. Unlike them the formal framework presented in this paper is integrated with automatic translators that obtain the specification that represent a model from the model depicted with a CASE-tool.

Some research [11] has been done on formalization of the UML Statechart diagram. This approach uses the SPIN model checker to perform the verification. The tool verifies several properties and generates a sequence diagram that shows how to reproduce the error in the model. However, this work suffers from some problems, such as a poor efficiency of the implementation. In the Maude design, efficiency has been considered from its beginning, resulting in a fast execution of the reductions and rewrites.

5   Conclusions

This work presents a formal approach to improve the inter-diagrams consistency. The specifications created in [1, 2, 3, 4] offer a good framework to guarantee the model consistency, because all the specifications are integrated.

Furthermore, this formal framework has been revealed as a useful instrument to realize verification of properties, both intra-model and inter-model. Modifying the semantics of existing property specifications and/or adding new property specification is very easy, once the formal framework (basic sorts, operations and equations) is available. Another possible application is the realization of precise transformations that help to find better models.

As further work, we will continue to implement properties to improve the diagrams' consistency. Moreover, we are searching for real case studies to justify the use of algebraic specification within MDA. The integration of the Communication and Statechart Diagram is another of the research lines that we are working on, in order to verify that each object that takes part in the communication diagram has a consistent state when the interaction finishes. Another future work is the application of this approach on tools for the development of Web Information Systems (WIS), like MIDAS-CASE[12].

References

1. Fernández Alemán, J.L., Toval Álvarez, A.: Improving System Reliability via Rigorous Software Modeling: The UML Case. Proceedings of the 2001 IEEE Aerospace Conference (track 10: Software and Computing), Montana, USA IEEE Computer Society (2001)
2. Lucas Martínez, F.J., Toval Álvarez, A.: Formal Verification of Properties in the UML Collaboration Diagram. ICSSEA 2004: 3rd Workshop on SYSTEM TESTING AND VALIDATION. Paris (2004)
3. Fernández Alemán, J.L., Toval Álvarez, A.: Can Intuition Become Rigorous? Foundations for UML Model Verification Tools. International Symposium on Software Reliability Engineering, Published by IEEE Press (2000)
4. Whittle, J., Araújo, J., Toval Álvarez, A., Fernández Alemán, J.L.: Rigorously Automating Transformations of UML Behavior Models. Dynamic Behaviour in UML Models: Semantic Questions in conjunction with UML 2000 York, UK , ACM SIGSOFT, IEEE Computer Society (2000)
5. OMG: Object Management Group. UML Superstructure 2.0. Draft adopted Specification. Retrieved from: http://www.omg.org/uml. (2004)
6. Huzar, Z., Kuzniarz, L., Reggio, G., Sourrouille, J.L.: Consistency Problems in UML-based Software Development. In UML Modeling Languages and Applications. UML 2004 Satellite Activities Lisbon, Portugal, October 11-15, 2004 Revised Selected Papers. Jardim Nunes, N. Selic, B., Silva, A. Toval, A. (Eds.) **Springer Verlag vol. 3297 of LNCS** (2004)
7. Clavel, M., Durán, F., Eker, S., Lincoln, P., Martí-Oliet, N., Meseguer, J., Talcote, C.: Maude 2.0 Manual. Versión 1.0, http://maude.csl.sri.com/. (2003)
8. Lucas Martínez, F.J., Toval Álvarez, A.: Algebraic Specification of the UML Collaboration Diagram and Applications. Tech. Rep. LSI 3, Department of Informatics and Systems. University of Murcia. (2004)
9. Truong, N.T., Souquieres, J.: An approach for the verification of UML models using B. 11th IEEE International Conference and Workshop on the Engineering of Computer-Based Systems (ECBS'04) (2004)
10. Kim, S.K., Carrington, D.: A Formal Object-Oriented Approach to defining Consistency Constraints for UML Models. 2004 Australian Software Engineering Conference (ASWEC'04) (2004)
11. Litius, J., Porres Paltor, I.: vUML: a Tool for Verifying UML Models. 14th IEEE International Conference on Automated Software Engineering (1999)
12. Vara, J., de Castro, V., Cáceres, P., Marcos, E.: Arquitectura de MIDAS-CASE: una herramienta para el desarrollo de SIW basada en MDA. 4ª Jornadas Iberoamericanas de Ingeniería del Software e Ingeniería del Conocimiento. Madrid (2004)

A UML 2 Profile for
Business Process Modelling*

Beate List and Birgit Korherr

Women's Postgraduate College for Internet Technologies,
Institute of Software Technology and Interactive Systems,
Vienna University of Technology
{list, korherr}@wit.tuwien.ac.at
http://wit.tuwien.ac.at

Abstract. Current UML Profiles for Business Process Modelling realise a narrow focus of the process, and capture the process flow on a low level of detail. They do not provide a comprehensive coverage of theoretical aspects. In this work, we have designed a UML 2 Profile for Business Process Modelling that provides two complementary perspectives, focussing on the business process context (e.g. goals, measures, products, customers, etc.) as well as on the detailed business process flow. Therefore, the profile presents a business process in a very comprehensive way. It is tested with an example business process.

1 Introduction

Business processes are often the starting point for software development and define requirements for software systems to be designed. Research and industry have addressed the alignment of business processes and IT only marginally. Most software developers are not aware of business processes or are not able to read the models, as different modelling languages with different diagrams and notations are used in both domains. In order to overcome this gap, we have developed a UML 2 Profile for Business Process Modelling (BPM), with the goal:

- To provide business process models to software developers in a well-known notation (Reuse of the UML notation).
- To present business process models to software developers through UML tools (Reuse of UML tools).
- To develop a meta-model that covers comprehensive aspects of business process theory, including business process context.

UML profiles provide an extension mechanism for building UML models for particular domains. None of the existing UML profiles for business process modelling [1, 6, 7, 14, 15] cover business process theory systematically. They realise a narrow focus of the process, and capture the process flow on a low level of detail. In contrast, we cover business process theory in the UML 2 profile for BPM comprehensively. As a

---

* This research has been funded by the Austrian Federal Ministry for Education, Science, and Culture, and the European Social Fund (ESF) under grant 31.963/46-VII/9/2002.

J. Akoka et al. (Eds.): ER Workshops 2005, LNCS 3770, pp. 85–96, 2005.

basis for the UML 2 profile, we have developed a meta-model that provides two complementary perspectives: the business perspective and the sequence perspective. The business perspective presents the business process from a wide angle. Software developers will get a full understanding of the process without working through the complex process logic. The business perspective provides a model that gives a comprehensive understanding of the process, as it describes the major business process characteristics, such as goals and their measures, the deliverables, the owner, the type and the customer.

The sequence perspective refines the business perspective and describes the detailed flow of the process. It can utilise any business process modelling language, e.g. the Event-driven Process Chain [13], or the Business Process Modelling Notation [2]. A detailed description of the sequence perspective is out of the scope of this paper.

The contribution of the UML 2 profile for BPM is:

- It provides business process models to software developers in UML notation. As software systems support the business processes of an organisation, the profile represents the business context and business requirements to software developers in a formal and well-known modelling notation.
- The profile can support the elicitation of requirements from the business process models for the software systems to be developed. Deriving requirements from the business process models ensures a business-goal oriented software development.
- The profile can be integrated into the Computation Independent Model (CIM) of the Object Management Group's (OMG) Model Driven Architecture (MDA) [11] approach. Because the CIM model is a business model capturing the requirements of the software systems and is traceable to code, the integration of the UML profile can improve the quality of the requirements and the design of the software. The profile is raising the level of abstraction at which software development starts.
- The UML 2 profile for BPM can be easily extended and mapped to Business Process Execution Languages (BPEL). Mapping tools are able to take the business processes models developed in a UML tool and convert them to the correct BPEL, and vice versa. Thus, high productivity will be resulting, even if the underlying technology changes.
- The profile facilitates the seamless integration of already available business process models into a UML tool, because there is no additional modelling effort required.
- It could abandon Business Process Modelling tools, as almost all UML tools support UML profiles.
- The profile integrates business process models into the standard software development environment and can be seen as a further step towards bridging the gap between business process engineering and software engineering.

Based on the requirements of business process models in Section 2, we have developed a meta-model for the UML 2 profile for BPM in Section 3. The profile is described in Section 4 and tested with an example business process in Section 5. Related work is presented in Section 6.

2   Requirements - What Must a Business Process Model Capture?

Davenport, Hammer and Champy created with business process reengineering a new discipline at the beginning of the 1990ies and provided the theoretical concept for business process modelling. So far, in the business process modelling community attention has only been given to the modelling of certain aspects of processes (e.g. roles, activities, interactions). These aspects are represented for example in the Business Process Modelling Notation [2], the Event-driven Process Chain [13], the UML 2 Activity Diagram [9], or the Role Activity Diagram [12]. But there are a lot of other aspects including the business process context that should be also represented in a model.

A business process is defined as a "group of tasks that together create a result of value to a customer" [3]. Its purpose is to offer each customer the right product or service (that is, the right deliverable), with a high degree of performance measured against cost, longevity, service and quality [5]. The term customer should be used in an extended meaning. It can literally be simply a customer, but it can also be another process in the environment that is external to the company, such as a partner or subcontractor. Thus, a customer can be an internal or external role that receives products or services from a business process. An external customer is outside of the organisation, while the internal customer is part of the organisation.

Three types of business process are differentiated: core, support and management processes [12]. Core processes concentrate on satisfying external customers. They directly add value to the organisation. They respond to a customer request and generate customer satisfaction. Support processes concentrate on satisfying internal customers. They might add value to the customer indirectly by supporting a core business process. Management processes concern themselves with managing the core processes or the support processes, or the concern themselves with planning at the business level.

Functional structures have functional managers, and business processes have a similar concept for management, namely the process owner. He or she is an individual concerned with the successful realisation of a complete end-to-end process, the linking of tasks into one body of work, and making sure that the complete process works together [3]. Often these tasks are delegated and the process owner is focused on measuring the achievement of goals and initiate actions if necessary. Therefore, this person should be in a very powerful position, especially when she / he is the owner of a core business process.

Business processes support the achievement of enterprise goals in an organisation. Process goals support enterprise goals. The achievement of goals must be measured either by qualitative or quantitative measures. Measures aim at reaching a to-be-value or target value and are very important for business process improvement. "Measurements are the key. If you cannot measure it, you cannot control it. If you cannot control it, you cannot manage it. If you cannot manage it, you cannot improve it." [4].

Thus, beside activities and roles there are a lot of other aspects to be included into a business process model e.g. customers, process owners, process types, deliverables, goals and measures. And if there is a need for more things to be integrated into a business process model, feel free to make an extension!

3 The Business Process Modelling (BPM) Meta-model

Business process models need to express a lot of aspects. Capturing all available process characteristics in one model will completely overload it. Therefore, we have developed a business process meta-model that provides two complementary perspectives: a business perspective and a sequence perspective. The *Business Perspective* is an external perspective of the business process. It presents a general view of the process and describes its major characteristics e.g. the goals, the deliverables, or the customer at a glance. The *Sequence Perspective* provides the details of the process. It is an internal perspective and describes the process flow in detail, e.g. the order of tasks, the roles that perform the tasks, or the resources created. The perspectives of this business process modelling meta-model were inspired by Jacobson et al. [5], who proposed two perspectives for business process modelling as well. All other concepts of our approach are fully different.

3.1 Business Perspective

The business perspective provides the most important characteristics for describing a business process without showing its detailed flow. Software developers, who do not know or do not need to know the process in detail, will get a full understanding of the process without working through the complex process logic. The business perspective can be seen as a starting point for getting to know a business process or as an end in itself. Such a perspective is not provided by any state-of-the art process modelling language. Therefore, we have developed a meta-model (Fig. 1) capturing the major characteristics of a business process.

The meta-model of the business perspective (Fig. 1) presents the business process in relation to other process characteristics. A *Business Process* can be either a *Core Process*, a *Support Process* or a *Management Process*. A core process is either independent from support processes or supported by one or more support processes, which in turn support one or more core processes. A business process satisfies one or more *Customers*, who can be either *External* or *Internal*. External customers are outside of an organisation. Internal customers are part of the organisation and represent other groups or departments. External customers relate to core processes and internal customers relate to support or management processes. A *Detailed Process Diagram* describes each business process and can be seen as a link to the sequence perspective. A business process may be also composed of other business processes (or subprocesses), which in turn may be part of other business processes.

A *Process Owner* is responsible for one or more business processes. Each business process generates one or more *Deliverables,* which are either *Services* or *Products*. Each business process must achieve one or more *Process Goals*, which in turn support one or more *Enterprise Goals*. Concrete *Measures* describe the achievement of *Goals*, both process and enterprise goals. Each measure has a *To Be Value* assigned, which is sometimes also called a target value, and should be reached by each process instance. A *Unit* is also assigned to one or more measures. Measures judge the quality of goals and can be either *Qualitative* or *Quantitative*.

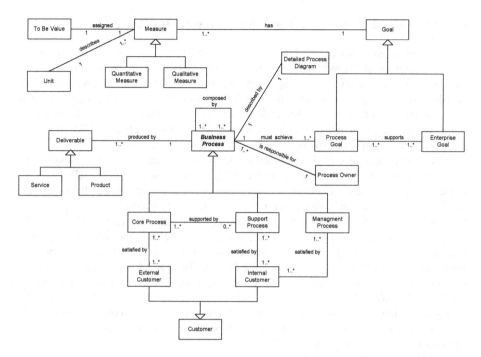

Fig. 1. Meta-Model of the Business Perspective

3.2 Sequence Perspective

The sequence perspective provides a detailed flow of the business process. Today, there are a lot of meta-models and diagrams (based on meta-models) for business process or workflow modelling available, e.g. the Business Process Definition Meta-Model [10], the Business Process Modelling Notation [2], the Event-driven Process Chain [13], the UML 2 Activity Diagram [9], the Role Activity Diagram [12]etc. Therefore, we have decided not to design 'yet another meta-model', but rather to integrate a well-established one. This can foster the acceptance of this profile in general, and simplify the integration of already available process models into a UML tool in particular. Thus, we have decided to keep the meta-model of the sequence perspective open for the special requirements and the choice of the process modeller. Process models are often available in a lot of modelling tools (e.g. ARIS Toolset, MS Visio, Adonis, etc). This facilitates the integration of already available process models into a UML tool without additional modelling effort. In contrast, meta-models that have no sufficient tool support require modelling from scratch. A detailed description of the sequence perspective is out of the scope of this paper, as it depends on the specific preference for a business process modelling diagram of the organisation.

4   The UML 2 Profile for Business Process Modelling

In this section, we describe the UML 2 profile for BPM. It is based on the business perspective of the BPM meta-model. UML offers a possibility to extend and adapt its meta-model to a specific area of application through the creation of profiles. UML profiles are UML packages with the stereotype «profile». A profile can extend a meta-model or another profile [9] while preserving the syntax and semantic of existing UML elements. It adds elements which extend existing classes. UML profiles consist of *Stereotypes*, *Constraints* and *Tagged Values*.

A stereotype is a model element defined by its name and by the base class(es) to which it is assigned. Base classes are usually meta-classes from the UML meta-model, for instance the meta-class *«Class»*, but can also be stereotypes from another profile. A stereotype can have its own notation, e.g. a special icon.

Constraints are applied to stereotypes in order to indicate restrictions. They specify pre- or post conditions, invariants, etc., and must comply with the restrictions of the base class [9]. Constraints can be expressed in any language, such as programming languages or natural language. We use the Object Constraint Language (OCL) [8] in our profile, as it is more precise than natural language or pseudocode, and widely used in UML profiles.

Tagged values are additional meta-attributes assigned to a stereotype, specified as name-value pairs. They have a name and a type and can be used to attach arbitrary information to model elements.

The UML 2 profile for BPM creates an overview model in order to describe the major characteristics of business processes. It extends the meta-class *«Class»*, the meta-class *«Property»* and the meta-class *«Actor»*. In Fig. 2 we show a part of the UML 2 meta-model related to the classifier concept of UML 2 (light) to illustrate how the stereotypes we designed (dark) fit into to the existing UML 2 meta-model.

In the UML 2 profile for BPM, we use the classes *Class* and *Actor* as base classes for all stereotypes. The OMG has defined a class as "a set of objects that share the same specifications of features, constraints, and semantics. The purpose of a class is to specify a classification of objects and to specify the features that characterize the structure and behavior of those objects." [9]. Therefore, classes are appropriate for

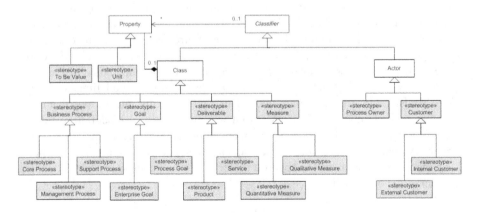

Fig. 2. Extending the UML2 Meta-Model with Stereotypes for BPM

business processes and their characteristics. An actor is used in UML 2 for modelling roles interacting with a system and is defined as a "specification of a role which is played by a user or any other system that interacts with the subject." [9]. Therefore, actors are suited for the purpose of showing process owners and customers.

The relationships between the stereotypes defined are described with the *Association* between classes. The OMG specifies an association as "a semantic relationship that can occur between typed instances" [9]. So there is no need to define an additional stereotype for the relationships between stereotypes.

Table 1. Customers and Process Owners: Specification of Stereotypes

Name	Customer	
Base Class	Class	
Description	A customer is an internal or external role that receives products or services from a business process [5].	
Constraints	None	
Name	**External Customer**	
Base Class	Customer	«External Customer»
Description	An external customer is outside of the organisation.	

Constraints	An external customer has to be satisfied by at least one core process which has to produce one or more deliverables.
	`context External inv:`
	`self.CoreProcess->size() >= 1 implies`
	`self.Deliverable->size() >= 1`
		Name

Name	**Internal Customer**	
Base Class	Customer	«Internal Customer»
Description	The internal customer is part of the organisation.	

Constraints	An internal customer has to be satisfied by at least one support process or management process which has to produce one or more deliverables.
	`context Internal inv:`
	`self.SupportProcess->size() >= 1 or`
	`self.ManagementProcess->size() >= 1 implies`
	`self.Deliverable->size() >= 1`
		Name

Name	**Process Owner**	«Process Owner»
Base Class	Actor	
Description	The process owner is concerned with the successful realisation of a process, (...) and making sure that the complete process works together [3], and with measuring it against target values.	

Constraints	A process owner is responsible for one or more processes.
	`context ProcessOwner inv:`
	`self.BusinessProcess->size() >= 1`
		Name

As described in Fig. 2, a business process and its characteristics can be defined with six abstract top-level stereotypes, «Process Owner» and «Customer» (the specification of the stereotypes is listed in Table 1), «Business Process» (see Table 2), «Goal» (see Table 3) «Deliverable» (see Table 4) «Measure» (see Table 5). The semantics of the individual elements were described in greater detail in Section 3.

Table 2. Business Processes: Specification of Stereotypes

Name	**Business Process**	
Base Class	Class	
Description	A business process is a group of tasks that together create a result of value to a customer [3]. Three types of business process are differentiated: core, support and management processes [12].	

Constraints	A business process is composed by at least one business process.
	`context BusinessProcess inv:`
	`self.BusinessProcess->size()>=1`
	A business process is designed by one or more detailed process diagrams.
	`context BusinessProcess inv:`
	`self.DetailedProcessDiagram->size()>=1`
	A business process is in the responsibility of one process owner.
	`context BusinessProcess inv:`
	`self.ProcessOwner->size()>= 1`
	A business process wants to achieve at least one process goal.
	`context BusinessProcess inv:`
	`self.ProcessGoal->size()>= 1`	

Name	**Core Process**	
Base Class	Business Process	
Description	Core processes concentrate on satisfying external customers. They directly add value to the organisation [12].	«Core Process» Name

Constraints	A core process has to produce one or more deliverables to satisfy one ore more external customers.
	`context CoreProcess inv:`
	`self.External->size()>= 1 implies`
	`self.Deliverable->size() >= 1`	

Name	**Support Process**	
Base Class	Business Process	
Description	Support processes concentrate on satisfying internal customers. They might add value to the customer indirectly by supporting a core business process [12].	

Constraints	A support process has to produce one or more deliverables to satisfy one or more internal customers.
	`context SupportProcess inv:`
	`self.Internal->size() >= 1 implies`
	`self.Deliverable->size() >= 1`
	A support process has to support one or more core processes.
	`context SupportProcess inv:`
	`self.CoreProcess->size() >= 1`
		«Support Process» Name

Name	**Management Process**	
Base Class	Business Process	
Description	Management processes concern themselves with managing the core processes or the support processes, or the concern themselves with planning at the business level [12].	«Management Process» Name

Constraints	A management process has to produce one or more deliverables to satisfy one or more internal customers.
	`context ManagementProcess inv:`
	`self.Internal->size() >= 1 implies`
	`self.Deliverable->size() >=1`	

Table 3. Goals: Specification of Stereotypes

Name	**Goal**
Base Class	Class
Description	Goals express intentions and capture reasons for a system to be built.

Constraints	A goal has at least one measure.
	`context Goal inv:`
	`self.Measure->size() >= 1`

Name	**Process Goal**	
Base Class	Goal	
Description	Process goals support enterprise goals and represent the intentions of a certain business process.	«Process Goal»

Constraints	A process goal has to be achieved by one business process.
	`context ProcessGoal inv:`
	`self.BusinessProcess->size() = 1`
	A process goal has to be supported by one or more enterprise goals.
	`context ProcessGoal inv:`
	`self.EnterpriseGoal->size() >= 1`
		Name

Name	**Enterprise Goal**	«Enterprise Goal»
Base Class	Goal	
Description	Enterprise goals represent the intentions of an organisation.	

Constraints	An enterprise goal supports one or more process goals.
	`context EnterpriseGoal inv support:`
	`self.ProcessGoal->size() >= 1`
		Name

Table 4. Deliverables: Specification of Stereotypes

Name	**Deliverable**
Base Class	Class
Description	A business process creates a deliverable that is a service or a product for a customer [5].

Constraints	If the deliverable is produced by a core process, then it has to satisfy at least one external customer.
	If the deliverable is produced by a support or management process, then it has to satisfy at least one internal customer.
	`context Deliverable inv:`
	`  if self.CoreProcess->size() = 1`
	`    then self.External->size() >= 1`
	`  endif`
	`  if self.SupportProcess->size() = 1 or`
	`    self.ManagementProcess->size() = 1`
	`    then self.Internal->size() >= 1`
	`  endif`

Name	**Service**	«Service»
Base Class	Deliverable	Name
Description	A service is a non-tangible good.	
Constraints	None	

Name	**Product**	«Product»
Base Class	Deliverable	Name
Description	A product is a tangible good.	
Constraints	None	

Table 5. Measures: Specification of Stereotypes

Name	**Measure**	
Base Class	Class	
Description	A measure is a basis for comparison, a reference point against which other things can be evaluated.	

Constraints	A Measure has to be assigned to a to be value, and is described by a unit. Both are attributes.
	`context Measure inv:`
	`self.Measure.AllAttributes->includes(unit) and`
	`self.Measure.AllAttributes->includes(tobevalue)`
	A measure has to measure one goal.
	`context Measure inv:`
	`self.Goal->size() = 1`	

Name	**Qualitative Measure**	«Qualitative Measure» Name
Base Class	Measure	
Description	The measurement of descriptive elements (e.g. age).	
Constraints	None	To Be Value Unit
Name	**Quantitative Measure**	«Quantitative Measure» Name
Base Class	Measure	
Description	Quantitative Measures are expressed in numerical values.	
Constraints	None	To Be Value Unit

5 Example

We demonstrate the practical applicability of the business perspective of the UML 2 profile for BPM in Fig. 3 with the example business process of an insurance company: the *Processing of Claims* business process. Fig. 3 shows that a business process based on the UML 2 profile for BPM can be grasped at a glance.

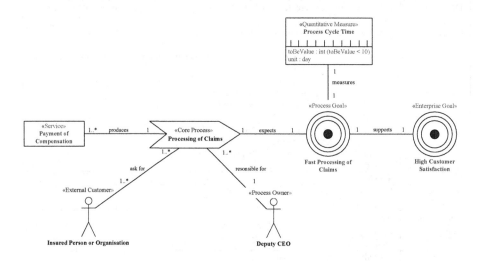

Fig. 3. Processing of Claims Business Process

The *Processing of Claims* business process is a *Core Process* and therefore the *Customer* is *External*. The customer is an *Insured Person or Organisation*. Beside the proposal business process, the processing of claims business process is the most important process in an insurance company. Therefore, a very powerful position is required for the *Process Owner*. In this example it is the *Deputy CEO*. *Fast Processing of Claims* is the *Process Goal* that supports the *Enterprise Goal, High Customer Satisfaction*. The *Quantitative Measure* is the *Process Cycle Time* and measures the achievement of the process goal. The *To Be Value* of the cycle time is *less than 10 Days*, the *Unit* of the measure.

6   Related Work

There are already some UML profiles for business process modelling available in the current literature. The profiles focus primarily on the sequential flow of the business process and represent comprehensive business process concepts only partly. All of the existing profiles are based on UML 1.4., whereas the UML profile developed in this paper is based on UML 2.

The UML profile for business modeling of the OMG [7] is defined in the UML 1.4 specification and embodies the object-oriented approach for business engineering developed by Jacobson et al. [5]. The model consists of two views, an external and an internal view. The external view is described by the use case model, the internal view by the object model. The model lacks a detailed process flow with a sequence of activities, but also business context.

Johnston extended the UML profile for business modeling of the OMG [7] with goals and events, and an activity diagram to represent the process flow. It is called the rational UML profile for business modeling [6]. This UML profile is a component of the Rational Unified Process® (RUP®). It presents a UML language for capturing business models and is supported by the business modelling discipline in the RUP.

The UML profile for business modelling in [14] proposes a basic meta-model covering a business process, its resources, and goals. The profile gives a basic overview of the process, and provides a detailed description of the process flow with Action-States (from ActivityGraphs).

The UML profile for business modelling in [15] focuses on the integration of business processes into software development. The profile maps between business concepts and software artefacts. Therefore, the profile describes the process flow in a very detailed way, and in addition adds goals, measures and resources, but lacks more advanced concepts like customers, process types or process owners.

The UML profile for modeling workflow and business processes in [1] is closely related to UML activity diagrams. The profile targets the modelling of business process architectures and other activities of concurrent processes. It provides dynamic semantics for these modelling concepts which can be used as a basis for the construction of automated analysis tools which provide performance simulations for the established models. The profile contains only basic concepts like activities and participants, such as actors and resources.

7 Conclusion

In this paper, we have developed a UML 2 profile for BPM targeting software developers to view business processes in their own notation. The profile provides two complementary perspectives, the business perspective and the sequence perspective. The sequence perspective refines the business perspective and describes the detailed flow of the process. The business perspective presents the business process from a wide angle by integrating aspects like goals, customers, deliverables, or process types etc. In order to capture these characteristics, we have developed a meta-model for the business perspective, described with stereotypes. The UML 2 profile for BPM was tested with an example business process.

References

1. Bochmann, G. v.: A UML profile for modeling workflow and business processes. http://beethoven.site.uottawa.ca/dsrg/PublicDocuments/Publications/Boch00d.pdf (8/8/05).
2. Business Process Management Initiative (BPMI): Business Process Modeling Notation (BPMN), Specification Version 1.0, May 3, 2004, http://www.bpmn.org/Documents/BPMN%20V1-0%20May%203%202004.pdf (8/8/05).
3. Hammer, M.: Beyond Reengineering – How the process-centered organization is changing our work and our lives. Harper Collins Publishers, 1996.
4. Harrington, J.H.: Business Process Improvement – The breakthrough strategy for total quality, productivity, and competitiveness. McGraw-Hill, 1991.
5. Jacobson, I., Ericson, M., Jacobson, A.: The Object Advantage – Business Process Reengineering with Object Technology. ACM Press, Addison-Wesley Publishing, 1995.
6. Johnston, S.: Rational UML Profil for Business Modeling. http://www-128.ibm.com/developerworks/rational/library/5167.html#author1 (8/8/05).
7. Object Management Group: UML 1.4 Specification http://www.omg.org/docs/formal/01-09-75.pdf (8/8/05).
8. Object Management Group: UML 2.0 OCL Specification http://www.omg.org/docs/ptc/03-10-14.pdf (8/8/05)
9. Object Management Group: UML 2.0 Superstructure http://www.omg.org/cgi-bin/doc?ptc/2004-10-02 (8/8/05).
10. Object Management Group: Business Process Definition Metamodel http://www.bpmn.org/Documents/BPDM/OMG-BPD-2004-01-12-Revision.pdf (8/8/05).
11. Object Management Group: MDA Guide Version 1.0.1 http://www.omg.org/docs/omg/03-06-01.pdf (8/8/05).
12. Ould, M.: Business Processes – Modelling and Analysis for Re-engineering and Improvement. John Wiley & Sons, 1995.
13. Scheer, A.-W.: ARIS – Business Process Modeling. Springer Verlag, 1999.
14. Sinogas, P., Vasconcelos, A., Caetano, A., Neves, J., Mendes, R., Tribolet, J.: Business Processes Extensions to UML Profile for Business Modeling. Proceedings of the International Conference on Enterprise Information Systems, 2001.
15. Tyndale-Biscoe, S., Sims, O., Wood, B., Sluman C.: Business Modelling for Component Systems with UML. Proceedings of the 6th International Enterprise Distributed Object Computing Conference (EDOC '02), IEEE Press, 2002.

Preface to AOIS 2005

Paolo Bresciani and Manuel Kolp

Information systems have become the backbone of all kinds of organizations today. In almost every sector—manufacturing, education, health care, government, and businesses—large and small information systems are relied upon for everyday work, communication, information gathering, and decision-making. Yet, the inflexibilities in current technologies and methods have also resulted in poor performance, incompatibilities, and obstacles to change. As many organizations are reinventing themselves to meet the challenges of global competition and e-commerce, there is increasing pressure to develop and deploy new technologies that are flexible, robust, and responsive to rapid and unexpected change.

The Agent Oriented computing paradigm adopts concepts and techniques which hold great promise for responding to the new realities of information systems. They offer higher level abstractions and mechanisms which address issues such as knowledge representation and reasoning, communication, coordination, cooperation among heterogeneous and autonomous parties, perception, commitments, goals, beliefs, intentions, etc. These features are promising at the system implementation level as well as at the methodological one. The concrete implementation of these concepts can lead to advanced functionalities, e.g., in inference-based query answering, transaction control, adaptive workflows, brokering and integration of disparate information sources, and automated communication processes. At the methodological level, since agent based representational capabilities allow more faithful and flexible treatments of complex organizational processes, more effective requirements analysis and architectural and detailed design can be produced, which also allows for improved communication between the stakeholders and the software engineers.

The bi-conference workshop on Agent Oriented Information Systems (AOIS), which is at its seventh edition in 2005, focuses on how agent-based concepts and techniques may contribute to meeting information systems needs. The workshop represents a unique opportunity for the information systems community and the agent systems community to meet and interact on this theme of common interest. As well, the AOIS edition at ER-2005 also include a *Special Track on Agent Oriented Methodologies and Conceptual Modeling*, with the aim of promoting a deeper understanding of the agent-based conceptual notions and the impact they have on conceptual modeling languages and methods adopted or created for agent-oriented methodologies.

J. Akoka et al. (Eds.): ER Workshops 2005, LNCS 3770, p. 97, 2005.

Agent Oriented Data Integration

Avigdor Gal and Aviv Segev

Technion - Israel Institute of Technology,
avigal@ie.technion.ac.il, asegev@tx.technion.ac.il
Christos Tatsiopoulos, Kostas Sidiropoulos, Pantelis Georgiades
European Profiles SA
ctatsio@hol.gr

Abstract. Data integration is the process by which data from heterogeneous data sources are conceptually integrated into a single cohesive
data set. In recent years agents have been increasingly used in information systems to promote performance. In this work we propose a modeling
framework for agent oriented data integration to demonstrate how agents
can support this process. We provide a systematic analysis of the process
using real world scenarios, taken from email messages from citizens in a
local government, and demonstrate two agent oriented data integration
tasks, email routing and opinion analysis.

1 Introduction

Data integration is the process of combining two or more data sets together
for sharing and analysis to support information management. Agents are autonomous, or semi-autonomous proactive and reactive computer software. Although there is a vast corpus of research of data integration, this research has
little impact on the state-of-the-art in agent oriented systems. We believe this
chasm can be attributed to the fact that most approaches rely on semantic reconciliation to be resolved first (probably manually), before attending to the more
"technical" aspects of the integration. However, researchers and practitioners
alike are coming to realize that there can be no solution to the delivery of integrated information unless the semantic heterogeneity problem is tackled head-on
[20]. This research works towards this goal through the use of ontologies.

This approach of agent oriented data integration was recently adopted in
QUALEG, a European Commission project aimed at increasing citizen participation in the democratic process.[1] In QUALEG, contexts are used to specify the
input from citizens and then to provide services - routing emails to departments
and performing opinion analysis on topics at the forefront of public debates.

We present the QUALEG approach towards agent oriented data integration.
We first propose a modeling framework for agent oriented data integration. We
then provide a systematic analysis of the process using real world scenarios,
taken from email messages from citizens in a local government, and demonstrate
two agent oriented data integration tasks, email routing and opinion analysis.

---

[1] http://www.qualeg.eupm.net/

J. Akoka et al. (Eds.): ER Workshops 2005, LNCS 3770, pp. 98–108, 2005.

2   Background and Motivation

2.1   Data Integration

Data integration has become a common theme in the information technology world. As information is increasingly becoming more complex and vast and the management of information more critical, the need to ascertain data integrity and replication is key to the reliable operation of the information system.

Many techniques are employed to promote the data integration process, such as event-based software integration [1], database schema integration [2], Web-based information integration [11], and semantic integration [3]. The field has seen the development of many tools, such as DIKE [21], Clio [17], Cupid [13], and OntoBuilder [18], to name a few.

Although there has been extensive research performed on data integration, the use of agents to promote this process has not been addressed adequately. This paper therefore presents an agent oriented approach to data integration.

2.2   Agents

In today's world, with the proliferation of computers, agents are necessary to promote the user's effective exploitation of software systems. Agents are used to initiate communications, monitor events, and perform tasks to assist users to understand the technically complex world. Agent concepts and techniques already appear in many information system architectures.

There are agents for accessing Web Information Systems (WIS) through Mobile Devices [22]. There are also multi-agent systems that assist individual investors with stock market investments [26]. Text mining agents for net actions have been also extensively analyzed [12]. Agents have been introduced into digital libraries, such as in University of Michigan Digital Library [4] and the ZUNO Digital Library (ZUNODL) [7], a commercial framework for building digital libraries. In this work we propose a framework and architecture of agents for data integration.

2.3   Context and Ontology

Contexts and ontologies are defined and used in various research areas, including philosophy, artificial intelligence, information sciences, knowledge representation, object modeling, and most recently, eCommerce applications.

Context is defined as a first class object [15]. McCarthy defines a relation $ist(\mathcal{C}, P)$, asserting that a proposition P is true in a context $\mathcal{C}$. Previous work on contexts [24] uses metadata for semantic reconciliation. It has been proposed to use a multilevel semantic network to represent knowledge within several levels of contexts [27]. This paper employs an agent based, fully automated context recognition algorithm that uses the Internet as a knowledge base and as a basis for clustering [23].

Ontology is defined as a world of systems [6]. Bunge in his seminal work provides a basic formalism for ontologies. Typically, ontologies are represented using a Description Logic [5, 8], where subsumption typifies the semantic relationship between terms; or Frame Logic [10], where a deductive inference system provides access to semi-structured data.

The realm of information science has produced an extensive body of literature and practice in ontology construction [28], ontology management [25], ontology learning [14, 9], and the use of ontology in knowledge representation source [16, 19].

This paper presents a agent oriented model for the integration of data into an ontological structure. The data structures are represented by the ontology concepts. Each ontology concept represents a possible topic or a possible opinion.

2.4 Example

To illustrate agent oriented data integration, consider the following example of the local government of Saarbrücken.

Example 1. Two ontology concepts in the ontology of Saarbrücken are:
(Perspectives du Theatre, $\{\{\langle$Öffentlichkeitsarbeit, 2$\rangle\}$, $\{\langle$Multimedia, 1$\rangle\}$,
$\{\langle$Kulturpolitik, 1$\rangle\}$, $\{\langle$Musik, 6$\rangle\}$, ...$\})$
and
(Long Day School, $\{\{\langle$Förderbedarf, 1$\rangle\}$, $\{\langle$Mathematik, 2$\rangle\}$, $\{\langle$Musik, 2$\rangle\}$,
$\{\langle$Interkulturell, 1$\rangle\}\})$
The set of descriptors define possible contexts with appropriate weights defining the importance of each descriptor in the context. There are also two ontology concepts that represent a positive opinion and a negative opinion. Each of these opinions can be ascribed to each of the above fields of interest.

The following email is received in the local government of Saarbrücken:
Eine leerer und verwaister Festivalclub, Regen und eine lustlose Band prägten das Bild der Auftaktveranstaltung des diesjährigen Festivals.

An agent can extract the following context of the email message using the algorithm in [23] (to be described later): $\{\{\langle$Musik, 8$\rangle\}$, $\{\langle$Open Air, 1$\rangle\}\}$.

An agent can map the email to the correct ontology topic which represents a field of interest and can forward the email to the correct local government representative handling this topic.

Another agent can identify the opinion of the email and store the information. This information can be statistically analyzed, integrated, and displayed as the citizens opinions on each of the fields of interest of the local government.

3   Model

Agents can be used to automatically extract context from a given text and then map context to ontology. We propose an agent oriented method for the management of data integration involved in automatic knowledge extraction and context-to-ontology mapping.

3.1   Context Recognition Algorithm

A *context* $\mathcal{C} = \left\{ \{\langle c_{ij}, w_{ij} \rangle\}_j \right\}_i$ is a set of finite set of descriptors c_{ij} from a domain $\mathcal{D}$ with appropriate weights w_{ij}, defining the importance of c_{ij}. For example, a context $\mathcal{C}$ may be a set of words (hence, $\mathcal{D}$ is a set of all possible character combinations) defining a document *Doc*, and the weights could represent the relevance of a descriptor to *Doc*. In classical Information Retrieval, $\langle c_{ij}, w_{ij} \rangle$ may represent the fact that the word c_{ij} is repeated w_{ij} times in *Doc*.

Several methods have been proposed in the literature for extracting context from text. One method proposed in the IR community is based on the principle of counting the number of appearances of each word in the text, assuming that the words with the highest number of appearances serve as the context. Variations on this simple mechanism involve methods for identifying the relevance of words to a domain and using methods such as stop-lists and inverse document frequency.

This agent oriented model employs a context recognition algorithm that uses the Internet as a knowledge base to extract multiple contexts of a given situation, based on the streaming in text format of information that represents situations [23]. This algorithm has been extensively tested and was found to obtain similar cobtexts to those proposed by human experts. This algorithm is currently part of the QUALEG solution.

The input to the algorithm is a stream, in text format, of information. The context recognition algorithm output is a set of contexts that attempts to describe the current scenario most accurately. The set of contexts is a list of words or phrases, each describing an aspect of the scenario. The context recognition algorithm consists of the following major phases: collecting data, selecting contexts for each text, ranking the contexts, and declaring the current contexts. The phase of data collection includes parsing the text and checking it against a stoplist. To improve this process, the text can be checked against a domain-specific dictionary. The result is a list of keywords obtained from the text. The selection of the current context is based on searching the Internet for relevant documents according to these keywords and on clustering the results into possible contexts. The output of the ranking stage is the current context or a set of highest ranking contexts. The set of preliminary contexts that has the top number of references, both in number of Internet pages and in number of appearances in all the texts, is declared to be the current context.

Up to this stage, the agent has achieved a set of contexts describing the given scenario. In the next stage, the agent maps these contexts to ontology concepts to achieve the automatic data integration.

3.2   Data Integration Using Contexts and Ontologies

When a context is extracted automatically from some information source (*e.g.*, an email message), the assumption is that it is correct, although it may not be extracted accurately and context descriptors may have been erroneously added or eliminated. Moreover, there may be inaccuracies in the definition of ontologies.

Therefore, to integrate the data it is necessary for the agent to map the extracted contexts to the relevant ontology concepts - a set of sets of contexts.

A context can belong to multiple context sets, which in turn can converge to different ontology concepts. Thus, one context can belong to several ontology concepts simultaneously.

For example, a context $\langle$Musik, 2$\rangle$ can be shared by many ontology concepts with interest in culture (such as schools, after school institutes, non-profit organizations, *etc.*) yet it is not in their main role definition. Such overlap of contexts in ontology concepts affects the task of email routing. The appropriate interpretation of a context of an email, when the context is part of several ontology concepts, is that the email is relevant to all such concepts. Therefore, it should be delivered to multiple departments in the local government.

A good algorithm for context extraction generates contexts in which false negatives and false positives are considered to be the exception, rather than the rule. Therefore, we would like to measure some "distance" between an extracted context and various ontology concepts, assuming a "closer" ontology concept to be better matched. To that end, we define a metric function for measuring the distance between a context and ontology concepts, as follows.

We first define distance between two weighted context descriptors $\langle c_i, w_i \rangle$ and $\langle c_j, w_j \rangle$ to be:

$$d(c_i, c_j) = \left\{ \begin{array}{ll} |w_i - w_j| & i = j \\ \max{(w_i, w_j)} & i \neq j \end{array} \right.$$

This distance function assigns greater importance to descriptors with larger weights, assuming that weights reflect the importance of a descriptor within a context. To define the best ranking concept in comparison with a given context we use Hausdorff metric. Let A and B be two contexts and a and b be descriptors in A and B, respectively. Then,

$$d(a, B) = \inf\{d(a, b) | b \in B\}$$
$$d(A, B) = \max\{\sup\{d(a, B) | a \in A\}, \sup\{d(b, A) | b \in B\}\}$$

The first equation provides the value of minimal distance of an element from all elements in a set. The second equation identifies the furthest elements when comparing both sets.

Of particular interest are ontology concepts that are considered "close" under some distance metric. As an example, consider the task of opinion analysis. With opinion analysis, a system should not only judge the relevant area of interest of a given email but also determine the opinion that is expressed in it. Consider an opinion analysis task, in which opinions are partitioned into categories (*e.g.*, "for" and "against"). We can model such opinions using a common concept ontology (say, that of Perspectives du Theatre, see Example 1), with the addition of words that describe opinions. An email whose context fit with the theme of the ontology concept will be further analyzed to be correctly classified to an opinion category.

Example 2. (Email Routing) Returning to our case study example, the context $\{\{\langle \text{Musik}, 8\rangle\}, \{\langle \text{Open Air}, 1\rangle\}\}$ may be relevant to both Perspective du Theatre and Long Day School, since in both, a descriptor Musik is found, albeit with different weights. The distance between $\langle \text{Musik}, 8\rangle$ and $\langle \text{Musik}, 6\rangle$ in Perspective du Theatre is 2, and to $\langle \text{Musik}, 2\rangle$ in Long Day School is 6. Assume that $\{\langle \text{Open Air}, 1\rangle\}$ is a false positive, which does not appear in either Perspective du Theatre or in Long Day School. Therefore, its distance from each of the two points accumulation is 1 (since $\inf\{d(a,b)|b \in B\} = 1$, *e.g.*, when comparing $\{\langle \text{Open Air}, 1\rangle\}$ with $\{\langle \text{Kulturpolitik}, 1\rangle\}$). We can therefore conclude that the distance between the context and Perspective du Theatre is 2, which is smaller than its distance from Long Day School (computed to be 6). Therefore, Perspective du Theatre will be ranked higher than Long Day School.

4 Architecture

This agent oriented method for integrating the context into the ontology concept according relevance is applied in the tasks of email routing and opinion analysis.

Email routing: The user provides QUALEG with a distance threshold t_1. Any ontology concept that matches with a context, automatically generated from an email, and its distance is lower than the threshold $(d(A, B) < t_1)$ will be considered relevant, and the email will be routed accordingly.

Opinion analysis: A relevant set of ontology concepts is identified, similarly to email routing. Then for each ontology concept, the relative distance of the different opinions of that concept is evaluated. If the difference in distance is too close to call (given an additional threshold t_2), the system refrains from providing an opinion (and the email is routed accordingly). Otherwise, the email is marked with the opinion with minimal distance.

These tasks are achieved through the implementation of the agent oriented Qualeg architecture, which consists of the following main seven components: (1) Agora - A Web interface to the system through which the citizen interacts via emails, chats and forums with the civil servant. (2) Datamart - The component that stores all the Qualeg data. (3) Qualeg ontology - A multilingual ontology describing the public and e-government issues. (4) Knowledge Extractor - The previously described context extraction algorithm used by the agents. (5) Qualeg Workflow - The component that handles the flow of processes relevant to the public servants and administrations. (6) A set of Intelligent Agents, which in the backstage handle the main control of the Qualeg system, acting asynchronously and handling the data to be communicated among various modules and passing control this way. (7) A set of Web Services offered for seamless data handling to and from the Datamart.

There are five different agents in the system, classified according to their task as follows: Knowledge Extraction, Opinion Analysis, Off-line Questionnaires, Email, and Email Handler.

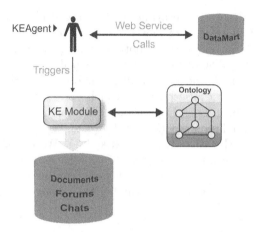

Fig. 1. Agent Architecture

The main focus of this part is on the use of data integration intelligent agent interactions with the rest of the Qualeg modules as a means for both asynchronous and synchronous control. The intelligent agents have been developed in the JADE platform and in line with the FIPA specifications for interoperable intelligent multi-agent systems. The following agents are provided in the Qualeg Architecture solution:

Knowledge Extraction Agent. The Knowledge Extraction Agent (KE Agent) illustrated in Figure 1 has the responsibility to trigger the Knowledge Extraction Module so that the context of the stored information is regularly analyzed. There are four types of documents that should be analyzed: documents uploaded to AGORA, text in forums, chats, and incoming e-mail messages. In particular, the KE Agent performs periodical searches in the platform's databases for new information to be analyzed. Every transaction with the database is carried out by means of Web services. If new documents are found, the agent triggers the previously described knowledge extraction algorithm on them. Hence, the KE Agent parses all the required information - such as document id, document name, document url - to the KE module. The KE module performs the mapping with reference to an ontology, which defines the set of concepts and their relationships. After the KE process is completed, a set of keywords is stored in a database.

Opinion Analysis Agent. Similarly to the KE Agent, the Opinion Analysis Agent (OAAgent) regularly searches in QUALEG's databases to find which documents have to be analyzed by the Opinion Analysis Module (OA Module). Once again, all the agent's database transactions are carried out through Web service calls. If documents requiring analysis are found, the agent triggers the opinion analysis algorithm on them in the same way as the KE agent. Opinion Analysis output is an ontology concept and a list of words.

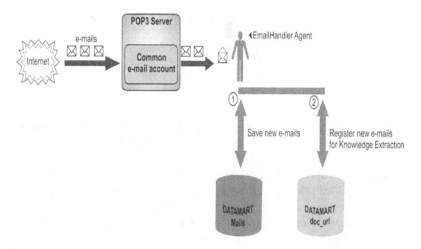

Fig. 2. Agent System Architecture

Email Handler Agent. All incoming e-mails that have been sent to a common e-mail account of the local government are gathered by an agent. The EmailHandlerAgent illustrated in Figure 2 disassembles each e-mail into its parts and distils the contained text. Next, the agent registers the new e-mail to a designated database in DATAMART. In particular, the EmailHandlerAgent, using Web services, saves information concerning the e-mail, *eg. sender, subject, body etc.*

5   Results

The aim of the QUALEG project is to support the electronic interactions between civil servants and citizens. Our experiment domain was the Perspectives Festival of May 15-21, 2005 in Saarbrücken (http://www.perspectives-sb.de/) along with similar data from the previous year's festival, which included films, theatre, street events, music, etc. Given the daily communications (in German) about this event, which consisted primarily of emails from citizens to the city hall or press releases and announcements from the city outward, our challenge was to analyze this material and provide a useful set of classifications so that the materials could be rapidly understood and sent to the appropriate people for response. Two different agent systems were developed, separating the task of knowledge extraction from that of opinion analysis. The main difference between the two agents is that the knowledge extraction agent avoids the language specific implementation and bases its analysis techniques on the use of a large corpus of relevant documents taken from the Internet, while the opinion analysis agent uses techniques from IR and NLP to improve content understanding. The systems analyzed the materials by topic (ticket/travel information, finances, organization,

Table 1. Context Recognition / Knowledge Extraction Agent

Precision	85.37 %
Recall	84.34 %
F-Score	84.85 %

etc.) and opinion (positive, negative, etc.). The system's average performance achieved high correspondence to human results for the different classes.

Our first experiment included 104 different emails to analyze the knowledge extraction agent. Table 1 summarizes the comparison of the results of the context recognition knowledge extraction agent to the human judgements. Our second experiment included 72 different emails to analyze the opinion analysis agent. Table 2 summarizes a similar comparison of the results of the opinion analysis agent. Both tables contain the precision, recall, and the weighted toward Precision F-score obtained. These results show the promising ability of our agents to integrate the data from the citizens with the local government specifications.

Table 2. Opinion Analysis Agent

Precision	78.95 %
Recall	69.23 %
F-Score	73.77 %

6   Conclusion

Data integration is a key field in the management of information systems today. The use of autonomous or semi-autonomous agents can effectively promote the process of data integration. The paper presents a modeling framework for agent oriented data integration to demonstrate how agents can support this process. The agent architecture and the analysis of the empirical results are based on real life scenarios, taken from email messages from citizens in a local government, and demonstrate two agent oriented data integration tasks, email routing and opinion analysis.

Initial tests show that the algorithm also achieves high performance compared to manually integrated data. Future directions of research include automatic agent responses to incoming data based on the previously integrated data.

Acknowledgments

The work of Gal was partially supported by two European Commission 6th Framework IST projects, QUALEG and TerreGov, and the Fund for the Promotion of Research at the Technion.

References

1. D. Barret, L. Clarke, P. Tarr, and A. Wise. A framework for event-based software integration. *ACM Transactions on Software Engineering and Methodology*, 5(4), 1996.

2. C. Batini, M. Lenzerini, and S. Navathe. A comparative analysis of methodologies for database schema integration. *ACM Computing Surveys*, 18(4):323–364, December 1986.

3. S. Bergamaschi, S. Castano, M. Vincini, and D. Beneventano. Semantic integration of heterogeneous information sources. *Data & Knowledge Engineering*, 36(3), 2001.

4. W. P. Birmingham, E. H. Durfee, T. Mullen, and M. P. Wellman. The distributed agent architecture of the university of michigan digital library. In *AAAI Spring Symposium on Information Gathering*, 1995.

5. A. Borgida and R. J. Brachman. Loading data into description reasoners. In *Proceedings of the 1993 ACM SIGMOD international conference on Management of data*, pages 217–226, 1993.

6. M. Bunge. *Treatise on Basic Philosophy: Vol. 4: Ontology II: A World of Systems.* D. Reidel Publishing Co., Inc., New York, NY, 1979.

7. D. Derbyshire, I. A. Ferguson, J. P. Muller, M. Pischel, and M.Wooldridge. Agent-based digital libraries: Driving the information economy. In *Sixth IEEE Workshops on Enabling Technologies: Infrastructure for Collaborative Enterprises*, 1997.

8. F.M. Donini, M. Lenzerini, D. Nardi, and A. Schaerf. Reasoning in description logic. In G. Brewka, editor, *Principles on Knowledge Representation, Studies in Logic, Languages and Information*, pages 193–238. CSLI Publications, 1996.

9. V. Kashyap, C. Ramakrishnan, C. Thomas, and A. Sheth. Taxaminer: An experimentation framework for automated taxonomy bootstrapping. *International Journal of Web and Grid Services, Special Issue on Semantic Web and Mining Reasoning*, September 2005. to appear.

10. M. Kifer, G. Lausen, and J. Wu. Logical foundation of object-oriented and frame-based languages. *Journal of the ACM*, 42, 1995.

11. C.A. Knoblock, S. Minton, J.L. Ambite, N. Ashish, I. Muslea, A. Philpot, and S. Tejada. The Ariadne approach to web-based information integration. *International Journal of Cooperative Information Systems (IJCIS)*, 10(1-2):145–169, 2001.

12. Y. Kusumura, Y. Hijikata, and S. Nishida. Text mining agent for net auction. In *ACM Symposium on Applied Computing*, 2004.

13. J. Madhavan, P.A. Bernstein, and E. Rahm. Generic schema matching with Cupid. In *Proceedings of the International conference on very Large Data Bases (VLDB)*, pages 49–58, Rome, Italy, September 2001.

14. A. Maedche and S. Staab. Ontology learning for the semantic web. *IEEE Intelligent Systems*, 16, 2001.

15. J. McCarthy. Notes on formalizing context. *In Proceedings of the Thirteenth International Joint Conference on Artificial Intelligence*, 1993.

16. D.L. McGuinness, R. Fikes, J. Rice, and S. Wilder. An environment for merging and testing large ontologies. In *Proceedings of the Seventh International Conference on Principles of Knowledge Representation and Reasoning (KR2000)*, 2000.

17. R.J. Miller, M.A. Hernàndez, L.M. Haas, L.-L. Yan, C.T.H. Ho, R. Fagin, and L. Popa. The Clio project: Managing heterogeneity. *SIGMOD Record*, 30(1):78–83, 2001.

18. G. Modica, A. Gal, and H. Jamil. The use of machine-generated ontologies in dynamic information seeking. In C. Batini, F. Giunchiglia, P. Giorgini, and M. Mecella, editors, *Cooperative Information Systems, 9th International Conference, CoopIS 2001, Trento, Italy, September 5-7, 2001, Proceedings*, volume 2172 of *Lecture Notes in Computer Science*, pages 433–448. Springer, 2001.

19. Fridman N. Noy and M.A. Musen. PROMPT: Algorithm and tool for automated ontology merging and alignment. In *Proceedings of the Seventeenth National Conference on Artificial Intelligence (AAAI-2000)*, pages 450–455, Austin, TX, 2000.

20. A.M. Ouksel and A.P. Sheth. Semantic interoperability in global information systems: A brief introduction to the research area and the special section. *SIGMOD Record*, 28(1):5–12, March 1999.

21. L. Palopoli, L.G. Terracina, and D. Ursino. The system DIKE: Towards the semi-automatic synthesis of cooperative information systems and data warehouses. In *PADBIS-DASFAA*, pages 108–117, 2000.

22. A. C. Ramos, J. Gensel, M. Villanova-Oliver, and H. Martin. Adapted information retrieval in web information systems using pumas. In *Workshop on Agent-Oriented Information Systems (AOIS)*, 2005.

23. A. Segev. Identifying the multiple contexts of a situation. In *Proceedings of IJCAI-Workshop Modeling and Retrieval of Context (MRC2005)*, 2005.

24. M. Siegel and S. E. Madnick. A metadata approach to resolving semantic conflicts. In *Proceedings of the 17th International Conference on Very Large Data Bases*, pages 133–145, 1991.

25. P. Spyns, R. Meersman, and M. Jarrar. Data modelling versus ontology engineering. *ACM SIGMOD Record*, 31(4), 2002.

26. S. C. Sundararajan, S. Sankarlal, and A. Kumar. Inca (investor network collaborative architecture) a method in the madness of wall street. In *Workshop on Agent-Oriented Information Systems (AOIS)*, 2005.

27. V. Terziyan and S. Puuronen. Reasoning with multilevel contexts in semantic metanetwork. In R. Nossun P. Bonzon, M. Cavalcanti, editor, *Formal Aspects in Context*, pages 107–126. Kluwer Academic Publishers, 2000.

28. B.C. Vickery. *Faceted classification schemes*. Graduate School of Library Service, Rutgers, the State University, New Brunswick, N.J., 1966.

AOSE and Organic Computing – How Can They Benefit from Each Other? Position Paper

Bernhard Bauer and Holger Kasinger

University of Augsburg, 86135 Augsburg, Germany
{bauer, kasinger}@informatik.uni-augsburg.de

Abstract. Organic computing is an upcoming research area with strong relationships to the ideas and concepts of agent-based systems. In this paper, we therefore will have a closer look at agent systems, organic computing systems (as well as autonomic computing systems) and state commonalities and divergences between them. We then propose a common view on these technologies and show, how they can benefit from each other with regard to software engineering.

1 Introduction

Over the past few years technical systems as airplanes, vehicles, telecommunication networks or manufacturing installations became more and more complex. This is not only a result of the continuing evolution in microelectronics but also of the immense embedding of huge hardware and software complexes into these systems. But the producer's painful experiences show that these systems already today are difficult to manage. Thus, with respect to the future evolution, new advanced management principles have to be developed. A feasible principle is an autonomic behavior of the systems which is addressed by two research directions, namely agent technology and organic/autonomic computing.

Agent technology is believed to be able to play a key role in this "revolution", e.g. by automating daily processes, enriching higher level communication or enabling intelligent service provision. An intelligent agent is "a computer system, situated in some environment that is capable of flexible autonomous actions in order to meet its design objectives" [1]. The real strength of agents is based on the community of a multi-agent system and the negotiation mechanisms and coordination facilities. A multi-agent system is "a dynamic federation of software agents that are coupled through shared environments, goals or plans and that cooperate and coordinate their actions" [2]. It is this ability to migrate, communicate, coordinate and cooperate that makes agents and multi-agent systems a worthwhile metaphor in computing and that makes them attractive when it comes to tackling some of the requirements in next-generation systems.

Another worthwhile metaphor is provided by organic computing (OC) systems [3] that can be seen as an extension to autonomic computing (AC) systems [4]. The latter – driven by IBM since 2001 – draw analogies from the human

J. Akoka et al. (Eds.): ER Workshops 2005, LNCS 3770, pp. 109–118, 2005.

body, in particular from the autonomic nervous system where all reactions occur without explicit override by the human brain – so to say autonomous. By embedding this behavior into technical systems, the administrative complexity of next-generation systems can be left to the systems themselves. IBM refers to this autonomy as "self-management" that includes four so-called "self-x properties", namely self-configuration (configuration and reconfiguration according to policies), self-optimization (permanent improvement of performance and efficiency), self-healing (reactive and proactive detection, diagnostics and reparation of localized SW/HW-problems) and self-protection (defense of the system as a whole). Furthermore, AC systems are self-aware, context-sensitive, non-proprietary, anticipative and adaptive. OC systems instead draw analogies from the biological world and try to use perceptions about the functionality of living systems for the development and management of artificial and technical systems respectively. In addition to the properties of AC systems they are defined as being self-organizing (hence they do not necessarily have to be self-aware).

As OC systems basically have the same objectives and concepts as AC systems, we will mostly treat them as one single technology for the rest of the paper, which is organized as follows: In section 2 we present the concepts of agents as well as autonomic/organic computing and the existing software-engineering approaches for these technologies. Section 3 relates the technologies and presents a common view on them. Based on this view, in section 4 we present a development process, which helps to benefit AOSE and OC from each other before we conclude with open issues and an outlook for further research in section 5.

2   Concepts

In this section we give an overview on agent technology as well as on autonomic/organic computing and consider the associated methodologies.

2.1   Agents

Software agents are software components characterized by autonomy (to act on their own), reactiveness (to process external events), proactiveness (to reach goals), cooperation (to efficiently and effectively solve in common tasks), adaptation (to learn by experience) and mobility (migration to new places). For further details on agent technology see e.g. [5] or [6].

Often, agents are subdivided into three functional sections: The *agent body* wraps a software component (e.g. a database, a calendar or an external service) and controls it through the software API. Connected to external software, the agent acts as an application agent by transforming the application API into agent communication language (ACL) and vice versa. Messages of such ACLs are highly structured and must satisfy standardized communicative (speech) acts which define the type and the content of the messages (like FIPA-ACL [7] or KQML [8]). The order of exchanged messages is fixed in protocols according to the relation of agents or the intention of the communication.

The *agent head* is responsible for the agent's intelligence. It is connected to the agent body on one side and to the agent communicator on the other side. The agent head contains knowledge bases storing knowledge of certain types like facts, beliefs, goals or intentions, preferences, motivations and desires concerning the agent itself or associated ones. Further, it contains a world model as an abstraction of relevant states of the real world. It is updated by information from other agents or through real world interfaces, e.g. sensors. The agent head is able to evaluate incoming messages with respect to its goals, plans, tasks, preferences and to the world model.

The *agent communicator* converts logical agent addresses into physical addresses and delivers messages on behalf of the agent head through appropriate channels to the receivers. Furthermore, the communicator listens for incoming messages (e.g. by running an event loop) and forwards them to the agent head. The agent behavior should be benevolent, which means that an agent at least understands the interaction protocols and reacts accordingly.

2.2 Autonomic/Organic Computing

According to [9], AC systems are composed of four levels: On the lowest level *managed resources (MR)*, e.g. HW/SW-components as servers, databases or business applications, are located, together making up the complete IT infrastructure. So-called *touchpoints* on the next level provide a manageability interface – similiar to an API – for each MR by mapping standard sensor and effector interfaces on the sensor and effector mechanisms (e.g. commands, configuration files, events or log files) of a specific MR. The next level is composed of so-called *touchpoint autonomic managers (TAM)* directly collaborating with the MRs and managing them through their touchpoints.

An *autonomic manager (AM)* in general implements an intelligent control loop (closed feedback loop) called *MAPE loop*. The latter is composed of the components *monitor* (collects, aggregates, filters and reports MR's details), *analyze* (correlates and models complex situations), *plan* (constructs actions needed to achieve goals) and *execute* (controls execution of a plan). Additionally, a knowledge component provides the data used by the four components, including policies, historical logs and metrics. Together with one or more MRs, an AM represents an *autonomic element (AE)* (see Fig. 1). A TAM also provides a sensor and an effector to *orchestrating autonomic managers (OAM)* residing on top level. The latter achieve system-wide autonomic behavior, as TAMs are only able to achieve autonomic behavior for their controlled MRs.

As (strong) self-organizing systems (like OC systems) are defined as systems "that change their organization without any explicit – internal or external – central control" [10], there can be no single instance within an OC system that is aware of all system's components or states. From our point of view, system-wide autonomic behavior in OC systems is in contrast to AC systems therefore an emergent behavior of the system's component interactions and not the achievement of a single OAM. This issue has significant impact on software engineering but not on the concepts mentioned above which are also used in OC systems.

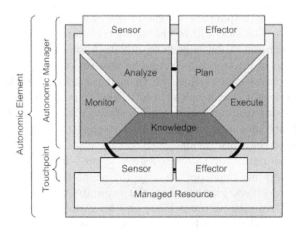

Fig. 1. Logical structure of an autonomic element

2.3 Software Engineering Methodologies

Agent-oriented Software Engineering Methodologies. A considerable number of AOSE methodologies and tools are available today (see our work in [11] or [12] for a more detailed survey), and the agent community is facing the problem of identifying a common vocabulary to support them.

The knowledge engineering community inspired most early approaches supporting the SE of agent-based systems: The CommonKADS [13] was developed to support knowledge engineers in modeling expert knowledge and developing design specifications in textual or diagrammatic form. To consider agent-specific aspects CoMoMAS [14] and MAS-CommonKADS [15] were developed.

Gaia [16] is a methodology designed to deal with coarse-grained computational systems, having static organization structures and agents with static abilities and services. ROADMAP [17] extends Gaia by adding elements to deal with the requirements analysis in more detail by using use cases, handling open system environments and specification of interactions. SODA [18] addresses aspects like open systems or self-interested agents, based on the analysis and design of agent societies (exhibiting global (emergent) behavior not deducible from the behavior of the individual agents) and agent environments.

One of the first methodologies for the development of BDI agents based on OO technologies was presented in [13] and [19]. The methodology distinguishes between the external viewpoint – the system is decomposed into agents, modeled as complex objects characterized by their purpose, their responsibilities, the services they perform, the information they require and maintain, and their external interactions – and the internal viewpoint – the elements required by a particular agent architecture must be modeled for each agent, i.e. an agent's beliefs, goals and plans.

MESSAGE [20] is a methodology that extends UML by agent-related concepts (inspired e.g. by Gaia). TROPOS [21] uses UML for the development of

BDI agents. Prometheus [22] it is an iterative methodology covering the complete SE process and aiming at the development of intelligent agents using goals, beliefs, plans and events, resulting in a specification which can for example be implemented with JACK [23]. MaSE [24] has been developed to support the complete software development life cycle. PASSI [25] is an agent-oriented iterative requirement-to-code methodology for the design of multi-agent systems mainly driven from experiments in robotics.

Autonomic / Organic Computing Methodologies. Continuous and consistent SE methodologies for AC/OC systems are more or less not available now, since most of the research activities are in the area of algorithms, middleware, hardware concepts as well as application areas. Nevertheless, the objective in particular of OC has to be on the control of such systems by engineering methods. Traditional SE methods are strictly hierarchic and follow a top-down approach by transforming the entire specification into detailed modules. For emergent and self-organizing systems this strict approach is abandoned. System states have to be reached that are not imagined beforehand. This is a fundamental contradiction between a top-down-control and a creative bottom-up-behavior.

Today it is not clear, how to combine these opposite tendencies. However, there are some approaches based on constraint propagation, the use of assertions and so-called observer/controller architectures. Assertions can be used for monitoring values of special variables. Yet, the limitation of emergent behavior of OC systems will be crucial for their technical application. Thus, constraints play an important role to the limitation of learning in self-organizing systems as constraint violations result in warnings.

3 Relating Agents and Organic Computing

Based on the presented concepts we try to relate agents and OC in this section and propose a common view on these technologies.

Both technologies incorporate managed objects, either software components wrapped in the agent body or managed resources on the OC-side. In addition, both technologies have an institution for intelligent and autonomic behavior, namely the agent head and the autonomic manager respectively. Moreover an agent communicator is in a sense comparable to a touchpoint in OC.

Thus, in order to bring the technologies together, we view an autonomic element from now on as the combination of agents and organic computing with the following properties: Having a BDI mental model about other autonomic elements; using a MAPE loop similar to the control loop of agents, with monitoring and analyzing the environment and messages, consulting the knowledge base, planning and execution; managing the internal behavior automatically, like OC does it, without interaction with the environment; interacts with its environment, not only via direct messages but also via e.g. stigmergy – therefore the environment has to be modeled explicitly, like for swarm intelligence, or ant algorithms. Moreover, an autonomic element community consists of cooperating

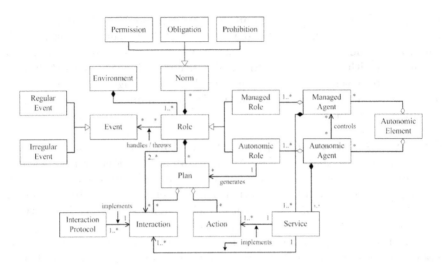

Fig. 2. The meta model for organic computing systems

autonomic elements explicitly communicating based on speech acts and interaction protocols or implicitly via the environment. Additionally these cooperating elements have to satisfy global system rules such that no unintentional behavior of the system takes place.

Having this in mind we propose a meta model for both a MASs with OC properties and OC systems as MASs (see Fig. 2). Therefore, we have combined different proved concepts of existing agent architectures and their SE methodologies as well as AC/OC concepts.

Similar to many existing agent methodologies a *role* is the central architectural concept. The complete set of roles builds up the *environment*. The life cycle of a role is traditionally: A role or rather the enacting agent recognizes a situation, makes a decision based upon it and executes appropriate activities. The recognition of situations is based on *events*. *Regular events* are familiar to a role, e.g. by design or by adaption, whereas *irregular events* are new to a role, e.g. by failure appearance. *Norms* regulate the behavior of a role and are a generalization of either a *permission*, an *obligation* or a *prohibition* and consist of a goal and activation as well as deactivation events. The decision making is based on *plans* that fire certain events at the end (as notification of being in a certain state) which may correspond to a norm's goal or event respectively. A plan consists of actions (internal activities of a role) and interactions (external activities between different roles) and are chosen accordingly to a goal of an activated norm. *Interactions* are implemented by specific *interaction protocols*. The relation between interactions and interaction protocols is the same as between interfaces and their implementations. Thus, according to diverse requirements, an interaction may be implemented by different kinds of protocols for direct (e.g. by auctions) or indirect (e.g. by stigmergy) communication. Interactions and actions are both implemented by *services* with different visibilities.

Roles are logically divided into *managed roles (MR)* and *autonomic roles (AR)* (similar to the AC concepts). MRs are responsible for the business logic of a system and reside on versatile resources. They are controlled by one or more ARs that are responsible for the self-management of a system. ARs do not necessarily have to be located at the same resource as its MRs. In contrast to MRs the ARs are able to generate new plans based on the received data of their MRs. The latter do not have to generate new plans as they communicate the occurrence of irregular events to their monitoring ARs and mostly are not in possession of further required information. Both roles are taken over dynamically by *managed agents* and *autonomic agents* respectively. *Autonomic elements* contain one or more autonomic agents and managed agents at the same time.

4   Software Engineering for OC and AO Systems

As a result of the common view presented in the previous section, we propose a development process in this section which can be used for both AOSE and OC. The process is based on the Model Driven Architecture (MDA), a framework for software development driven by the Object Management Group (OMG). It comprises a *Computation Independent Model (CIM)* (model of a system that abstracts from any computation), a *Platform Independent Model (PIM)* (model of a system that abstracts from any specific platform) and a *Platform Specific Model (PSM)* (model of a system that is tailored to one or more specific implementation platforms). For a more detailed description see [26].

The process consists of 19 activities and encompasses an analysis phase (activities 1-5) and a design phase (activities 6-19). Each activity results in a specific model either in the CIM (analysis phase) or the PIM (design phase) (see Fig. 3). An implementation phase is not considered yet, but can be added smoothly in the future. Notice, the process does not prescribe a process model.

The analysis phase consists of the activities (1) 'Definition of the business context', (2) 'Definition of business processes being supported', (3) 'Characteri-

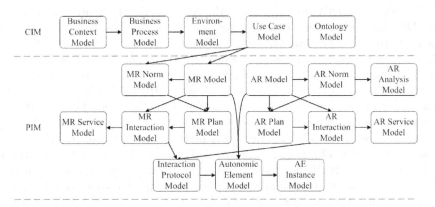

Fig. 3. MDA-based development process models for agent and OC systems

zation of the environment', (4) 'Assembly of potential use cases' and (5) 'Assembly of common vocabulary'. The resulting models are: *Business Context Model:* As a result of (1) the business context of the future system is modeled by an UML activity diagram. This model only considers higher level correlations and abstracts from concrete business processes; *Business Process Model:* As a result of (2) the business processes supported by the later system are modeled by an UML activity diagram; *Environment Model:* As a result of (3) important environment objects of all types are modeled by an UML class diagram; *Use Case Model:* As a result of (4) the system application is declared abstractly in an UML use case diagram. The model is supported by an UML sequence diagram to explain the message flow of the system clearly; *Ontology Model:* As a result of (5) all important knowledge blocks and common vocabulary are categorized in an UML class model.

The design phase consists of the activities (6) 'Identification of MRs', (7) 'Specification of norms for MRs', (8) 'Development of plans for MRs', (9) 'Derivation of interactions between MRs', (10) 'Specification of services of MRs', (11) 'Identification of ARs', (12) 'Specification of norms for ARs', (13) 'Development of an analysis for ARs', (14) 'Development of plans for ARs', (15) 'Derivation of interactions between ARs', (16) 'Specification of services of ARs', (17) 'Development of interaction protocols', (18) 'Identification of AE' and (19) 'Deployment of AE'. The resulting models of this phase are: *Managed Role Model:* As a result of (6) the MRs are identified and modeled similar to a class in an UML composition structure diagram; *MR Norm Model:* As a result of (7) the norms (containing goals, activation and deactivation events) of MRs are specified and modeled similar to a class in an UML class model; *MR Plan Model:* As a result of (8) the plans (containing input and output parameters, actions and interactions, and events) of MRs are modeled in an UML activity diagram; *MR Interaction Model:* As a result of (9) the interactions between MRs are derived and the exchanged objects (information carriers) are modeled in an UML sequence diagram; *MR Service Model:* As a result of (10) the signature of provided services (containing visibility, input and output parameters) of a MR are specified and modeled similar to a class in a UML class diagram again.

The results of activities (11), (14), (15) and (16), the *Autonomic Role Model,* the *AR Plan Model,* the *AR Interaction Model* and the *AR Service Model* are similar to the corresponding MR models. Further resulting models are: *AR Norm Model:* As a result of (12) and parallel to (11) the norms for ARs are specified according to desired self-x properties. Notice, a norm of an AR realizes a part of a certain self-x property of a system; *AR Analysis Model:* As a result of (13) the monitoring and analysis of events and data by an AR is modeled in an UML activity diagram as a premise for the right choosing of a plan; *Interaction Protocol Model:* As a result of (17) the interaction protocols for the (direct/indirect) interactions between all types of roles are specified in an UML sequence diagram; *Autonomic Element Model:* As a result of (18) MRs and ARs are combined into AEs that are modeled similar to a class in an UML composition structure diagram again; *Autonomic Element Instance Model:* As a result of (19) the deployment of the AEs onto resources is defined similar to an UML deployment diagram. Note, activities (11)-(16) are logically separated and represent the way of self-x property development.

5   Conclusion, Open Issues and Outlook

As described in this paper, agent systems and OC systems have many conceptual commonalities which result in benefits for both AOSE and OC: On the one side open agent systems can be developed that exhibit OC properties, on the other side OC can make use of the experiences in AOSE and adopt existing concepts.

The open issues in this context for us are: Where are the borders between an autonomic element, an agent or multi-agent system? How to deal with the emergent behavior of the system such that no unintentional behavior of the system occurs? How to define emergency strategies if the system is out of control, with regard to the emergent behavior? Should we have an hierarchical composition, like grouping autonomic elements to autonomic communities, view these communities as autonomic elements and grouping them to autonomic communities, etc.? How to model self-x properties in the local as well as in the global sense and how does the local behaviors result in a global behavior? How to integrate interaction (communication protocols) in such OC systems? What is the appropriate middleware/platform for OC systems (web services, grid computing middleware, agent platforms, . . .)?

In this context our vision is to combine different but related technologies, like grid computing, semantic web, (semantic) web services and web service composition, P2P, business processes and OC with its self-x properties, since these technologies deal with similar aspects (service provisioning, service access, service and data distribution, service and resource work loading, processes in distributed environments) and use similar standards.

References

1. Jennings, N.R., Sycara, K., Wooldridge, M.J.: A Roadmap of Agent Research and Development. Autonomous Agents and Multi-Agent Systems, 1(1) (1998) 7–38
2. Huhns, M.N.: Multiagent Systems. Tutorial at the European Agent Systems Summer School (EASSS 99) (1999)
3. Organic Computing website: http://www.organic-computing.org
4. Horn, P.: Autonomic Computing: IBM's Perspective on the State of Information Technology. http://www.research.ibm.com/autonomic/manifesto/ autonomic_computing.pdf (2001)
5. Müller, J. P.: The design of intelligent agents. A layered approach. Lecture Notes of Artificial Intelligence, Volume 1177. Springer-Verlag (1996)
6. Huhns, M.N., Singh, M.P.: Agents and Multiagent Systems: Themes, Approaches, and Challenges. Readings in Agents, Morgan-Kaufmann (1998), 1–24
7. FIPA: http://www.fipa.org
8. Finin, T., Fritzson, R., McKay, D., McEntire, R..: KQML as an Agent Communication Language. Proceedings of the Third International Conference on Information and Knowledge Management (CIKM'94). ACM Press (1994) 456–463

9. IBM: An architectural blueprint for autonomic computing. http://www-03.ibm.com/autonomic/pdfs/ACBP2_2004-10-04.pdf (2004)
10. Di Marzo Serugendo, G., Gleizes, M.-P., Karageorgos, A.: Self-Organisation in Multi-Agent Systems. AgentLink News (16) (2004) 23–24
11. Bauer, B., Müller, J.P.: Methodologies and Modeling Languages. In: Luck M., Ashri R. D'Inverno M. (eds.): Agent-Based Software Development. Artech House Publishers, Boston, London (2004)
12. Iglesias, C.A., Garijo, M., Centeno-González, J.: A Survey of Agent-Oriented Methodologies. In Proceedings of Fifth International Workshop on Agent Theories, Architectures, and Languages (ATAL 98) (1998) 317–330
13. Kinny, D., Georgeff, M., Rao, A.: A Methodology and Modeling Technique for Systems of BDI Agents. 7th European Workshop on Modelling Autonomous Agents in a Multi-Agent World (MAAMAW 96), LNAI 1038, Springer (1996) 56–71
14. Glaser, N.: Contribution to Knowledge Modelling in a Multi-Agent Framework (the Co-MoMAS Approach). PhD thesis, L'Universtité Henri Poincaré, Nancy I, France (1996)
15. Iglesias, C.A., Garijo, M., Centeno-González, J., Velasco, J.R.: A methodological proposal for multiagent systems development extending CommonKADS. In Proceedings of 10th Knowledge Acquisition for Knowledge-Based Systems Workshop (KAW 96), Banoe, Canada (1996)
16. Wooldridge, M., Jennings, N.R., Kinny, D.: The Gaia Methodology for Agent-Oriented Analysis and Design. Journal of Autonomous Agents and Multi-Agent Systems, 3 (3) (2000) 285–312
17. Juan, Th., Pearce, A., Sterling, L.: ROADMAP: Extending the Gaia Methodology for Complex Open Systems. In Proc. of the First Int. Joint Conf. on Autonomous Agents and Multiagent Aystems (AAMAS 02), ACM Press (2002) 3–10
18. Omicini, A.: SODA: Societies and Infrastructures in the Analysis and Design Of Agent-based Systems. In Proceedings of Agent Oriented Software Engineering (AOSE 00), LNCS 1957, Springer (2000) 185–193
19. Kinny, D., Georgeff, M: Modelling and Design of Multi-Agent Systems. Intelligent Agents III: Proceedings of Third International Workshop on Agent Theories, Architectures, and Languages (ATAL 96), LNAI 1193, Springer (1996)
20. Caire, G., Coulier, W., Garijo, F., Gomez, J., Pavon, J., Massonet, P., Leal, F., Chainho, P., Kearney, P., Stark, J., Evans, R.: Agent Oriented Analysis using MESSAGE/UML. In Proceedings of the Second International Workshop on Agent-Oriented Software Engineering II (AOSE 01), Springer (2002) 119–135
21. Bresciani, P., Giorgini, P., Giunchiglia, F., Mylopoulos, J., Perini, A.: Tropos: An Agent-Oriented Software Development Methodology. Journal of Autonomous Agent and Multi-Agent Systems, 8 (3) (2004) 203–236
22. Padgham, L., Winikoff, M.: Developing Intelligent Agent Systems: A Practical Guide. John Wiley & Sons (2004)
23. Busetta, P., Rönnquist, R., Hodgson, A., Lucas, A.: JACK Intelligent Agents - Components for Intelligent Agents in Java. AgentLink News (2) (1999) 2–5.
24. DeLoach, S.A., Wood, M.F., Sparkman, C.H.: Multiagent Systems Engineering. The International Journal of Software Engineering and Knowledge Engineering, 11 (3) (2001) 231–258
25. Cossentino, M., Potts, C.: A CASE tool supported methodology for the design of multi-agent systems.. In Proceedings of the 2002 International Conference on Software Engineering Research and Practice (SERP'02), Las Vegas, USA (2002)
26. Model Driven Architecture website: http://www.omg.org/mda

Modeling Dynamic Engineering Design Processes in PSI

Vadim Ermolayev[3], Eyck Jentzsch[1], Oleg Karsayev[2], Natalya Keberle[3],
Wolf-Ekkehard Matzke[1], and Vladimir Samoylov[2]

[1] Cadence Design Systems, GmbH, Feldkirchen, Germany
{wolf, jentzsch}@cadence.com
[2] SPII RAS, Saint Petersburg, Russia
{ok, samovl}@iias.spb.su
[3] Zaporozhye National Univ., Zaporozhye, Ukraine
{eva, kenga}@zsu.zp.ua

Abstract. One way to make engineering design effective and efficient is to make its processes flexible – i.e. self-adjusting, self-configuring, and self-optimizing at run time. This paper presents the descriptive part of the Dynamic Engineering Design Process (DEDP) modeling framework developed in the PSI[1] project. The project aims to build a software tool to assist managers to analyze and enhance the productivity of the DEDPs through process simulations. The framework incorporates the models of teams and actors, tasks and activities as well as design artifacts as the major interrelated parts. DEDPs are modeled as weakly defined flows of tasks and atomic activities which may only "become apparent" at run time because of several presented dynamic factors. The processes are self-formed through the mechanisms of collaboration in the dynamic team of actors. These mechanisms are based on several types of contracting negotiations. DEDP productivity is assessed by the Units of Welfare collected by the multi-agent system which models the design team. The models of the framework are formalized in the family of DEDP ontologies.

1 Introduction

It is widely accepted that the processes of engineering design differ from manufacturing processes by the fact that they "… are frequently chaotic and non-linear, and have not been well served by project management or workflow tools" (cf. [1]). The primary reason is that the ability to design is one of the signatures of human intelligence which can hardly be framed by the rigid and static bounds of pre-defined business processes. Therefore one of the promising ways to make engineering design effective and efficient is to manage its processes in a flexible manner – i.e. make them self-adjusting, self-configuring, and self-optimizing at run time. By doing so we may enhance the degree of coherence between the interrelated activities and make them better coordinated and therefore more productive. Hence, the model of a DEDP should be at least capable to account for the constellation of the factors which make a

---

[1] Productivity Simulation Initiative (PSI) is the R&D project of Cadence Design Systems GmbH.

J. Akoka et al. (Eds.): ER Workshops 2005, LNCS 3770, pp. 119–130, 2005.
© Springer-Verlag Berlin Heidelberg 2005

DEDP "chaotic and non-linear" and, at most, to eliminate them as much as possible. Provided that we have built such a fine-grained DEDP modeling framework, we may expect implementing software tools allowing to assess a process and, ultimately, to optimize DEDPs in terms of engineering design productivity.

Improving DEDPs in terms of engineering design productivity is the focus of PSI project. The project has prototyped a software tool which provides for the assessment of the accomplished DEDPs and the prediction of the characteristics of the planned DEDPs through their simulations. This simulation prototype has been implemented as a multi-agent system [2][2] which models designers' teams working on projects by dynamically formed teams of software agents, DEDPs performed by these teams – by tasks, and the results of these processes – by design artifacts. The knowledge on the performed processes is formalized and stored to PSI testbed in terms of DEDP family of ontologies presented in this paper. Thus we obtain the incremental collection of actors' experience which is further on used to make simulation results more reliable.

The paper is structured as follows. Section 2 discusses modeling requirements justifying the necessity to cope with the dynamic character of DEDPs. Section 3 outlines our approach to assessing the productivity of DEDPs. Then PSI Actor model, Task-Activity model and Design Artifact model are presented in Sections 4 – 6. Section 7 deals with the epistemological aspects of DEDP ontologies family and the usage of these ontologies in PSI in the form of DEDP-lite ontologies. Section 8 surveys the related work and analyses the contributions of DEDP modeling framework.

2   The Model of a Dynamic Engineering Design Process

A DEDP is a process of aiming a weakly defined engineering design workflow to achieve its goal in an optimal way in terms of result quality and gained productivity. It is therefore clear that the following entities are involved in the process: actors, who form design teams and collaboratively do the work in the flow; activities which are the atomic parts of a workflow defined by the technology used in the house; tasks which are subjective actors' representations of activities' compositions and choreography[3]; and design artifacts which are the results of engineering design activities. Hence, only engineering design activities are defined by the design technology and are well known before a DEDP starts. Other elements may only "become apparent" at run time because:

- The treatment of a task as atomic or composite is different by the actors having different capabilities. A task which is perceived as an atomic activity by one actor may be recognized as composite by another actor.

---

[2] [2] is the parallel paper which reports on the implementation and the evaluation experiments with PSI simulation prototype. In this paper we omit the description of this important part of our research due to space limitations.

[3] Choreography in the mentioned context is understood similarly to Web Services choreography and means the way of arranging material input – output communication among the dependent activities.

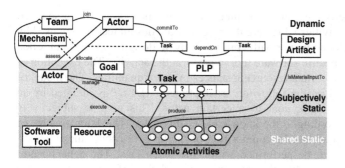

Fig. 1. Static and dynamic components of the modeling framework

- The composition [3] of the activities is defined only subjectively and partially. Tasks in our model may be composed of the activities and other tasks in different ways by different actors having different knowledge. It implies that the sequence of activities and sub-tasks in a task may be understood differently in the partial local plans of different actors.

- The number of activity loops is not defined in advance. It depends on the quality checks at intermediate steps. Changing the number of activity loops may cause the changes in its duration. In turn, it may cause the delays of the dependent tasks and activities with associated penalties for, e.g., deadline violation.

- The duration of activity execution is not defined in advance. Different actors possess different capacities to be spent for the activity at a certain time. They may perform the same activity with different efficiency (productivity – Section 3). An activity may remain idle while waiting until the pre-conditions have been triggered. Idle state duration can't be computed in advance because the preconditions may be formed by the other activities executed by other actors.

- The actors are not assigned specific activities in advance. An actor is chosen by the Task Manager when s/he decides to assign the activity. In PSI framework contracting negotiations are the means to optimally choose the actor to perform the task. DEDP model should therefore incorporate the actor model and the means to arrange actors' collaboration through peers' assessment and negotiations.

Mentioned factors provide certain degrees of freedom[4] in DEDP planning, re-planning, scheduling, re-scheduling, and execution. In PSI a DEDP is never rigidly planned before it starts. The decisions on how to continue its execution are taken each time it reaches a certain state in the state space. These decisions are taken by the design team members (Section 4) who manage the tasks which continue the process. According to the aforementioned properties of a DEDP different paths through the state space may be more productive or less.

As shown in Fig. 1 a DEDP has components which differ along the dimensions of their changeableness. The first dimension is the dynamic character ranging from static, i.e. pre-defined for all possible DEDPs, to dynamic, i.e., subjected to changes in a DEDP. Another dimension is the sphere of visibility or commitment. It ranges from shared, i.e., having the same meaning and instances for all DEDP participants, to subjective, i.e. having specific instances for different actors (though in the terms of a common ontology). Static shared DEDP components are atomic activities, associated

---

[4] It should be noted here that this freedom implies more complications in planning, scheduling and the necessity to deal with finer grained DEDP model.

software tools, and resources. It is assumed that the processes are assembled (ultimately) of atomic activities which are the pieces of the design technology used by the company. The technology normally provided by a design support unit often suggests the usage of a specific software tool to perform the activity. The execution of a given activity consumes certain resource instances in given quantities. The model of a design process is based on the following assumptions. A DEDP is initiated by an external influence providing a goal to a certain actor. This goal is subjectively transformed to a task according to the knowledge of this actor. The actor uses his or her subjective knowledge about the composition of the task, i.e. about the sub-tasks and the atomic activities to be performed within the given task. The dependencies between different tasks are also the subjective knowledge of an actor and are formalized in his or her Partial Local Plans (PLP). The actor may decide to perform a sub-task or to execute an activity of a decomposed task himself or to hire (for the price) another actor through the available collaboration mechanism (contract net negotiations in PSI). In the latter case the sub-task becomes the goal of another peer-actor who commits himself to perform the corresponding task by striking the contract deal. Hence the appearance of actor-task combinations in a DEDP is subjectively dynamic. The mechanism of incorporating new actors to the process and the model of the design team are subjectively dynamic as well since they depend on the decisions and choices taken at run time by the actors which states change in the process. The rules of encounter of the mentioned mechanism are shared static and provide the horizontal laws for the system [3], [4]. A design artifact is a subjectively dynamic outcome of the process since it is formed out by subjectively dynamic collaborative team of actors. However, the proposed layering allows reaching this effect through applying shared static atomic activities, though in subjectively dynamic combinations. For an activity a design artifact is both the material input and the result of its execution.

The actors who perform a task and initiate collaboration are Task Managers. Their rational goal with respect to the performed task is to choose the next step on the process path as productive as possible. Of course an actor needs a sort of productivity assessment model for that (Section 3).

3 Assessing Productivity by the Earned Units of Welfare

Productivity by its very nature is one of the most important economic metrics and is defined by the ratio of the produced output (value) to the consumed input (value). As such it is an integral characteristic of any transformation process, e.g. a DEDP. This neo-classical definition of productivity imposes rigid requirements on the process under consideration. The homogeneity of inputs and outputs is the most severe one with respect to engineering design. Known productivity measurement methodologies in engineering design ground themselves on the assessment of design complexity characteristics in the creation of homogeneous input- and output-measures. They pretend to do it by applying heuristic weights to compared parameters (e.g., the normalized transistor count[5] in Semiconductor and Electronic Systems (SES) design,

---

[5] Measuring IC and ASIC Design Productivity. White Paper. Numetrics Management Systems, 5201 Great America Parkway, Suite 320 Santa Clara, CA 95054, 2000.

FP, KSLOC counts[6] in software design, etc.). The fundamental problem of this approach is that the complexity characteristics need to be invariant both to the type of a process and to the transformed design artifact. If those characteristics are not invariant, measurement scales tend to lack well-defined units. Consequently the properties of the measurement scale, the labeling of the units, and the interpretation of the values derived are of very limited practical use. Furthermore in non-deterministic environments such measures are not very reliable even if proposed. It is therefore important to build a measure which addresses the homogeneity requirement with respect to inputs and outputs and which is invariant to the dynamic characteristics of a process (Section 2). Such a measure may be based on the integral process success indicators like for example the ratio of the Earned Value to the Planned Value or to the Actual Cost at a Sign-off Stage of the process. This implies that productivity of a DEDP may be assessed by the value produced and accumulated by designers in a team. The more value produced by a designer – the more relatively productive s/he is. It is also true in a longer run if several DEDPs are taken into consideration. Hence more productive designers are characterized by the higher volume of accumulated Units of Welfare (UoW) if designers are incentivized adequately to their produced value (assumed in PSI). This characteristic is invariant to all aforementioned dynamic features of an engineering design process. UoW is a normalized scalar measure which by its semantics is similar to the notion of a Utility which is used in Distributed Rational Decision Making. UoW earning mechanisms in PSI are based on contracting deals stricken through several types of negotiations [4].

4   Actors, Teams, Beliefs, and Negotiations

Actors and related concepts are denoted by the DEDP Actor ontology which is outlined in Fig. 2. An Actor is the abstraction of a person who performs Tasks and executes atomic Activities which result in the transformation of Design Artifacts. An Actor as the part of an organization plays Organizational Roles which are regulated by organizational Policies. An Organizational Role is the subclass of an abstract Role. A Role specifies the set of requirements to an Actor with respect to his or her capability to execute Activities. Thus Organizational Roles and Policies constitute the organizational framework of DEDP model. A Collaboration Role is another subclass of a Role specifying the Roles of Actors with respect to their encounters governed by PSI Negotiation Mechanisms defined by interaction protocols, negotiation sets, and negotiation strategies. Therefore another important aspect covered by DEDP Actor ontology is Collaboration and Team Formation framework. Chosen collaboration mechanisms based on contracting negotiations (full details are in [4]) imply the appearance of the following subclasses of an Actor: a Task Manager and a Believed Performer. A Task Manager intends to out-source a Task to one of his or her peers. The following two aspects constrain the set of peers to the sub-set of the Believed Performers: a Task Manager believes that the Believed Performers are (1) Capable to perform the Task and (2) Credible enough to trust the performance of the Task to them. These Beliefs in PSI are (1) formalized by Belief sub-ontology and (2) adjusted by the Post-Effects of Activities (Section 5) through capability and credibility

---

[6] FP stands for Functional Point, KSLOC – for kilo lines of source code.

assessment mechanisms adopted from RACING [3]. A Contractor is the sub-class of a Believed Performer as s/he is the only one of Believed Performers who receives the negotiated Task and commits him- or herself to perform it according to the commitment-convention framework [4]. If a Contractor according to his subjective knowledge decides that the received Task comprises only one atomic Activity[7] then s/he becomes an Activity Executor. S/he also becomes the member of the design Team by committing him- or herself to the Activity execution. A Team is therefore formed of Task Managers and Activity Executors through contracting negotiations. Conceptually a Team is the bridge providing the relationship of a DEDP to the Project which is implemented through this DEDP.

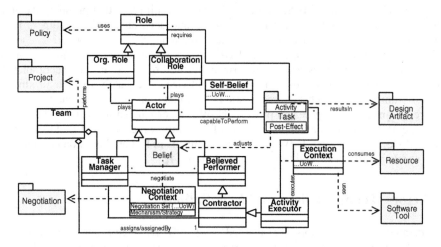

Fig. 2. The outline of DEDP **Actor** ontology

It is assumed in PSI that collaboration mechanisms are based on three types of negotiations which use one basic protocol (extended FIPA CNP) but differ by negotiation sets and strategies [4]: (1) on Task allocation; (2) on Design Artifact re-use; (3) on Software Tool provision. PSI Negotiation ontology based on [5] is used as the namespace for formalizing Negotiation Contexts in all negotiation types.

The central property of an Actor is the capability to perform Tasks. An Actor is capable to perform Tasks in frame of the Organizational Role he plays in the sense that s/he has the subjective knowledge on the following: (1) if the certain Task is a *composite* one or it contains only one an *atomic* Activity; (2) if s/he can perform this Task *by himself* or he should *allocate* it to another Actor paying a price in UoW. This knowledge constitutes Actor's Self-Beliefs. Another portion of subjective Task-related knowledge is formalized by the DEDP Task ontology (Section 5). However the Actor ontology provides for the clear separation between the notions of a Task and an Activity. A Task is *performed* – i.e. arranged and managed by Task Managers. An Activity is *executed* by Activity Executors – physically: using Design Artifacts as material Inputs and Software Tools as instruments, consuming Resources, producing

---

[7] As the Contractor believes.

material Outputs in the form of Design Artifacts. These aspects are captured by Execution Context concept of the Actor ontology. UoW are spent by Activity Executors for lending Software Tools and using Resources.

5   Tasks, Activities, and Partial Local Plans

The DEDP Task-Activity model provides formal shared static description framework (Fig. 3) used in the knowledge models of Actors[8] to form their subjective static knowledge on Task compositions, Activity choreography, and Task dependencies. This knowledge according to the Task-Activity model is tightly linked to the Belief and Self-Belief parts of DEDP Actor ontology.

An Activity is the basic process building element which is shared static (defined by the design technology) and is treated as objectively atomic by all DEDP participants. Material Inputs and Outputs of Activities are also fixed by the technology and are Design Artifacts. Task-Activity model provides corresponding relationships. An Activity as the basic building element is the sub-class of a Task concept. In difference to a Task an Activity is the only piece of a DEDP which is executed and produces material Outputs. A Task is subjectively static as the representations of the compositions of the same Task may differ from Actor to Actor. This is one of the explicit reasons to introduce a TaskByActor concept as the sub-class of an abstract Task. A Task is linked to an Actor by the capability relationship with the associated Self-Belief context. Unlike an abstract Task a TaskByActor is associated with a Task Manager. Thus its seman tics become even more subjective in the sense that it is the Task which is managed and, therefore, can not comprise only one atomic Activity. A TaskByActor is the Task to which the Task Manager has committed him- or herself by striking the deal in the contracting negotiations. Hence, a TaskByActor (but not a Task) has UoW property

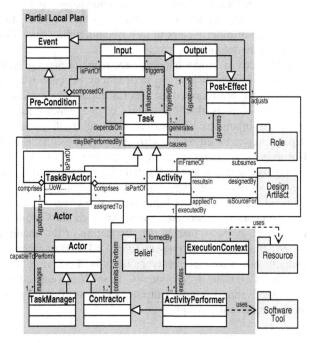

Fig. 3. The outline of DEDP **Task-Activity** ontology

---

[8] Actors are modeled by economically rational software agents in PSI DEDP Simulation Prototype [2].

associated with it. UoW property of a TaskByActor reflects the result of the negotiations on this very task providing the Contractor with the budget figure.

A Task in contrast to an Activity is managed. Task management comprises the proper scheduling of its sub-Tasks which requires the knowledge on the dependencies among these sub-Tasks. In the Task-Activity model Tasks may be *independent* or *strongly dependent* on other Tasks. The model also indirectly allows coping with *facilitations* (or *weak dependencies*). Task t_1 is said to be independent of Task t_2 if the performance of t_2 does not depend of t_1 performance or of the results of t_1 and vice versa. The task t_1 is said to be *strongly dependent* of task t_2 if the results of t_2 are essential to start the performance of t_1. Finally, task t_1 is said to be *facilitated* by task t_2 if the performance of t_2 or the results of t_2 may help to execute t_1 in less time, with less resources consumed or obtaining better quality with the same resource consumption. From productivity viewpoint facilitation means UoW savings. For example getting a proper Design Artifact from a fellow for re-use may facilitate to the design of the similar Design Artifact resulting in effort, resource (and therefore UoW) savings. Hence, fine-grained knowledge on Task dependencies allows making the process properly coordinated and therefore more productive.

One more important aspect captured by the discussed model is the subjectivism of dependencies' representations in the PLPs of different Actors. Dependency plans are denoted as partial local because different Actors: (1) have different knowledge on Task dependencies – these pieces of knowledge are the subjective parts of the whole picture possibly leading to alternative paths in DEDP state space; (2) do not use the knowledge of other Actors in task planning and scheduling – i.e. take their decisions locally or autonomously.

Task-Activity model handles dependencies among Tasks based on the assumption that the existence of a strong dependency between t_1 and t_2 implies that the material Outputs of t_2 are required as material Inputs to t_1 before t_1 starts. Therefore the Pre-condition of t_1 is that the events of the appearance of all the necessary Inputs (eventual Inputs to be shorter) have all took place thus triggering t_1. A weak dependency is based on the same triggering mechanism through the eventual Inputs. However in the latter case the trigger just lowers the amount of UoW required for managing the dependent task reflecting that the facilitation has occurred. PLP part of the Task-Activity ontology frames out the sets of eventual Inputs as Pre-conditions. It is stated that an eventual Input is the sub-class of an eventual Output because only some outputs may become inputs. An eventual Output is in turn the sub-class of a Post-effect. A Post-effect is the abstraction of the changes implied by the Task onto its environment. With respect to a DEDP Post-effects are not only the eventual Outputs but also the events caused by Task re-planning and re-scheduling like deadline violations. Consequently, Post-effects cause the changes in Actors' Beliefs (Fig. 3). Eventual Inputs, Outputs, and Post-effects are ultimately the sub-classes of an abstract Event concept.

6   Design Artifacts and Project Memory

The purpose of PSI Design Artifact model is twofold: (1) it provides the grounding to SES design domain and (2) it reflects the project-oriented nature of engineering design. DEDP Design Artifact ontology is outlined in Fig. 4. From the point of view of domain grounding the model specifies that a Design Artifact comprises the hierarchy of Functional Blocks as the structural elements of designed functionality. Functional Blocks are generally viewed as "gray boxes" with functional subdivision to digital, analog and mixed-signal blocks according to the function and components used in their design. Therefore the Interfaces of Functional Blocks are of type Digital or Analog. A Functional Block of mixed functionality may have Interfaces of both types. The Functional Block of the topmost level is finally materialized in the corresponding Chip. The description of a Chip ready for production is considered the terminal output of a DEDP. Functional Blocks are complemented by TestBenches and Verification Runsets – the means to test and verify designs according to the provided engineering design technology.

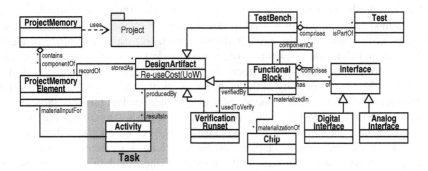

Fig. 4. The outline of DEDP **Design Artifact** ontology

The Design Artifact model provides the formal frame for handling material Inputs and Outputs of DEDP Activities. It is considered that a Design Artifact is the material Output of an Activity (through resultsIn – producedBy relationship) and is stored to the design Project Memory as a Project Memory Element. A Project Memory Element (but not a Design Artifact) is therefore the material Input to an Activity. Hence a Design Artifact may be rightfully used as the material Input for an Activity only after properly stored to the Project Memory. PSI mechanisms assume that a Project Memory is a shared tuple space used for activity run-time coordination based on blackboard principles.

7 DEDP Ontologies: Epistemology and Usage

The descriptive part of DEDP modeling framework has been initially designed as the family of DEDP ontologies and coded in the set of UML class diagrams (further on referenced as DEDP-full). Further formalization and implementation work has been performed in the way aligned with scenarios of ontology usage identified by Uschold and Jasper [6]. DEDP ontologies are used [2] for authoring DEDP logs recorded to PSI testbed (neutral authoring), for specifying the designs of DEDP-MAS simulator software (ontology as software specification), and as shared ontologies for agent communication at run time (common access to information). Ontology usage aspects influenced the choice of the formal languages for coding the ontologies. The ontologies were first coded in OWL-lite[9] (further on referenced as DEDP-lite). This language was chosen because it is one of the de-facto ontology specification standards. The second reason for choosing OWL-lite was that its expressive power is similar to that of the internal mental model specification language (MMSL) of MASDK [7] which has been used for specifying the design and prototyping of PSI prototype – DEDP-MAS. From epistemological viewpoint the transformation of DEDP-full ontologies to DEDP-lite required the changes of UML associations to the constructs with binary relationships with restrictions. This transformation has been performed manually with the help of Protégé 3.0[10] ontology editor.

DEDP-MAS has been implemented to evaluate the modeling framework and to assess the feasibility of building a software tool for DEDP optimization through their productivity assessment. In the performed evaluation experiments [2] the simulator is used in two application modes: playback and "freestyle" (predictive) simulation. In playback mode the simulation is used to assess the performance of the DEDPs which have been accomplished in the past. Predictive simulation supports project managers in planning and dynamic re-planning of running design projects in cases of several kinds of events which are out of their direct control: late changes to the design objective, sudden unavailability of the team members, the changes in the workload of the designers according to the influence of the other independent projects, etc.

Simulations performed on the With DEDP records stored to PSI testbed demonstrated that the simulator develops DEDPs very closely to what happened in reality. Observed fluctuations were caused by the changes in the parameters of the availability of the team members in the course of the simulation experiments by "screwing" their available capacities. This fact confirms the adequacy of the developed framework to the industrial requirements in SES.

8 Related Work and Discussion

The projects which pioneered R&D in agent-based engineering design process modeling, support and automation appeared about a decade ago, e.g. [8], [9], [10]. Some projects of the "second wave" [11], [12] helped to specify the focus of PSI in automating the near-optimal arrangement of DEDPs in terms of their productivity. In difference to e.g. [22] the objective of PSI is not to automate the design process itself, but to automate the arrangement of its activities in the most productive way. In PSI the activities resulting in the elaboration of design artifacts are performed manually by human designers.

The DEDP modeling framework in its part of organizational and actor-related knowledge representation is based on the frameworks [13], [14], [5]. PSI contribution in this part is the incorporation of roles and actors with its specific subclasses, teams of actors, negotiation context in one coherent ontologies' family and its binding to the engineering design domain through incorporating Design Artifacts and Software Tools ontologies. The main contribution of the family of DEDP ontologies is the model of a dynamic team of designers which is formed through contracting negotiations and performs dynamically orchestrated processes. Hence DEDPs in PSI

---

[9] OWL Web Ontology Language Reference. http://www.w3.org/TR/owl-ref/.
[10] Protégé ontology editor and knowledge acquisition system http://protege.stanford.edu/.

are understood as socially performed processes in the sense close to [15]. For example the notions of a Role or a Policy of PSI Actor ontology are semantically close to that of the normative multi-agent framework.

In the part of process modeling PSI bases its approach on [16], [3], [17]. In the family of DEDP ontologies engineering design processes are modeled as tasks composed of sub-tasks and atomic activities. Similarly to [18] subtasks and activities may have weak and strong dependencies. However, in PSI the knowledge on these dependencies is local and differs from actor to actor as specified in their partial local plans. Similarly to [17] activities have pre-conditions and post-effects. However, DEDP Task-Activity ontology constrains the semantics of pre-conditions and post-effects by making them sub-classes of an event concept. Material inputs and outputs belong semantically and structurally to DEDP Design Artifacts ontology.

Examples of theoretical frameworks for solving planning tasks are Decision Theoretic Planning (DTP) [19] and Hierarchical Task Networks (HTN) [20]. PSI framework is built upon the conceptual denotation of the planning task shared by the mentioned frameworks. Planning is understood as the process of cascade decomposition of the goal, transformation of the sub-goals to Tasks and committing Actors to Tasks. However PSI framework extends the capabilities of the classical AI approaches to planning by accounting the dynamic character of the process and by the capability to collaborative distributed planning through negotiation mechanisms. The latter feature also distinguishes our descriptive framework from the plan-task ontology of KMI [21]. Moreover, the family of DEDP ontologies provides conceptual means for dynamic re-scheduling based on the concepts of Self-Beliefs and Beliefs.

9 Conclusions

The paper presented the descriptive part of the DEDP modeling framework developed in the PSI project. The project is aimed to build a software tool assisting in analysis and optimization of DEDPs' productivity through agent-based simulations. The framework incorporates the models of teams and actors, tasks and activities, design artifacts as the major interrelated parts. DEDPs are modeled as weakly defined flows of tasks and atomic activities which may only "become apparent" at run time because of several factors which are beyond the control of the design team members. The processes are self-formed through the mechanisms of collaboration in the dynamic team of actors. These mechanisms are based on several types of negotiations. DEDP productivity is assessed by the Units of Welfare collected by the multi-agent system which models the design team. The models of the framework are formalized in the family of DEDP ontologies. These ontologies are used in the implemented simulator software prototype. Initial evaluation experiments have been performed using the PSI testbed [2].

References

1. Neal, D., Smith, H. and Butler, D.: The evolution of business processes from description to data to smart executable code – is this the future of systems integration and collaborative commerce? *Research Services Journal*: March 2001, 39-49
2. Gorodetsky, V., et al.: Agent-Based Framework for Simulation and Support of Dynamic Engineering Design Processes in PSI. In: Proc. CEEMAS'05, 15-17 Sept. 2005, Budapest, Hungary. (draft: http://eva.zsu.zp.ua/eva_personal/PS/PSI-CEEMAS.pdf)

3. Ermolayev, V., et al: Towards a framework for agent-enabled semantic web service composition. *Int. J. of Web Services Research*, 1(3): 63-87, 2004.
4. Ermolayev, V. et al: Agent-Based Dynamic Engineering Design Process Modeling Framework. Technical Report. Cadence Design Systems, GmbH, 29 p., 2004, http://eva.zsu.zp.ua/eva_personal/PS/PSI-DEDP-MF-v10-Feb-2004.pdf
5. Ermolayev, V. Keberle, N. and Tolok, V.: OIL Ontologies for Collaborative Task Performance in Coalitions of Self-Interested Actors. In: H. Arisawa, Y. et al (Eds.): Conceptual Modeling for New Information Systems Technologies ER 2001 Workshops, Yokohama Japan, November 27-30, 2001. LNCS vol. 2465, 390-402, 2002.
6. Uschold, M. and Jasper, R.: A Framework for Understanding and Classifying Ontology Applications. In: 12-th Workshop on Knowledge Acquisition, Modeling and Management (KAW'99), Banff, Alberta, CA, 16-21 Oct., 1999
7. Gorodetsky, V. et al.: Multi Agent System Development Kit: MAS software tool implementing GAIA Methodology. In: Z. Shi and Q. He (eds.) Int. Conf. on Intelligent Information Processing (IIP2004), Beijing, Springer, 69-78. 2004.
8. Cutkosky, M.R. et al: PACT: An Experiment in Integrating Concurrent Engineering Systems. *IEEE Computer* 26(1), 28-38, 1993
9. Darr, T. P., Birmingham, W. P.: An Attribute-Space Representation and Algorithm for Concurrent Engineering. CSE-TR-221-94, University of Michigan, Department of Electrical Engineering and Computer Science, Ann Arbor, Michigan 48109-2122, 1994.
10. Balasubramanian, S. and Norrie, D. H.: A multi-agent intelligent design system integrating manufacturing and shop-floor control. In: Proc. First Int. Conf. on Multi-Agent Syst., San Francisco, 3-9, 1995
11. Parunak, H.V.D. et al: The RAPPID Project: Symbiosis between Industrial Requirements and MAS Research. *Autonomous Agents and Multi-Agent Systems* 2: 111-140, 1999.
12. Danesh, M. R. and Jin, Y.: An Agent-Based Decision Network for Concurrent Engineering Design. *CERA* 9(1), 37-47, 2001.
13. Fox, M.C. and Gruninger, M.: Enterprise Modelling. *AI Magazine* 19(3): 109–121, 1998.
14. Uschold, et al: The Enterprise Ontology. Knowledge Engineering Review, 13(1), 1998
15. Boella, G. and van der Torre, L.: An Agent Oriented Ontology of Social Reality. In: Varzi, A., Vieu, L. (Eds.) Proc. 3-d Int. Conf on Formal Ontology in Information Systems (FOIS'04), Turin, Nov. 3-6, 199-209, 2004
16. Buhler, P. and Vidal, J.M.: Enacting BPEL4WS specified workflows with multiagent systems. In Proc. of the Workshop on Web Services and Agent-Based Engineering, 2004
17. Fensel, D. and Bussler, C.: The Web Service Modeling Framework WSMF. *Electronic Commerce Research and Applications* 1(2): 113-137, 2002.
18. Nagendra Prasad, M. V., and Lesser, V. R. (1999) Learning situation-specific coordination in cooperative multi-agent systems. *Autonomous Agents and Multi-Agent Systems*. 2(2): 173-207, 1999
19. Blythe, J.: Decision-Theoretic Planning. *AI Magazine*, 20 (2), 1999.
20. Erol, K., Hendler, J. and Nau, D. S.: Semantics for Hierarchical Task-Network Planning. Technical report CS-TR-3239, University of Maryland at College Park, 1994.
21. Rajpathak, D. and Motta, E.: An Ontological Formalization of the Planning Task. In: Varzi, A., Vieu, L. (Eds.) Proc. 3-d Int. Conf. on Formal Ontology in Information Systems (FOIS'04), Turin, Nov. 3-6, 2004.
22. Capera, D., Picard, G., and Gleizes, M.-P.: Applying ADELFE Methodology to a Mechanism Design Problem. In: Proc. 3-d Int. Joint Conf. AAMAS'04, 1508-1509, 2004

Preliminary Basis for an Ontology-Based Methodological Approach for Multi-agent Systems

Ghassan Beydoun[1], Numi Tran[1], Graham Low[1], and Brian Henderson-Sellers[2]

[1] School of Information Management and Technology Management,
University of New South Wales
{g.beydoun, g.low, numitran@unsw.edu.au}
[2] Faculty of Information Technology,
University of Technology of Sydney, Sydney, Australia
{brian@it.uts.edu.au}

Abstract. The influence of ontologies in Knowledge Based Systems (KBS) methodologies extends well beyond the initial analysis phase, leading in the 1990s to domain-independent KBS methodologies. In this paper, we reflect on those lessons and on the roles of ontologies in KBS development. We analyse and identify which of those roles can be transferred towards an ontology-based MAS development methodology. We identify ontology-related inter-dependencies between the analysis and design phases. We produce a set of six recommendations towards creating a domain-independent MAS methodology that incorporates ontologies beyond its initial analysis phase. We identify its essential features and sketch the characteristic tasks within both its analysis and design phases.

1 Introduction

We argue in this paper that a methodology that is ontology-based (i.e., it uses ontologies as a central model beyond the analysis phase) will have at least the following two advantages: firstly, such a methodology can serve as a unification platform between varying concerns of existing methodologies that do not individually address all MAS applications. Secondly, it can better produce reusable MAS designs and components (beyond the ontologies themselves). Here, we map out the software engineering requirements to create such an ontology-based MAS methodology. Our survey in [27] shows that only in a very few of the most prominent Agent Methodologies have ontologies been used; and none in the design phase. We use as a guide the roles of ontologies of reuse and domain-independent development in modern Knowledge Based Systems (KBSs) rooted in the situated view of knowledge (as advocated for example in [25]). This leads us to highlight the often-overlooked ontology-related interactions between the analysis and design software development phases for MAS (with intelligent knowledge-based agents as advocated by the popular BDI agent model [31]).

Using the domain independence of KBS methodologies as a guide, we believe that what is required in order that ontologies are effectively accommodated in the MAS architectures and throughout their design phases as ready components is a domain-independent methodological approach founded on ontological analysis throughout the

J. Akoka et al. (Eds.): ER Workshops 2005, LNCS 3770, pp. 131–140, 2005.

whole development cycle. The realization of this requires a combination of all domain-dependent concerns of existing and future methodologies. We investigate the way forward towards this. We use ontologies beyond the analysis phase to the design phase. We analyse the interplay between analysis and design and suggest the requirements that an ontology-based domain-independent methodology needs to fulfil.

2 Traditional KBS Development and the Role of Ontologies

Decoupling problem-solving knowledge from domain knowledge was the breakthrough needed to address usability-reusability limitations faced by developers of intelligent single agent systems (or KBS) [16]. This decoupling was born out of investigation of techniques necessary to use knowledge specified at the knowledge level [23] and to turn it into a working KBS [8]. It has been realised that no simple generic techniques, such as deductive reasoning techniques, are sufficient to utilise knowledge in order to address every kind of problem. Rather, specific techniques for different kinds of problems are necessary in order to build relatively complete and competent systems. This resulted in collections of problem-solving methods to be used in conjunction with domain ontologies as well as the relevant domain knowledge (see, e.g. [1, 22, 24]). Ontologies provided domain-dependent reusable encapsulation of the structural basis of domain knowledge. Problem-solving Methods (PSMs) [17] provided a reusable and domain-independent encapsulation of problem-solving knowledge. The reliance on ontologies and PSMs produced reusable and effective components for building robust KBSs more economically. Moreover, this led to methodologies founded on ontological analysis that are consequently domain-independent. Examples include Ibrow3 [2], KAMET II [7], KADS [30] and CommonKADS [26]. In contrast, in the current state of the art in MAS Software Engineering (usually referred to as Agent Oriented Software Engineering (AOSE)), many agent-oriented methodologies are being published, most acknowledging their own suitability for a given class of applications. For example, Adelfe [3] targets adaptive systems; Passi [11] is limited to a class of communication architectures.

Few MAS methodologies include ontologies in their models and processes e.g. [13, 18]. The inclusion of ontologies in such works is confined to the analysis phase of the development. For instance, [18] distinguishes between an initial ontology and a domain model geared towards designing an MAS and these authors specify how a *domain model* that includes goal and role analyses is developed from an initial ontology. Similarly, in [13], the MaSE methodology is extended to incorporate the use of an ontology to mediate the transition between the goal and the task analyses (both being within the analysis phase). Our work in this paper is perhaps closest to recent work in [6] which recognizes the usefulness of ontologies for verification of models during the analysis phase. Outside the analysis phase, ontologies currently are mainly used to express a common terminology for agent interactions in an MAS e.g. [15]. These interactions have no parallel within a single agent KBS (since an agent does not usually need to interact with itself!). We find that the initial motivation for using ontologies (for single agent systems), that of enhancing reuse (cf. [20]) of system architectures and components, is absent in AOSE.

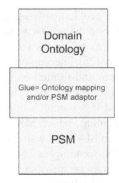

Fig. 1. KBS architecture-based on an ontology and a Problem Solving Method

Reusability of system design is recognised as a key concern in single agent knowledge-based systems [9, 28] and is the impetus for the ontology-based architectural view of a KBS as being formed from two components: a PSM and a suitable ontology (Figure 1). This view is central to many KBS methodologies e.g. [2, 26, 30]. It was the impetus for most of the KBS research in the '80s and '90s, with the aim of reducing KBS analysis and design to ontology engineering coupled with a suitable choice of a PSM from some existing library of PSMs [1]. Ontologies were used to support the reuse of PSMs in different problem areas. Alternatively, PSM components permitted reuse of ontologies to address different problems within the same area.

Domain ontologies aim to reuse part of the domain knowledge in different systems i.e. a domain is characterised by a set of objects referred to by a set of terms deemed relevant and that can be used by different systems to handle different types of tasks. The development of reusable ontologies create the problem that a general-purpose ontology is very rich, while for a particular task only a small part of it will actually be needed. To compensate, KBS developers carefully choose a suitable problem-solving method and adapt the ontology used to a suitable level of refinement. With this idea, it has eventually become possible for a single methodology to address the development of any system (e.g. CommonKADS). With current MAS development approaches, this idea cannot have a direct parallel from single agents to multi-agents, nor do we have a universal MAS methodology. We advocate in this paper a similar ontology-centric development process-based on that initial *reuse* motive. Towards this, we examine how assumptions, about the way knowledge is used, vary between a single agent system and an MAS. We highlight what changes should be embraced by methodological approaches of MAS development in order to accommodate reuse together with ontology-oriented MAS analysis and design phases. We argue that a unified domain-independent methodology that is ontology-based is required in order to create easily reusable MAS architectures and components. Availability of universal methodologies for single agent systems was made possible partly because of ontologies.

3 Ontologies for MAS Development

In an MAS, two or more agents interact or work together to achieve a set of goals [31]. Agents have their own localised knowledge bases. The coordination between agents possessing diverse knowledge and problem-solving capabilities usually enables the achievement of global goals that cannot be otherwise achieved by a single agent working in isolation. MASs are thought to be an answer to a number of shortcomings of general problem-solving limitations [25]: incomplete knowledge requirement specification, incomplete PSM requirement and limited computational resources. MASs are particularly useful in the engineering of open, dynamic and adaptive systems. Agents in an MAS are usually smaller and less complex than a standalone single agent in a KBS. Associated with these shortcomings which MASs address, we note the following differences between agents within an MAS and a single agent KBS: an MAS may have different PSMs for different agents, some agent ontologies may be incomplete in an MAS, individual PSMs for agents may be insufficient for their own goals in an MAS and agents within an MAS may have limited execution resources. In what follows, we present an overview of how these differences characterize the way agents may interact within an MAS, noting that six potential influences their way of interacting has on how ontologies should be utilised in an MAS.

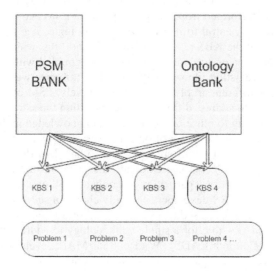

Fig. 2. As new problems arise, the PSM and the ontology banks are used to construct suitable KBSs. An ontology from the *ontology bank* strengthens a given PSM from the *PSM bank* to suit the domain.

An MAS may have different kinds of PSMs: In the case of multi-agent systems, different problem solvers operate on the same domain. Using ontologies in an MAS is complicated by having to provide knowledge requirements to different PSMs at the same time. Whilst individual PSM may operate at different levels of abstraction of the domain, they still need to share their results using a common terminology. PSMs may be complementary and may have different degrees of strength. How much specificity they exhibit to a given domain may vary. In contrast, within a single agent KBS, ontologies were conceived and used to strengthen a single PSM for a given domain. Their use for KBS was never intended to *simultaneously* strengthen different PSMs for the same domain (see Figure 2). Therefore, in developing MASs, we may additionally need the following requirements:

Requirement 1: Ontology mappings allow individual problem solvers (of individual agents) to interact and use a common domain conceptualization.

Requirement 2: Verification of individual PSM knowledge requirements against allocated ontologies is required at design time.

Individual agent ontologies may be incomplete in an MAS: A domain ontology underlying knowledge requirements of all agents is available. However, the version available to an individual agent, matching its PSM, is not necessarily complete (as is assumed to be the case for single agent systems). In addition to 1 and 2, we add:

Requirement 3: Knowledge extensibility is required at the agent level at least to accommodate any new ontological units added to the system about the domain. This can often create inconsistencies [5].

Requirement 4: Associated with 3, a structured and understood knowledge representation is required to resolve inconsistencies.

Individual PSM may be incomplete in an MAS: An agent PSM is not assumed to be powerful enough to respond to all events it encounters during its lifetime within an MAS. It usually negotiates cooperation from other agents. Current practices often assume that functional goal analysis is sufficient to specify the knowledge requirement for agents [19], and any deficiencies in its later problem-solving capacity is assumed to be offset by that cooperation. However, in our view, without consideration of its actual PSM (or other available PSMs within the system), there is no guarantee that this cooperation would ultimately work. This suggests:

Requirement 5: Iteration between the PSM design and the goal analysis is required to ensure that the chosen problem solver for a given agent is capable of meeting its specified goals.

Requirement 6: A consideration of the total PSMs of all agents is required to ensure that system goals are achievable.

Agents within an MAS may have limited external resources: Agents are limited by their resources e.g. computation, storage and response time. It is often assumed that agents cooperate through sharing of their processing resources. This requires synchronisation. Common agent platforms such as Jade can resolve this.

4   Ontologies for MAS Reuse

Similar to KBS development, we assume that the choice of PSM may be made independently of domain analysis. Moreover, we assume that a domain ontology describing domain concepts and their relationships is available e.g. from an existing repository e.g. [12] or a domain analysis may be considered the first stage in developing the system. The purpose of such a domain analysis would only be to identify concepts and their relationships as proposed in [10]. Given such a domain ontology and our six SE requirements, we sketch features of the analysis and design phases for an ontology-based MAS methodology.

There is inter-play between the role of reuse and other roles of ontologies in an MAS. Various reuse roles cannot be smoothly accommodated (e.g. interoperability at run-time) without careful consideration of run-time temporal requirements. For example, the role of ontologies in reasoning at run-time are based on fulfilling PSM knowledge requirements at design time. This requires scoping domain analysis for each individual agent at design time (towards requirement 2).

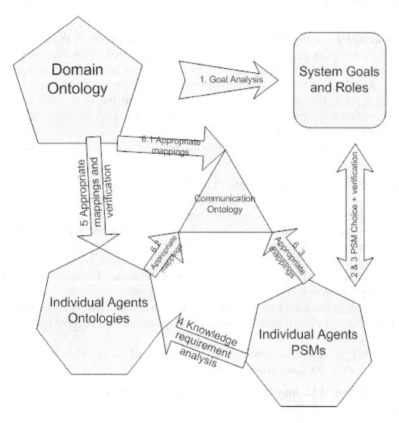

Fig. 3. 1. Ontology-based MAS development: Domain Ontology produces Goal Analysis 2. Goal analysis produces a collection of PSMs (using a PSM bank) 3. Knowledge requirement analysis (4). can then be used to delineate local ontologies that can be verified against the domain ontology (step 5). Finally, in step 6 the communication ontology (language) can then be derived using appropriate mappings.

Requirement 2 recognizes that the key to ontology-based design of an MAS is the appropriate allocation of a PSM to individual agents in order to match system requirements. Towards this, we note that goal analysis is the usual way to express requirements e.g. see [19, 32], and we suggest associating PSMs (using PSM libraries) and system goals in the early stages of an MAS design. The rest of the system can then be developed with appropriate ontological mappings (Figure 2). The collection of all PSMs for local goals should also be verified for completeness against stated system goals. These goals should also be checked against cooperation potential (a form of distributed goal interaction evaluation could be done using approaches such as [29]). Most current methodologies view the decision of problem-solving mechanisms as a low level design step. In our current view, paralleling KBS development, ontology-based design and development requires elevating this to an early design phase and making it central to later decisions on the communication and interface requirements of each agent (rather than the other way around as in many other methodologies e.g. [19, 32]). This elevation of reasoning and iterative verification with goal analysis is one way to satisfy Requirements 5 and 6 (see Figure 2).

Chosen problem solving capabilities for different agents in a given MAS do not necessarily have the degree of domain dependence. Hence, for a PSM chosen for some agents, their ontology may need to be adapted. For this, the domain ontology is again the most convenient reference point. Ontology mapping (between portions of domain ontology and local agent's knowledge) is required to ensure that all PSMs have their knowledge requirement available to their reasoning format. Agents need to communicate their results and instigate cooperation using a common language. For this purpose, we recommend a global communication ontology (as in [15]), rather than many-to-many individual mappings between agents. Such a communication ontology is most conveniently based on the domain ontology available, and it depends on the individual ontology of each agent. In some cases, an ontology mapping may be required between PSM ontologies and the communication ontology. The same adaptation between the reasoning and domain ontology can be used to map the result of reasoning back to a common communication ontology (based on the domain ontology). In the case of open systems, introducing new agents may require extending the communication ontology or some local ontologies to allow cooperation with new agents.

Towards requirements 3 and 4, hierarchical ontologies are one way to have flexible domain ontology refinements for agents according to their PSMs, and to accommodate differences in the strength of the PSM of agents. A common hierarchical domain ontology can be used as a starting point for verification during development and for multiple access at multiple abstraction levels depending on the individual knowledge requirement of each agent PSM. For this purpose, Multiple Hierarchical Restricted Domain (MHRD) ontologies, employed by many authors (e.g. [14]), are well understood and expressive for most domains. MHRD models are sets of inter-related concepts that are defined through a set of attributes, so the presence of axioms between these attributes is not considered. There can be *part-of* and taxonomic relations among the concepts so that attribute (multiple) inheritance is permitted.

Figure 3 provides a methodological sketch accommodating the observations of this section. The MAS development process starts with a domain ontology, used to identify goals and roles that index an appropriate set of PSMs from a bank of PSMs (similar to Figure 2). Appropriate individual ontologies for each PSM are extracted from the initial ontology. These ontologies are used for reasoning by individual problem solvers and may be used to represent results communicated by the individual problem solver. They are next verified against the knowledge requirement of chosen PSMs. The collection of these ontologies is then used to develop a common communication ontology. Appropriate mappings may be needed between individual local ontologies and the communication ontology, in order to facilitate communicating results between individual agents. Verification between problem solvers and the communication ontology is undertaken, which may result in further localized ontology mappings.

5 Discussion, Conclusion and Future Work

In this paper, we have evaluated the goal of long term reuse of software engineering knowledge and effort involved in developing MASs. This reuse may take the form of extending functionality of an existing system, reusing components of an existing system in an entirely different context or creating a new system using the design (in whole or part) of an existing system. We have argued that an ontology-founded MAS methodology can produce reusable MAS components and designs, an issue often overlooked in the MAS software engineering community. Moreover, we have argued that an ontology-based MAS methodology can truly become a domain-independent methodology by combining domain-dependent concerns of existing methodologies.

The current use of ontologies in MAS methodologies is limited to the early analysis phases and, in other cases, to express the communication languages for agents within the system. Current usage ignores the impact of using ontologies for the late design phase where components of the system begin to emerge. Taking into account this impact, we have highlighted software engineering requirements for ontology-based multi-agent systems development. We have drawn from lessons of the knowledge based systems (KBS) and engineering communities to use the separation of problem solving methods and ontology as a basis for reuse. As a conclusion of our analysis, we have sketched an MAS ontology-based methodology which assumes that an initial domain ontology is available. This methodology guides the allocation of individual ontologies and problem solving capabilities to individual agents in the system.

To complete our sketched methodology, domain-dependence of some of its steps described in Section 4 should be recognized. An example of where this may occur is during the step producing goal analysis from the initial domain ontology, in order to index individual agent's PSM. In other words, we acknowledge that it is not wise to assume that all domain dependencies are bundled in the PSM bank. Cordi *et al.* [10] explain the best way to undertake such domain-dependent model conversions. Our sketched methodology also requires developing and adapting appropriate interfaces to PSM and ontology banks.

As for the later phases of our sketched methodology, there are many existing agent-oriented methods with differing concerns and assumptions that can be combined to produce a broad domain-independent unified approach. The results in a comprehensive framework addressing the ontology concerns elucidated here and combining all domain-dependent techniques. This would produce the equivalent of PSM banks, but for the MAS software development process itself. We are currently examining different ways to unify all domain-dependent concerns of existing methodologies and interleave the domain-independent ontological SE guidelines as outlined in this paper. Metamodelling-based method engineering as outlined in [4] and [21] is particularly promising.

References

1. R. Benjamins: Problem solving methods for diagnosis and their role in knowledge acquisition. *International Journal of Expert Systems: Research and Applications*, 1995. **2**(8): p. 93-120.
2. V.R. Benjamins, E. Plaza, E. Motta, D. Fensel, R. Studer, B. Wielinga, G. Schreiber and Z. Zdrahal: IBROW3 - An Intelligent Brokering Service for Knowledge-Component Reuse on the World Wide Web, in *Banff Knowledge Acquisition Workshop (KAW98)*. 1998. Canada.

3. C. Bernon, M.-P. Gleizes, S. Peyruqueou and G. Picard: ADELFE, a Methodology for Adaptive Multi-Agent Systems Engineering, in *Engineering Societies in the Agents World*. 2002. Spain.

4. G. Beydoun, C. Gonzales-Perez, G. Low and B. Henderson-Sellers: Synthesis of a Generic MAS Metamodel, in *Software Engineering for Large Scale Multi Agent Systems 2005 (SELMAS2005)*. 2005.

5. G. Beydoun, A. Hoffmann, J.T.F. Breis, R.M. Béjar, R. Valencia-Garcia and A. Aurum: Cooperative Modeling Evaluated. *International Journal of Cooperative Information Systems, World Scientific*, 2005. **14**(1): p. 45-71.

6. A.A.F. Brandao, V.T.d. Silva and C.J.P.d. Lucena: Ontologies as Specification for the Verification of Multi-Agent Systems Design, in *Object Oriented Programmings, Systems, Languages and Applications Workshop (2004)*. 2004. California.

7. O. Cairo and J.C. Alvarez: The KAMET II Approach for Knowledge-Based System Construction, in *8th International Conference on Knowledge-Based Intelligent Information and Engineering Systems (KES 2004)*. 2004. New Zealand: Springer.

8. B. Chandrasekaran: Generic tasks in knowledge-based reasoning: High level building blocks for expert system design. *IEEE Expert*, 1986. **3**(1): p. 23-30.

9. B. Chandrasekaran, T. Johnson and J. Smith: Task Structure Analysis for Knowledge Modelling. *Communications of ACM*, 1992. **35**(9): p. 124-137.

10. V. Cordi, V. Mascardi, M. Martelli and L. Sterling: Developing an Ontology for the Retrieval of XML Documents: A Comparative Evaluation of Existing Methodologies, in *AOIS2004 @CaiSE04*. 2004.

11. M. Cossentino and C. Potts: A CASE tool supported methodology for the design of multi-agent systems, in *International Conference on Software Engineering Research and Practice (SERP'02)*. 2002. Las Vegas (NV), USA.

12. DARPA: Ontology Repository. 2000, http://www.daml.org/ontologies/.

13. J. Dileo, T. Jacobs and S. Deloach: Integrating Ontologies into Multi-Agent Systems Engineering, in *4th International Bi-Conference Workshop on Agent Oriented Information Systems (AOIS2002)*. 2002. Italy.

14. C. Eschenbach and W. Heydrich: Classical mereology and restricted domains. *International Journal of Human-Computer Studies*, 1995. **43**: p. 723-740.

15. M. Esteva, D.d.l. Cruz and C. Sierra: ISLANDER: an electronic institutions editor, in *International Conference on Autonomous Agents & Multiagent Systems (AAMAS02)*. 2002. Italy: ACM.

16. D. Fensel: Using Ontologies for Defining Tasks, Problem-Solving Methods and Their Mappings, in *European Knowledge Acquisition Workshop*. 1997. Spain: Springer-Verlag.

17. D. Fensel, V.R. Benjamins, E. Motta and B. Wielinga: UPML: A framework for knowledge system reuse, in *Sixteenth International Joint Conference on Artificial Intelligence (IJCAI99)*. 1999. Sweden: Morgan Kaufmann Publishers.

18. R. Girardi, C.G.d. Faria and L. Balby: Ontology-based Domain Modeling of Multi-Agent Systems, in *OOPLSA Workshop*. 2004.

19. F. Giunchiglia, J. Mylopoulos and A. Perini: The Tropos Software Development Methodology: Processes, Models and Diagrams, in *Agent-Oriented Software Engineering III: Third International Workshop, AOSE 2002*, F. Giunchiglia, J. Odell, and G. Weiß, Editors. 2003, Springer. p. 162-173.

20. T.R. Gruber: A Translation Approach to Portable Ontology Specifications. *Knowledge Acquisition*, 1993. **5**: p. 199-220.

21. B. Henderson-Sellers: Creating a comprehensive agent-oriented methodology - using method engineering and the OPEN metamodel, in *Agent-Oriented Methodologies*, B. Henderson-Sellers and P. Giorgini, Editors. 2005, Idea Group. p. 368-397.

22. E. Motta and Z. Z.: Parametric design problem solving, in *10th Banff Knowledge Acquisition for Knowledge Based System Workshop*. 1996. Canada.

23. A. Newell: The knowledge level. *Artificial Intelligence*, 1982. **18**: p. 87--127.
24. F. Puppe: *Systematic Introduction to Expert Systems: Knowledge Representation and Problem-Solving Methods*. 1993, Berlin: Springer-Verlag.
25. S. Russell and P. Norvig: *Artificial Intelligence, A modern Approach, the intelligent agent book*. 2003: Prentice Hall.
26. G. Schreiber, H. Akkermans, A. Anjewierden, R.d. Hoog, N. Shadbolt, W.V.d. Velde and B. Wielinga: *Knowledge Engineering And Management: The CommonKADS Methodology*. 2001, London: The MIT Press.
27. Q.N.N. Tran, G.C. Low and M.A. Williams: A Feature Analysis Framework for Evaluating Multi-Agent System Development Methodologies, in *14th International Conference on Methodologies for Intelligent Systems*. 2003. Japan.
28. M. Uschold and M. Grueninger: Ontologies: Principles, Methods and Application. *Knowledge Engineering Review*, 1996. **11**(2): p. 93-195.
29. A. van Lamsweerde, R. Darimont and E. Letier: Managing Conflict in Goal-Driven Requirements Engineering. *IEEE Transaction on Software Engineering*, 1998. **24**(11).
30. B. Wielinga, G. Schreiber and J. Breuker: KADS: a modelling approach to knowledge engineering. *Knowledge Acquisition*, 1992. **4**: p. 5-53.
31. M. Wooldridge: *Reasoning About Rational Agents*. 2000: MIT Press.
32. M. Wooldridge, N.R. Jennings and D. Kinny: The Gaia Methodology for Agent-Oriented Analysis and Design, in *Autonomous Agents and Multi-Agent Systems*. 2000. The Netherlands: Kluwer Academic Publishers.

DDEMAS: A Domain Design Technique for Multi-agent Domain Engineering

Rosario Girardi and Alisson Neres Lindoso

Federal University of Maranhão, Portugueses Av., Campus do Bacanga,
65.080-480, São Luiz-MA, Brazil
rgirardi@deinf.ufma.br, alissonlindoso@uol.com.br

Abstract. Multi-agent Domain Engineering is a process for the construction of domain-specific agent-oriented reusable software artifacts, like domain models representing the requirements of a family of multi-agent systems, and frameworks, implementing an agent-oriented solution to those requirements. This work describes DDEMAS, an ontology-based technique for the architectural and detailed design of multi-agent frameworks providing a solution to the requirements of a family of multi-agent software systems specified in a domain model. DDEMAS is part of MADEM, a methodology for domain analysis and design of a family of multi-agent systems in a domain. Domain models and multi-agent frameworks are part of a knowledge base constructed through the instantiation of ONTOMADEM, an ontology that represents the knowledge of MADEM. Some examples from a case study on the application of DDEMAS on the construction of a multi-agent framework for the development of usage mining-based Web personalization systems are also described.

1 Introduction

Considerable advances on the systematization of the agent-oriented development paradigm have been achieved and several techniques, methodologies and software development environments are already available for the development of multi-agent applications [3][4][6][8][27][28][31]. Some methodologies promote the reuse of software patterns [6], however, little work has been done on the development of techniques and methodologies for the construction of high-level reusable software abstractions in this development paradigm.

Domain Engineering and Application Engineering [1][7][22] are two complementary software processes. Domain Engineering, also known as Development FOR Reuse, is a process for creating software abstractions reusable on the development of a family of software applications in a domain, and Application Engineering or Development WITH Reuse, the one for constructing an specific application using reusable software abstractions available in the approached domain(s).

The process of Domain Engineering is composed of the phases of analysis, design and implementation of a domain. Domain analysis activities identify reuse opportunities and determine the common and variable requirements of a family of applications. The product of this phase is a domain model. Domain design activities

J. Akoka et al. (Eds.): ER Workshops 2005, LNCS 3770, pp. 141–150, 2005.
© Springer-Verlag Berlin Heidelberg 2005

look for a documented solution to the problem specified in a domain model. The product of this phase is composed of one or more frameworks and, possibly, a collection of design patterns, documenting good solutions in that domain. Reusable components integrating the framework are constructed during the phase of domain implementation. This is the compositional approach of Domain Engineering. In a generative approach, Domain Engineering produces Domain Specific Languages (DSLs), which can be used as application generators to construct a family of applications in a domain. Knowledge of the domain and design patterns are encoded in DSLs [7][18].

Ontologies [5] are knowledge representation structures particularly useful for the specification of high-level reusable software abstractions [15][20][21]. They provide an unambiguous terminology that can be shared by all involved in a development process. Ontologies can also be as generic as needed allowing its reuse and easy extension.

A collection of ontology-based reusable software abstractions is being developed in the context of a Multi-Agent Domain and Application Engineering research project [15][16][17][18][20][21]. The multi-agent paradigm has been adopted because of its effectiveness to approach software complexity.

This work describes the DDEMAS technique for the architectural and detailed domain design of multi-agent systems. The technique is part of MADEM ("Multi-Agent Domain Engineering Methodology"), an ontology-based methodology that provides support for all the phases of the Multi-agent Domain Engineering process. MADEM integrates GRAMO [15][17], a technique for domain analysis of multi-agent systems. Previous work on the DDEMAS technique has been already published [12].

The paper is organized as follows. Section 2 summarizes the modeling phases and respective tasks of MADEM. ONTOMADEM, an ontology that is been used as a tool for capturing and representing the products of the Multi-agent Domain Engineering process is also briefly described. Section 3 details the architectural and detailed design phases of DDEMAS. Section 4 discusses related work on this research topic. Section 5 concludes the paper with some remarks on further work being conducted.

2 The MADEM Methodology

The knowledge of the MADEM methodology has been represented in ONTOMADEM, an ontology that is been used as a tool for capturing and representing the products of the Multi-agent Domain Engineering process, created through the instantiation of their hierarchy of classes. Therefore, domain models and frameworks are embedded in a knowledge base where concepts are semantically related and where inferences can be made thus facilitating the understanding and reuse of the common and variable features of both the requirements and design solutions of a family of multi-agent applications. ONTOMADEM has been developed with the Protégé ontology editor [13].

Fig. 1 illustrates the MADEM methodology in the context of the Multi-agent Domain Engineering process and Table 1 summarizes the modeling phases and respective tasks and products of MADEM.

Domain analysis supported by the GRAMO technique approaches the specification of current and future requirements of a family of multi-agent applications in a domain model by considering domain knowledge and development experiences extracted from specific applications already developed in the domain (Fig. 1). Existing analysis patterns can also be reused in this modeling task.

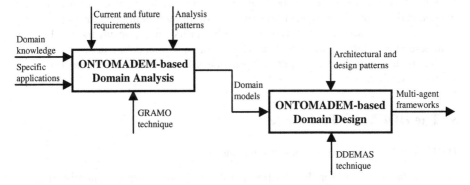

Fig. 1. The MADEM methodology in the context of the Multi-agent Domain Engineering process

Table 1. Summary of the modeling phases and tasks of the MADEM methodology

Phases		Tasks	Products	
Domain Analysis		Concept Modeling	Concept Model	Domain Model
		Goal Modeling	Goal Model	
		Role Modeling	Role Model	
		Variability Modeling	*in the models above*	
		Modeling of Role Interactions	Role Interaction Models	
		Specification of non-functional requirements	Non-functional requirement specification	
Domain Design	Architectural design	Mapping of the Role Model into a first draft of an Agent Society Model	Agent Society Model	Architectural Model
		Mapping of the Role Interaction Models into first drafts of the Agent Interaction Models	Agent Interaction Models	
		Reorganization of the agent society through cooperation and coordination mechanisms	Coordination and Cooperation Model	
	Detailed design	Identification of a detailed design pattern	Agent Template	Agent Models
		Definition of the agent type		
		Modeling the agent behavior	Agent Behavior Model	
	Modeling the knowledge of the multi-agent society		Model of the Multi-agent Society Knowledge	

Domain analysis can focus either the formulation of a problem (e.g. User Modeling) or the representation of a knowledge area (e.g. the Tourism area) and is based on the following modeling tasks. For the formulation of a problem, the tasks

Goal Modeling, Role Modeling, Variability Modeling and Modeling of Role Interactions are performed. A Domain Model is obtained through the composition of a Goal Model, a Role Model and a set of Role Interaction Models. For the representation of a knowledge area only the Concept and Variability Modeling tasks are performed. A Domain Model is obtained, which consists of the developed Concept Model.

Domain design supported by the DDEMAS technique approaches the architectural and detailed design of multi-agent frameworks providing a solution to the requirements of a family of multi-agent software systems specified in a domain model. Existing architectural and design patterns can also be reused in this modeling task.

The modeling tasks and generated products of DDEMAS are described in the following section.

3 The DDEMAS Technique

The DDEMAS technique consists of three sub-phases:

- *Architectural design*, for the construction of the architecture of a family of multi-agent systems.
- *Detailed design,* for the construction of the internal architecture of each agent in the society.
- *Modeling of the knowledge of the multi-agent society*, for representing the meaning of concepts which agents in the society needs to understand in order to communicate with each others.

Next sub-sections detail the tasks performed in each sub-phase of DDEMAS illustrated with examples extracted from a case study on the development of ONTOWUM, a family of multi-agent systems for Web personalization based on usage mining [26].

A collection of architectural and detailed agent-oriented design patterns approaching both general purpose and specific problems has been developed [14][19]. An ontology-based knowledge base has been developed with this collection to facilitate the localization and reuse of patterns in MADEM [21].

3.1 Architectural Design

The purpose of this sub-phase is to develop an *architectural model* representing an agent-oriented solution to the problem specified in the domain model. This architectural model is composed of three sub-products: an *agent society model*, *agent interaction models* and a *coordination and cooperation model* developed through the following tasks:

- *Mapping of the Role Model into a first draft of an Agent Society Model*. Here the purpose is to identify the agents that will compose the multi-agent society. The agents are identified from the roles specified in the *Role Model* of the *Domain Model*. Initially, a mapping of one "role" to one "agent" is done, as well as responsibilities, activities, inputs, outputs, pre and post-conditions and resources. The *Agent Society Model* is represented graphically in a three level organizational chart. Agents and resources are represented in the first and third level, respectively; responsibilities, activities, inputs, outputs, pre and post-conditions in the second one;

- *Mapping of the Role Interaction Models into first drafts of Agent Interaction Models.* Here the purpose is to identify the interactions between agents needed to accomplish their responsibilities. For that, initially, they are extracted from the interactions between roles in the *Role Interaction Models* of the *Domain Model* and represented in *Agent Interaction Models* whose graphical representation is inspired in the interaction diagrams of AUML [27]. An *Agent Interaction Model* provides the dynamic view of an *Architectural Model*.

- *Organization of the agent society through appropriate cooperation and coordination mechanisms.* From the first drafts of the Agent Society Model and Agent Interaction Models, and according to both the functional and non-functional requirements (e.g. performance) specified in the Domain Model; well-known design rules like functional cohesion and considering available architectural patterns [10][14][19], and/or appropriate mechanisms of cooperation and coordination [9][11][23][24][25], the agent society is organized in a *Cooperation and Coordination Model*. For the identification of an architectural pattern, the descriptions of general and specific goals in the *Goal Model* of the *Domain Model* are matched with the description of the problem attribute in a pattern description. Obviously, selected patterns must have a context related with the architectural design of multi-agent systems. If a total or partial matching is obtained, the solution described in the pattern is considered on the execution of the other tasks of the architectural design. Through this reorganization one or more agents can fusion or one agent can be divided in two or more agents. Therefore, some interactions, responsibilities, activities, inputs, outputs, pre and post-conditions can disappear and new ones can emerge. These changes are represented in a new *Agent Society Model* and new *Agent Interaction Models*, which are also detailed according to a particular Interaction protocol (e.g. KQML). An example of a *Coordination and Cooperation Model* is shown in Fig. 2, where a two-layer architecture is adopted to organize the agents that compose the framework ONTOMUW [26]. The upper layer is responsible for processing user information, while the lower one leads with the discovery of navigational patterns. The model follows the architectural design of a multi-agent layer pattern [19]. A *Coordination and Cooperation Model* provides the static view of an *Architectural Model*.

3.2 Detail Design

The purpose of this sub-phase is to perform the detailed design of each agent in the framework, resulting in a set of *agent models*, each one composed by an *agent template* and an *agent behavior model*. For that, the following tasks are performed:

- *Identification of a detailed design pattern and definition of the agent type.* Design patterns describing solutions for the detailed design of each agent in the framework are identified by first selecting patterns whose *context* description refers to the design of the internal architecture of an agent; then the *problem* and *forces* in the selected patterns are matched with the description of the responsibilities in the agent template of each *Agent Model*. After the selection of a pattern, the agents are structured according to the solution proposed by the pattern.

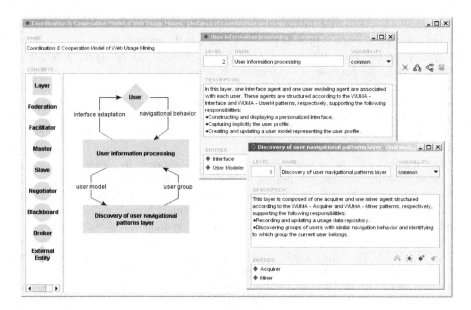

Fig. 2. Coordination and Cooperation Model of the ONTOWUM architectural based on a multi-agent layer pattern

- *Definition of the agent type.* If there is not a reusable solution available for the design of an agent, a specific one must be constructed. In this case, the type of agent (reactive or deliberative) should be selected, establishing the mechanisms for mapping perceptions to agent actions [29] by considering non-functional requirements (e.g. performance);
- *Modeling the agent behavior.* The purpose of this subtask is to specify the behavior of each agent according to the activities specified in the *Agent Society Model* and representing them in an *Agent Behavioral Model* of each agent composed of a *State Model* and an *Activity Model*. Behaviors that can be reused in the implementation phase from a particular implementation platform can also been specified. The graphical notation of the *State* and the *Activity Models* is similar to the corresponding diagram of UML [2].

3.3   Modeling the Knowledge of the Multi-agent Society

The purpose of this sub-phase is to represent the meaning of concepts that agents of the society need to understand in order to communicate with each other. This is done through the construction of a *model of the multi-agent society knowledge*, represented in a semantic network. For that, the techniques specified as resources for the execution of agent activities in the *Agent Society Model* are analyzed and a basic vocabulary is defined for each resource. Each term in the vocabulary is represented as a node in the semantic network. The relationships between the different concepts of techniques are also defined and represented as links in the semantic network.

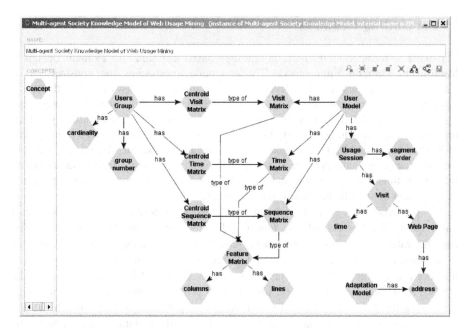

Fig. 3. Multi-agent Society Knowledge Model of the ONTOWUM framework

Note that alternative resources originate alternative semantic networks. When a domain is related to a knowledge area besides a problem-solving area, the concept model specified in the domain analysis phase must also be mapped to the semantic network representing the knowledge of the multi-agent society.

Fig. 3 shows an example of *Multi-agent Society Knowledge Model*, which is part of the framework ONTOMUW [26] and specifies the semantics of concepts involved in a web usage mining process.

For the discovery of groups of similar users over unlabeled usage data, a clustering technique is used [30]. For that, it is necessary to match each individual user model with each other in order to construct groups with the most similar users. For the construction of the individual user models to be further matched, it is used the Feature Matrix model [30]. Note that these concepts are captured through an analysis of the Feature Matrix model specified as a resource of the User Modeler and Miner agents of the *Agent Society Model* of the domain model of ONTOWUM [20].

4 Related Work

Several techniques for Domain Engineering [1][7][22] and development of multi-agent systems [3][4][6][8][27][28][31] were analyzed and have influenced in different aspects the definition of the DDEMAS technique.

Two main features distinguish DDEMAS from other existing approaches. First, it provides support for the construction of reusable agent-oriented software artifacts, and second, it is a knowledge-based technique where models of agents and frameworks are represented as instances of the ONTOMADEM ontology.

5   Concluding Remarks

This work introduced DDEMAS, an ontology-based technique for Domain Design in Multi-agent Domain Engineering. The technique approaches the construction of frameworks to be reused on the development of multi-agent software applications.

Frameworks are embedded in a knowledge base and created through the instantiation of the hierarchy of classes of ONTOMADEM, an ontology which represents the knowledge of MADEM. This is a methodology which integrates DDEMAS and GRAMO, a technique for Domain Analysis in Multi-agent Domain Engineering. Thus, concepts are semantically related allowing effective searches and inferences thus facilitating the understanding and reuse of the models during the development of specific applications in a domain.

Using MADEM, a case study has been developed where a domain model and a multi-agent framework of a family of multi-agent applications for Web personalization based on usage mining have been constructed [20][26]. From this experience, a system of architectural and design patterns for that problem-solving area has been extracted [19] and classified in the ONTOPATTERN ontology [21]. ONTOPATTERN collects general and specific problem solving patterns for agent-oriented software development. These models and patterns are being reused on the development of recommendation systems for the juridical domain according to the techniques for Multi-Agent Application Engineering we are currently developing.

Acknowledgments

This work is supported by CNPq, an institution of Brazilian Government for scientific and technologic development.

References

1. Arango, G.: Domain Engineering for Software Reuse. Ph.D. Thesis. Department of Information and Computer Science, University of California, Irvine (1988).
2. Booch, G., Rumbaugh, J. and Jacobson, I.: Unified Modeling Language User Guide. Reading: Addison Wesley (1999).
3. Bresciani, P., Giorgini, P., Giunchiglia, F., and Mylopoulos, J., and Perini, A.: TROPOS: An Agent-Oriented Software Development Methodology. In Journal of Autonomous Agents and Multi-Agent Systems, Kluwer Academic Publishers Volume 8, Issue 3, May (2004) 203 - 236
4. Caire, G., et al.: Agent-Oriented Analysis using MESSAGE/UML. In: Second International Workshop on Agent-Oriented Software Engineering, AOSE 2001 (2001) pp. 101-108.
5. Chandrasekaran, B., Josephson, J. R., and Benjamins, V. R.: What are Ontologies, and why do we need them? IEEE Intelligent Systems, Vol. 14 no.1, January (1999) pp. 20-26.
6. Cossentino, M., Sabatucci, L., Sorace, S. and Chella, A.: Patterns reuse in the PASSI methodology. In: Proceedings of the Fourth International Workshop Engineering Societies in the Agents World (ESAW'03), Imperial College London, UK. October (2003) pp. 29-31.

7. Czarnecki, K., Eisenecker, U. W.: Generative Programming: Methods, Tools, and Applications. ACM Press/Addison-Wesley Publishing Co., New York, NY (2000).
8. Dileo, J., Jacobs, T. and Deloach, S.: Integrating Ontologies into Multi-Agent Systems Engineering. Proceedings of 4th International Bi-Conference Workshop on Agent Oriented Information Systems (AOIS 2002), pp. 15-16, Bologna (Italy), July (2002).
9. Demazeu, Yves; Muller, Jean-Pierre. Descentralized Artificial Intelligence. North-Holland: Elsevier Science Publishers (1990) pp.3-131.
10. Deugo, Dwight; Weiss, Michael; Kendall, Elizabeth. Reusable Patterns for Agent Coordination. Computer Systems Engineering, Royal Melbourne Institute of Technology. Australia, (1998).
11. Ferber, Jacques. Multi-Agent Systems: An Introduction to Distributed Artificial Intelligence, ed.Addison-Wesley (1999).
12. Ferreira, S. and Girardi, R.: Specification of a Generic Ontology for the Domain Design of Multi-agent Applications. In the Proceedings of the 3° Software Engineering Workshop (WIS 2003), Chilean Journeys of Computer Science 2003, University of Bio-Bio. Chilán, Chile. November (2003). (In Portuguese)
13. Gennari, J., Musen, M. A., Fergerson, R. W. et al.: The Evolution of Protégé: An Environment for Knowledge-Based Systems Development. Technical Report SMI-2002-0943. (2002). Available at http://smi.stanford.edu/pubs/SMI_Abstracts/SMI-2002-0943.html
14. Girardi, R., Oliveira, I, and Bezerra, G.. "Towards a System of Patterns for the Design of Agent-based Systems", Proceedings of The Second Nordic Conference on Pattern Languages of Programs" (VikingPLoP 2003). Bergen, Norway. (2003).
15. Girardi, R. and Faria, C.: A Generic Ontology for the Specification of Domain Models, Proceedings of 1st International Workshop on Component Engineering Methodology (WCEM'03) at Second International Conference on Generative Programming and Component Engineering, Ed. Sven Overhage and Klaus Turowski, pp. 41-50. Erfurt, Germany. September 22-25 (2003).
16. Girardi, R., Faria, C. and Marinho, L.: Ontology-based Domain Modeling of Multi-Agent Systems. Proceedings of the Third International Workshop on Agent-Oriented Methodologies at International Conference on Object-Oriented Programming, Systems, Languages and Applications (OOPSLA 2004), Ed. Cesar Gonzalez-Perez, pp. 51-62. Vancouver, Canada. October 24th to 28th, (2004).
17. Girardi, R., and Faria, C.: An Ontology-Based Technique for the Specification of Domain and User Models in Multi-Agent Domain Engineering. CLEI Electronic Journal, V. 7, N. 1, Pap. 7, June (2004).
18. Girardi, R, Serra, I.: Using Ontologies for the Specification of Domain-Specific Languages in Multi-Agent Domain Engineering. Proceedings of the Sixth International Bi-Conference Workshop on Agent-oriented Information Systems (AOIS-2004) at The 16th International Conference on Advanced Information Systems Engineering (CAISE'04), Ed. Janis Grundspenkis and Marite Kirikova (Eds.), pp. 295-308. Riga, Latvia. June 07 -11 (2004).
19. Girardi, R., Marinho, L. and Oliveira, I.: A System of Agent-based Patterns for User Modeling based on Usage Mining. Interacting with Computers, Elsevier, v. 17, n. 5, pp. 567-591, Sept. (2005).
20. Girardi, R. and Marinho, L., An Ontology-based Domain Model for Usage Mining-based Web Personalization Systems, submitted paper (2005).
21. Girardi, R. and Lindoso, A.. Using ontologies for the representation and reuse of software patterns. In: ECOOP 2005 1ST WORKSHOP ON BUILDING A SYSTEM USING PATTERNS: EXAMINE THE ILLUSTRIOUS CLAIM, Glasgow (2005). (to appear)

22. Harsu, M.: A Survey of Domain Engineering. Report 31, Institute of Software Systems, Tampere University of Technology, December (2002).
23. Huhns, N., and Stephens, L. M. Multi-Agent Systems and Societies of Agents, In: Multiagent Systems - A Modern Approach to Distributed Artificial Intelligence, G. Weiss (ed.), The MIT Press (1999).
24. Jennings, N. R., "Commitments and Conventions: The Foundation of Coordination in Multi-Agent Systems", The Knowledge Engineering Review, v. 8, n. 3 (1993) pp. 223—250.
25. Jennings, Nicholas R. Coordination Techniques for Distributed Artificial Intelligence, in O'Hare G M P and Jennings N R (Eds): Foundations of Distributed Artificial Intelligence, London, Wiley (1990) pp 187-210.
26. Marinho, Leandro B.: A Multi-Agent Framework for Usage Mining and User Modeling-based Web Personalization. Master dissertation, Federal University of Maranhão - UFMA – CPGEE (2005). (In Portuguese)
27. Odell, J., Parunak, H.V.D. and Bauer, B.: Extending UML for Agents. Proc. of the Agent-Oriented Information Systems Workshop at the 17th   National Conference on Artificial Intelligence, accepted role, AOIS Workshop at AAAI (2000) pp. 3-17.
28. Omicini, A.: SODA Societies and Infrastructures in the Analysis and Design of Agent-based Systems. Proceedings of the First International Workshop, AOSE 2000 on Agent-Oriented Software Engineering, Limerick, Ireland, January (2001) pp. 185-193,.
29. Russell, S. and Norvig, P.: Artificial Intelligence: A Modern Approach. Prentice-Hall (1995).
30. Shahabi, C. and Banaei-Kashani, F.: Efficient and Anonymous Web Usage Mining for Web Personalization. INFORMS Journal on Computing - Special Issue on Data Mining, Vol.15, No.2, Spring (2003).
31. Wooldridge, M., Jennings, N. and Kinny, D.: The Gaia Methodology for Agent-Oriented Analysis and Design. In the International Journal of Autonomous Agents and Multi-Agent Systems, v. 3 (2000).

An Agent-Oriented Meta-model for Enterprise Modelling

Ivan Jureta and Stéphane Faulkner

Information Management Research Unit,
University of Namur,
8 Rempart de la vierge, B-5000 Namur, Belgium
{ivan.jureta, stephane.faulkner}@fundp.ac.be

Abstract. This paper proposes an agent-oriented meta-model that provides rigorous concepts for conducting enterprise modelling. The aim is to allow analysts to produce an enterprise model that precisely captures the knowledge of an organization and of its business processes so that an agent-oriented requirements specification of the system-to-be and its operational corporate environment can be derived from it. To this end, the model identifies constructs that enable capturing the intrinsic characteristics of an agent system such as autonomy, intentionality, sociality, identity and boundary, or rational self-interest; an agent being an organizational actor and/or a software component. Such an approach of the concept of agent allows the analyst to have a holistic perspective integrating human and organizational aspects to gain better understanding of business system inner and outer modelling issues. The meta-model takes roots in both management theory and requirements engineering. It helps bridging the gap between enterprise and requirements models proposing an integrated framework, comprehensive and expressive to both managers and software (requirements) engineers.

1 Introduction

Business analysts and IT managers have advocated these last fifteen years the use of enterprise models to specify the organizational and operational environment (outer aspects of the system) in which a corporate software will be deployed (inner aspects of the system) [20]. Such a model is a representation of the knowledge an organization has about itself or of what it would like this knowledge to be. This covers knowledge about functional aspects of operations which describe what and how business processes are to be carried and in what order; informational aspects that describe what objects are to be processed; resource aspects that describe what or who performs these processes according to what policy; organizational aspects that describe the organization architecture within which processes are to be carried out ; and finally, strategic aspects that describe why processes must be carried out. The specification of these key aspects of the core business of an enterprise is an effective tool to consider for gathering and eliciting software requirements. It may be used to [1, 4]:

- analyze the current organizational structure and business processes in order to reveal problems and opportunities;
- evaluate and compare alternative processes and structures;

J. Akoka et al. (Eds.): ER Workshops 2005, LNCS 3770, pp. 151 – 161, 2005.

- achieve common understanding and agreement between stakeholders (e.g., managers, owners, workers, etc.) about different aspects of the organization;
- reuse knowledge available in the organization.

This paper proposes an integrated agent-oriented meta-model for enterprise modelling. The agent paradigm is a recent approach in software engineering that allows developers to handle the life cycle of complex distributed and open systems required to offer open and dynamic capabilities in the latest generation enterprise software (see e.g., [22]).

The proposed meta-model takes inspiration from research works in requirements engineering frameworks (see e.g., [3, 5]), management theory concepts found to be relevant for enterprise modelling (see e.g., [11, 12, 13]) and agent oriented software engineering (see e.g., [22]). It leads to reduce the semantic gap between enterprise and requirements representations, providing a modelling tool that integrates the outer specification of the system together with its inner specification. Our proposal implicitly suggests a holistic approach to integrate human and organizational issues and gain better understanding of business processes and organisations representation. To this end, we introduce new concepts to enterprise modelling, related to authority, power and interest.

The rest of this paper is organized as follows. Section 2 describes the main concepts of our meta-model. Sections 3 and 4 detail some elements of the meta-model and discuss their relevance for enterprise modelling. Section 5 gives an overview of related works and Section 6 summarizes the results and points to further work.

2 An Agent-Oriented Enterprise Meta-model

The motivation of our proposal is to understand precisely the semantics of the organizational environment of the system and to produce an agent-oriented requirements specification for the software to build. The framework described in this section provides modelling constructs that enable the representation of the autonomy, intentionality, sociality, identity and boundary, and rational self-interest of actors, i.e., agents in the real world and/or software agents. Actors are autonomous as their behaviour is not prescribed and varies according to their dependencies, personal goals and capabilities. They are intentional since they base their actions and plans on beliefs about the environment, as well as on goals they have to achieve. Being autonomous, actors can exhibit cooperative behaviour, resulting from similar goals and/or reciprocal dependencies concerning organizational roles they assume. The dependencies can either be direct or mediated by other organizational roles. Actors can have competing goals which lead to conflicts that may result from competing use of resources. Actors have varying power and interest in the ways in which organizational goals contribute to their personal ones. Boundary and identity are closely related to power and interest of actors. We model variations in boundary and identity as resulting from changes in power and interest since these vary with respect to the modifications in the roles an actor assumes and the dependencies involving these roles. Actors can act according to their self-interest, as they have personal goals to achieve. They have varying degrees of motivation to assume organizational roles, according to the degree of contribution to personal goals these roles have in achieving organizational ones. Actors apply plans according to the rationale described in terms of personal, organizational goals, and capabilities. The rationale of our actors is not perfect, but bounded [10, 15], as they can act based on beliefs that are incomplete and/or inconsistent with reality. We provide constructs such as AndOr relationships, non-functional requirements [22]… to evaluate alternative deployments of the software in the organizational environment.

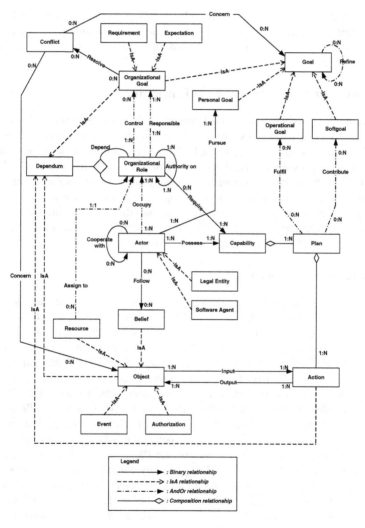

Fig. 1. The agent-oriented meta-model

Fig. 1 introduces the main entities and relationships of our meta-model. For clarity, we have subdivided it into five sub-models:

- *Organizational sub-model*, describing the actors of the organization, their organizational roles, responsibilities and capabilities.
- *Goals sub-model*, describing enterprise and business process purposes, i.e. what the actors are trying to achieve and why.
- *Conflict sub-model*, indicating inconsistencies in the business process.
- *Process sub-model*, describing how actors achieve or intend to achieve goals.
- *Objects sub-model*, describing non-intentional entities and assumptions about the environment of the organization and the business processes.

Due to a lack of place, the paper only details the organizational and goal sub-models, their integration and discusses their relevance for enterprise modelling. We first sketch the meta-model from the point of view of system developers and of organization managers.

2.1 Information System Development Perspective

The meta-model provides widely-used constructs for specifying the architecture of an agent-oriented information system: *Actors* are agents of the system. They *possess Capabilities* composed of *Plans*, each *Plan* representing a sequence of atomic *Actions*. When applying *Plans*, *Actors fulfil* or *contribute* to system *Goals*. *Actors follow Beliefs* which represent assertions about aspects of the organization and/or its environment. *Actions* can take *Objects* as *input* from the system or its environment. New *Objects* can be produced or existing ones modified by carrying out *Actions*, i.e., they can be *output* from *Actions*. *Objects* represent any thing of interest for the system: *Resources, Beliefs, Authorizations* or *Events*.

2.2 Management Perspective

The meta-model provides common terms used to describe an organization. *Organizational Roles* are *responsible* of *Organizational Goals*, which may be either *Operational* (i.e. can be actually fulfilled) or *Softgoals* (such as e.g., broadly specified business objectives). *Organizational Roles* can *depend* on one another for the provision of *Dependums - Actions, Objects*, or *Organizational Goals*. An *Actor*, being a *Legal Entity* or a *Software Agent*, can *occupy Organizational Roles*, as long as it *possesses* the *required Capabilities* to do so. *Actors* exhibit intentional behaviour as they act according to *Goals* and *Beliefs* about their environment. As *Beliefs* may be incoherent, and as they pursue *Personal Goals*, *Actors* can exhibit competitive behaviour. They will exhibit cooperative behaviour when they are responsible of identical *Organizational Goals*. *Actors* execute *Plans*, composed of *Actions*, in order to fulfil and contribute to *Goals*. By doing so, they comply with the responsibilities of *Organizational Roles* they *occupy*. As a matter of organizational policy, *Resources* in the organization are assigned to *Organizational Roles*. The allocation of Resources is determined by both *authority* among *Organizational Roles* and *Authorizations* that may be *input* or *output* of specific *Actions*.

Common ground between both points of view resides in the sense that the information system can be developed to automate some (part of) business processes (e.g., administrative tasks) or to radically modify ways in which *Goals* are fulfilled (e.g., reorganizing customer relationship management by deploying e-commerce facilities). The model provides an unambiguous representation serving both software staff and organization strategic management.

Primitives of our framework are of different types: meta-concepts (*Goal, Actor, Object*, etc.), meta-relationships (*possess, require, pursue*, etc.), meta-attributes (*Power, Interest, Motivation*, etc.), and meta-constraints (e.g., "*an actor occupies a position if that actor possesses all the capabilities required to occupy it*").

All meta-concepts, meta-relationships and meta-constraints have the following mandatory meta-attributes:

- *Name*, which allows unambiguous reference to the instance of the meta-concept (e.g. "European Commission" for the Actor meta-concept).
- *Description*, which is a precise and unambiguous description of the corresponding instance of the meta-concept. The description should contain sufficient information so that a formal specification can be derived for use in requirements specifications for a future information system.

3 Organizational Sub-model

The Organizational sub-model is used to identify the relevant *Actors* of the organization, the *Organizational Roles* they *occupy*, the *Capabilities* they *possess*, and the *Dependums* for which *Actors depend on* one another.

3.1 Actor

An Actor applies *Plans* (which are part of his *Capabilities*) to *fulfil* and/or *contribute* to *Organizational Goals* for which the *Organizational Role* he *occupies* is *responsible*, and *Personal Goals* he *pursues* (i.e. wishes to achieve). As the *Actor* exists in a changing environment, it *follows Beliefs* about the environment in order to adapt its behaviour to environmental circumstances.

An *Actor* is either a *Legal Entity* or a *Software Agent*. A *Legal Entity* is used to represent any person, group of people, organizational units or other organizations that are significant to the organization we are modelling, i.e., that have an influence on its resources, its goals, etc. A *Software Agent* is used to represent a software component of an information system(-to-be). An *Actor* can *cooperate with* another *Actor* to fulfil and/or contribute to *Organizational Goals* common to the *Organizational Roles* that each of these *Actors* occupies.

Besides standard meta-attributes, an *Actor* possesses the *Motivation* meta-attribute, whose values describe the degree of motivation of an *Actor* to *occupy* an *Organizational Role*. Values are functions of the degree of contribution to *Personal Goals* the Actor's *Organizational Role* have in achieving *Organizational Goals* and of functions of the conflicts involving this *Actor*.

A *Legal Entity* is characterized with two specific meta-attributes: *Interest* and *Power* [11]. *Interest* is the degree of satisfaction of an actor to see *Organizational Goals* positively contributing to its *Personal Goals*. *Power* is the degree to which the actor is able to modify the objectives of the organization or its business processes through its *Capabilities*. For instance, when automating a business process, the values of *Interest* and *Power* meta-attributes of *Legal Entities* change: in the new configuration of the process, some actors will gain decision power while maintaining the same level of interest; others that previously benefited from high power in the initial process structure might become less powerful. It is crucial to take these changes into account when eliciting software requirements. It may lead otherwise to introducing *Goals* not identified during the initial requirements analysis, and/or changing *Priority* of already specified *Goals*. *Interest* and *Power* help to find *Legal Entities* that will play a crucial role in the software-to-be. For example, focus in some business process might shift to *Legal Entities* which were not considered very significant during the inception phase, and whose needs were not specified in depth. This would result in that these now crucial processes would not be fully exploited, and would lead to the overall failure of the requirements specification efforts.

3.2 Organizational Role

Actors occupy Organizational Roles. They can take many forms: a unique functional position (e.g. the Project Manager), a unique functional group (e.g. the Marketing Department), a rank or job title (e.g. the CIO), a class of persons (e.g. Customer), etc. *Organizational Roles* are responsible of *Organizational Goals*. They cannot be responsible of *Personal Goals*. Each *Organizational Role* requires a set of *Capabilities* which can be used to fulfil or contribute to *Organizational Goals* for which it is responsible. *Organizational Roles* can be attributed only to those *Actors* that possess all the capabilities required to occupy these *Organizational Roles*.

Organizational Roles can have different levels of authority. Consequently, an *Organizational Role* can have *authority on* another *Organizational Role*. The *authority on* meta-relationship specifies the hierarchical structure of the organization.

3.3 Capability

A *Capability* is a set of *Plans* an *Actor* can execute. An *Actor possesses Capabilities*.

When exploring possible alternative business processes or organizational structures, newly identified *Organizational Roles* can *require Capabilities* that no *Actor possess*. These *Capabilities* have to be confronted to those available in the organization, in order to evaluate the proposed alternatives with respect to the current *Roles* and the way they use existing *Capabilities*. This is significant to determine which and how proposed *Capabilities* and *Roles* will be finally introduced through the system-to-be.

3.4 Dependum

An *Organizational Role depends* on another *Organizational Role* for a *Dependum*, so that the latter may provide the *Dependum* to the former. A *Dependum* can be an *Organizational Goal*, an *Object*, or an *Action*. In the *depend* meta-relationship, the *Organizational Role* that depends on is called the depender, and the *Organizational Role* being depended upon is called the dependee. We define the following dependency types:

- *Organizatonal Goal*-dependency: the depender *depends* on the dependee to *fulfill* and/or *contribute* to an *Organizational Goal*. The dependee is given the possibility to choose *Plans* through which it will *fulfill* and/or *contribute* to the *Organizational Goal*.

- *Action*-dependency: The depender *depends* on the dependee to accomplish some specific *Action*.
- *Object*-dependency: The depender *depends* on the dependee for the availability of an *Object*.

The *depend* on relationship differs from the dependency relationship in *i** [3] in several aspects. In our meta-model, dependencies are not among *Actors*, but among *Organizational Roles*. *Organizational Roles* are independent concepts and separated from *Actors*. They only *occupy* them, in order to enable changing of *Actors* in *Organizational Roles* without reviewing the entire process or the organizational structure. For example, an *Actor* a_1 can *occupy* some *Organizational Role* r; but if in some point in the future the organization has access to some *Actor* a_2 which can provide better performance in terms of fulfilling and/or contributing to *Organizational Goals* for which *Organizational Role* r is *responsible*, then *Actor* a_2 might be chosen to *occupy* the *Organizational Role* r instead of *Actor* a_1. This replacement would be done without reviewing the entire business process and/or organizational structure – it is sufficient to replace a_1 by a_2 in the *occupy* relationship of r.

The *Object*-dependency allows us to represent any specialization of the *Object* meta-concept as a *Dependum*. For example, an *Organizational Role* r_1 might *depend* on another *Organizational Role* r_2 for an *Authorization*. This has implications on the *authority* on meta-relationship, as this dependency means that $r2$ must have *authority on* r_1.

4 Goals Sub-model

A *Goal* describes a desired or undesired state of the environment. A state of the environment can be described through the states of *Objects* (*Beliefs*, *Authorizations*, *Resources*, etc.). In addition to standard meta-attributes, a *Goal* is characterized by the optional *Priority* meta-attribute, which specifies the extent the goal is optional or mandatory.

A *Goal* can be *refined* [7] into alternative sets of other *Goals*. Each such set is identified through goal refinement. Informally, goal refinement consists of asking "how" questions about a *Goal* G in order to find alternative sets of *Goals*. Each alternative set of *Goals* that *refine* G provides an alternative way of *fulfilling* and/or *contributing* to G in such a way that the *fulfilment* and/or *contribution* to all of the *Goals* in the set *fulfils* and/or *contributes* to G. Goal refinement is introduced in the model using the *refine* meta-relationship.

The *refine* meta-relationship is an *AndOr* relationship, making it possible to show (directly in the model) alternative refinements of a *Goal*. The *refine* meta-relationship is characterized with an *Alternative Name* and *Alternative Status* meta-attributes. *Alternative Status* indicates whether the alternative is sufficient or not to *fulfil* the *Operational Goal* it *refines*.

Goal types are defined along two axes: *Operational Goals vs. Softgoals* and *Organizational Goals vs. Personal Goals*. In addition, we use patterns to specify the temporal behaviour of *Goals*.

Operational Goals vs. Softgoals. An *Operational Goal* is a set of *Objects* (*Beliefs, Resources*, etc.) describing the environment state that can be achieved by *Plans*. We can always determine whether an *Operational Goal* has been *fulfilled* or not by verifying whether the environment state described by the *Operational Goal* has or has not been achieved.

An *Operational Goal* has *State* and *Status* optional meta-attributes. *State* explicitly describes (in terms of *Objects*) the environment in which the *Operational Goal* is fulfilled. *Status* indicates whether the *State* of the *Operational Goal* has been reached, i.e. whether the *Goal* has been fulfilled or not.

A *Softgoal* describes the environment state which can never be achieved since its achievement criteria are not objective. This makes it impossible to formally verify whether a *Softgoal* has been achieved. *Plans* that are otherwise applied to *fulfil Operational Goals* can only *contribute* (positively or negatively) to *Softgoals*. For example, "increase customer satisfaction", "implement a flexible IS", "improve productivity of the workforce", are *Softgoals*.

Organizational Goals vs. Personal Goals. An *Organizational Goal* describes the state of the environment that should be achieved by cooperative behaviour of *Actors*. An *Organizational Goal* is either a *Requirement* or an *Expectation*. A *Requirement* is an *Organizational Goal* under the responsibility of an *Organizational Role occupied* by a *Software Agent*. An *Expectation* is an *Organizational Goal* under the responsibility of an *Organizational Role occupied* by a *Legal Entity*.

Organizational Goals can solve *Conflicts* by specifying the state of the environment in which the *Conflicts* cannot be true.

A *Personal Goal* describes the state of the environment that an *Actor pursues* (wants to obtain) and which can require competitive behaviour among *Actors*.

Organizational Roles are *responsible* of *Organizational Goals*, and *Actors pursue Personal Goals*, i.e., we distinguish what is expected from an actor's participation in the process (through the *Organizational Role* it *occupies*), from what the *Actor* expects from his participation in the process (*fulfilment* of or *contribution* to its *Personal Goals*). In reality, consistency between the *Organizational Goals* and *Personal Goals* is not necessarily ensured. Consequently, it is important to reason about *Conflict* that may arise between *Personal* and *Organizational Goals*, as well as about the degree to which an *Organizational Goal* assists in the *Actor's* pursuit of *Personal Goals*. We use *fulfil* and *contribute* meta-relationships to show how *Plans fulfil* and *contribute* to both *Personal Goals* that the *Actor pursues* and *Organizational Goals* for which its *Organizational Roles* are *responsible*.

5 Related Works

Process-Oriented Approaches such Activity Diagrams, DFDs, IDEF0, workflows (see e.g., [2, 14, 15, 16]) describe enterprise's business processes as sets of activities. Strong emphasis is put on the activities that take place, the order of activity invocation, invocation conditions, activity synchronization, and information flows. Among these approaches, workflows have received considerable attention in the literature. In such kind of process-oriented approaches, agents have been treated as a computational paradigm, with focus on the design and implementation of agent systems, not analysis on enterprise models.

Actor-Oriented Approaches emphasize the analysis and specification of the role of the actors that participate in the process [17]. The $i*$ modelling framework [3] has been proposed for business process modelling and reengineering. Processes, in which information systems are used, are viewed as social systems populated by intentional actors which cooperate to achieve goals. The framework provides two types of dependency models: a strategic dependency model used for describing processes as networks of strategic dependencies among actors, and the strategic rationale model used to describe each actor's reasoning in the process, as well as to explore alternative process structures. The diagrammatic notation of $i*$ is semi-formal, and proved useful in requirements elicitation (see e.g., [18, 12, 19]). In this context, actor-oriented approaches provide significant advantages over other approaches: agents are autonomous, intentional, social, etc. [21] which is of particular importance for the development of open distributed information systems in which change is ongoing. However, actors have served mostly as requirements engineering modelling constructs for real-world agents, without assuming the use of agent software as the implementation technology nor the use of organizational actors for enterprise modelling.

Goal-Oriented Approaches focus on goals that the information system or a business process should achieve. Framework like KAOS [5, 8] provides a formal specification language for requirements engineering, an elaboration method, and meta-level knowledge used for guidance while the method is applied [6]. The KAOS specification language provides constructs for capturing the various types of concepts that appear during requirements elaboration. The elaboration method describes steps (i.e. goal elaboration, object capture, operation capture, etc. [6]) that may be followed to systematically elaborate KAOS specifications. Finally, the meta-level knowledge provides domain-independent concepts that can be used for guidance and validation in the elaboration process.

Enterprise Knowledge Development (EKD) [17] is used primarily in modelling of business processes of an enterprise. Through goal-orientation, it advocates a closer alignment between intentional and operational aspects of the organization and links re-engineering efforts to strategic business objectives. EKD describes a business enterprise as a network of related business processes which collaboratively realise business goals. This is achieved through several sub-models: enterprise goal sub-model (expressing the causal structure of the enterprise), enterprise process sub-model (representing the organizational and behavioural aspects of the enterprise), and information system component sub-model (showing information system components that support the enterprise processes) [17]. Agents appear in the EKD methodology but without explicit treatment of their autonomy and sociality [21]. In KAOS, agents interact with each other non-intentionally, which reduces the benefits of using agents as modelling constructs.

6 Conclusion

Modelling the organizational and operational context within which a software system will eventually operate has been recognized as an important element of the engineering process (e.g., [20]). Such models are usually founded on primitive concepts such as those of *actor* and *goal*. Unfortunately, no specific enterprise modelling framework really exists for engineering modern corporate IS. This paper proposes an integrated agent-oriented meta-model for enterprise modelling. Moreover, our approach differs primarily in the fact that it is founded on ideas from in requirements engineering frameworks, management theory concepts found to be relevant for enterprise modelling and agent oriented software engineering.

We have only discussed here the concepts that we consider the most relevant at this stage of our research. Further classification of, for instance, goals is possible and can be introduced optionally into the meta-model. For example, goals could be classified into further goal categories such as Accuracy, Security, Performance, etc. We also intend to define a strategy to guide enterprise modelling using our meta-model as well as to define a modelling tool à la Rational Rose to visually represent the concepts.

References

1. Koubarakis M., Plexousakis D.: A formal framework for business process modelling and design. Information Systems 27, 2002, pp. 299-319.
2. Kamath M., Dalal N.P., Chaugule A., Sivaraman E., Kolarik W.J.: A Review of Enterprise Process Modelling Techniques. In Prabhu V., Kumara S., Kamath M.: Scalable Enterprise Systems: An Introduction to Recent Advances. Kluwer Academic Publishers, Boston, 2003.
3. Yu E.: Modelling Strategic Relationships for Process Reengineering. Ph.D. thesis, Dept. of Computer Science, University of Toronto, 1994.
4. Bernus P.: Enterprise models for enterprise architecture and ISO9000:2000. Annual Reviews in Control 27, 2003, pp. 211–220.
5. Dardenne A., van Lamsweerde A., Ficklas S.: Goal-directed requirements acquisition. Sci. Comput. Programming 20, 1993, pp. 3-50.
6. van Lamsweerde A., Darimont R., Letier E.: Managing Conflicts in Goal-Oriented Requirements Engineering. IEEE Transactions on Software Engineering, Special Issue on Managing Inconsistency in Software Development, 1998.
7. van Lamsweerde A.: Goal-Oriented Requirements Engineering: A Guided Tour. Proceedings RE'01, 5th IEEE International Symposium on Requirements Engineering, Toronto, 2001, pp. 249-263.
8. van Lamsweerde A.: The KAOS Meta-Model –Ten Years After. Technical report, (2003).
9. Simon H. A.: Administrative Behavior : A Study of Decision-Making Processes in Administrative Organization. New York: The Free Press 3rd ed., 1976.
10. Johnson G., Scholes K.: Exploring Corporate Strategy, Text and Cases. Prentice Hall, 2002.
11. Brickley J.A., Smith C.W., Zimmerman J.L.: Managerial Economics and Organization Architecture. McGraw-Hill Irwin 2nd ed., 2001.
12. Faulkner S., Kolp M., Coyette A., Tung Do T.: Agent-Oriented Design of E-Commerce System Architecture. Proceedings of the 6th International Conference in Enterprise Information Systems Engineering, Porto, 2004.
13. Simon H. A.: Rational Decision Making in Business Organizations. The American Economic Review vol.69 no.4, 1979, pp. 493-513.
14. Elmagarmid A., Du W.: Workflow Management: State of the Art Versus State of the Products. In Dogac A., Kalinichenko L., Tamer Ozsu M., Sheth A.: Workflow Management Systems and Interoperability. NATO ASI Series, Series F: Computer and Systems Sciences 164, Springer Heidelberg, 1998.

15. Mentzas G., Halaris C., Kavadias S.: Modelling business processes with workflow systems: an evaluation of alternative approaches. International Journal of Information Management 21, 2001, pp. 123-135.
16. Sheth A.P., van der Aalst W., Arpinar I.B.: Processes Driving the Networked Economy. IEEE Concurrency 7, 1999, pp. 18-31.
17. Kavakli V., Loucopoulos P.: Goal-Driven Business Process Analysis Application in Electricity Deregulation. Information Systems 24, 1999, pp. 187-207.
18. Liu L., Yu E.: Designing information systems in social context: a goal and scenario modelling approach. Information Systems 29, 2004, pp. 187–203.
19. Briand L., Melo W., Seaman C., Basili V.: Characterizing and Assessing a Large-Scale Software Maintenance Organization. In Procedings of the 17th International Conference on Software Engineering, Seattle, WA, 1995.
20. Castro J., Kolp M. and Mylopoulos J.: Towards Requirements-Driven Information Systems Engineering: The Tropos Project. In Information Systems (27), Elsevier, Amsterdam, 2002.
21. Yu E.: Agent-Oriented Modelling: Software Versus the World. Proceedings of the Agent-Oriented Software Engineering AOSE-2001 Workshop, Springer Verlag, 2001.
22. Chung L.K., Nixon B.A., Yu E., Mylopoulos J.: Non-Functional Requirements in Software Engineering. Kluwer Publishing, 2000.

An Approach to Broaden the Semantic Coverage of ACL Speech Acts

Hong Jiang and Michael N. Huhns

Department of Computer Science and Engineering,
University of South Carolina, Columbia, SC 29208, USA
{jiangh, huhns}@engr.sc.edu

Abstract. Current speech-act based ACLs specify domain-independent information about communication and relegate domain-dependent information to an unspecified content language. This is reasonable, but the ACLs cover only a small fraction of the domain-independent information possible. As a key element of modern ACLs, the set of communicative acts needs to be as complete as possible to enable agents to communicate the widest range of information with agreed-upon semantics. This paper describes a new approach to broaden the semantic coverage of ACL speech acts. It provides agents with the ability to express more of the possible meanings in human languages and yields a more powerful ACL. Specifically, we first compare Austin's and Searle's classifications, and Ballmer and Brennenstuhl's comprehensive classification of speech acts. The main meaning categories and their semantics are given next. Finally, a multifaceted evaluation of our approach is presented, which points out that the approach potentially can combine the benefits of the FIPA ACL with Ballmer and Brennenstuhl's speech act classification, resulting in a more expressive ACL.

1   Introduction

Agent communication languages (ACLs) are a critical element of multiagent systems and a key to the successful application of agents in commerce and industry. Modern ACLs, such as the FIPA ACL, provide a standardized set of performatives denoting types of communicative actions. Such ACLs have been designed as general purpose languages to simplify the design of multiagent systems. However, recent research shows that these ACLs do not support adequately all relevant types of interactions. Serrano and Ossowski [1] report a need for new *ad hoc* sets of performatives in certain contexts, which the FIPA ACL does not support. Singh [2] points out that agents from different venders or even different research projects cannot communicate with each other. In [3], Kinny shows that FIPA reveals a confusing amalgam of different formal and informal specification techniques whose net result is ambiguous, inconsistent, and certainly underspecified communication. He proposes a set of requirements and desiderata against which an ACL specification can be judged, and briefly explores some of the shortcomings of the FIPA ACL and its original design basis.

J. Akoka et al. (Eds.): ER Workshops 2005, LNCS 3770, pp. 162–171, 2005.
© Springer-Verlag Berlin Heidelberg 2005

Therefore, a complete set of speech acts as communicative acts in an ACL would be desirable in order to improve understanding among the agents in a multiagent system. Recognizing that the ~4800 speech acts in [8] would be desirable but impractical to use individually, we describe a feasible approach to broaden the semantic coverage of ACLs by formalizing speech act categories that subsumes the ~4800, enabling the meanings of all the speech acts to be conveyed. Different from [11], we focus on the standard messages used for communication instead of designing a conversation protocol.

Specifically, Section 2 compares Austin's, Searle's, and Ballmer and Brennenstuhl's classifications of speech acts. Based on an abstract model that separates protocols, agent types, and decision mechanisms from communications, Section 3 describes the main meaning categories. In Section 4, we use FIPA's formal semantic language to represent the semantics of our speech act categories. This enables our approach to combine the benefits of the FIPA ACL with a broader set of speech acts. An evaluation of this approach is discussed in Section 5.

2   Comparison of Austin's, Searle's and Ballmer's Classification

Current ACLs derive their language primitives from the linguistic theory of speech acts. The original speech act theory was developed by Austin [4]. The most important part of Austin's work was to point out that human natural language can be viewed as *actions* and people can perform things by saying. According to his theory, a speech act has three aspects, as summarized in [7]: Locution, Illocution, and Perlocution. Austin also tried to classify speech acts. He classified illocutionary acts as verfictives, exercitives, commissives, behabitives, and expositives [4]. The classification has been criticized for overlapping categories, too much heterogeneity in categories, ambiguous definitions of classes, and misfits between the classification of verbs and the definition of categories [8, 13, 12].

Austin's work was extended by Searle [5, 14, 6, 13], who posited that an illocutionary speech act forms the minimum meaningful unit of language. He classified speech acts into five categories: Assertives, Directives, Commissives, Declaratives, and Expressives. Searle's speech act theory focuses on the speaker. The success of a speech act depends on the speaker's ability to perform a speech act that should be understandable and successful.

Ballmer and Brennenstuhl [8, 12] criticize the clarity of Searle's classification, definition of declaratives as a speech act type, principles used in the classification, selection of illocutionary verbs from all verbs, vague definition of the illocutionary point, and vagueness of the line between illocutionary force and propositional content. Based on their criticism, they propose an alternative classification of speech acts, which contains both simple linguistic functions such as expression and appeal, and more complex functions such as interaction and discourse. Models for alternative actions are formed and verbs are classified according to the phases of the model.

Ballmer and Brennenstuhl's classification motivates us to rethink the speech acts used in ACLs. Since the classification is based on an almost complete domain (~4800 speech acts) and the authors claim they provide a "theoretically justified" classification that is "based explicitly and systematically on linguistic data," we believe that to generate a speech act set for ACLs based on their classification will be a powerful way to represent meaning. However, this classification is not perfect: the classification for English is obtained by translating the verbs of the German one, the names of the categories are not systematically chosen, and there are no formal semantic representation for the categories. However, by rebuilding the categories, most of above problems can be fixed. Thus, we endeavor herein to derive a reasonable set of categories for agent communication from their theory, and to give a formal semantics using more typical English names.

3   Method Description

3.1   Abstract Communication Model

Based on current popular communication models, we generalize an abstract communication model for agents consisting of agents and environments. An environment constrains the agents and affects the communications among them via a message control mechanism that provides protocols, routing, and message delivery. This gives flexibility to a multiagent system at an abstract level, in which the message control mechanism could be any one of many possibilities.

Following similar choices made for KQML and the FIPA ACL, the mechanism specifies from whom did a message originate, to whom should the message be sent, what information is being communicated, and how should the message be delivered. The "what" is separated into a communicative act and content, where the communicative act reflects domain *independent* information to be communicated, and content reflects domain *dependent* information. The message may also separately specify the language used in the content, encoding and decoding functions, ontology, and protocol.

We focus on the communicative acts. They should be in a formal form, as FIPA provides well. Further, they should cover all the possible domain-independent meanings in human communication, as we address next.

3.2   Overview of Meaning Categories

This section describes semantic categories for a relatively complete set of human speech activity verbs, derived from the classification in [8]. The categories reflect an ontological and a conceptual structuring of linguistic behavior. The main categories and their relationships are represented in Figure 1. The topmost node, Speech Acts, represents the entire set of speech acts in human language. The four major groups—Emotion Model, Enaction Model, Interaction Model, and Dialogic Model—represent four basic functions of linguistic behavior.

The *Emotion Model* is the least hearer-oriented and least extroverted function of the four, and focuses on representing the kinds of emotional states of a human or agent.

The *Enaction Model* is a function clearly directed toward a hearer. In other words, the speaker tries to get control over the hearer.

The *Interaction Model* is a function involving speaker and hearer in mutual verbal actions. Among this group are three sub-categories to represent different degrees of the mutual competition: Struggle Model, Institutional Model, and Valuation Model. In the Struggle Model, the speaker tries to get control over the hearer, or the speaker is more competitive in controlling mutual verbal actions. In contrast, the hearer is more competitive in the Valuation Model. In the Institutional Model, the hearer and speaker are equally competitive.

The *Dialogic* Model covers a kind of reciprocal cooperation where there is a better-behaved and more rigidly organized verbal interaction. Its three subcategories focus on different types of the content and the organization: the Discourse Model focuses on the organization and types of discourse, the Text Model focuses on the textual assimilation and processing of reality, briefly, specific knowledge involved, and the Theme Model focuses on the process of thematic structuring and its results, in other words, the structure or organization of some knowledge system.

Fig. 1. Ontology of the Main Speech Act Categories

In the above ontology, the four basic models can be divided into unilateral and multilateral models. The Emotion Model and Enaction Model are unilateral, because they focus on a single speech action. The Interaction Model and Dialogic Model are multilateral because they consider the response from a hearer. The Emotion Model and Interaction Model are more original and racy, and the Enaction Model and Dialogic Model are more institutionalized and controlled. Practically, these four basic models may be combined.

4   Semantics of Meaning Categories

4.1   Formal Semantic Model Notations

The semantic model used in representing the categories in this paper follows the formal semantic language as described for the FIPA ACL [9]. Components of the formalism are

- $p, p_1, ...$ are closed formulas denoting propositions;
- ϕ, ψ are formula schemes, which stand for any closed proposition;
- i, j are schematic variables denoting agents.

The mental model of an agent is based on four primitive attitudes: belief (what the agent knows or can know); desire (what the agent desires); intention (which is defined as a persistent goal that could lead to some actions); and uncertainty. They are respectively formalized by operators B, D, I, and U:

- $B_i p$ agent i (implicitly) believes (that) p;
- $D_i p$ agent i desires that p currently holds;
- $I_i p$ agent i intends a persistent goal p;
- $U_i p$ agent i is uncertain about p;

To enable reasoning about action, we also introduce operators *Feasible*, *Done*, and *Agent*:

- *Feasible*(a, p) means that an action a can take place and, if it does, then p will be true.
- *Done*(a, p) means that when p is true, then action a takes place.
- *Agent*(i, a) means agent i performs action a.

Generally, the components of a speech act model involved in a planning process should contain both the conditions that have to be satisfied for the act to be planned and the reasons for which the act is selected. The former is named *FP* (feasibility preconditions), and the latter *RE* (rational effect) in FIPA ACL. We use the same model here, which is represented as follows:

$$< i, act \ (j, C) >$$
$$FP : \phi_1 \tag{1}$$
$$RE : \phi_2$$

where i is the agent of the act, or speaker, j the recipient or hearer, act the name of the speech act, C stands for semantic content, and ϕ_1 and ϕ_2 are propositions.

4.2   Emotion Model

The Emotion Model focuses on representing the emotional states of a human or agent. We assume there is a finite set of emotions, E, represented as

$$E = \{e_+, e_0, e_-\} \tag{2}$$

where e_+ is an emotion in the set of positive emotions, which is characterized by or displaying a kind of certainty, acceptance, or affirmation (about the content involved), such as $\{happy, love, ...\}$; e_0 is in the set of neutral emotions, which does not show any tendency, such as $\{hesitate, ...\}$; e_- is in the set of negative emotions, which intend or want to express a kind of negation, refusal, or denial, such as $\{angry, sad, afraid, ...\}$.

Table 1. Foundational Meaning Units of Emotional Speech Acts

+	0	-
happy	N/A	sad
love	N/A	hate
excited	nervous	angry
desire	hesitate	fear
N/A	shocked	N/A

The Emotion Model can be represented as follows:

$$< i,\ em\ (j, \phi) >$$
$$FP : \neg B_i\ (B_j Agent(i, em(\phi))) \wedge D_i(B_j Agent(i, em(\phi))) \qquad (3)$$
$$RE :\ B_j\ Agent(i, em(\phi))$$

where $em \in E$, and the semantic content ϕ could be empty. This model represents that agent i sends a message to j that i has emotion em about ϕ or i is in status of em when ϕ is empty. The FP shows that, when agent i does not believe agent j knows about ϕ, i is currently in emotion em about ϕ, and i desires that j knows it, then this message could be sent. The RE shows that the desired result is that agent j believes that i is in emotion em about ϕ.

To simplify usage of this model, we could directly use e_+, e_0, or e_- as communicative acts. In this case, we focus on the effect of the emotion speech act on the content ϕ. That is, for a positive effect, i hopes j will increase its intention on ϕ; for a negative one, i hopes j will decrease its intention on ϕ; for a neutral one, i shows its attitude is uncertain about ϕ. Just as for human interactions, we do not have to know the precise value of an attitude. Instead, we just need to know that something is viewed favorably, unfavorably, or neutrally.

However, detailed emotions are also desirable in some cases. To make this usable, we generate a set of foundational meaning units from 155 emotion speech acts listed in [8]. Table 1 gives the foundational meaning units of emotion with consideration of positive, neutral, and negative values.

In Table 1, each row represents one kind of meaning unit. In the first row, *sad* has the opposite meaning of *happy*. *Hate* has the opposite meaning of *love* in the second row. *Excited* represents an opposite attitude to something with strong feeling, *nervous* represents a strong uncertain feeling about something, and *angry* represents a strong negative feeling about something. In the fourth row, *desire* shows a feeling to get something, *hesitate* shows no intentions, and *fear* shows a feeling to avoid something. In the last row, *shocked* shows a neutral feeling about surprise.

4.3 Enaction Model

In the Enaction Model, the speaker more or less coercively attempts to get the hearer to do something by expressing an idea, wish, intention, proposal, goal, etc. There are many speech acts in this group. To organize them and simplify the usage, we define the set of enactions as:

$$EN = \{en_+, en_-\} \tag{4}$$

Unlike the Emotion Model, which focuses on presenting a kind of description or knowledge, the Enaction Model tries to make the hearer do something. Thus, there are no neutral enactions: if agent i does not want j to do anything, i does not have to send any message to j. en_+ is an action in the set of positive enactions, such as $\{intend, desire, askfor, encourage, ...\}$; en_- is an action in the set of negative enactions, such as $\{warning, cancel, ...\}$.

The Enaction Model can be defined as:

$$
\begin{aligned}
&< i, \quad en_\pm \quad (j, \phi) > \\
&FP: \neg B_i\phi \wedge D_i\phi \wedge B_i(B_j\phi \wedge \neg D_j\phi) \\
&RE: Agent\ (j, en_\pm(\phi))
\end{aligned}
\tag{5}
$$

where $en_\pm \in EN$. This model represents that agent i sends a message to j to ask j to do $en_\pm$ on ϕ. The FP shows that this message could be sent when i does not believe that i can do ϕ and it desires ϕ. On the other hand, i believes that j can do it, but j does not want to do it. The expected result is j does $en_\pm$ on ϕ. Practically, j could just add the action to its action queue for a positive enaction, or delete it from its action queue for a negative enaction.

4.4   Interaction Model

The Interaction Model is a function involving speaker and hearer in mutual verbal actions. First of all, we assume an interaction set IN, and for some $in_1, in_2 \in IN$, $\exists rule: in_1 \rightarrow in_2$, such that:

$$
\begin{aligned}
&< i, \quad in_1 \quad (j, (a, goal)) > \\
&FP: I_igoal \wedge \neg B_ia \wedge D_ia \wedge B_i(B_ja \wedge \neg D_ja) \\
&RE: Agent\ (j, a) \wedge \\
&\quad (< j, \ in_2(i, (a', goal - a) > \vee < j, in_2(i, +) > \vee < j, in_2(i, -) >)
\end{aligned}
\tag{6}
$$

where a, a' are actions, and $goal$ can be looked as a plan or a sequence of actions. This model represents that agent i sends a message to j to ask j to do action a for some $goal$. The FP shows that i intends to achieve the $goal$, so i desires to do a but can't do it itself, and i believes that j can do it. However, j does not desire to do it. The expected result is j does a first, and then generates another message back to i. This reply message follows the rule $in_1 \rightarrow in_2$. Generally, the message has the form $< j, in_2(i, (a', goal - a) >$, which mentions that after j has done a, it generates another action a' and reduces the $goal$. In some special case, for example after j has done a, the $goal$ is already achieved, then j sends back message $< j, in_2(i, +) >$, where $+$ means the $goal$ is achieved. Another extreme case is that j finds out that the $goal$ is impossible to be achieved, then it sends back message $< j, in_2(i, -) >$, where $-$ means the $goal$ is unachievable.

Among this model group, there are three subcategories to represent different degrees of the mutual competition: Struggle Model, Institutional Model, and Valuation Model. In the Struggle Model, the speaker tries to get control over the hearer, or the speaker is more competitive in controlling mutual verbal actions. In this case, the rule $in_1 \rightarrow in_2$ is decided by the speaker or sender i.

In the Institutional Model, the hearer and speaker are equally competitive. For example, the establishment of a behavior in an institution equally affects the upholders of and the participants in the institution, especially when entering an institution and thereby adopting its norms, following its norms and rules, violating them, and being pursued by the upholders of the institution. Thus, the agents i and j should have some common rule system defined in advance.

In the Valuation Model, the hearer is more competitive, so it decides which communication act will be replied. That is, the rule $in_1 \rightarrow in_2$ is decided by agent j after its evaluation of the previous message. Details of the Valuation Model cover both positive and negative valuations of actions, persons, things, and states of affairs.

4.5 Dialogic Model

The Dialogic Model tends to a kind of reciprocal cooperation, and is a better-behaved and more rigidly organized verbal interaction. For this model, we at first assume a dialogic speech act set DS, and for some actions $d_1, d_2 \in DS$, $\exists rule : d_1 \rightarrow d_2$, such that

$$
\begin{aligned}
&< i, \ d_1 \ (j, \phi) > \\
FP &: \neg B_i \, B_j \phi \wedge D_i B_j \phi \\
RE &: B_j \phi \wedge < j, d_2(i, \phi') >
\end{aligned}
\tag{7}
$$

This model represents that agent i sends a message to j about ϕ. For this message to be sent, agent i does not believe j believes ϕ, and i desires j to believe it. The expected result is that j believes ϕ and j replies to i with another message about a new ϕ, which is the reasoning result of agent j, and the communicative act used in the message follows the rule $d_1 \rightarrow d_2$.

According to the three sub-categories, which focus on different types of content and organization, we can define three types for ϕ:

- The Discourse Model focuses on the organization and types of discourse. In other words, ϕ points to some kind of type or organization that is predefined. For example, according to the status of a discourse, it could be { *beginning discourse, being in discourse, discourse inconvenience, reconciliation of discourse, ending discourse* }; according to the attitude for some content, it could be { *accept, refuse, cancel* }; according to the number of agents involved in the discourse, it could be { *discourse with several speakers, discourse with one speaker, ...* }; or a kind of irony, joke, etc.
- The Text Model focuses on the textual assimilation and processing of reality, briefly, the specific knowledge involved. Or, ϕ focuses on some description

of specific knowledge. For example, it could be perceiving reality, producing texts, promulgating texts, systematically searching for data, etc.
- The Theme Model focuses on the process of thematic structuring and its results, in other words, ϕ points to some structure or organization about some knowledge system.

5   Evaluation

For evaluation of an ACL based on our extended classification of speech acts, we focus on the following five aspects to compare with current ACLs:

Better coverage: By including speech acts with approximately 4800 verbs, while current ACLs include speech acts with 20 to 30 verbs, our approach provides better coverage.

Precise semantics: Precise semantics is an important property for an ACL, and one of the nice features of FIPA is that it provides one for its ACL. We adapt it for the four basic categories and subcategories of our approach.

Easy usage: For practical reasons the ACL must be easy to use. The FIPA ACL already has many successful uses. Instead of replacing it, we substitute our speech acts and keep its message structure. We organize the speech acts as an ontology with different abstract levels, so that a user can more easily navigate through them to choose the desired ones.

Better understood: Easy usage requires that the ACL be well understood. However, the original categories given by Ballmer and Brennenstuhl's classification are very poor on this point, because the classification is obtained by translating German verbs and the names of the categories are not systematically chosen. In our research, we modified their classification by using typical English names, which should be more understandable.

Efficiency: Efficiency is desirable for usage of an ACL. Since we have not yet deployed our approach, we can not evaluate this aspect.

For evaluation, we will have a small group of users encode the conversations among the agents in some scenarios, for example:

"Agent Bob wants to ask agent Sue to a dance, but he doesn't want to call directly, so he decides to find out Sue's intention in advance. If Bob knows that Sue would say 'yes', then he would call; otherwise, he wouldn't bother to call. To avoid embarrassment, he decides to ask Sue's friend agent Jill to find out if Sue is available. Based on what Jill finds out, he will decide to call or not. Meanwhile, Sue wants to go to the dance. She prefers to go with Bob, but will go with agent Jack if he asks her before Bob does."

The users will encode the conversations using both the FIPA ACL and our ACL. We will then survey the users to find out which is easier and preferred, and analyze the resulting agents to see which are better understood, etc.

6   Conclusion and Future Work

In this paper, we first compare Austin's, Searle's, and Ballmer's classification of speech acts. Then we provide an abstract model, which separates protocols, agents types, and decision mechanisms from the communications, so that we can focus on common messages for communication. Based on this model, we describe the semantic categories that are derived from Ballmer and Brennenstuhl's classification, which attempts to represent all possible meanings in human language. We also give a formal representation for each category and describe the subcategories. This formal representation follows the formal semantic language used for the FIPA ACL. Thus, our approach could combine the benefits from FIPA ACL and Ballmer and Brennenstuhl's speech act classification.

Above all, our approach is theoretically more expressive in representing a broader range of domain-independent communication semantics, while remaining consistent with current approaches to ACLs. However, a comprehensive evaluation is needed, and much work remains to be done to make this approach complete and practically applicable.

References

1. Serrano, J.M., Ossowski, S.: An Organizational Metamodel for the Design of Catalogues of Communicative Actions. PRIMA. (2002) 92–108.
2. Singh, Munindar P.: Agent Communication Languages: Rethinking the Principles. IEEE Computer, vol.31(12). (1998) 40–47.
3. Kinny, David: Reliable Agent Communication – A Pragmatic Perspective New Generation Computing, Vol.19(2). (2001)139–156.
4. Austin, J. L.: How to do Things with Words. Clarendon, Oxford, UK, 1962.
5. Searle, J.: Speech Acts: An Essay in the Philosophy of Language. Cambridge U. Press, 1970.
6. Searle, J.: Expression and Meaning. Cambridge U. Press, Cambridge, 1979.
7. Wooldridge, M.: Introduction to MultiAgent Systems. JohnWiley and Sons, 2002.
8. Ballmer, Th., Brennenstuhl, W.: Speech Act Classification – A Study in the lexical Analysis of English Speech Activity Verbs. Springer-Verlag Berlin Heidelberg New York, 1981.
9. FIPA: FIPA Communicative Act Library Specification. Foundation for Intelligent Physical Agent, Geneva, Switzerland, 2002.
10. Bellifemine, F., Caire, G., Poggi, A., and Rimassa, G.: JADE: A White Paper. TILAB "EXP in search of innovation" Journal. Vol.3(3), 2003.
11. Chang, M.K., Woo, C.C.: A Speech-Act-Based Negotiation Protocol: Design, Implementation, and Test Use. ACM Transactions on Information Systems, Vol.12(4), 360–382, 1994.
12. Esa A., Lyytinen K.: On the Success of Speech Acts and Negotiating Commitments. LAP'96, Tilburg, The Netherlands, 1996.
13. Searle J., Vanderveken D.: Foundations of Illocutionary Logic, Cambridge U. Press, London, 1985.
14. Searle J.: A Taxonomy of Illocutionary Acts. 1971. Language, Mind, and Knowledge: Minnesota Studies in the Philosophy of Science. Keith Gunderson(Ed.). Minneapolis: U of Minnesota P, 1975.

Normative Pragmatics for Agent Communication Languages

Rodrigo Agerri and Eduardo Alonso

Dept. of Computing, City University,
EC1V 0HB London, UK
{rag, eduardo}@soi.city.ac.uk

Abstract. The ability to communicate is one of the crucial properties of agents. In this paper a normative approach to the pragmatics of Agent Communication Languages (ACLs) is proposed. In an open environment, like the Internet, in which agents are designed in many different ways, it is important to clearly establish the meaning of a standard language for artificial agents. Traditionally, the pragmatics of ACLs take the form of interaction protocols, which only specify the order in which messages occur without taking into account the content of the message, or the role of the agents. We present a unified ACL which includes the semantics and pragmatics of ACLs, focusing on a pragmatic level based on the social and normative notion of right. The framework is developed extending CTL with modal and deontic operators, and the pragmatics are expressed by means of a prolog-like declarative language.

1 Introduction

The adoption of a standard Agent Communication Language (ACL) is crucial for artificial agents to interact in open environments. Communication is a kind of interaction that should not affect the autonomy or heterogeneity of the agents. This is particularly true in open environments, such as electronic commerce applications based on the Internet, where agents are designed by different constructors and work for their individual interests.

Most of the approaches to ACLs are based on speech act theory [16]. According to this theory, linguistic communication is just a special type of action which consists of three components. An *illocution* is the central component of a communicative action and it corresponds to what the action is intended to achieve. This goal should be distinguished from the effect that the communicative action is meant to produce on the receiver (*perlocution*), as well as from how the actual communication is physically carried out (*locution*). We argue that it is possible to define a pragmatics for ACLs by means of conversation policies (CPs), which account for the social effects of performing a communicative action and thereby facilitate the achievement of perlocutionary effects.

Current approaches on ACLs respond to three different views. KQML (Knowledge Query Manipulation Language [8]), and FIPA ACL (Foundation for Intelligent Physical Agents [9]), based on the *mentalistic* approach, are the most widely used ACLs. The meaning of the performatives is defined in terms of the mental states of the agents.

J. Akoka et al. (Eds.): ER Workshops 2005, LNCS 3770, pp. 172–181, 2005.

It has been argued that, in open environments, in which agents are heterogeneous and competitive, it is not sensible for agents to trust their opponents in a negotiation process by making assumptions about their current beliefs or intentions ([18, 10]). Besides, the perlocutionary effects of the communicative actions are difficult to specify. Dealing with autonomous agents, it is not possible to guarantee that the perlocutionary action is satisfied, because its fulfillment depends on the receiving agent. The second approach, known as *procedural*, focuses on the design of pattern conversation templates. ACLs are defined in terms of message sequences. Examples of this approach can be found in [11] and [14], among others. It has been claimed that procedural accounts over-constrain the behaviour of the agents, transforming communication in a meaningless exchange of ordered tokens ([18]). Finally, the *social* approach takes into account the social consequences derived from performing a communicative action. For instance, the commitments that agents acquire by sending a particular message. Some authors take *commitment* as the core social notion to define the meaning of the performatives [18, 10]. We understand that commitment-based approaches fail to capture the illocutionary aspect of agent communication, in which agents perform speech acts in order to achieve a particular goal.

We agree with the above criticisms but we also believe that some of them are the result of a misconception, namely, that the semantic level of an ACL should guarantee the social and public character of communication, or that it should achieve the perlocutionary effects. Although the social approach facilitates the verification of ACLs [21], the rest of the issues still need to be solved. This paper proposes a normative-based pragmatics to define meaningful conversations (and CPs) which constrain agents' behaviour with the purpose of achieving the perlocutionary effects.

The remainder of the paper is structured as follows: In the next section, we introduce the main concepts of our ACL framework. In Section 3 a formal definition for the semantics of ACLs is given. Section 4 defines CPs and interaction protocols (IPs) using the framework provided. Section 5 discusses how our proposal compares to related approaches and presents some conclusions and further research.

2   General Framework

In open environments, agents work on behalf of the interest of their designers. Because agents are designed by different vendors, their internal structures are different. Thus, to guarantee that interactions between agents are successful, we need agents to behave according to their roles within a normative system.

A well defined semantics is a central component of the specification of an ACL. However, to this date, most of the ACLs do not include a pragmatic component to regulate the use of the semantics. Traditionally, ACL specifications would just consist of a set of communicative actions, and several interactions protocols would then separately define conversational templates for specific scenarios (e.g. auctions). Conversely, our approach places the social aspect of communication at the pragmatic level, maintaining the illocutionary aspect as the central feature of the semantics (communicative actions). This allows us to define conversational policies (pragmatics) to guide and constrain the use of the performatives. The concept of right plays a central role in the definition of

CPs, allowing us to express the social consequences of performing a particular communicative action.

Our ACL framework includes both the social and mental aspects of communication. Following [21], a complete ACL will consist of a set of communicative actions, that is, the language $\mathcal{L}_c$ expressing the semantic meaning of the performatives, a set of CPs restricting its use, $\mathcal{L}_r$, and the semantic languages $\mathcal{L}_s$ and $\mathcal{L}_p$ for $\mathcal{L}_c$ and $\mathcal{L}_r$ respectively. Note that IPs would also be defined using $\mathcal{L}_r$. In short, the set of communicative actions defined by $\mathcal{L}_c$ represent the semantics of the ACL, whereas the set of policies defined by $\mathcal{L}_r$ represent the pragmatics. The semantics encode the illocutionary character of communication between autonomous agents. The pragmatic level takes into account the social consequences of performing a communicative action. Thus, an ACL is defined as a the tuple

$$\mathcal{ACL} = \langle \mathcal{L}_c, \mathcal{L}_s, \mathcal{L}_r, \mathcal{L}_p \rangle$$

The syntax of the communication language $\mathcal{L}_c$ is based on the FIPA ACL [9]. The semantics of the modal and temporal operators will be given by $\mathcal{L}_s$ in the next section. The language $\mathcal{L}_s$ is based on Computation Tree Logic (CTL [7]) extended with operators for beliefs, goals and intentions. Using a type of temporal logic would facilitate to relate the language $\mathcal{L}_s$ to a computational model and, as a consequence, its verification [21]. Mental states are not understood in this paper as private mental states of the agents; as in theory of planning, goals represent states of the world which are desirable for the agents. When an agent expresses the intention to execute an action, it is expressing publicly its willingness to perform such an action. Intentions refer to the actions that the agent is committed to perform in order to achieve one or more goals. Besides, holding an intention to execute an action presupposes the ability to perform it.

CTL extended with deontic operators for rights and obligations is used to define the language $\mathcal{L}_p$, which provides the semantics for the normative operators needed to specify a set of CPs (pragmatics). We believe that the CPs can be expressed by a declarative language like prolog. An interesting lead is to investigate how our CPs could be imported as reasoning rules in an programming environment for cognitive agents like 3APL [3].

In the following section the language $\mathcal{L}_s$ provides a complete set of communicative actions, reformulating the corresponding FIPA ACL performatives. It is generally agreed that the FIPA Communicative Actions Library (CAL) accounts only for the illocutionary act since the perlocution (effects) cannot be guaranteed. We deal with the (social) effects in the pragmatic level limiting the semantics to the illocutions.

3   Illocutionary Actions

CTL is a type of branching temporal logic which describes properties of a computation tree [7]. Temporal model checking using CTL is one of the most used techniques for verifying properties of state-transition systems. The logic presented here, $\mathcal{L}_s$, extends CTL by adding modal operators for beliefs, goals and intentions. The main difference with other extensions (e.g. [15] [18]) is the use of a goal operator which cannot conflict with beliefs, that is, we distinguish between goals and desires. The objects of goals are states (ϕ) whereas the objects of intentions are actions (α).

The recursive definition of $\mathcal{L}_s$ formulae is as follows:

$$\varphi := AP|\neg\varphi|\varphi_1 \wedge \varphi_2|B_i\varphi|G_i\varphi|I_i\varphi|E\varphi|A\varphi|X\varphi|\varphi_1 U\varphi_2$$

The boolean operators are standard. E and A are quantifiers over paths, meaning "there exists an execution" and "for all executions" respectively; $\phi U\psi$ means that ψ does eventually hold and that ϕ will hold everywhere until ψ holds. The rest of the formulae can be introduced as abbreviations: $EF\phi$ for $E\ true\ U\phi$; $AF\phi$ for $A\ true\ U\phi$; $EG\phi$ for $\neg(A\ true\ U\neg\phi)$; finally, $AG\phi$ abbreviates for $\neg(E\ true\ U\neg\phi)$. Beliefs are represented by a $KD45$ axiomatization relative to each agent. For goals and intentions, we assume a minimal KD axiomatization to ensure consistency.

A structure is a tuple $M = \langle S, R, L, T, A, \mathcal{B}, \mathcal{G}, \mathcal{I}, \rangle$ where: S is a set of states, R is a total binary relation $\subseteq S \times S, \forall s \in S s.t.(s, t) \in R$, $L : S \to PowerSet(\Phi)$ is an interpretation $L : S \to 2^\Phi$ where Φ is a set of atomic expressions, $T : S \to P$ gives the real path conveyed by a state, where P is the set of paths derived from L; PP gives the powerset of P. A is a set of agents, $\mathcal{B} : S \times A \to S$ gives the accessibility relation for beliefs; $\mathcal{G} : S \times A \to PP$ and $\mathcal{I} : S \times A \to PP$ are interpretations for goals and intentions respectively.

For a Kripke structure M and a state s_0, we write $M, s_0 \models \phi$, for a state formula ϕ. For a structure M and a full path χ, we say that $M, \chi \models \phi$ for a path formula ϕ. The semantics of $\mathcal{L}_s$ is as follows:

$M, s_0 \models \phi$ iff $\phi \in L(s_0)$, for $\phi \in AP$
$M, s_0 \models \phi \wedge \psi$ iff $M, s_0 \models \phi$ and $M, s_0 \models \psi$
$M, s_0 \models \neg\phi$ iff it is not the case that $M, s_0 \models \phi$
$M, s_0 \models E\phi$ iff $\exists$ a full path $\chi = (s_0, s_1, s_2, \ldots)$ in M and $M, \chi \models \phi$
$M, s_0 \models A\phi$ iff $\forall$ full paths $\chi = (s_0, s_1, s_2, \ldots)$ in M, and $M, \chi \models \phi$
$M, s_0 \models B_i(\phi)$ iff $\forall s_1 : s_1 \in \mathcal{B}(i, s_0) \Rightarrow M, s_1 \models \phi$
$M, s_0 \models G_i(\phi)$ iff $\forall \chi : \chi \in \mathcal{G}(i, s_0) \Rightarrow M, \chi \models \phi$
$M, s_0 \models I_i(\phi)$ iff $\forall \chi : \chi \in \mathcal{I}(i, s_0) \Rightarrow M, \chi \models \phi$
$M, \chi \models \phi U\psi$ iff $\exists i, M, s_i \models \psi$ and $\forall_j < i, M, s_j \models \phi$
$M, \chi \models X\phi$ iff $M, s_1 \models \phi$

We can use now $\mathcal{L}_s$ to express the meaning of the relevant classes of illocutionary actions. Following Searle's taxonomy [16], we classify the actions into assertives, commissives, directives, declarations and expressives. The last category is not relevant for the purposes of this paper, so it will not be included. The syntax of $\mathcal{L}_c$ is based on FIPA

Table 1. A complete list of communicative actions

$\langle i, inform(j, \phi) \rangle$	$\langle i, request(j, \alpha) \rangle$
$FP : B_i(\phi) \wedge G_i(B_j(\phi))$	$FP : G_i(I_j(F\alpha))$
$RE : B_j\phi$	$RE : EF\alpha$
$\langle i, promise(j, \alpha) \rangle$	$\langle i, declare(j, \phi) \rangle$
$FP : I_i F\alpha$	$FP : G_i(\phi)$
$RE : F\alpha$	$RE : \phi$

ACL. In Table 1 some performatives for each of the remaining four types of categories are defined.

The two performatives at the top, *inform* and *request*, represent the assertives and directives respectively. *Declare* is an action of the declarative class and *promise* is a commissive.

4 Normative Pragmatics

Interaction protocols (IPs) define the sequences in which communicative actions can be performed, so that agents can engage in a meaningful conversation. We believe that IPs can be modelled as the right agents have to use a performative based on previous speech acts. Thus, IPs provide the set of performatives that can be used at a given time. IPs are, in turn, constrained by the pragmatics of the language specified as a set of CPs.

Informally, if an agent has the right to execute a set of actions, then: (i) it is permitted to perform it (under certain obligations); (ii) the rest of the agents are not allowed to perform any action that violates the right-holder's action; and (iii), the rest of agents, the group, has the obligation to sanction any inhibitory action. The function of rights for agent communication is to stabilize social interactions by making the behaviour of agents predictable to the other agents of the system. Permissions are usually defined as the dual of obligation, meaning that an agent that is not obliged not to do α is permitted to do α. Rights are not simply the absence of obligations. For an agent to have the right to execute α, is must be given permission to do so. Not being obliged not to do α does not mean that the agent has the right to do α.

To define the language $\mathcal{L}_p$, we extend CTL to express obligations and rights within an organizational structure in which agents have a role assigned [20]. Thus, special propositions i rr j, g_i r_i are introduced to mean that agents i and j are role-related by rr, i is a member of group g, and i plays the role r, respectively. A role is a set of constraints that should be satisfied when an agent plays the role. For example, the role of auctioneer constrains the goals, obligations, permissions and rights of the agent that plays that role. The scope of the role depends on the institutional reality in which it is defined (e.g., auction). A group is a set of agents (roles) that share a specific feature (i.e., being auctioneers). Finally, role relations constrain the relations between roles (e.g., the auctioneer-bidder relation). An example involving passengers and ticket controllers in the London Tube will illustrate this point when defining the notion of right.

We also need to speak about agents performing actions, such as giving or cancelling rights. $Done_i(\alpha)\phi : i \in A, \alpha \in \Phi$, where A is a set of agents and Φ a set of atomic expressions, means that the performance of action α by agent i makes the proposition ϕ true.

The resulting language $\mathcal{L}_p$ is an extension of CTL with deontic and dynamic operators. The combination is straightforward, the models include, as independent layers, temporal transition functions and a deontic accessibility relation.

Thus, given a finite set of agents A, a finite set of group names GN, a finite set RN of role names, a finite set RR of role relations, and a countable set AP of primitive propositions, the syntax of $\mathcal{L}_p$ is given the following BNF expression:

$$\varphi := AP|\neg\varphi|\varphi_1 \wedge \varphi_2|G_i\varphi|O_{ij}\varphi|Done_i\varphi|E\varphi|A\varphi|X\varphi|\varphi_1 U\varphi_2$$

In our framework, the deontic operators are directed, e.g., $O_{ij}\phi$ means that agent i has the obligation to bring about ϕ towards agent j [6, 17, 20]. As usual, permissions are defined as the dual of obligations. The axiomatization of obligation is given by the system KD. The semantics of $\mathcal{L}_p$ inherit from the semantics of $\mathcal{L}_s$. The truth of $O_i\varphi$ and $Done_i\varphi$ is defined below. We add a deontic accessibility relation $\mathcal{O}$, $\mathcal{O} : 2^A \rightarrow 2^{S \times S}$, and a function $\mathcal{D}$, $\mathcal{D} : S \times \Phi \rightarrow 2^S$ that gives us the state transitions caused by the achievement of an action.

1. $M, s_0 \models O_i\phi$ iff $\forall s_1$ such that $s_0 \mathcal{O}_i s_1$ we have $M, s_1 \models \phi$
2. $M, s_0 \models Done_i(\alpha)\phi$ iff $\forall\chi$ such that $\chi \in \mathcal{D}(i, s_0) \Rightarrow M, \chi \models \phi$

Note that in our framework, an expression such as $r_i \rightarrow B_i\phi$ will not be expressed by an axiom but by a formula. The reason is that role related conditions should not be substituted by another proposition.

It remains to define our notion of right. In order to do so, we need to introduce the violation predicate V [19].

Definition 1 (Violation).

From each literal built from a variable α, $V \in A$, $V(\alpha)$ means that some agent A determines that α is a violation. Then $\neg\alpha$ is a violation under state ϕ for some $m \in NS$, such that NS is a set of norms, iff

$$O_{ij}(\alpha)\, U\phi \rightarrow \neg Done_i(\alpha)\psi\, U\phi$$

Rights are considered here exceptions to obligations [19]. An agent has the right to do α under some condition ϕ if it has the goal not to believe α as a violation ($\neg V(\alpha)$) when the agent that gives the right believes that ϕ.

Definition 2 (Right).

Let NS be a set of norms $\{m_1, \ldots, m_n\}$, and let the variables of agent A contain a set of violation variables $V = \{V(\alpha)$ such that $\alpha \in AP\}$. Agent i believes that it has the right given by agent j to do α, $R_{ij}\alpha$, under situation ϕ, $\phi \in S$ iff for some $m \in NS$

$$G_j(\neg V\alpha)\, U B_j\phi$$

Therefore, if the agent j giving the right to do α believes that ϕ then j will not want to consider α as a violation $\neg V\alpha$ until $B_j\phi$. We can now use the semantics defined by $\mathcal{L}_p$ to specify a set of IPs and CPs for agent communication. This is done by a prolog-like declarative language $\mathcal{L}_r$.

It is possible to define the notion of sanction as a type of obligation (e.g., [20]). For instance, agent i wants to travel by tube from Oxford Circus to London Bridge (its role is passenger, $p \in RN$) and has not paid a ticket before the journey started. The London Tube has given the right to some agent j to impose fines while playing the role of ticket controller, $tc \in RN$.

$$p_i \wedge tc_j(\neg Done_i(\alpha)\phi\, U\psi \rightarrow O_{i,j}\psi)$$

4.1   Conversational Policies

IPs aim to constrain the interaction to facilitate the desired outcomes (for example, that the highest bidder is found in an auction). Besides, CPs make coordination easier since they assign rights and obligations on the participants, and specify which communicative actions are appropriate at certain states (the agent playing the role of auctioneer establishes the rights and permissions of the participants). The conformity of the participants to the protocol is based on the content of the performatives used.

The language $\mathcal{L}_p$ provides the semantics of the normative notions needed to specify CPs for agent communication ($\mathcal{L}_r$).

Using a declarative language to express CPs is inspired by work related to 3APL [3] and within the Semantic Web framework [12].

A policy consists of the following components: communicative actions, domain actions, normative rules and facilitator actions. The first component is given by $\mathcal{L}_c$ and the second is defined by $\mathcal{L}_p$; the kind of domain actions agents can perform will depend on the abilities agents have; like communicative actions, domain actions are expressed in terms of goals, preconditions and effects.

$$done(Agent, Goal, Precondition, Effect)$$

There are two types of domain actions: those actions specific of the institution (e.g., bidding) in which the interaction is taking place, that we call context-dependent actions, and those rights-related actions aimed to create and cancel rights. The right-related actions are defined in the general framework, whereas the context-dependent actions will depend on the specific scenario.

Definition 3 (Create).
The creation of a right is satisfied in the model M along a path χ iff the performance of an action α, that is, $Done_i(\alpha)$, makes true the right. That is, if the agent issuing the right does not consider executing φ to be a violation.

$$M, s_0 \models Create_j(R_{ij}\varphi) \text{ iff } \exists \alpha \in \Phi \text{ in } M \text{ and } M, \chi \models Done_j(\alpha)R_{ij}\varphi$$

Definition 4 (Cancel).
An agent j cancels the the right of agent i to do φ iff j performs an action so that the right does not hold.

$$M, s_0 \models Cancel_j(R_{ij}\varphi) \text{ iff } \exists \alpha \in \Phi \text{ in } M \text{ and } M, \chi \models Done_i(\alpha)\neg R_{ij}\varphi$$

Facilitator actions depend on the platform in which agents run. That is, facilitator actions are defined by the programming language in which agents are built. For example, in Java built platforms like JADE, sending messages is simply a case of creating an ACLMessage, setting the parameters (sender, receiver, reply-to, performative, etc.) and then sending it using the send() method in the agent object.

Finally, normative rules consist of a deontic operator (obligations, rights) and a condition that has to be true for the rule to be applicable:

$$right(X, request(X, Y, Condition)$$

Agents hold the right to do α as long as α does not constitute a violation. An obligation rule states that an agent must perform an action before its applicability condition becomes false; a permission rule establishes that the agent can perform an action α if its condition(s) is true. We can now use the language $\mathcal{L}_r$, whose semantics are specified by $\mathcal{L}_p$, to model IPs of FIPA ACL in terms of the rights of the agents to use the performatives.

In the FIPA interaction protocol for query-if, agent X queries agent Y whether or not a proposition P is true. The receiver has the right to either agree or refuse to send and *inform* message providing an answer (a definition of agree and refuse is provided in [1]). In the case that agent Y agrees, then it has obligation to send a notification which can be an inform stating the truth of falsehood of the proposition P. If agent Y sends a refuse message the protocol ends there. We can complement this by specifying the roles of the participating agents as follows:

```
role(X, customer).
role(Y, seller).

right(X, query-if(X, Y, P), _).

right(Y, agree(Y, X, P));
right(Y, refuse(Y, X, P)) :-
   receive(query-if(X, Y, P).

obligation(Y, inform(Y, X, P));
obligation(Y, inform(Y, X, not P)) :-
   send(agree(Y, X, P), _).
```

Policies can then be defined to constrain the agents' use of the performatives in virtue of their content. For example, agent Y, acting on behalf of an airline company serving flights to European countries, could have a CP that states that it should agree to every query regarding flight tickets to Europe (i.e., answering about flight times and providing the best offer for a potential buyer) and another one specifying that it has the obligation to refuse every query about flights to non European countries.

```
role(X, customer).
role(Y, seller).

obligation(Y, agree(Y, X, P) :-
   receive(query-if(X, Y, P)), europeanFlight(P)).

obligation(Y, refuse(Y, X, P) :-
   receive(query-if(X, Y, P), nonEuropeanFlight(P)).
```

Similarly, other CPs can be defined to state that an agent can deceive, or that it has the right to do so in particular circumstances. It can be specified that an agent X will always answer to every message it receives, etc.

5   Concluding Remarks

Our approach analyzes agent communication in terms of the social consequences of executing an action. The idea of using rights to constrain agents' communicative behaviour is inspired by [13] and [2]. The characterization of the normative and organizational concepts is inspired by the work of van der Torre [19, 20]. Our proposal is also related to [12], but here we present a complete semantics for the deontic operators. Other authors ([5]), have also presented a temporal deontic logic with dynamic operators.

Two semantic languages based on CTL that give the semantics for the communicative actions and the conversational policies of an ACL are presented. We understand that using a temporal logic would facilitate the compliance testing of our ACL [21], something not possible for KQML and FIPA ACL.

For standardization reasons, the ACL specification proposed here intends to be as close as possible to the FIPA ACL specification. With this purpose, we provide definitions for the actions absent in FIPA CAL: commissives and declaratives. We understand that in FIPA CAL some of the definitions are unnecessarily complex. This is partially due to the multimodal language used as the semantic language. Besides, unlike CTL, it is not a language that can be grounded in a computational model.

We have shown with an example, how our proposal can be used to define IPs using a declarative language. Unlike KQML and FIPA ACL, our normative-based ACL is not confined to a meaningless sequential exchange of tokens. By defining normative CPs, we facilitate the fulfillment of the perlocutionary effects of a communicative action. In this sense, our ACL can be applied in open environments, such as e-commerce taking place in the Web, in which agent interaction should not rely on agents trusting each other.

Future work involves its application in a programming environment such as 3APL [4]. It is possible to implement 3APL agents that employ external prolog files which can be loaded into the 3APL platform. The prolog files are part of the belief base, which can contain general rules to be applied for some problem domain. We believe that, for communicative agents, our right-based CPs can work as these general rules. Finally, we also plan to extend the right-related actions (create, cancel) to specify a set of inference rules that can be used for agents to take decisions about giving or cancelling rights and obligations.

References

1. Agerri, R., Alonso, E.: Semantics and Pragmatics for Agent Communication. In Proceedings of the 12th Portuguese Conference on Artificial Intelligence (EPIA 05). To be published
2. Alonso, E.: Rights and argumentation in open multi-agent systems. Artificial Intelligence Review **21** (2004) 3–24
3. Dastani, M., van der Ham, J., Dignum, F.: Communication for goal directed agents. In Huget, M.P., ed.: Communication in Multiagent Systems - Agent Communication Languages and Conversation Policies. Springer-Verlag (LNCS 2003) 239–252
4. Dastani, M., Riemsdijk, M., Dignum, F., Meyer, J.J.: A programming language for cognitive agents: Goal-directed 3apl. In Dastani, M., Dix, J., Fallah-Seghrouchni, A.E., eds.: Programming Multi-Agent Systems (LNAI 3037). Springer-Verlag, Berlin (2004) 111–130

5. Dignum, F., Kuiper, R.: Combining dynamic deontic logic and temporal logic for the specification of deadlines. In Sprague, J.R., ed.: Proceedings of thirtieth HICSS, Hawaii (1997)
6. Dignum, F.: Autonomous agents with norms. Artificial Intelligence and Law **7** (1999) 69–79
7. Emerson, E.A.: Temporal and modal logic. In van Leeuwen, J., ed.: Handbook of Theoretical Computer Science, volume B. North Holland, Amsterdam (1990) 995–1072
8. Finin, T., Fritzson, R., McKay, D., McEntire, R.: KQML as an Agent Communication Language. In Adam, N., Bhargava, B., Yesha, Y., eds.: Proceedings of the 3rd International Conference on Information and Knowledge Management (CIKM'94), Gaithersburg, MD, USA, ACM Press (1994) 456–463
9. FIPA ACL: FIPA Communicative Act Library Specification (2002) http://www.fipa.org/repository/aclspecs.html
10. Fornara, N., Colombetti, M.: A commitment-based approach to agent communication. Applied Artificial Intelligence an International Journal **18** (2004) 853–866
11. Greaves, M., Holmback, H., Bradshaw, J.: What is a Conversation Policy? In Dignum, F., Greaves, M., eds.: Issues in Agent Communication. Heidelberg, Germany: Springer-Verlag (2000) 118–131
12. Kagal, L., Finin, T., Joshi, A.: A policy language for a pervasive computing environment. In: IEEE 4th International Workshop on Policies for Distributed Systems and Networks (2003)
13. Norman, T.J., Sierra, C., Jennings, N.R.: Rights and commitment in multi-agent agreements. In: Proceedings of the Third International Conference on Multi-Agent Systems. (1998) 222–229
14. Pitt, J., Mamdani, A.: A protocol-based semantics for an agent communication language. In: Proceedings of the 16th International Joint Conference on Artificial Intelligence IJCAI'99, Stockholm, Morgan-Kaufmann Publishers (1999) 486–491
15. Rao, A., Georgeff, M.: Modeling rational agents within a bdi-architecture. In Allen, J., Fikes, R., Sandewall, E., eds.: 2nd International Conference on Principles of Knowledge Representation and Reasoning (KR'91), Morgan Kaufmann Publishers (1991) 473–484
16. Searle, J.R.: Speech Acts. An Essay in the Philosophy of Language. Cambridge: Cambridge University Press (1969)
17. Singh, M.P.: An ontology for commitments in multiagent systems: toward a unification of normative concepts. Artificial Intelligence and Law **7** (1999) 97–113
18. Singh, M.P.: A social semantics for agent communication languages. In: Issues in Agent Communication, volume 1916 of LNAI. Berlin: Springer-Verlag (2000)
19. van der Torre, L.: Contextual deontic logic: Normative agents, violations and independence. Ann. Math. Artif. Intell. **37** (2003) 33–63
20. van der Torre, L., Hulstijn, J., Dastani, M., Broersen, J.: Specifying multiagent organizations. In: Proceedings of the Seventh Workshop on Deontic Logic in Computer Science (Deon'2004), LNAI 3065. Springer (2004) 243–257
21. Wooldridge, M.: Semantic issues in the verification of agent communication languages. Journal of Autonomous Agents and Multi-Agent Systems **3** (2000) 9–31

Experimental Comparison of Rational Choice Theory, Norm and Rights Based Multi Agent Systems

Peter Kristoffersson and Eduardo Alonso

Department of Computing, City University, London EC1V 0HB, United Kingdom
{dn184, eduardo}@soi.city.ac.uk
Tel: +44 (0)20 7040 8552, Fax: +44 (0)20 7040 0244

Abstract. As utility calculus cannot account for an important part of agents' behaviour in Multi-Agent Systems, researchers have progressively adopted a more normative approach. Unfortunately, social laws have turned out to be too restrictive in real-life domains where autonomous agents' activity cannot be completely specified in advance. The idea of Rights is a halfway concept between anarchic and off-line constrained interaction. Rights improve coordination and facilitate social action in Multi-Agent domains, they allow agents enough freedom, and at the same time constrain them (prohibiting specific actions). So far rights have not been tested or proven experimentally. We are comparing experimentally the three mentioned interaction architectures in the domain of agent-based traffic simulation.

1 Introduction

The Rational Choice Theory (RCT) has been the most influential theory for designing agents in Artificial Intelligence and Distributed Artificial Intelligence. According to this approach to rationality, agents with complete knowledge make their decisions in order to maximise their own utilities. In this non-constrained approach agents are assumed to be `free': They act of their own accord and are not subject to any set of (social) rules. However fruitful this approach has been, there have been pointed out (e.g., [4]) some drawbacks in RCT:

- In real dynamic domains agents do not have enough information or time to perform complex, optimal utility calculus.
- The utilitarian approach fails to explain cooperation and social action.

In order to cope with these problems, the MAS community has adopted a more constrained approach to rationality including conventions, norms and/or social laws. It is well-known that agents working under norms do not need to calculate continuously their utilities and, consequently, do not need complete information. Agents are supposed to act in a somehow predetermined way according to the principle of `mutual expectation'. Besides, norms imply that the agents respect certain social constraints that deter them from breaking agreements. Unfortunately, research in this field has fallen into two extreme positions:

Shoham and Tennenholtz [5] have studied off-line social laws, which agents must comply with automatically. Here the agents are assumed to follow rules just because

J. Akoka et al. (Eds.): ER Workshops 2005, LNCS 3770, pp. 182–191, 2005.

they are designed to do so. Following this line of argumentation agents are not seen as autonomous any more. Proposals so formulated are thus closer to Distributed Problem Solving than to MAS.

Alternatively, conventions [8] have been introduced as rules emerging during repeated encounters in open normative systems. The problem here is that no notion of sanction is considered. Consequently, if the agents have the chance to calculate their utility each time they interact, conventions are continually under consideration. In other words, following a convention is not always a stable strategy.

It seems, therefore, that we need a concept that allows agents to reason and make decisions, but that implies enforcement at the same time. The idea of "right" has been proposed by Alonso [1] as such a concept and further explained and axiomatically represented in [2]. However it has still not been proven experimentally whether it works and how well it performs in real life situations. We will therefore explore and compare experimentally off-line designed rights with off-line designed social laws (focusing on obligations and prohibitions) and RCT architecture. Due to space constraints, we will not explain RTC or social norms in detail. The reader is assumed to be familiar with game theoretic and normative approaches to MAS coordination. Neither will we discuss other alternatives (such as bounded rationality etc) to RCT as Rights are more related to Norms than other solutions. The reminder of the paper is structured as follows. In the second section we present the concept of rights in more detail and what we gain by introducing tem. Section three present the system in which we test the architectures while section four describes how these architectures where implemented. Section five defines the experiment parameters while in sections six and seven show the results and analyse them. We finish with some conclusions and further research.

2 Rights

Roughly stated, a right is considered as a set of restrictions on the agents' activities which allow them enough freedom, but at the same time constrain them. Not surprisingly, some authors (e.g., [7]) have expressed the same idea from a RCT perspective, by introducing some constraints in the set of strategies available to the agents. In so doing, agents are free to converge on `stable social laws' (qualitative equilibrium). However interesting this approach may be, it presents a serious handicap: To make sure that the agents choose a stable and efficient strategy, the designer decides beforehand which strategies should be eliminated. The designer, therefore, manipulates the process and creates an `illusion of freedom'.

Generally speaking, if an agent has the right to execute a set of actions then (a) he is permitted to perform it (under certain constraints or obligations), (b) the rest of the group is not allowed to execute any action inhibiting the agent from exercising his right, and (c) the group is obliged to prevent this inhibitory action.

Rights can be modelled as norms but to do so is very difficult. A rights based system can be seen as a normative system in the instance the decision is being made. The difference is that not all agents will have to obey the norms and that every agent will have a different set of norms in the situation. The set of norms that governs each agent will also be different from one instance to another. For a more comprehensive description and a formal characterisation of rights using the language L and the axiomatic proof the readers are referred to Alonso's [2] work.

2.1 Gains from Using Rights

The idea of using rights is worthy of consideration because it makes easier to have agents coordinated. This has already been described and showed qualitatively by Alonso in [2]. As it has been repeatedly pointed out (e.g., [3, 6, 9]) coordination is mainly concerned with complexity, efficiency, stability, and flexibility. Rights aid all of these. For a more detailed explanation the readers are referred to Alonso's [2] work.

2.2 Evaluating Gains

To prove experimentally that rights make it easier to coordinate agents, we have decided to create traffic MAS simulation and to test the three mentioned coordination mechanisms in this environment. We are interested in testing the stability and efficiency of the system with regards to agent survival rate and average speed. Even though we will be comparing the outcomes of the three mechanisms it is important to understand that the results in themselves can always be challenged. Therefore, even though we are evaluating the results, we are more interested in the result patterns rather than the results themselves. The reason for this is that it is very difficult if not impossible to evaluate the mechanisms against each other. There is always a chance that one could design a better architecture that could outperform the others. If we however look at it as finding patterns in behaviour of the coordination architecture we will gain a better understanding of the outcomes and how these are achieved.

3 Experiment Environment

The reason for using traffic simulation is that this domain is intuitively easy to understand. The created system is based on a microscopic traffic simulation system developed by Tom Fotherby. Our redesign changed most of the internal working of the system with the exception of the time engine, graphics and road design ability. The agent architecture, information provided by the system to the agents and users, data saving, statistics and interaction between the agents (crashes) have been created by us.

3.1 Internal Architecture

The system is built in java making it portable between operating systems. The internal engine of the system is based on two main methods, a "pretick" and a "tick" in each agent. The system alternates invoking the "pretick" and the "tick" methods in all registered agents. Firstly all "pretick" methods are run after which all the "ticks" are run, this going in a loop. This allows the agent to firstly calculate what to do next (in the "pretick") without any risk that the environment will change before the actions can be implemented. Then in the "tick" all these actions can be implemented so that they happen simultaneously from the agents' perspective. The time measurement in the system is done through steps where one time step is defined as one loop of "preticks" and "ticks".

3.2 System Features

The system allows agents to perceive their environment forward, backwards, and to the sides back and forth. It gives full information about the distance to other agents as long as the other agent is on the same stretch of the road. It also gives their speeds and direction. In the system the agent can only see one agent ahead meaning that if we have three agent-cars driving in a row in front of us, we will only see the closest one. The agents can change their speed and position on the road (lane) in order to go past obstacles. Each car's maximum speed is set randomly with minimum of 44 and maximum of 82. The system enables defining the rate of new incoming agents, where new agents enter (are created) the system every N time steps (one car every N time units => 1/N) at the spot where the lane touches the border of the simulation window (see Fig. 1) and are removed from the system when crashed (after 10 time units) or when they reach the end of the lane. The entry per time unit is connected to each lane (so two lanes in the same direction will have 2* 1 entries every N time units). Every car that crashes will be immovable for 10 time steps after which it will disappear.

4   Agent Architecture

In our experiments the agent plays the role of a car that wants to survive (not crash) the trip and get through the system as fast as possible. The agents are homogenous. The main goal for the agents is obviously survival. In order to ensure that it was the architecture and not the coding that created better performance, the normative architecture is basically enhanced free rider architecture and the rights architecture is an enhanced normative architecture.

4.1   Free Rider Architecture

The free rider architecture is a simple deliberative architecture. It allows to choose the best action for any given situation by evaluating which one would allow the agent to perform best (drive faster and not crash). At each time steps the agents are re-evaluating their choices. The agent can only perform one action at a time. The actions are arranged in a hierarchy. The possible actions are (according to their hierarchy): accelerate, do nothing, switch to left lane, switch to right lane, de-accelerate. This architecture was selected as it is simple to implement, easy to understand, easy to extend with new choices and allows prioritisation between actions when two actions have same utility figure.

> For each time step
> > Evaluate all possible actions and assign them utility values
> > Discover the highest obtained utility value for this time step
> > Perform the action with highest value and hierarchy

The free agents are using this to decide what to do next from their own selfish perspective. They are allowed to do whatever they want. They can drive on the wrong lane if they so choose to do.

4.2   Normative Architecture

Normative agents are using the selfish agent architecture with an added filter. The method evaluates whether performing (or not performing) an action would violate the norms. If that is the case, the method then changes the utility value of the affected action to either 0 or to the highest possible depending on the violation and the norm. In the experiment we are using three norms.

- Cars are not allowed to drive on the wrong side (lane in the wrong direction)
- Cars must drive on the left lane unless they are overtaking
- The maximum allowed speed is 55

These were selected from the norms governing English roads. There was no particular norm analysis or selection process involved. It was however intuitively felt that these would minimise the amount of crashed cars. The maximum speed norm was introduced after some preliminary experiments showed that this minimizes number of crashed agets. The architecture of the normative system looks as follows:

For each time step
   Evaluate all possible actions and assign them functionality values
   Adjust the functionality values according to the norms
   Discover the highest obtained functionality value for this time step
   Perform the action with highest value and hierarchy

4.3 Rights Architecture

Rights-based agents are using normative architecture as a base. Here however we are now using rights instead of norms. Looking at this as a right- hierarchy we have:

- Right to live – do not do anything that could put you or others in danger
- Right to drive on your side – an agent on correct side has the right not to be obstructed by agents going in opposite direction
- Right to overtake – if the agent in front is slower than this agent then this agent has the right to overtake
- Right to use the full speed – if this is not in conflict with previous rights
- Right to drive on the road – if this is not in conflict with previous rights

These rights (except the first one) are corresponding to the norms defined earlier although not perfectly as it is not possible to make a perfect translation. The top right is here the most important one as it states that safety is paramount and thus allows or disallows invoking of any other rights. It will also force slowing down or accelerating in dangerous situations.

As stated earlier we wanted the systems to be based on the same basic architecture. We have therefore decided that even this one should be based on the previous one. We have done this by adding a new method that together with the norm-method created earlier, evaluates the situation and from a agent rights perspective (with safety as the main right) and either allows or disallows certain actions (depending on the rights). The idea is that this will lead to evaluation of architectures rather than the code itself as the basic code is the same.

For each time step
   Evaluate all possible actions and assign them functionality values
   Adjust the functionality values according to the norms
   Evaluate each adjusted action and change the value according to rights
   Discover the highest obtained functionality value for this time step
   Perform the action with highest value and hierarchy

5 Experiment Parameters

All the experiment results are based on 100k time steps for each experiment, where the data for each 10 steps is averaged and saved for analysis making 10k data points for each experiment. The reasoning behind this number is that it should give enough information about the performance of each architecture in each scenario. If the number was smaller, one could argue that random statistical errors could have affected the results, while a larger number would have not changed the results and therefore would be pointless. In total we have 8 experiment scenarios. Each experimental scenario is tested three times. Table 1 shows how the parameters change in each scenario. If the data for each experiment (in a single scenario) is consistent with remaining two, the experiment results are then averaged into one set. In a situation where the results would not be consistent more experiments would have been run. Fortunately this has not happened and supports our choice for 100k time steps experiments. To simplify these experiments we introduced a single road without any junctions. On this road the lanes run in both directions and the agents are not penalised for taking a curve at full speed. Since we intend to compare the three methods we need to understand how they perform in different situations. To do so we decided on to use two different parameters:

- **The number of lanes in each direction.** This parameter was chosen as the number of lanes does affect both behaviour and through flow (efficiency) in a traffic system. It is therefore interesting to find out how it affects the architectures and the efficiency of the system.
- **The rate of incoming agents.** The number of entering cars affects obviously traffic flow (efficiency). Same architecture might perform differently depending on the complexity of the situation. Having more agents increases the complexity thus testing the stability of the system.

Table 1. Experiment Layout for each Architecture

Entry Rate	500	300	100	50
Single Lane	3 exp	3 exp	3 exp	3 exp
Double Lane	3 exp	3 exp	3 exp	3 exp

The parameters were selected after considering real life traffic scenarios. What often changes is the number of lanes and the number of incoming cars. In order to analyse the results we have decided to measure the following facts about the system: average speeds and number of crashed cars. Both of these figures describe the system and how it is performing in a specific situation. As in any scenario we want to avoid the crashes as much as possible while maintaining as high speed as possible. Our main goal is to survive! In situations where survival rates are comparable we will compare agents' average speeds.

6   Results

This chapter presents the results. The tables give basic information about the system performance. Entry rate states an agent will enter the system every 500, 300, 100 etc time steps depending on the experiment. The graphs show the average speeds for the systems and are used to show graphically system behaviour.

7 Analysis

The results show that both Normative and Rights based architectures outperform the Free-rider in each case. Even though the average speed of Free-rider system is higher than the corresponding Rights and Normative systems we can clearly see that when it comes to efficiency these systems still outperforms the RCT as the number of crashed agents is a lot smaller.

Table 2. Single Lane Experiments. AS=Average speed, CC=Crashed Cars, GL=Grid Lock.

Entry Rate	500		300		100		50		10	
	AS	CC	AS	CC	AS	CC	AS	CC	AS	CC
RCT	53	112	53	364	GL	GL	GL	GL	GL	GL
Normative	44	0	44	0	43	0	41	0		
Rights	52	0	50	0	44	0	41	0		

Table 3. Double Lane Experiments. AS=Average speed, CC=Crashed Cars, GL=Grid Lock.

Entry Rate	500		300		100		50		10	
	AS	CC	AS	CC	AS	CC	AS	CC	AS	CC
RCT	54	79	53	224	51	2335	49	7048	GL	GL
Normative	48	0	47	70	47	1560	45	4545		
Rights	51	0	49	28	47	1210	47	4707		

In single lane 1/500 experiments (Table 2) the number of entering cars is 400 (200 from each side). Using RCT the number of crashed cars makes 20% of the total number of entered cars. One fifth of all the agents will not accomplish their goal meaning that even though average speed is higher than in the other two systems, the efficiency is 1/5 lower. At the same time neither in the Normative or the Rights systems have any crashed cars. The Rights systems average speed is almost 20% more than the Normative agents and very close to the RCT. In the next scenario, single lane 1/300, the number of entered cars is 667. In the RCT system over 50% of the agents crash. The Normative and Rights systems have again 0 crashes. Even in this case the average speed of the Rights system is well above the speed of the Normative system.   The third single lane scenario, 1/100, we see a smaller difference between the Normative and Right based systems. The RCT agent based system cannot handle the number of agents and ends up in grid lock situations.

The final single lane scenario shows no difference between Rights and Normative systems. In the Free rider scenario we see again grid locks blocking the whole system.

In the double lane scenarios (Table 3) we see similar results. In 1/500 experiments the RCT system is faster than both Normative and Rights system. However ca 10% of all Free rider agents crash whereas both other architectures have no crashes.

Next double lane scenario, 1/300, sees the rise of dead agents to 16% for the RCT system. At the same time we also notice agents crashing in the Normative and the Rights systems. 5% Normative agents never reach their destination while only 2% of Right based agents crash. This is still outperforming the free-rider agents as we are interested in stability and efficiency and having 16% crashes is a lot more than 5% or

2%. In the third double lane scenario, 1/100, we continue to see similar trends to the previous two situations. The RCT system has now over 60% failure rate while the Normative system has a failure rate of 39% and the Rights system 30%. RCT fails twice as often as the Rights system! Speed wise Norm based and Right based systems are doing the same (Fig. 2). Failure wise Right system exceeds Norms by 25%.

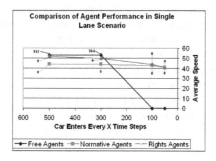

Fig. 1. Agent Performance in Single Lane scenario. The numbers beside the lines show the number of crashed cars for each architecture.

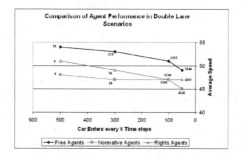

Fig. 2. Agent Performance in Double Lane scenario. The numbers beside the lines show the number of crashed cars for each architecture.

In the final scenario, 1/50, the vast majority (88%) of RCT agents crash while the same figures for Normative architecture is 57% and 59% for the Rights system. This rather unexpected result will have to be looked into in more detail.

As we are looking for efficient and stable results, any experiments with large number of crashed agents will automatically be assumed as underperforming. When we compare the three architectures we can clearly see that most cases the Rights system is the best with regards to efficiency and stability. The free agent system if faster only because many agents crash. However since we are interested is the survival of as many agents as possible the average speed of the system becomes less interesting and is only used for comparison when agent failure rates are the same.

In single lane scenarios at levels of 1/100 the RCT agents end up in situations that cannot be resolved and the whole system locks with throughput 0. This obviously leads to the conclusion that the stability and efficiency is a lot more difficult to obtain in RCT systems. At the same time both the normative and rights system keep on functioning. As the failure rates are the same (0) we then compare the average speeds.

In the beginning the difference between the two is quite large (up to 20%, Fig. 1) in favour of the Rights system. As the number of cars entered per time unit increases the rights agents results (the average speed) are converging toward the results of normative system. This is expected and explained by the fact that when the environment becomes more hostile (more cars using the road simultaneously) the right to "not being obstructed by other agents" is used a lot more. The fact remains however that the Rights system is more, and in worse case scenario just as efficient as Normative system. In double lane scenarios we see this even more clearly as failure rate for the Rights system is significantly lower than in the Normative one. In dynamic MA Systems, we want autonomous agents to obtain the best stable results using as few resources as possible. Any agent that fails is a waste of resources. We are therefore interested in as high survival rate as possible. The results clearly state that in most scenarios the Rights system will be the most successful one.

8   Discussion and Insights

So what do these results mean? We have already established earlier that a result comparison should not be taken as it is but rather a behaviour pattern needs to be discovered. If we look at the graphs and the behaviour of the systems we can clearly see some patterns emerging. In the RCT system, the more complex the scenario becomes the worse does the system perform. In non complex scenarios the RCT system will on average perform better that the normative one. Reason being that RCT does not have to follow any behaviour constraining rules. In a situation with only one agent there is no risk for crashes and the agent does not have to take into account anything else. It can therefore use its full potential. In a complex situation however the free choice means that agents cannot have full knowledge of how others will behave. This results in crashes. For non complex systems RCT will perform extremely well. On the other hand in a normative system we see little difference between very complex and non complex scenarios. The system performance worsens only marginally when the complexity becomes higher. In a non complex scenario the agents will not perform at the top of their capabilities and the system efficiency will not be utilised to maximum. The norms make sure that the agents always perform the same. The rights based system behaves differently to the other two. In non complex situations it behaves like a RCT system and in very complex situations it behaves like a normative one. As complexity goes up the behaviour of a rights system converges towards a normative one. This can be illustrated with a single car driving on a road. When there is no one else that could be affected by a car's actions, the car will drive as fast as possible. In a very complex the rights of others might outweigh the rights of this agent. It will therefore adjust its behaviour to others just as agents do in normative systems with the difference that for each time step the particular norm set might be different. In the rights system, the rights are flexible. Different rights will be applied depending on the complexity of the situation. The more complex a situation is, the higher hierarchy rights will be used. This means that the system as a whole changes its behaviour depending on what is best for it. A rights based system can be a RCT or a normative system depending on the circumstances.

9   Conclusions and Further Work

We have now presented an empirical comparison of free, normative and rights based agent coordination mechanisms in a simple car traffic simulation scenario. Rights give a system flexibility to perform more efficiently. In non complex situation it allows the agent to behave like RCT and in very complex scenarios the agent will behave like a normative one. This flexibility between the two extremes and a range of in between stages and the fact that the system adjusts itself make Rights a very promising alternative to RCT and norms. Further work will focus on more complex scenarios with better defined behaviours, more norms and more rights as well as other types of social norms.

References

1. Alonso E. (2004), Rights and Argumentation in Open Multi-Agent Systems. Artificial Intelligence Review, 21(1), 3-24.
2. Alonso E. (2004), A Formal Theory of Rights and Argumentation in Open Normative Multi-Agent Systems. In W. Zhang and V. Sorge, (Eds.), Distributed Constraint Problem Solving and Reasoning in Multi-Agent Systems, Frontiers in Artificial Intelligence and Applications 112, IOS Press, 153-167.
3. O'Hare G.M.P. and Jennings N.R. (Eds.). Foundations of Distributed Artificial Intelligence. John Wiley and Sons. New York, 1996
4. Reiner R. Arguments against the possibility of perfect rationality. Minds and Machines, 5:373-389, 1995.
5. Shoham, Y. & Tennenholtz, M. (1992). On the synthesis of useful social laws for artificial agents societies. In Proceedings of the Tenth National Conference on Artificial Intelligence, AAAI-92, 276{281. Menlo Park, CA: AAAI Press.
6. Sycara K.P. Multiagent Systems. AI Magazine, 19:79-92, 1998.
7. Tennenholtz, M. (1998). On stable social laws and qualitative equilibria. Artificial Intelligence 102: 1-20.
8. Walker, A. & Wooldridge, M. (1995) Understanding the emergence of conventions in multi-agent systems. In Proceedings of the First International Conference on Multi-Agent Systems, ICMAS-95, 384-389. Cambridge, MA: MIT Press.
9. Weiss G. Multiagent Systems: A Modern Approach to Distributed Artificial Intelligence. MIT press, Cambridge, MA, 1999

Preface to CoMoGIS 2005

Michela Bertolotto

CoMoGIS 2005 was the second International Workshop on Conceptual Modeling in GIS held in conjunction with the annual International Conference on Conceptual Modeling (ER'05) on October 27th, 2005 in Klagenfurt, Austria. Following the success of CoMoGIS 2004 (held in Shanghai, China), its aim was to bring together researchers investigating issues related to conceptual modeling for geographic and spatial information handling systems, and to encourage interdisciplinary discussions including the identification of emerging and future issues.

The call for papers attracted 31 papers of which 12 were selected for presentation at the workshop and inclusion in the proceedings. The accepted papers relate to topics that range from spatial and spatio-temporal data representation, to the management of spatial relations, to spatial queries, analysis and data mining, to 3D data modelling and visualisation. Our key note presentation, given by Prof. Andrew Frank, discussed the definition of functors for extending map algebra to deal with temporal data.

The workshop would not have been a success without the efforts of many people. I wish to thank the authors for the high quality of their papers and presentations, and the International Program Committee members for their invaluable contribution. Many thanks to the ER'05 and local organisers for their help. Furthermore I would like to thank Prof. Andrew Frank for accepting to be our keynote speaker.

J. Akoka et al. (Eds.): ER Workshops 2005, LNCS 3770, p. 193, 2005.
© Springer-Verlag Berlin Heidelberg 2005

Map Algebra Extended with Functors
for Temporal Data

Andrew U. Frank

Dept. of Geoinformation and Cartography,
Technical University, Vienna
frank@geoinfo.tuwien.ac.at

Abstract. This paper shows how to extend and generalize Tomlin's Map Algebra to apply uniformly for spatial, temporal, and spatio-temporal data. A specific data layer can be seen as a function from location to a value (Goodchild's geographic reality). Map layer but also time series and other similar constructions are functors, mapping local operations to layers, time series, etc. Tomlin's Focal Operations are mostly convolutions and the zonal operations are summaries for zones. The mathematical framework explained justifies polymorphic overloading of operation names like + are made to work for layers, time series, etc. There is also a uniform method to apply user-defined local functions to them. The result is a consistent extension of Map Algebra with a simplified user interface. The implementation covers raster operations and demonstrates the generality of the concept.

1 Introduction

The integration of temporal data into GIS is arguably the most important practical problem currently posed to the GIS research and development community (Frank 1998). Temporal data is collected for administration and scientific applications of GIS. Geographic data gives nearly always a snapshot of the state of our ever changing environment (Snodgrass 1992). These collections of snapshots contain information about changes and processes, but users are left to invent their own methods for temporal analysis. Many *ad hoc* extensions to commercial systems to handle spatial data from different epochs are reported (for example, at recent ESRI user conferences).

The central concept in GIS is the overlay process: Data from different sources are combined (figure 1). This is a computational version of the traditional physical overlaying of maps on a light table (McHarg 1969). Map Algebra is a strong conceptual framework for this method of spatial analysis and has changed little in the 20 years since its "invention" (Tomlin 1983b; Tomlin 1983a), which demonstrates its conceptual clarity. Dana Tomlin's Ph.D. thesis described Map Algebra in a semi-formal way providing all the information necessary for others to use and to produce implementations of Map Algebra (Tomlin 1983a). Several public domain or low cost implementations were around since 1980 (Tomlin's IBM PC version, OSUmap, IDRISI, to name but a few I have used). All the commercial GIS today organize geographic data in layers or themes (ESRI 1993) and contain map algebra operations;

J. Akoka et al. (Eds.): ER Workshops 2005, LNCS 3770, pp. 194–207, 2005.

OGC and ISO standards include them as well (sometimes overshadowed by a multitude of operations for the maintenance of the data and administrative queries).

I demonstrate in this paper that Map Algebra can be extended to include temporal data analysis. I start from the observation that processing of time series is similar to Map Algebra overlay operations. The clarification of the underlying theory—category theory and functors —leads to a generalization that extends Map Algebra to include processing of time series. Extended Map Algebra applies to spatial, temporal, and spatio-temporal data and generalizes the current implementations. It simplifies the user interface:

- The same operations apply to spatial, temporal, and spatio-temporal data, which reduces learning of commands, and makes the experience users have with Map Algebra valuable to solve spatio-temporal problems.
- Irrelevant detail is removed from the user interface.
- Users can define new functions without learning a special language.
- Map layers and snapshots are typed and errors are detected before starting lengthy processing to produce non-sense results.

The theory produces the consistency in the approach and justifies the solution. In addition, it gives guideline for the implementation and optimization of execution. The implementation in a very high-level language (Peyton Jones, Hughes et al. 1999) is only a few pages of code. It uses the second order concepts built into modern languages and demonstrates feasibility. The translation into imperative languages is straight forward.

The paper is structured as follows: the next section reviews map overlay and processing of time series. The following sections prepare the mathematical background, first discussing functions and then mappings between collections of functions seen as categories. The next two sections apply these concepts to extend processing of single values in formulae to local operations applied to collections of values, like map layer or time series. We then introduce summary functions and show, how focal operations fit. Zonal operations are a special case of summary operations and introduce comparable regional operations. At the end we list the improvement of Extended Map Algebra compared to current solutions and review the solution from an abstract point of view. The paper closes with a suggestion for future work and a summary.

2 Map Algebra

Map Algebra realizes the central tenet of GIS: the analysis of data related to the same location in space (figure 1). Tomlin has organized the operations with map layers in three groups (Tomlin 1990):

1. Local operations combine values from the same location,
2. Focal operations combine values from a location and the immediate surroundings, and
3. Zonal operations combine values from all locations in a zone.

In this paper I will first concentrate on the local operations and then show how focal and zonal operations fit in the framework.

Local operations in map algebra are intuitively understandable: the values of corresponding (homological) cells are combined to produce a new value (figure 1 left). The implementation is straightforward with loops over the indices of the raster array:

```
procedure overlay (a, b: layer; out result: layer; op:
function)
for i:= 1 to ymax do
     for j:= 1 to xmax do
          c [i,j] := op (a[i,j], b[i,j]);
```

The examples for map layers are male and female population per cell (*mpop* and *fpop*). They can be combined as shown in figure 1 left. Classification separates urban from rural areas, using a threshold of 300 persons/cell (figure 1 right).

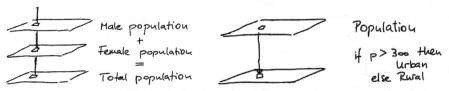

Fig. 1. (left) Total population is male population plus female population; (right) City areas are cells with total population higher than 300

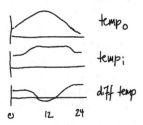

Fig. 2. The temperature difference between outside and inside

The processing of time series is similar (figure 2). Given two time series for the inside and outside temperature (*tempo* respectively *tempi*), the difference between the two at any given moment in time gives a new time series (*difftemp*). The computation for adding map layers or computing the difference between two time series is similar: combine homological, respective synchronous, values. What is the general rule? What are the limitations?

3   Computations are Functions Transforming Sets

A time series can be seen as a function $f(t) = \ldots$ Computation with functions is usual in electrical engineering; image processing uses a concept of images as a 2 dimensional function $g(x,y) = ..$(Horn 1986). Goodchild (1990; 1992) has suggested that geographic reality is a function of location and time $f(x, y, h, t)$. Can this viewpoint contribute to Map Algebra?

3.1 Definition of Function

"A function *f: S -> T* on a set *S* to a set *T* assigns to each *s* ∈ *S* an element *f(s)* ∈ *T*. ... The set *S* is called the domain of *f*, while *T* is the codomain. "... A function is often called a 'map' or a 'transformation' " (Mac Lane and Birkhoff 1967 p. 4). For example the function + takes pairs of values and maps them to a single value (+ :: (Int, Int) -> Int). The classification function *ur* takes a single value from the domain *population count* and maps it to the set *U* with the values *Urban* or *Rural*.

> *ur (p) = if p > 300 then Urban else Rural.*

3.2 Application Functions

A function takes a value from a cell—or corresponding cells—and produces the value for the corresponding cell in the new layer. In figure 1 (left), this is the operation +, in figure 1 (right) the classification function *ur(p)*. In figure 2 the two time series are combined with the function *diff (t1, t2) = t1 – t2*.

```
ur x = if x > (300::Float) then 'U' else 'R'
diff a b = a - b
```

These functions operate on values from domains of relevance to the user: Population density (in a discrete case: population count per cell), temperature, etc. They are sets of values and operations that map between them with rules like: *a + b = b + a* (commutative law). The domains, the operations, and the rules are applicable to these operations form algebras (Couclelis and Gale 1986; Loeckx, Ehrich et al. 1996).

Users *compose* more complex formulae, for example to compute the percent difference between male and female population:

> *(mpop – fpop) * 100/ pop*

or convert the temperature from degree Centigrade to degree Fahrenheit (*c2t*). For other datasets, one could compute the value for the 'universal soil loss formula'.

```
reldiff a b = abs (a - b) * (100.0::Float) / (a + b)
c2f t = 32 + (t * 9 / 5 )
f2c t = 5 * (t - 32)/ 9
```

3.3 Data Management Functions

The local operations for map layers or time series apply the application functions uniformly to every value (or pair, triple, etc. of homological values). This is encoded as a loop. Templates in C++ (Stroustrup 1991) separate the management of the data storage and access to the data from the processing with the application functions. The data management functions (in C++ called "iterators") are closely related to the data constructors.

A layer can be seen as a function that gives to each pair of indices the value at that position. For example, the layer population is a function:

> *pop :: (Int x Int) -> p.*

In the code above these access functions are written as *x[i,j]*. The generalization to volumes and combinations of times series with spatial data is immediate. Comparable 1, 2 and 3 dimensional snapshot and temporal data constructors are:

timeSeries:: t -. v
layer :: xy -> v
stack of layer :: h -> xy -> v
volume :: xyh -> v
timeSeries of layer:: t -> xy -> v
timeSeries of volume :: t -> xy -> h -> v

4 Morphism

Functions (mappings, transformations) that preserve algebraic structure are called morphism; for example the function double is a morphism of addition. A morphism $M:: C \to D$ maps the element of the domain to the codomain and maps the operation f $:: C \to C$ on the domain to a corresponding operation $f':: D \to D$ on the codomain, such that $M(f(x)) = f'(M(x))$ (figure 3) (Mac Lane and Birkhoff 1967 p. 37)

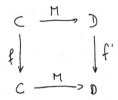

Fig. 3. Morphism M maps f to f'

In many cases we use the same name for the two operations despite the fact that they apply to different domains (polymorphism). Calculation with logarithms using the rule $log(a * b) = log\ a + log\ b$ provides an example, where an operation is mapped to a seemingly very different operation (* becomes +). The function log is a group morphism.

5 Construction of New Concepts from Existing Ones

The commonly available mathematical methods are not sufficient to construct a theory of map algebra. Functions operate on values, not layers. How to construct an extension?

5.1 Extension by Functors

Constructions in mathematics follow often the same pattern: A representation together with operations is found to be insufficient; the example from high school is that integers are insufficient to represent the solution to the problem of dividing 2 pies between 3 people, giving *2/3* to each.

Fractions are introduced as pairs of integers (numerator and denominator), such that the rules for addition and multiplication 'carry over'; integers map by the functor F to fractions by adding the denominator 1; rules like $a + 0 = a$ are preserved: $a/1 +$ $0/1 = a/1$; addition and multiplication of fractions is commutative, etc. The same "trick" has been used to construct point coordinates, imaginary numbers, polynomials, etc. How to apply to geography?

"Many constructions of a new algebraic system from a given one also construct suitable morphism of the new algebraic system from morphism between the given ones. These constructions will be called 'functors' when they preserve identity morphism and composites of morphisms." *(Mac Lane and Birkhoff 1967 p.131)*. Functors map domains to domains and morphism to morphism, i.e., they map categories to categories. Categories are sets of morphism, for example, the application domains used to encode map layers and the operations applicable to them form a category *V*. In our examples these are *integers, reals,* and the set *{Urban, Rural}* and the operations *+, -, *, ur, c2f,* etc. In each category, there is a special morphism, called *identity* morphism, which maps every value into the same value; it is the 'do nothing' operation. Categories have a single operation, namely *composition* of functions, written as '.' : $f(g\ (x)) = (f.g)\ x$.

6 Map Layers and Time Series are Functors

The representation of a snapshot of geographic reality for an area with a single number is insufficient. We may approximate the properties with a collection of values describing small areas (cells). A sampled image can be represented as a regular grid of values of type *v*, which we describe as a function

> *vlayer:: (Int x Int) -> v.*

Local operations in map algebra correspond to the operations on values: we add two layers and understand this operation to add homological cell values. For example, adding two layers $C = A + B$ means $(c[i, j] = a\ [i,j] + b[i,j])$.

The constructor for map layers of a fixed size, which makes a "map layer of type *r*" from values of type *r*, is a functor *M*. The function *M*, which takes a single value *v* from the category *M* and maps it to a map layer produces layers with a all cells of this value $(a[i, j] = v)$. Rules valid for the domain of cell values apply also for layers: for example the commutative law is valid for operations with layers: $A + B = B + A$). A similar argument—restricted to 1 dimension—shows that 'time series' is a functor as well (for historic reasons, it could be called *fluent* (Lifschitz 1990)). Indeed, all constructors listed above are functors.

6.1 Lifting Operation with Functors

Functors map operations on single values to operations on layers, time series, etc. It is customary to use the (polymorphic) operation *lift* to describe this mapping; we say the operation + is lifted to apply to layers (figure 1 left). If *M :: C -> D* is a functor and *op :: C -> C* is an operation then

> $op' = M\ (op) = lift\ op :: D -> D.$

Because functors preserve composition $(M\ (f\ .\ g) = M\ f\ .\ M\ g)$ not only single operations can be lifted, but also complex formulae like *ur, c2f,* or, e.g., the 'universal soil loss equation'.

Functors preserve the rules applicable to the operations. It is therefore appropriate to use the names of the ordinary functions directly for the corresponding operations on layers, time series, etc. (polymorphism). This simplifies the user interface: in stead of commands like *localSum, localDifference* (Tomlin 1990 p. 72) one uses directly the familiar +, - and writes:

```
pop = mpop + fpop
              -- adding male and female population
pop' = fpop + mpop
difftemp = tempi - tempo
              -- computing the difference in temp
tempif = lift c2f tempi
              -- conversion of time series to ºC
cities = lift ur pop
              -- classify in urban/rural
```

This works with time series of Population layers (e.g., population layers from 1950, 1960, 1970…). Applying the condition for *urban* to a time series of population layers gives a time series of a growing number of urban cells.

```
popTS = mpopTS + fpopTS
-- constructing a time series of total population
citiesTS = lift1 (lift1 ur) popTS
-- classify the time series of total population
```

If a formula contains a constant, the constant is lifted and becomes a layer with all values equal to the constant. Constant layers are useful to compute distances between locations: construct a layer where each cell contains the coordinate for its central location (*coord* layer) and then apply a "distance to point *p*" to it gives a layer with the distance of each cell from *p* (here *p*= 2, 3).

```
distance (x1,y1) (x2,y2)
              = sqrt (sqr (x1 - x2) + sqr (y1 - y2))
distTo23 = lift (distance (2,3)) coord
```

7 Summary Operations

For a time series, one might ask, what the maximum was or the minimum value, for example, for the temperature during a day, but the same questions is valid for a map: maximum or minimum population per cell, maximum or minimum height. For values on a nominal scale, one might ask what the most often occurring value is. From these primary summary values, other summary values are derived, e.g., average or higher statistical moments.

There are only a small number of functions that can serve to compute a summary. I propose the hypothesis, that only binary functions $f :: a \rightarrow a \rightarrow a$ with a zero such that $f(0, a) = a$ and which are additive can be used. Additive for a function in this context means that for two datasets A and B, the summary of the summary $(f(A))$of A and the summary $(f(B))$of B must be the same as the summary of A merged with B $(A + B)$

$$f(f(A), f(B)) = f(A + B).$$

A summary operation has no corresponding operation on a single value; a summary with the + operation is comparable to an integral over the time series, the map layer or the data volume. The operation *fold* converts a summary function like + to an operation on a data collection; for example *fold (+)* computes the sum for the data collection, *fold min* the minimal value. The second order function fold depends only on the data type.

Summary operations can apply to the complete data collection (e.g., a time series of data volume) or be applied to 'slices', for example the summary for each data cube in a time series gives a time series of single (summary) values (figure 4). Another example: a summary in a data cube can be for each layer giving a value for each height, or can be vertical for each location, summing all the values in the different heights.

```
popsum   = sum pop              -- total population
tempimax = fold max tempi  -- max temperature
tempomin = fold min tempo
popTSsum = lift sum popTS
              -- time series of total population
```

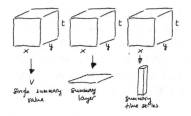

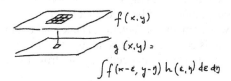

Fig. 4. Different summaries for a time series of layers

Fig. 5. A focal operation combines values in a neighborhood

8 Focal Operations in Map Algebra: Convolutions

Focal operations combine values in the neighborhood of a value to a new value (figure 5). Such operations do not have an equivalent operation for single values. Considering a layer as function reveals that Tomlin's zonal operations (or nearly all of them) are convolutions. A convolution of a function f with a weight function h (kernel) is the integral of the product of the layer f times the kernel function h (this is comparable to a weighted average).

Focal operations can be applied to functions of any number of variables. The 2-, 3-, or 4- dimensional convolutions are constructed analogously as double, triple, and quadruple integrals. Convolution is linear and shift-invariant (Horn 1986 p.104). Best known are the discrete forms of convolution, used as filter operations in image processing software. A 'focal average' operation is close to a filter to smooth an image, with a special weighting function (a Gaussian), but convolution can be used to detect edges, etc.

Applying a convolution to a multi-dimensional dataset, say a data volume (*vol: z -> x -> y -> v*) allows some options, which must be selected according to the application requirements: A *3d*-convolution can be applied to the volume (*conv k3 v*) (figure 6 left). Considering the data volume as a stack of layers, a *2d*-convolution can be applied to each layer in the stack (*lift (conv k2) v*) (figure 6 middle). Applying a *1d*-convolution to the data in each vertical data set is *conv k1 v* (where *k1* is a *1-d* kernel to select by polymorphism the *1d* convolution, which is then applied as a local operation to homologous values in the stack) (figure 6 right). In each case, different neighborhoods are used in the convolution.

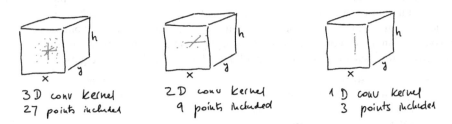

Fig. 6. 3-, 2-, and 1-dim convolution

Focal operations are very flexible if we permit functions that include conditions. For example, the Game of Life of John Conway can be expressed as a convolution or a convolution can be used to compute the next step in a simulation of a forest fire, given wind, amount of fuel, etc. The kernel could be the rule: if the central cell contains fuel and is burning then it continues to burn (with reduced fuel left); if it is not burning and one of the neighbor cells upwind from this cell then this cell starts burning as well.

9   Zonal Operations in Map Algebra

For time series, it is often desirable to compute the summary values for each day; for example a daily maximum and minimum. Such summaries for a regular aggregation are rare for space; here the summary should be computed for an irregular area. The area can be given as a spatial layer with value *True* for the cells included in the area; this is in image processing known as the characteristic function. Combining this area layer with the value layer by a function $f(a, v) = $ *if a then v else 0* and then apply the summary function to the result gives the summary for the desired area. This works for all dimensions of data sets. The combination of the layers and the summary operation can always be combined in a single operation (for a proof (Bird and de Moor 1997 p. 10)).

Tomlin introduced summary functions that compute a single a summary value for a zone. Tomlin's definition of a zone is an "area which has some particular quality that distinguishes it from other geographic areas" (Tomlin 1990 p. 10). Practically, zones are areas that have the same value in a layer; zones are not necessarily connected.

For many applications of spatial analysis zones are not suitable and simply connected regions are required. For example, to assess the suitability of a habitat the size of a connected wood area (a wood region) can be important and cannot be replaced with the size the wood zone, which gives the total wood area in the area (Church, Gerrard et al. 2003). To identify regions requires a function to form the connected components of a zone.

Zonal operations compute for each zone (or for each region) a summary value. In Tomlin's Map Algebra, the zonal operations are defined such that they produce again a map layer, where every cell in the layer has the value obtained for the zone (or region) it is part of. This is necessary to make Map Algebra closed—operations have layers as inputs and produce layers as results.

The zonal operations apply to all dimensions of data collections. For time series, one might ask: what is the maximum temperature for the 'rain zone'? (The 'rain zone' can be defined, for example, as days for which more than 5 mm precipitation was measured).

```
condUrban uOr popVal = if uOr == 'U' then popVal else
zero
popCity = lift condUrban cities pop
-- restrict the population values to the urban zone
citypop = sum popCity
-- total population in urban zone

-- the urban zones and the their total population
--          for the population time series
popCityTS = lift (lift condUrban) citiesTS popTS
citypopTS = lift sum popCityTS
```

10   What Is Achieved?

10.1   Consistent Operations for Maps, Time Series, Stacks of Maps, etc.

Map layers and time series, their combination and their extensions are all functors and instances of a single concept, namely a class of functors *layerN*, where *n* can be 1 for time series, 2 for Tomlin's Map Algebra, and 3 for either *2d* space and time or *3d* space and 4 for *3d* space and time (figure 6). This covers Godchild's concept of 'geographic reality' $f(x,y,h,t)$ and reveals it as a functor.

10.2   Polymorphic Application gives Local Operations "for Free"

Polymorphism and automatic lifting of functions makes operations and functions that apply to single values to apply to data collections as local operations. In stead learning a command like *localSum* or *localMinimum* one can directly use the operations + or *min* and write l*ayerC* = *layerA* + *layerB* or *layerD* = *min (layerA, layerB)*. Functors preserve the rules applicable to operations; users know that $C = A + B$ is the same as $C' = B + A$.

10.3   Extensibility Without Special Language

If an application requires the value for a formula, for example, how much is the difference between the highest and the lowest value in relation to the average, one could search in the manual if such a Map Algebra function exists or combine it from available functions, computing intermediate layers. Knowing that layer is a functor, we just write the function and apply it (with *lift*) to the layers. This is valuable, as sequences of routine Map Algebra operations can be combined in a single formula and users need not learn a special scripting language.

```
r a b = 2.0 * (max a b - min a b )/(a + b)
rpop = lift r mpop fpop
```

10.4 Typing

The description of map algebra given here is compatible with a typed formalism (Cardelli and Wegner 1985; Cardelli 1997) and requires it to make polymorphism work. It is therefore possible to identify errors and advise users to avoid non-sense operations: if A is a layer with values *True* or *False*, then $C = A + B$ is ill typed. The operation + is not applicable to layers with Boolean values, because + is not applicable to Boolean values and can not be lifted to layers of Booleans.

10.5 Implementation

The strong theory helps us with the implementation. In a modern language, constructs like Functor or lift are directly expressible (in Haskell directly, in C++ with templates).This gives an implementation for local operations of map algebra in few lines of high level code.

The use of the GHCi (available from www.haskell.org) interpreter gives a full environment, in which functions can be loaded from files and computations executed interactively and results displayed. All the examples included here have been run in this environment. Functions that are used often could be compiled with GHC to improve performance. The concepts explained here are not tied to this or any other programming languages and can be implemented with any current programming language.

10.6 Optimization

Using Map Algebra leads to the computation of many intermediate layers, which are then used as inputs to other operations. This is useful when exploring a question; once a sequence of operations that lead to the desired result is found, the intermediate layers are a nuisance, because they require careful naming to avoid confusions. In the categorical framework, the production of intermediate layers can be avoided and all the operations done in one sweep of the data, for example using rules like:

$$lift f (lift g \ la \ lb) = lift (f . g) \ la \ lb.$$

This rule should remind programmers of methods to merge sequences of loops into a single one! Having a strong theory gives us guidelines when and how optimizations are possible; understanding when optimization is not permitted may be more important to avoid producing wrong results. For example, it is not evident how to combine local and focal operations; understanding that a focal operation is a convolution and as such linear answers the question.

11 Why Does This Work

Category theory is the abstraction from algebra, it is 'one step up the abstraction ladder' (Frank 1999).Why is it useful here? Category theory is unifying very large parts of mathematics and brings them into a single context. It integrates in the same framework the theory of computing: "The main methodological connection between programming language theory and category theory is the fact that both are essentially 'theories of functions'." (Asperti and Longo 1991 p.ix).

All computing is in a single category, namely the category of sets with functions. The state of a computer is the collection of the values in all memory cells, CPU, output channels, etc. All computing is a—program controlled—transformation of the state of the machine to a new state. Every program, but also every step in a program is a function that transforms the state of the machine; application programs are composition of functions.

In the categorical framework, a mathematical treatment of computing becomes feasible. Besides standard mathematics, the construction of data types and the control of flow in a program with its statements are expressible. Functions with *if_then_else* logic can be written with the McCarthy conditional function and compose with other functions.

In general, programming an application means—independent if the analyst and programmer understand this theoretically or not—as the formalization of a suitable category for the concepts and operations of the application and then a mapping of this category to the category of sets because this is the category in which the program works.

12 Future Work

The description here as well as Tomlin's original text, does not depend on a discrete, regular raster. The theory is developed in terms of continuous functions in 2, 3, or more variables. The translation to a discrete, regular raster based implementation is immediate; the use of data represented by irregular subdivision (so-called vector data) is more involved. I suggest as future work a comprehensive investigation resulting in a theoretically justified implementation of Extended Map Algebra covering both regular and irregular subdivision data with the same operations.

13 Conclusion

As expected, there exists a strong theory behind Map Algebra. The identification of this theory has lead us to see commonalities between the methods we use to process spatial and temporal data and to understand that the general concept of map algebra can be applied to spatial data of 2 or 3 dimension, to temporal data, and to spatio-temporal data of $2d + t$ or $3d + t$; it can be used to combine data collections of any dimension.

The formal theory gives us rules for the reorganization of processing geographic data with Map Algebra operations. Descriptions of processing steps can be formalized and optimized using rules that guarantee that the same result is obtained with less effort.

The user interface can be reduced; the number of operation names required is drastically reduced exploiting polymorphism when it is appropriate. This should reduce the instruction time necessary for future users of GIS. The introduction of typed data can advise users when they try to compute nonsensical operations ahead of time (Cardelli 1997).

My esteemed PhD advisor Rudolf Conzett used to quote Boltzmann: "There is nothing more practical than a good theory". I believe that formal investigations lead to a deeper understanding and, as a consequence, a simplified solution. This paper demonstrates how theory leads to more powerful Extended Map Algebra with at the same time a simplified user interface.

Acknowledgements

Many of my colleagues and students have helped me to advance to this point. Particular thanks go to the Haskell community (www.haskell.org) for their continuous efforts to produce tools to play with category theory and apply it to practical problems. Funding from the European Commission for projects DigitalSEE, Geocompass, Georama and REVIGIS is acknowledged. Special thanks to my former student Dorenbeck for his work on an MSc. Thesis "Spatial Algebras in Geographic Information Systems" in 1991. The discussions with Max Egenhofer, Werner Kuhn and Gilberto Camara since then have contributed substantially to my current understanding.

References

Asperti, A. and G. Longo (1991). Categories, Types and Structures - An Introduction to Category Theory for the Working Computer Scientist. Cambridge, Mass., The MIT Press.

Bird, R. and O. de Moor (1997). Algebra of Programming. London, Prentice Hall Europe.

Cardelli, L. (1997). Type Systems. Handbook of Computer Science and Engineering. A. B. Tucker, CRC Press: 2208-2236.

Cardelli, L. and P. Wegner (1985). "On Understanding Types, Data Abstraction, and Polymorphism." ACM Computing Surveys 17(4): 471 - 522.

Church, R. L., R. A. Gerrard, et al. (2003). "Constructing Cell-Based Habitat Patches Useful in Conservation Planning." Annals of the Association of American Geographers 93(4): 814-827.

Couclelis, H. and N. Gale (1986). "Space and Spaces." Geografiska Annaler 68(1): 1-12.

ESRI (1993). Understanding GIS - The ARC/INFO Method. Harlow, Longman; The Bath Press.

Frank, A. U. (1998). GIS for Politics. GIS Planet '98, Lisbon, Portugal (9 - 11 Sept. 1998), IMERSIV.

Frank, A. U. (1999). One Step up the Abstraction Ladder: Combining Algebras - From Functional Pieces to a Whole. Spatial Information Theory - Cognitive and Computational Foundations of Geographic Information Science (Int. Conference COSIT'99, Stade, Germany). C. Freksa and D. M. Mark. Berlin, Springer-Verlag. 1661: 95-107.

Goodchild, M. F. (1990). A Geographical Perspective on Spatial Data Models. GIS Design Models and Functionality, Leicester, Midlands Regional Research Laboratory.

Goodchild, M. F. (1992). "Geographical Data Modeling." Computers and Geosciences 18(4): 401- 408.

Horn, B. K. P. (1986). Robot Vision. Cambridge, Mass, MIT Press.

Lifschitz, V., Ed. (1990). Formalizing Common Sense - Papers by John McCarthy. Norwood, NJ, Ablex Publishing.

Loeckx, J., H.-D. Ehrich, et al. (1996). Specification of Abstract Data Types. Chichester, UK and Stuttgart, John Wiley and B.G. Teubner.

Mac Lane, S. and G. Birkhoff (1967). Algebra. New York, Macmillan.

McHarg, I. (1969). Design with Nature, Natural History Press.

Peyton Jones, S., J. Hughes, et al. (1999). Haskell 98: A Non-strict, Purely Functional Language.

Snodgrass, R. T. (1992). Temporal Databases. Theories and Methods of Spatio-Temporal Reasoning in Geographic Space (Int. Conference GIS - From Space to Territory, Pisa, Italy). A. U. Frank, I. Campari and U. Formentini. Berlin, Springer-Verlag. **639**: 22-64.

Stroustrup, B. (1991). The C++ Programming Language. Reading, Mass., Addison-Wesley.

Tomlin, C. D. (1983a). Digital Cartographic Modeling Techniques in Environmental Planning, Yale Graduate School, Division of Forestry and Environmental Studies.

Tomlin, C. D. (1983b). A Map Algebra. Harvard Computer Graphics Conference, Cambridge, Mass.

Tomlin, C. D. (1990). Geographic Information Systems and Cartographic Modeling. New York, Prentice Hall.

A Formal Model for Representing Point Trajectories in Two-Dimensional Spaces

Valérie Noyon, Thomas Devogele, and Christophe Claramunt

Naval Academy Research Institute, BP 600, 29240, Brest Naval, France
{noyon, devogele, claramunt}@ecole-navale.fr

Abstract. Modelling moving points is a subject of interest that has attracted a wide range of spatio-temporal database research. Efforts so far have been oriented towards the development of database structures and query languages. The preliminary research presented in this paper introduces a formal analysis of spatio-temporal trajectories, where the objective is to complement current proposals by a categorization of the underlying processes that characterize moving points. The model introduced identifies the semantic exhibited by point versus point, point versus line and point versus region trajectories.

1 Introduction

The integration of the temporal dimension within GIS is an area that has attracted a wide range of spatio-temporal databases research [1], [2]. This is largely favoured by the constant evolution of geo-referencing systems such as transponders for airplanes and ships, and telecommunication and GPS for terrestrial navigation. Applications range from monitoring to simulation systems, prediction and planning studies where the objective is to control and analyse fast changing phenomena such as urban or navigation traffic and transportation systems. Achievements and prototypes realised so far include developments of spatio-temporal data types and query languages [3], [4], [5], [6] and physical storage structures [7], [8]. These research achievements call for further development of exploration interfaces and languages that will help to identify processes, trends, and patterns that characterize the dynamic properties of spatio-temporal applications. In particular, such systems should help traffic and transportation planners to observe and understand the evolution of a given dynamic system at different levels of granularity.

The objective of the research presented in this paper is to explore and develop a trajectory manipulation model that supports not only the representation of mobile trajectories, but also an intuitive data manipulation language that facilitates the understanding of the underlying behaviours, processes and patterns exhibited by moving points. Our objective is not to develop a new spatial query language, but rather to explore the semantics revealed by point trajectories in space and time, and to identify a topological language that allows for a derivation of a complete set of orthogonal and plausible processes.

The remainder of the paper is organised as follows. Section 2 introduces basic principles of trajectory modelling, using either absolute or relative views

J. Akoka et al. (Eds.): ER Workshops 2005, LNCS 3770, pp. 208–217, 2005.

of space, and motivates the need for an integration of absolute and relative representations. Section 3 develops a formal model of mobile trajectories, that is, points moving in a two-dimensional space with respect to points, lines and regions, and their continuous transitions. Finally section 4 concludes the paper and draws some conclusions.

2 Modelling Background

Recent developments in Geographical Information Science tend to provide some alternative models of space that surpass the conventional Euclidean vision of space. Spatial cognition, naive geography are some of the domains that contribute to the emergence of cognitive representations of space ([9], [10]) particularly in the case of "egocentric" views where the objective is to model and materialize space the way it is perceived from a mobile observer active in the environment [11].

Such aspects in the representation of space directly relate to the main difference between an absolute and a relative representation, where in the former the origin and system of reference are fixed, while in the latter they depend of the location of the observer, or the observers when multiple points of view are required. In fact, absolute and relative representations provide complementary views, they can be also partially derived one from the other. Absolute representations are very much adapted and applied to the analysis of global patterns and trends, using conventional spatial data analysis and statistics, while relative representations have not been very much used, to the best of our knowledge, for data analysis and mining. Taking a relative view of space, it might be possible to analyse the behaviour of a moving object in space, with respect to other regions of space with which this object is in interaction. However, this implies to explore in further details the semantics revealed by a trajectory, when perceived from a relative point of view.

At the perception level, cognitive and spatial reasoning studies [12], [13], [14], [15] have shown than distance and speed differences are amongst the relationships and processes that are intuitively perceived and understood by human beings when acting in the environment and perceiving other mobile actors. This leads us to explore and design a trajectory data model, where distance and speed differences are modelled over time. One of our research objectives is to explore to which degree such a model completes the conventional Euclidean view of a spatial trajectory. From a formal point of view, this should lead to design a model made of a complete set of physically plausible and orthogonal process primitives, that also makes sense as far as possible at the cognitive level.

3 Trajectory Modelling

Let us study the trajectory of a moving point, modelled relatively with respect to some parts of space (i.e. point, line or region). Let us consider a two-dimensional space, a moving region A and a moving point B. We say that the moving region

A is the reference of the relative view and then the origin of the relative frame of reference, the moving point B being the target (or the reverse). The two-dimensional relative view is given by a coordinate system where the abscissa axis represents the speed difference between the target point and the origin region. An object materialized by either a line or a region is considered having an homogeneous speed and not deformed through time. The difference of speed $(\Delta v_{A,B}(t))$ between a target point and an origin region (or the reverse) is given as follows :

$$\Delta v_{A,B}(t) = \begin{cases} +\sqrt{(v_{x_A}(t) - v_{x_B}(t))^2 + (v_{y_A}(t) - v_{y_B}(t))^2} \\ \text{If } \|\overrightarrow{v_A}(t)\| \geq \|\overrightarrow{v_B}(t)\| \\ -\sqrt{(v_{x_A}(t) - v_{x_B}(t))^2 + (v_{y_A}(t) - v_{y_B}(t))^2} \\ \text{Otherwise} \end{cases} \tag{1}$$

Per convention, we say that when the target point B is slower than the origin region A, $\Delta v_{A,B}(t)$ is negative, positive on the contrary. The ordinate axis represents, if one of the objects is a region, the minimum Euclidian distance between the region and the point when the point is outside the region, the negative of this value when the point is inside the region, and the null value when the point is in the boundary of the region. When the origin and the target are both points or a line and a point, the Euclidean distance is considered. These distance are given as follows:

$$d_{A \times B}(t) = \begin{cases} +(\sqrt{(x_A(t) - x_B(t))^2 + (y_A(t) - y_B(t))^2}) \\ \text{If the point } B \text{ is outside the region or not within the line A, or} \\ A \text{ and } B \text{ are two points ;} \\ \text{where } (x_A(t), y_A(t)) \text{ denotes the coordinates of the point of } A \text{ the} \\ \text{closest to } B \text{ if } dim(A) = 2 \text{ or } dim(A) = 1, \text{ the coordinates of the} \\ \text{point otherwise} \\ -(\sqrt{(x_A(t) - x_B(t))^2 + (y_A(t) - y_R(t))^2}) \\ \text{If the point } B \text{ is into the region } A \\ \text{where } (x_A(t), y_A(t)) \text{ denotes the coordinates of the point in the} \\ \text{boundary of } A \text{ the closest to } B \end{cases} \tag{2}$$

A two-dimensional representation space is derived from the difference of speed and distance values. This two-dimensional space constitutes a modelling support for the exploration of (1) different dynamic states, (2) possible transitions between them and (3) physically plausible transitions. We hereafter successively study these three aspects.

Firstly, this representation space gives a partition that supports the characterization of the different spatio-temporal configurations, and the underlying processes than can be derived from moving point trajectories (Figure 1):

- $K = \{(\Delta v(t), d(t)) | d(t) > 0 \wedge \Delta v(t) < 0\}$
- $L = \{(\Delta v(t), d(t)) | d(t) > 0 \wedge \Delta v(t) > 0\}$
- $U = \{(\Delta v(t), d(t)) | d(t) < 0 \wedge \Delta v(t) < 0\}$
- $V = \{(\Delta v(t), d(t)) | d(t) < 0 \wedge \Delta v(t) > 0\}$

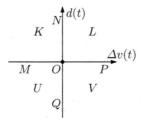

Fig. 1. Partition of the relative space

- $O = \{(\Delta v(t), d(t)) | d(t) = 0 \wedge \Delta v(t) = 0\}$
- $M = \{(\Delta v(t), d(t)) | d(t) = 0 \wedge \Delta v(t) < 0\}$
- $N = \{(\Delta v(t), d(t)) | d(t) > 0 \wedge \Delta v(t) = 0\}$
- $P = \{(\Delta v(t), d(t)) | d(t) = 0 \wedge \Delta v(t) > 0\}$
- $Q = \{(\Delta v(t), d(t)) | d(t) < 0 \wedge \Delta v(t) = 0\}$

Each of these spatio-temporal configurations denotes a spatio-temporal state of a target point B, with respect to a reference region A. We say that a state is valid over an interval of time $i \in I$, I being the set of temporal intervals.

Secondly, a state can be interpreted as a continuous event whose evolution is related to the way the two spatial dimensions considered evolve, i.e., distance and speed differences. Continuous transition between states can be formally studied using the notion of conceptual neighbours [16], [17] as distance and speed over time are continuous functions [18]. More formally, a continuous transition is defined as follows [16]:

Definition 1. *Continuous transition*
A continuous transition between two spatio-temporal states materializes a continuous change without any intermediary state.

The set formed by such conceptual neighbours relations is defined as in [16], where such transitions are orthogonal and form a complete set. Formally, two path-connected sets are conceptual neighbourhoods if there are some continuous transitions between them [19]. This allows us to refine the notion of continuous transitions between two given states:

Definition 2. *Continuous transitions versus path-connected sets*
There is a continuous transition between two states if their union is path-connected.

These definitions allowed us to identify the possible continuous transitions between the states identified [20]. These continuous transitions concern the changes of states from K to M, K and O, K and N, L and P, L and N, L and O, O and M, O and N, O and P, U and M, U and O, U and Q, V and P, V and O, V and Q and O and Q, and their reverses (Figure 2). One can remark that while some transitions between two states are continuous, other not (e.g. N and P or K and L). There is a continuous transition between two states if the

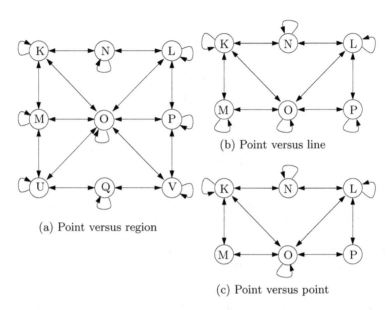

(a) Point versus region

(b) Point versus line

(c) Point versus point

Fig. 2. Continuous transitions

set composed by the two states is a connected set. Conversely, if a set composed by two states is disconnected [21], then there is no continuous transition between them.

Thirdly, let us study the soundness of these continuous transitions as previously defined in [22], and the constraints that materialize them. We still consider a target point B and a referent region, line or point A.

Constraint 1: *Changes of states*
Let us assume that the speeds of on origin region A and a target point B are constant and equal, and that A and B are located at a given constant distance. This represents the case where the point B follows the region A (or the reverse). If that distance is not null, B is in the state N or Q, otherwise in the state O. A change of distance implies a change of difference speed, then a change of state N (resp. Q) or O to state K or L (resp. U or V).

By a straight application of this constraint, it is immediate to derive that there is no sound continuous transition between the states N and O, and Q and O.

Constraint 2: *Stable states*
Let us consider B in a stable state over time with respect to A. If B and A are materialized by points, then a constant and null value of distance over this state should be valid for a given instant only, but not over an interval of time I. Also the distance $d_{A,B}(t)$ cannot be null for any instant $t \in I$ if the difference of speed $\Delta v_{A,B}(t)$ is not null.

Applying constraint 2, the states M and P cannot last over time when the target and origin objects are points. It is also worth noting that some states can be

Table 1. Change of speed processes (with $t_0 < t_1$)

1	$t_0 : v_A > v_B$ $t_1 : v_A = v_B$	Deceleration of A or/and acceleration of B to reach the same speed at t_1
2	$t_0 : v_A = v_B$ $t_1 : v_A > v_B$	Acceleration of A or/and deceleration of B standing from the same speed at t_0
3	$t_0 : v_A < v_B$ $t_1 : v_A = v_B$	Acceleration of A or/and deceleration of B to reach the same speed at t_1
4	$t_0 : v_A = v_B$ $t_1 : v_A < v_B$	Deceleration of A or/and acceleration of B standing the same speed at t_0
5	$t_0 : v_A > v_B$ $t_1 : v_A > v_B$	A is faster than B
6	$t_0 : v_A = v_B$ $t_1 : v_A = v_B$	A and B have the same speed
7	$t_0 : v_A < v_B$ $t_1 : v_A < v_B$	B is faster than A

stable although some others cannot. The application of constraints 1 and 2 make possible the stability over time of states K, N, L and O while states M and P are instantaneous states, only. If the origin or referent objects are materialized by a line (or a region), the states K, N, L, M, O, P, U, Q and V are stable states. Applying constraints 1 and 2, a complete and sound set of continuous transitions can be derived for all representations. Applied to the modelling of the relative trajectories of two mobile objects (i.e. point and region, point and line and two points) A and B, possible continuous transitions are shown in Figure 2. Note that point versus point and point versus line cases don't have the cases where the difference of speed is negative.

There are also no continuous transitions between O and Q, and O and N as these changes of state contradict constraint 1. Applying constraints 1 and 2, and for interactions between a point and a region, a set of twenty eight continuous transitions between states and nine stable states is identified and shown in table 2. For interactions between a point and a line, there are sixteen continuous transitions between states and six stable states (Table 3). The set of continuous transitions representing the interactions between two points is the set of continuous transitions for interactions between point and line without transitions $P \rightarrow P$ and $M \rightarrow M$ (Constraint 2)(Table 3). Table 2 and 3 materialize sound continuous transitions valid for a change of state between two time-stamps, that is, time instants denoted t_0 and t_1 with $t_0 < t_1$. The semantics of the processes that imply a change of speed are ilustred in table 1.

It is worth noting that each continuous transition tends to represent the semantic of a specific and relative process of a target object with respect to an origin object (Table 2). One can also remark that the continuous transitions exhibited by the model correspond to processes that can be discriminated by a natural language expression which has the advantage of being unambiguous and relatively short.

Table 2. Continuous transitions versus dynamic processes (with $t_0 < t_1$)

	From outside the region to the boundary $t_0 : d>0$ $t_1 : d=0$	From the boundary, to outside the region $t_0 : d=0$ $t_1 : d>0$	Outside the region $t_0 : d>0$ $t_1 : d>0$	On the boundary $t_0 : d=0$ $t_1 : d=0$	In the region $t_0 : d<0$ $t_1 : d<0$	From inside the region to the boundary $t_0 : d<0$ $t_1 : d=0$	From the boundary to inside the region $t_0 : d=0$ $t_1 : d<0$
1	$L \rightarrow O$	∅	$L \rightarrow N$	$P \rightarrow O$	$V \rightarrow Q$	$V \rightarrow O$	∅
2	∅	$O \rightarrow L$	$N \rightarrow L$	$O \rightarrow P$	$Q \rightarrow V$	∅	$O \rightarrow V$
3	$K \rightarrow O$	∅	$K \rightarrow N$	$M \rightarrow O$	$U \rightarrow Q$	$U \rightarrow O$	∅
4	∅	$O \rightarrow K$	$N \rightarrow K$	$O \rightarrow M$	$Q \rightarrow U$	∅	$O \rightarrow U$
5	$L \rightarrow P$	$P \rightarrow L$	$L \rightarrow L$	$P \rightarrow P$	$V \rightarrow V$	$V \rightarrow P$	$P \rightarrow V$
6	∅	∅	$N \rightarrow N$	$O \rightarrow O$	$Q \rightarrow Q$	∅	∅
7	$K \rightarrow M$	$M \rightarrow K$	$K \rightarrow K$	$M \rightarrow M$	$U \rightarrow U$	$U \rightarrow M$	$M \rightarrow U$

We illustrate the potential of the relative modelling approach using some simplified examples that support a schematic representation of the different continuous transitions and stability states supported by the trajectory model. These correspond to common dynamic processes exhibited by the behaviour of moving objects in space and time. Figure 3 presents two evolving configurations where a target object is getting close of the origin object at an increasing speed (Figure 3.b), while in Figure 3.d, a mobile object is getting close, but with a difference of speed that decreases. Taking an absolute view, these behaviours are ambiguous as changes of speed are not represented, whereas with the relative representation of difference of speeds, behaviours are unambiguous.

Table 3. Continuous transitions and dynamic behaviours (with $t_0 < t_1$)

	Connection	Separation	Same location	Different locations
	$t_0 : d > 0$ $t_1 : d = 0$	$t_0 : d = 0$ $t_1 : d > 0$	$t_0 : d = 0$ $t_1 : d = 0$	$t_0 : d > 0$ $t_1 : d > 0$
1	$L \to O$	$\varnothing$	$P \to O$	$L \to N$
2	$\varnothing$	$O \to L$	$O \to P$	$N \to L$
3	$K \to O$	$\varnothing$	$M \to O$	$K \to N$
4	$\varnothing$	$O \to K$	$O \to M$	$N \to K$
5	$L \to P$	$P \to L$	$P \to P$	$L \to L$
6	$\varnothing$	$\varnothing$	$O \to O$	$N \to N$
7	$K \to M$	$M \to K$	$M \to M$	$K \to K$

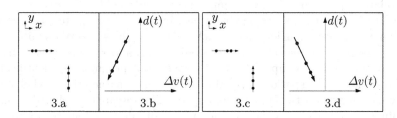

3.a   3.b   3.c   3.d

Fig. 3. One object getting close to the other

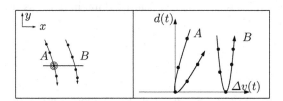

Fig. 4. Two points that touch a line with different behaviours

Figure 4 introduces an example where two target points cross a line with different behaviours. Point A stops on the line before moving away whereas point B crosses the line only, change which is apparent in the relative view only.

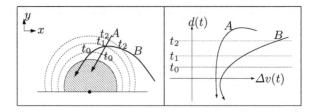

Fig. 5. Two points that enter into a region with different behaviours

Figure 5 presents another example of two target points with respect to an origin region. The distance between the region and the point, at t_0, t_1 and t_2, are the same for two target objects but the relative speeds are different. Speed differences are highlighted by the relative view in Figure 5.2.

4   Conclusion

Recent developments of database structures and languages for the modelling of moving points offer many opportunities for exploratory interfaces that will characterize the semantics exhibited by the underlying processes revealed by these trajectories. This paper introduces a qualitative representation of point trajectories where a relative-based view constitutes an alternative to the conventional Euclidean representation of space. The model is based on two trajectory primitives: relative speed and distance, that are commonly used and perceived as the basic relative constituents of a moving object in space. A complete set of orthogonal dynamic processes is identified, it characterizes the semantics exhibited by a moving point with respect to target points, lines and regions. Such a model complements the cartographical view of space and permits the identification and distinction of the spatio-temporal processes that characterize the behaviour of moving points in a two-dimensional space. The research developed so far is preliminary, further work concerns integration of additional spatial properties such as orientation and acceleration, and implementation of a prototype for the monitoring and analysis of mobile trajectories. We believe that this modelling approach can be applied to several application domains such as traffic monitoring in air, ground and sea.

References

1. Abraham, T., Roddick, J.: Survey of spatio-temporal databases. Geoinformatica **3** (1999) 61–99
2. Su, J., Xu, H., Ibarra, O.: Moving objects: Logical relationships and queries. In: Proceedings of the Seventh International Symposium on Spatial and Temporal Databases (SSTD). (2001) 3–19
3. Sistla, A., Wolfson, O., Chamberlain, S., Dao, S.: Modeling and querying moving objects. In: International Conference on Data Engineering (ICDE). (1997) 422–432

4. Güting, R., Böhlen, M., Erwig, M., Jensen, C., Lorentzos, N., Schneider, M., Vazir-giannis, M.: A foundation for representing and querying moving objects. ACM Transactions on Database Systems (TODS) **25(1)** (2000) 1–42

5. Porkaew, K.: Database support for similarity retrieveal and querying mobile objects. PhD thesis, University of Illinois (2000)

6. Huang, B., Claramunt, C.: STOQL: An ODMG-based spatio-temporal object model and query language. In Richardson, D., Oosterom, P., eds.: 9th Spatial Data Handling symposium, Springer-Verglag (2002) 225–237

7. Mokbel, M., Ghanem, T., Aref, W.: Spatiotemporal access methods. IEEE Data Engineering Bulletin **26** (2003) 40–49

8. Pfoser, D.: Indexing the trajectories of moving objects. IEEE Data Engineering Bulletin **25** (2002) 4–10

9. Tversky, B.: Cognitive maps, cognitive collages, and spatial mental models. In: Conference On Spatial Information Theory (COSIT). (1993) 14–24

10. Maass, W.: A cognitive model for the process of multimodal, incremental route descriptions. In: Conference On Spatial Information Theory (COSIT). (1993) 1–13

11. Imfeld, S.: Time, points and space - Towards a better analysis of wildlife data in GIS. PhD thesis, University of Zürich (2000)

12. Vieu, L.: Spatial Representation and Reasoning in Artificial Intelligence. In Stock, O., ed.: Spatial and Temporal Reasoning. Kluwer, Dordrecht (1997) 3–41

13. Gibson, J., ed.: The Ecological Approach to Visual Perception. Houghton-Mifflin (1979)

14. Levinson, S.: Studying spatial conceptualization across cultures: Anthropology and cognitive science. The Journal of the Society for Psychological Anthropology **26** (1998) 7–24

15. Morineau, T., Hoc, J.M., Denecker, P.: Cognitive control levels in air traffic radar controller activity. International Journal of Aviation Psychology **13** (2003) 107–130

16. Freksa, C.: Temporal reasoning based on semi-intervals. Artificial Intelligence **54** (1992) 199–227

17. Galton, A.: Qualitative Spatial Change. Oxford University Press (2000)

18. Galton, A.: Space, Time, and Movement. In: Spatial and Temporal Reasoning. Kluwer Academic Publishers (1997)

19. Worboys, M., Duckham, M.: GIS: A Computing Perspective. 2 edn. CRC Press (2004)

20. Noyon, V., Devogele, T., Claramunt, C.: A relative modelling approach for spatial trajectories. In: The Proceedings of the 4th ISPR Workshop on Dynamic and Multi-dimensional GIS. (2005) to appear

21. Egenhofer, M., Franzosa, R.: Point-set topological spatial relations. International Journal of Geographical Information Systems **5** (1991) 161–174

22. El-Geresy, B.: Qualitative representation and reasoning for spatial and spatio-temporal systems. PhD thesis, University of Glamorgan (2003)

A Logical Approach for Modeling Spatio-temporal Objects and Events

Cristian Vidal and Andrea Rodríguez

Department of Computer Science, University of Concepción,
Edmundo Larenas 215, Concepción, Chile
{cristianvidal, andrea}@udec.cl

Abstract. The formal specification of spatio-temporal information is essential to the definition of spatio-temporal database systems. The main contribution of this work is to provide a formal specification that uses object-oriented concepts associated not only with objects but also with events as primary classes of a model. The work is based on Event Calculus and C-logic to model objects and events and to provide a language for spatio-temporal queries. This work shows the possibility to combine the snapshot view with the event view of spatio-temporal information using a formal framework that serves for specifying information, checking consistency of specification, and being a reference for query languages.

1 Introduction

The formal specification of spatio-temporal entities is essential to the definition of spatio-temporal database systems. This work proposes a formal model based on object-oriented concepts for spatio-temporal entities. Its main contribution is to provide a formal specification that uses object-oriented concepts associated not only with objects but also with events as primary classes of the model. In this sense, this work agrees with [23] [8] about moving up the conceptual abstraction of spatio-temporal entities by having two entities existing in the world: *continuants or objects* and *occurrents or events*. Unlike previous studies, this work focused on a conceptual model upon which logical and physical implementations are possible and testable.

This work is embedded in a project that is currently implementing a spatio-temporal database for handling objects and events. Previous solutions to modeling and querying spatio-temporal data have proposed extensions to relational data models based on abstract data types that handle the state of objects' attributes in time. Such approaches focus, primarily, on the representation of spatio-temporal information as *snapshots* or temporally indexed objects [10] [7]. In this work, we use logic as the specification language of a conceptual model, since we believe that the deductive power of a formal specification can be integrated into a database to combine *snapshopts* with *events*. We consider an object-oriented view of the world that supports a rich collection of data modeling and manipulation concepts [22] [21] [20] [4] [6] [19] [9].

J. Akoka et al. (Eds.): ER Workshops 2005, LNCS 3770, pp. 218–227, 2005.

The focus of this work is on the specification of objects and events that have aspatial and spatial attributes changing in time. The work deals with events that concern discrete changes over objects. The work does not discuss the spatial representation of objects or events such that it decouples the modeling of spatio-temporal information from the underlying spatial representation (e.g., field versus object space). The approach in this paper uses already well known temporal logic [15] and logic for object-oriented concepts [5]. It aims to model objects and events as classes of an object-oriented or extended relational database, and on providing reasoning mechanisms to derive values of objects' attributes in time, time period of event occurrences, events' relationships, and time-varying objects' relationships. Considering that objects and events cannot be *situated* in more than one place at a given time, this work uses time as the independent data and space as the dependent data when modeling spatio-temporal information. This is similar to the way spatio-temporal settings were defined in [23], where settings are functions from time to space.

The structure of the paper is as follows. Section 2 reviews the two main concepts in modeling spatio-temporal information: objects and events. Section 3 describes the logic uses in the proposed framework. Section 4 presents the proposed spatio-temporal model. Final conclusions and future research directions are given in Section 5.

2   Spatio-temporal Entities: Objects and Events

A spatio-temporal object refers to objects situated in space and time. From the perspective of object orientation, spatial-temporal objects are complex or structured objects by the notion of complex types, where objects are instances (values) of a type [2]. Complex objects are obtained from atomic objects by using such constructors as *set* and *tuple*. Unlike the relational model or nonfirst normal form model, where each of the constructors can be applied only once or they have to be alternated, in a complex object model constructors can be applied arbitrarily deep [19]. Complex objects are grouped into classes whose semantics relations *is_a* and *is_part* are extracted from basic paradigms of object-oriented theory (inheritance and aggregation/composition, respectively) [13].

Common spatio-temporal database systems represent objects changing with respect to time as a collection of objects temporally indexed or as a collection of *snapshots* [3]. Snapshots represent images of the state of the world at particular instants in time. Such type of representation can only extract objects' changes by comparing different objects' states. Even more, snapshots do not allows us to know the exact time instant when changes occur or what originiate these changes. This motivates that recent investigation considers not only objects, but also events, when modeling spatio-temporal information [23] [8].

Object and event views for modeling and representing spatio-temporal information are complementary. In order to associate both views, relations between objects and events, can be used (e.g., involvement and participant) [8]. The semantics of these relations states that the occurrence of an event involves a set, empty or non-empty, of objects with some roles in the event.

In the context of spatio-temporal databases, entities have spatial and temporal locations, also called *settings* [23]. In same cases, the spatial dimension of an event may be derived from the spatial dimension of objects involved in such event. In the general case, however, the spatial dimension may be associated with objects, events or both. Spatial and temporal settings have also been considered as entities, not as attributes, in spatial and temporal ontologies [8]. In the context of databases, both the spatial and temporal settings are usually specified as complex types involved in the definition of object and event classes.

The modeling of spatial settings uses some representation primitives (e.g., points, lines, polygons and regions). For temporal settings, the most frequent temporal model for dynamic processes in the real world is the linear concept [3]. A conceptual issue in modeling spatio-temporal information is the debate between point versus intervals. A time point is a time instant, whereas a time interval is defined by a start and end point in time. Different types of temporal aspects that have been traditionally discussed in the database community are [11]:

- *valid time* is the time when a fact is true in a modeled reality;
- *transaction time* is the time when an element in the database, which is not necessarily a fact, is part of the current state in the database; and
- *existence time* of an object refers to the time when the object exists in reality.

3   Logic for Modeling Spatio-temporal Information

Few studies combine logic with object-oriented databases in the domain of spatio-temporal databases [1] [12]. This work presents a logical framework for spatial objects and events that highlights the importance of making an explicit modeling of events and relationship between objects and events.

3.1   Temporal Logic

There are a number of temporal logic in the literature, such as Situation Calculus [16] and Event Calculus [15]. This work follows the Event Calculus formalization, which has been previously used in combination with logic for object-oriented concepts [12]. Although this work concentrates on valid time of temporal database, extensions of the Event Calculus support both *valid time* and *transaction time* [18], as well, as *continuous* changes [17].

The Event Calculus was first introduced in [15] to formalize the reasoning about events and changes. It allows the explicit representation of events, events' occurrences and objects' changes. Since the first proposal of the Event Calculus, the Simplified Event Calculus [14] has been widely used for reasoning about events. In the simplified version of the Event Calculus, events initiate and terminate properties of the real world. Predicates that *initiate/terminate* properties are of the following form, where T is the valid time occurrence of the event Ev, and *Attr* is the property or attribute that is initiated or terminated by the event:

$$initiates\_at(Ev, Attr, T).$$
$$terminates\_at(Ev, Attr, T).$$

Updates are specified by facts of the form $happens\_at(Ev, T)$, and the effects of these updates are specified by the following general rules, which can be modified to handle changes over objects for time instants or time intervals [12]:

$$hold(Attr, T) \leftarrow happens\_at(Ev, T_s), T \geq T_s,$$
$$initiates\_at(Ev, Attr, T_s), \neg broken(Attr, T_s, T).$$
$$broken(Attr, T_s, T_e) \leftarrow happens\_at(Ev, T), T_s < T < T_e,$$
$$terminates\_at(Ev, Attr, T).$$

3.2 Logic for Complex Objects

This work uses C-logic [5] for the specification of a data model that includes objects and events. The main advantage of using C-logic is the direct mapping of its specification onto first-order formulas that use unary predicates for types and binary predicates for attributes. C-logic allows information about an entity to be specified and accumulated piecewise, which facilities the updates of subparts of objects' specifications independently. This type of specification can be easily implemented in logic programming language, such as PROLOG.

In C-logic, an object class (a group of objects with the same properties) is specified as

$$class : Object\_Class[attribute\_of \Rightarrow attribute_1, \ldots, attribute_n].$$

Using C-logic, complex objects are considered as collections or conjunctions of atomic properties. An object with several attribute labels can be described as a conjunction of several atomic formulas in first-order logic.

4 The Data Model

The model proposed in this paper defines classes for objects and events. A class in this model has a unique name and denotes a set of objects or events with the same attributes. Attributes of events or objects are spatial (e.g., geometry, location) or aspatial (e.g., name and population).

$$Object\_Class : Obj[attribute\_of \Rightarrow \{attribute_1, \ldots, attribute_n\}].$$
$$Event\_Class : Event[attribute\_of \Rightarrow \{attribute_1, \ldots, attribute_n\}].$$

Classes are organized into a class hierarchies by a is_a relation (a subclass or subset relation). The inheritance of attributes by the subclasses is guaranteed with the following predicate:

$$attribute\_of(Sub, Attr) \leftarrow is\_a(Sub, Class), attribute\_of(Class, Attr).$$

At the class level, and in addition to the is_a relation, interesting semantic relations are $part/whole$ between object classes. For space reasons, this paper just presents the specifications that involve class instances; that is, specification of objects, events, and their corresponding relations. Due to conceptual differences, specifications of instances of objects and events are described separately in the following subsections. A new predicate $involves(Ev, Obj, Role)$ establishes an *involvement* relation between an event and an object. This relation also indicates the role that the object plays in the event. An object may be involved in different events in time, and one or more than one object may be involved in an event. For example, objects can be destroyed or created, and more than one object are needed in *merge* events.

4.1   Objects

Values of objects' attributes are derived from the predicates *initiates* and *terminates* and predicates *holds_at* and *holds_for* of OEC [12].

$$initiates(Ev, Obj, Attr, Value).$$
$$terminates(Ev, Obj, Attr, Values).$$

The class of an object is specified by a predicate $instance\_of(Obj, Class, T)$, such that class membership of objects can be a time-varying relationship [12]. For example, a building may be a *school* at a given time and be a *hospital* at another time. To allow this change of membership, Kesim and Marek [12] introduces predicates $instance\_of(Obj, Class, T)$, $assigns(Ev, Obj, Class)$, $destroys(Ev, Obj)$, and $removes(Ev, Obj, Class)$. While the predicate *destroys* indicates that an object has been destroyed, a predicate *removes* indicates that an event removes an object from its class. Instead of predicates $destroys(Ev, Obj)$ and $removes(Ev, Obj, Class)$, we use the more general predicate $involves(Ev, Obj, Role)$ with two roles, *destroyed* and and *removed*, to define the predicate $instance\_of$:

$$instance\_of(Obj, Class, T_e) \leftarrow happens(Ev, T_s), T_s \leq T_e,$$
$$assigns(Ev, Obj, Class), \neg removed(Obj, Class, T_s, T_e).$$
$$removed(Obj, Class, T_s, T_e) \leftarrow happens(Ev, T^*),$$
$$T_s < T^* < T_e, involves(Ev, Obj, destroyed).$$
$$removed(Obj, Class, T_s, T_e) \leftarrow$$
$$happens(Ev[instance\_of \Rightarrow remove\_class, object\_class \Rightarrow Class], T^*),$$
$$T_s < T^* < T_e, involves(Ev, Obj, removed).$$

From the transitive property of subclasses, an instance of a subclass is also an instance of its superclass. This is defined by:

$$assigns(Ev, Obj, Class) \leftarrow is\_a(Sub, Class), assigns(Ev, Obj, Sub).$$

Another interesting semantic relation for objects is the part/whole relation, which is implemented by a particular predicate $is\_part(Obj_1, Obj_2, T)$. Unlike the membership relation (is_a) between classes, the is_part relation exists between classes and between objects. The is_part relations between classes define a semantic relation that all instances of the classes hold; however, one also wants to specify about the is_part relation between particular instances.

In this model, $is\_part(Obj_1, Obj_2, T)$ is also a time-varying relation. For example, a park can be part of a county and then, by a re-definition of the administrative boundaries of a country, be part of another county. A distinction in the relations between parts and wholes considers the concepts of *compositions* and *aggregations*. In a composition, the existence of parts is fully determined by the existence of their whole whereas, in an aggregation, the elimination of the whole does not imply the elimination of its parts.

In the implementation of a time-varying is_part relation, two new predicates that differentiate *part_whole* relations that are derived from compositions and aggregations are $compose(Ev, Obj_1, Obj_2)$ and $aggregate(Ev, Obj_1, Obj_2)$. The rules that specify the effects of such predicates are:

$$is\_part(Obj_1, Obj_2, T_e) \leftarrow happens(Ev, T_s), T_s \leq T_e,$$
$$compose(Ev, Obj_1, Obj_2), \neg destroyed(Obj_1, T_s, T_e),$$
$$\neg compose\_for(Obj_1, T_s, T_e), \neg destroyed\_by\_composition(Obj_2, T_s, T_e).$$
$$is\_part(Obj_1, Obj_2, T_e) \leftarrow happens(Ev, T_s), T_s \leq T_e,$$
$$aggregate(Ev, Obj_1, Obj_2),$$
$$\neg destroyed(Obj_1, T_s, T_e), \neg destroyed(Obj_2, T_s, T_e).$$
$$destroyed\_by\_composition(Obj, T_s, T_e) \leftarrow happens(Ev, T_s), T_s \leq T_e,$$
$$compose(Ev, Obj, Obj^*), \neg destroyed(Obj, T_s, T_e),$$
$$\neg compose\_for(Obj, T_s, T_e), \neg destroyed\_by\_composition(Obj^*, T_s, T_e).$$
$$destroyed(Obj, T_s, T_e) \leftarrow happens(Ev, T),$$
$$T_s \leq T \leq T_e, involves(Ev, Obj, destroyed).$$
$$compose\_for(Obj, T_s, T_e) \leftarrow happens(Ev, T), T_s < T \leq T_e, compose(Ev, Obj_1, \_).$$

The predicate *destroyed_by_composition* ensures that once a *whole* of a composition has been destroyed, all its *parts* are also destroyed. The predicate *compose_for* checks whether or not an object is involved as the *part* in a composition within a time interval. The use of this predicate enforces that an object can be only *part* of one whole by composition.

To make possible a transitive property of parts, we introduce the following rules:

$$is\_part\_at(Obj_1, Obj_2, T) \leftarrow is\_part(Obj_1, Obj_2, T).$$
$$is\_part\_at(Obj_1, Obj_3, T) \leftarrow is\_part(Obj_1, Obj_2, T), is\_part\_at(Obj_2, Obj_1, T).$$

The definitions of aggregated or composite objects also affect previous definitions respect to class membership and attributes' values. These rules should be modified to include a condition respect to the destruction of compositions.

4.2 Events

Unlike objects' attributes, this model assigns values to attributes of an event within the predicate of this event. While objects' attributes are derived through predicates in terms of events, values of events' attributes hold for the specific time in which the event occurs. In this sense, the attributes' values of events do not change with time, even when the event itself is temporal in nature. In this model, the predicate of events includes a specific attribute *instance_of*, which indicates the class to which the event belongs.

$$event : Ev[instance\_of \Rightarrow Class, Attribute_1 \Rightarrow Attr\_Value_1, \ldots,$$
$$Attribute_n \Rightarrow Attr\_Value_n].$$

Focusing on the spatial domain, this work describes 5 different types of events and their relationships with objects.

- **Creation.** An event that creates an object initiates, at least, a spatial attribute (G). As it was said before, such event of the creation of an object should also assign the object to a class (C), which is specified through an attribute *object_class* of the event. The specification of this event and the rules that show its effect follow:

$$assigns(Ev, Obj, Class) \leftarrow$$
$$event : Ev[instance\_of \Rightarrow creation, object\_class \Rightarrow C, geometry \Rightarrow R],$$
$$involves(Ev, Obj, created).$$
$$initiates(Ev, Obj, geometry, R) \leftarrow$$
$$event : Ev[instance\_of \Rightarrow creation, object\_class \Rightarrow C, geometry \Rightarrow R],$$
$$involves(Ev, Obj, created).$$

- **Destruction.** The inverse of creating an object is the elimination of an object and its attributes. This event also removes the object as an instance of its class. For example, the rules that describe the effect of the event *ev* that destroys the object *obj* are:

$$terminates(Ev, Obj, Attr, \_) \leftarrow$$
$$involves(event : Ev[instance\_of \Rightarrow destruction], Obj, destroyed).$$

An additional consideration when destroying an object is the case when the object that is destroyed is composed of parts. In such case the *is_part* relation ceases to exist and parts may (in case of composition) or may not (in case of aggregation) be destroyed. The effect of destroying the *is_part* relation has been already described when defining this relation. To explicitly handle the destruction of parts, and not only the destruction of the relation *is_part*, a new temporal rule *destroyed_at* indicates whether or not an object is destroyed at particular instant in time.

$$destroyed\_at(Obj, T) \leftarrow happens(Ev, T_s), T_s < T,$$
$$compose(Ev, Obj, Ob^*),$$
$$destroyed\_by\_composition(Obj^*, T_s, T_e).$$

- **Change.** Change is an event that changes a property of an object. This property can be aspatial or spatial. The effect of the event over properties of objects is specified with the *initiates* predicate. Consider, for example, an event *move* that changes an object's location to coordinates (x, y). The effect of such event over the location of the object is specified with the *initiates* predicates as follows

$initiates(Ev[instance\_of \Rightarrow move, location_x \Rightarrow X, location_y \Rightarrow Y],$
  $Obj, location_x, X) \leftarrow$
    $event : Ev(instance\_of \Rightarrow move, location_x \Rightarrow X, location_y \Rightarrow Y],$
    $involves(Ev, Obj, moved).$
$initiates(Ev[instance\_of \Rightarrow move, location_x \Rightarrow X, location_y \Rightarrow Y],$
  $Obj, location_y, Y) \leftarrow$
    $event : Ev[instance\_of \Rightarrow move, location_x \Rightarrow X, location_y \Rightarrow Y],$
    $involves(Ev, Obj, moved).$

- **Split.** Split is an event that creates a boundary to split an object. This event is spatial in nature, so its affect spatial attributes. In this model, we consider that the *split* event creates a new object of the same class from the splitting of an original object whose geometry is then modified by removing the region assigned to the new object. Thus, we have two objects involved in this event: a new object and an updated object. The effect of such event is shown in the following rules:

  $instance\_of(Obj, Class, T) \leftarrow$
   $happens(Ev[instance\_of \Rightarrow split, geometry_1 \Rightarrow R_1, geometry_2 \Rightarrow R_2], T_s),$
   $T \geq T_s, involves(Ev, Obj, created), involves(Ev, Obj^*, updated),$
   $instance\_of(Obj^*, Class, T_s),$
   $\neg removed(Obj, Class, T_s, T), \neg removed\_by\_composition(Obj, Class, T_s, T).$
  $initiates(Ev, Obj, geometry, R_2) \leftarrow$
   $event : Ev[instance\_of \Rightarrow split, geometry_1 \Rightarrow R_1, geometry_2 \Rightarrow R_2],$
   $involves(Ev, Obj, created).$
  $initiates(Ev, Obj, geometry, R_1) \leftarrow$
   $event : Ev[instance\_of \Rightarrow split, geometry_1 \Rightarrow R_1, geometry_2 \Rightarrow R_2],$
   $involves(Ev, Obj, updated).$

- **Merge.** The inverse case of the *split* event is the *merge* event. In this case, an object is merged into other object. The merged object is then destroyed. The effect of such event is shown in the following rules:

  $involves(Ev, Obj, destroyed) \leftarrow$
  $involves(Ev[instance\_of \Rightarrow merge, geometry \Rightarrow R], Obj, merged).$
  $initiates(Ev, Obj, geometry, R) \leftarrow$
   $event : Ev[instance\_of \Rightarrow merge, geometry \Rightarrow R], involves(Ev, Obj, updated).$

In our model, occurrences of events are basic facts upon which different derivations are possible. We use the relations between events (e.g., necessity, sufficiency, and so on) as constraints in our model. Consider the common case of a *necessity* relation; that is, if the entailing event occurs, the entailed event should have occurred before. For example, our model enforces that if something is moved_out, then it should have been moved_in before. The rules that check this constraint are specified as follows:

$consistent(Ev_1[instance\_of \Rightarrow move\_out]) \leftarrow$
  $happens(Ev_1[instance\_of \Rightarrow move\_out], T_1), involves(Ev_1, Obj, moved\_out),$
  $happens(Ev_2[instance\_of \Rightarrow move\_in], T), involves(Ev_2, Obj, moved\_in),$
  $\neg moved\_out(Obj, T, T_1), T < T_1.$
$moved\_out(Obj, T_s, T_e) \leftarrow$
  $happens(Ev[instance\_of \Rightarrow move\_out], T),$
  $involves(Ev, Obj, moved\_out), T_s \leq T < T_e.$

These constraints over events' occurrences should be included as conditions in the rules defined above (e.g., *hold_for* and *holds_at*). This is similar to the way impossible cases of events were handled in [17].

5   Conclusions

This paper presents an approach based on Event Calculus and C-Logic to model objects and events that vary in time. Whereas the spatial dimension of objects and events are handled as attributes, time is embedded in the formalism of the Event Calculus. The work describes types of events that are commonly found in the spatial domain, such as, split, merge, create, and destroy. Relations between objects are considered time varying and are specified in terms of events occurrences. Of particular interest are *is_part* and *is_a* relations between objects and classes. Relations between events are introduced as constraints in our model.

As further investigation, events with intervals, also called processes, could be also considered [17]. Such type of consideration may lead to the treatment of event granularity and temporal inclusion of relations between events. Likewise, the treatment of continuous changes is also a possible extension to Event Calculus [17]. Reasoning properties of the proposed model is related to the paradigm of deductive databases [12], which uses logic as computational formalism, database specification, and query language. The approach of deductive databases, however, face efficiency problems when handling large collections of data. Our current work is integrating the deductive power of the formalism proposed in this paper into an extended relational database that can handle both view of spatio-temporal information, namely, snapshots and events.

Acknowledgment. This work has been funded by CONICYT, Chile, under grant FONDECYT 1050944.

References

1. A. Abdelmoty, N. Paton, H. Williams, A. Fernandes, M. Barja, and A. Dinn. Geographic data handling in a deductive object-oriented database. In D.Karagiannis, editor, *International Conference on Database and Expert Systems Applications DEXA'94*, pages 445–454. Springer-Verlag, 1994.

2. S. Abiteboul and C. Beeri. The power of languages for the manipulation of complex values. *VLDB Journal*, 4(4):727–794, 1995.

3. T. Abraham and J. Roddick. Survey of spatio-temporal databases. *GeoInformatica*, 3:61–69, 1999.
4. K. Borges, C. Davis, and A. Laender. Omt-g: An object-oriented data model for geographic information applications. *GeoInformatica*, 5(3):221–260, 2001.
5. Weidong Chen and David Warren. C-logic of complex objects. *ACM SIGACT-SIGMOD-SIGART Symp. Principles of Database Systems*, pages 369–378, 1989.
6. Max Egenhofer and A. Frank. Object-oriented modeling for gis. *Journal Urban and Regional Information System Assoc.*, 4(2), 1992.
7. L. Forlizze, R. Güting, E. Nardelli, and M. Schneider. A data model and data structure for moving objects databases. In *ACM SIGMOD*, pages 319–330. ACM Press, 2000.
8. P. Grenon and Smith. SNAP and SPAN: Towards dynamic spatial ontology. *Spatial Cognition and Computation*, 4(1):69–103, 2004.
9. T. Griffiths, A. Fernandes, N. Paton, and R. Barr. The TRIPOD spatio-temporal data model. *Data and Knowledge Engineering*, 49(1):23–65, 2003.
10. R. Güting, M. Böhlen, M. Erwing, C. Jensen, N. Lorentzos, and E. Nardelli. *Spatio-Temporal Databases: The Chorochronos Approach*, chapter Spatio-Temporal Models and Languages, pages 117–176. Springer-Verlag, 2003.
11. C. Jensen and C. Dyreson. The concesus glossary of temporal concepts - february 1998 version. In *Lecture Notes in Computer Science 1399*, pages 367–405. Springer-Verlag, 1998.
12. F.N. Kesim and Marek Sergot. A logic programming framework for modeling temporal objects. *IEEE Transactions on Knowledge and Data Engineering*, 8(5):724–741, 1996.
13. S. Khoshafian and R. Abnous. *Object Orientation: Concepts, Language, Databases and User Interfaces*. John Wiley & Sons, New York, 1990.
14. R. Kowalski. Database updates in the event calculus. *Journal of Logic Programming*, 12(162):121–146, 1992.
15. R. Kowalski and M. Sergot. A logic-based calculus of events. *New Generation Computing*, 4:67–95, 1986.
16. J. McCarthy and P. Hayes. *Machine Intelligence*, chapter Some Philosophical Problems from the Standpoint of Artificial Intelligence, pages 463–502. Edinburg University Press, 1969.
17. R. Miller and M. Shanahan. The event calculus in classical logic - alternative axiomatisations. *Lonköping Electronic Articles in Computer and Information Science*, 4(16):1–30, 1999.
18. S. Sripada. *Temporal Reasoning in Deductive Databases*. PhD thesis, Dept. of Computing, Imperial College, 1991.
19. A. Voisard and B. David. A database pespective on geospatial data modeling. *IEEE Transactions on Knowledge and Data Engineering*, 14(2):226–246, 2002.
20. M. Worboys. Object-oriented approaches to geo-referenced information. *International Journal of Geographic Information Systems*, 8(4):385–399, 1994.
21. M. Worboys. A unified model of spatial and temporal information. *Computer Journal*, 37(1):26–34, 1994.
22. M. Worboys, H. Hearnshaw, and D. Maguire. Object-oriented data modelling for spatial databases. *International Journal of Geographic Information Systems*, 4(4):369–383, 1990.
23. M. Worboys and K. Hornsby. Geographic Information Science. *LNCS 3234*, chapter From Objects to Events: GEM, the geospatial event model, pages 327–343. Springer-Verlag, 2004.

Conceptual Neighbourhood Diagrams
for Representing Moving Objects

Nico Van de Weghe and Philippe De Maeyer

Department of Geography, Ghent University,
Krijgslaan 281 (S8), B-9000 Ghent, Belgium

Abstract. The idea of Conceptual Neighbourhood Diagram (CND) has proved its relevance in the areas of qualitative reasoning about time and qualitative reasoning about space. In this work, a CND is constructed for the Qualitative Trajectory Calculus (QTC), being a calculus for representing and reasoning about movements of objects. The CND for QTC is based on two central concepts having their importance in the qualitative approach: the theory of dominance and the conceptual distance between qualitative relations. Some examples are given for illustrating the use and the potentials of the CND for QTC from the point of view of GIScience.

1 Introduction

Since humans usually prefer to communicate in qualitative categories supporting their intuition and not in quantitative categories, qualitative relations are essential components of queries that people would like to run on a GIS. Temporal calculi, such as the Interval Calculus [1] and the Semi-Interval Calculus [2], have been proposed. In addition, spatial calculi, such as the RCC-calculus [3] and the 9-Intersection Model [4], both focusing on topological relations between regions, have been proposed. These topological relations form the basis of most current GIS[s]. Despite extensive research during the past decade, both from the area of spatio-temporal reasoning (e.g. [5-7]) and spatio-temporal databases (e.g. [8-11]), the representation of space-time is problematic, and a full temporal GIS is not available yet. We believe that qualitative spatio-temporal calculi should take a central place behind such a temporal GIS.

In [12], the Qualitative Trajectory Calculus is presented, being a theory for representing and reasoning about moving objects in a qualitative framework. Depending on the level of detail and the number of spatial dimensions, different types of this calculus are defined and studied in detail. In this paper, we focus on the basic Qualitative Trajectory Calculus in one dimension, QTC for short. In Section 2, QTC is presented. After a brief description concerning Conceptual Neighbourhood Diagrams (CND), this idea is applied to the domain of continuously moving objects in Section 3. We discuss two concepts, used for the construction of the CND[s] for QTC: *dominance space* and *conceptual distance*. In Section 4, some examples on how to use the CND for QTC are presented.

J. Akoka et al. (Eds.): ER Workshops 2005, LNCS 3770, pp. 228–238, 2005.

2 The Basic Qualitative Trajectory Calculus for One Dimension (QTC)

QTC handles the qualitative movement of a pair of point objects along a 1D line, such as cars driving on the same lane and trains moving on a railroad. We assume continuous time for QTC.

2.1 QTC of Level One (QTC$_{L1}$)

Because the movement is restricted to 1D, the velocity vector of an object is restricted to two directions, with the intermediate case where the object stands still. Hence, the direction of the movement of each object can be described by one qualitative variable, using the following conditions, resulting in 9 so-called *L1-relations* (Fig. 1).[1]

1. Movement of k, with respect to the position of l at time point t *(distance constraint)*
   $-$: k is moving towards l; $+$: k is moving away from l; 0: k is stable with respect to l.
2. Movement of l, with respect to the position of k at time point t *(distance constraint)*
   $-$: l is moving towards k; $+$: l is moving away from k; 0: l is stable with respect to k.

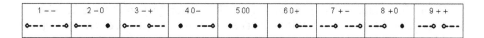

Fig. 1. L1-relation icons

2.2 QTC of Level Two (QTC$_{L2}$)

QTC$_{L1}$ can be extended with a third character giving the relative speed of the objects:

3. Relative speed of k at time point t with respect to l at t *(speed constraint)*
   $-$: v_k at $t < v_l$ at t; $+$: v_k at $t > v_l$ at t; 0: v_k at $t = v_l$ at t

This way, we can create all combinations for QTC$_{L2}$. 10 of the 27 theoretic possibilities are impossible, resulting in only 17 *L2-relations* (Fig. 2).[2]

Note that QTC can handle as well movements of objects along a curved line. This is done by considering the distance and speed constraints along the curved line. As a direct result, situations where, for example, two cyclists are moving along a curved cycle track, can be handled in QTC.

---

[1] The left and right dot represent respectively the positions of k and l. The dashed line segments represent the potential object movements; the lines are dashed since there is no direct information about the speed of the objects. The line segments represent whether each object is moving towards or away from the other. A dot is filled if the object can be stationary.

[2] Note that the lines can have different lengths giving the difference in relative speed.

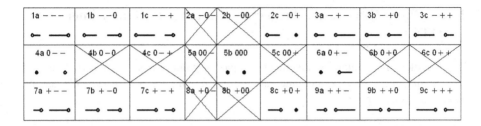

Fig. 2. L2-relation icons

3   Conceptual Neighbourhood Diagram for QTC

3.1   Definition of a Conceptual Neighbourhood Diagram (CND)

CND[s] have been introduced in the temporal domain [2], and have been widely used in spatial reasoning, e.g.: for topological relations [3,13]; cardinal directions [14], and for relative orientation [15]. CND[s] are typically used for qualitative simulation to predict what will happen in the future. Two relations between entities are conceptual neighbours, if they can be transformed into one another by continuously deforming, without passing another qualitative relation; a CND describes all the possible transitions between relations that can occur [2]. For clarification of these definitions, we use the CND for RCC (Fig. 3). The relations *DC* and *EC* are conceptual neighbours, since continuous deformation is possible by moving *k* and *l* towards each other. *DC* and *PO* are not conceptual neighbours, because a continuous deformation cannot transform from *DC* into *PO* without passing through *EC*.

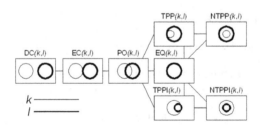

Fig. 3. CND for topological relations in the RCC-calculus

3.2   Construction Concepts

Based on [2], we define: two QTC relations are conceptual neighbours if they can directly follow each other during a continuous movement. We discuss two concepts for the construction of the CND[s] for QTC: *dominance space* and *conceptual distance*.

Theory of Dominance. Central in the *theory of dominance* [16,17] are the constraints imposed by continuity. Consider the qualitative distinction between −, 0 and +. A direct change from − to + is impossible, since such a change must pass the qualitative value 0, that only needs to hold for an instant. On the other hand, the + of a variable changing from 0 to + and back to 0, must hold over an interval. We say that 0 *dominates* − and + [18]. Now, let us focus on QTC. A change from the L1-relation (− +)$_{L1}$ to (+ +)$_{L1}$, must pass at least one QTC relation, since the first character cannot chance continuously from − to +. The shortest way is via (0 +)$_{L1}$. This relation only needs to hold for an instant. On the other hand, the (+ +)$_{L1}$ of the sequence of relations $\{(+\ 0)\leadsto(+\ +)\leadsto(+\ 0)\}_{L1}$, must hold over an interval.[3] In order to explain this, consider (+ 0)$_{L1}$ at t_1, (+ +)$_{L1}$ at t_2, and the speed of l at t_2 being 0.1 metres per second. One can always find a time point between t_1 and t_2 with the speed of l being somewhere in between 0 metres per second and 0.1 metres per second. In the words of Galton: "*When an object starts moving, there is a last moment when it is at rest, but no first moment when it is in motion*" [19, p.101]. Thus, (+ 0)$_{L1}$ dominates (+ +)$_{L1}$.

Now, one can construct a *dominance space*, being a space containing qualitative values and their according dominance relations [16]. Fig. 4 represents a basic example of the dominance space in 1D: a transition from − to 0 can occur and vice versa (with 0 dominating −); a transition from 0 to + can occur and vice versa (with 0 dominating +); a transition from − to + can only occur by passing through 0.

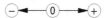

Fig. 4. Dominance space in 1D

It has been proved in [18] that *simple dominance spaces* can be combined for building *composite dominance spaces*. We use this theorem for the construction of the CND^s for QTC. Fig. 5a, for example, shows the composite dominance space in 2D, both dimensions containing the qualitative values 0 and +. Each dimension of this space, which can be seen as a subset of all L1-relations, contains two connections. Combining both dimensions, $\{(0\ 0)\leadsto(+\ +)\}_{L1}$ is constructed. This can be done by composing $\{(0\ 0)\leadsto(+\ 0)\}_{L1}$ with $\{(+\ 0)\leadsto(+\ +)\}_{L1}$, or by composing $\{(0\ 0)\leadsto(0\ +)\}_{L1}$ with $\{(0\ +)\leadsto(+\ +)\}_{L1}$. Note the importance of the direction of the connections. Based on [17] and Fig. 5b, we explain that it is impossible to construct $\{(0\ +)\leadsto(+\ 0)\}_{L1}$. (0 +)$_{L1}$ holds over an interval $]t_1,t_2[$, and (+ 0)$_{L1}$ holds over interval $]t_2,t_3[$. Thus, the first character changes from 0 to + at t_2, and the second character changes from + to 0 at t_2. Since 0 dominates +, the relation at t_2 will be (0 0)$_{L1}$.

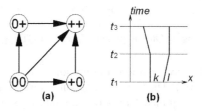

Fig. 5. Dominance space in 2D

---

[3] The transition from a to b is denoted by $a\leadsto b$.

3.2.2 QTC Distance (d_{QTC})

Based on the ideas of *topology distance* [13] and *distance measure between two cardinal directions* [20], we define *QTC distance* (d_{QTC}). The *conceptual distance* between qualitative relations is the shortest path between these relations in the CND, giving every arc a distance being equal to one. As a result, d_{QTC} is the conceptual distance between two QTC relations. Let us consider three examples:

- If R_1 and R_2 differ in one character that can change continuously between both states without passing through an intermediate qualitative value.
E.g.: if $R_1 = (0\ 0)_{L1}$ and $R_2 = (0\ +)_{L1}$ then $d_{QTC} = 1$, denoted by: $d_{QTC}(0\ 0,0\ +) = 1$.
- If R_1 and R_2 differ in one character that cannot change between both states without passing through an intermediate qualitative value, then d_{QTC} is composed of sub-distances.
E.g.: $d_{QTC}(0\ -,0\ +) = d_{QTC}(0\ -,0\ 0) + d_{QTC}(0\ 0,0\ +) = 1 + 1 = 2$
- If R_1 and R_2 differ in multiple characters, then d_{QTC} is a composition of the sub-distances determined for each individual character. If multiple compositions are possible, then the composition resulting in the lowest d_{QTC} is selected.
E.g.: $d_{QTC}(-\ -,+\ +) = d_{QTC}$ for 1st character + d_{QTC} for 2nd character $= 2 + 2 = 4$.

3.3   Conceptual Neighbourhood Diagram for QTC$_{L2}$ (CND$_{L2}$)[4]

Based on the theory of dominance and QTC distance, we construct CND$_{L2}$. As proposed in [16], we start with simple dominance spaces, and build these into composite dominance spaces. First, consider the three dominance spaces with $d_{QTC} = 1$ (*one-dominance space*) in Fig. 6a-c, representing three subsets of dominance relations: continuous changes of the first (a), second (b) and third (c) character.

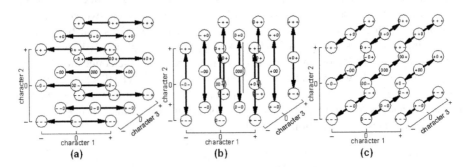

Fig. 6. One-dominance spaces for QTC$_{L2}$

There are three possible compositions of one-dominance spaces, resulting in a two-dominance space (Fig. 7a,b). There exists a second composition level in which each two-dominance space is combined with a one-dominance space (being orthogonal to both dimensions of the two-dominance space). Fig. 8a shows the three different composition possibilities resulting in the same three-dominance space (Fig. 8b).

---

[4] Due to space limitations, we will not handle the CND$_{L1}$, studied in depth in [6].

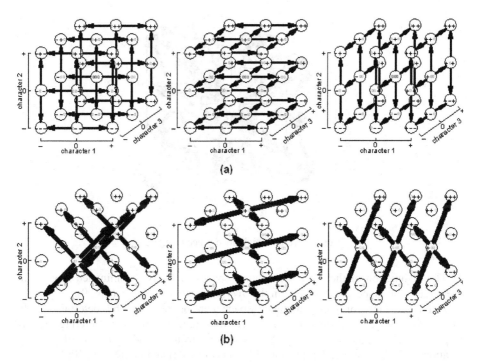

Fig. 7. Construction of two-dominance spaces for QTC$_{L2}$

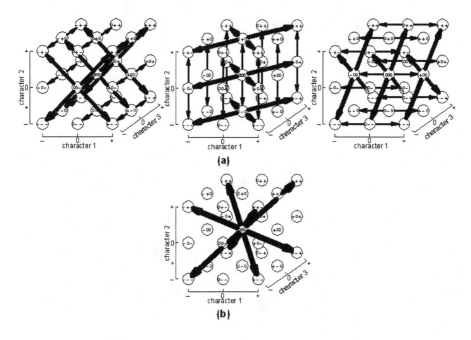

Fig. 8. Construction of three-dominance spaces for QTC$_{L2}$

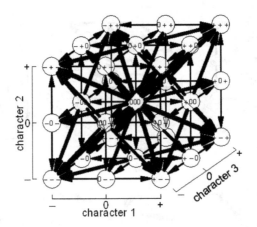

Fig. 9. Overall-dominance space for QTC$_{L2}$

The disjunction of the 3 one-dominance spaces, the 3 two-dominance spaces and the three-dominance space, results in the *overall dominance space* (Fig. 9).

A clearer view of the dominance space can be obtained by deleting all the 'impossible' nodes and rearranging those that remain [16]. By deleting the 10 impossible labels, and the according arcs starting from or ending in one of these 10 relations, CND$_{L2}$ is created, represented in 3D (Fig. 10a) and in 2D (Fig. 10b).

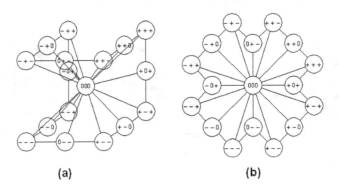

(a)   (b)

Fig. 10. CND$_{L2}$

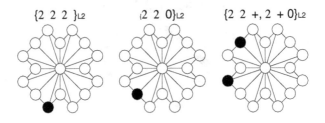

Fig. 11. Static CND$_{L2}$ icon

CND_{L2} forms the basis for two CND icons: the *static CND_{L2} icon* (Fig. 11) represents a set of L2-relations, the *dynamic CND_{L2} icon* (Fig. 13) represents a sequence of relations, and will be explained in detail in the next section.

4 Examples of CND for QTC

4.1 Multiple Time Points

Consider k and l moving in 1D, during a study period between t_1 and t_9 (Fig. 12). Nothing is known about the situations immediately before t_1 and immediately after t_9. Therefore, it is impossible to label t_1 and t_9. The L2-relations for i_1, t_2, i_2, ..., i_7, t_8, and i_8 can be created quite easily by use of some specific rules:

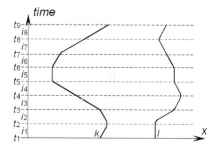

Fig. 12. History of multiple point objects in 1D

- **0 dominates + and –.** E.g.: t_7 is between $i_6(--0)_{L2}$ and $i_7(--+)_{L2}$. Thus, $t_7(--0)_{L2}$.
- **Transition from – to + (and vice versa) is impossible without passing 0.[5]** E.g.:

    Distance constraint: transition from $i_7(--+)_{L2}$ to $i_8(-++)_{L2}$ is impossible without passing through $t_8(-0+)_{L2}$.

    Speed constraint: transition from $i_2(++-)_{L2}$ to $i_3(+++)_{L2}$ is impossible without passing through $t_3(++0)_{L2}$.

Relations $t_8(-0-)_{L2}$ and $t_3(++0)_{L2}$ only hold for an instantaneous time point. One should be aware that the qualitative value 0 can also hold for an interval, e.g. $i_6(--0)_{L2}$. Be also aware that it is possible that multiple characters change simultaneously. This is for example the case for $\{t_6(0\ 0\ 0)\leadsto i_6(--0)\}_{L2}$.

- **Combination of both former rules.** E.g.: t_2 is between $i_1(-0+)_{L2}$ and $i_2(++-)_{L2}$. Thus, $t_2(0\ 0\ 0)_{L2}$.

Finally, CND_{L2} can be drawn (Fig. 13), with the labels and the directed arcs respectively referring to the temporal primitives and the transitions. If there is an absence of transition between multiple temporal primitives, all these units are written as has been done for '(0 0 0): t_5, i_5, t_6'. Note that the same node can be passed multiple times during a study period.

---

[5] This counts for the direction as well as the speed constraint.

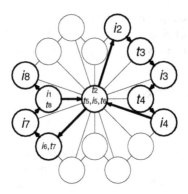

Fig. 13. Dynamic CND$_{L2}$ icon

4.2 Incomplete Knowledge About Moving Objects

This example starts from an expression (Ex1) forming fine knowledge concerning moving objects, and relaxes the constraints in order to get incomplete knowledge (Ex2 and Ex3). Thereafter, we work the other way round.

4.2.1 From Fine to Incomplete Knowledge
- *Ex1: k is moving towards l, which in turn is moving away from k, both objects having the same speed.* (Fig. 14a)

$$Ex1 \leftrightarrow (- + 0)_{L2}$$

- *Ex2: k is moving towards l, which in turn is moving away from k.* (Fig. 14b)

$$Ex2 \leftrightarrow (- + -, - + 0, - + +)_{L2}$$

- *Ex3: k is moving towards l.* (Fig. 14c)

$$Ex3 \leftrightarrow (- - -, - - 0, - - +, - 0 -, - 0 0, - 0 +, - + -, - + 0, - + +)_{L2}$$

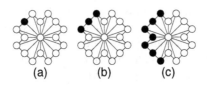

(a)   (b)   (c)

Fig. 14. Static CND$_{L2}$ icon for (a) Ex1, (b) Ex2 and (c) Ex3

4.2.2 From Incomplete to Fine Knowledge
Now, let us start from 3 expressions (Ex1a, Ex1b and Ex1c), which together form the fine compound expression Ex1:

- *Ex1a: k is moving towards l.* (Fig. 15a)

$$Ex1a \leftrightarrow (- - -, - - 0, - - +, - 0 -, - 0 0, - 0 +, - + -, - + 0, - + +)_{L2}$$

- *Ex1b: l is moving away from k.* (Fig. 15b)

$$Ex1b \leftrightarrow (-+-, -+0, -++, 0+-, 0+0, 0++, ++-, ++0, +++)_{L2}$$

- *Ex1c: k and l have the same speed.* (Fig. 15c)

$$Ex1c \leftrightarrow (--0, -+0, 0\,0\,0, +-0, ++0)_{L2}$$

The conjunction of the solutions of Ex1a, Ex1b and Ex1c gives $(-+0)_{L2}$ (Fig. 15d).

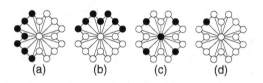

Fig. 15. Static CND_{L2} icon for (a) Ex1a, (b) Ex1a, (c) Ex1c, and (d) Ex1

By comparing Fig. 15d with Fig. 14a, one can state that the conjunction of the solution sets of the components of a compound expression is the same as the solution set of the compound expression.

5 Summary and Further Work

In this work, a conceptual neighbourhood diagram (CND) was constructed for the Qualitative Trajectory Calculus (QTC), being a calculus for representing and reasoning about movements of objects. After a brief description concerning CND^s, this idea was applied to the domain of continuously moving objects. We discussed two concepts, used for the construction of the CND^s for QTC: *dominance space* and *conceptual distance*. The CND for QTC forms the basis for the static CND icon (representing QTC relations) and the dynamic CND icon (representing a sequence of QTC relations). Finally, some examples on how to use the CND for QTC were presented. We believe that, apart from a neat visualisation, the CND and its icons can represent specific types of conceptual behaviour, which could lead to conceptual modelling and qualitative simulation of moving objects. The CND for QTC is specifically well-suited for reasoning about incomplete knowledge of moving objects.

This work is part of a larger research question that can be formulated as: 'how to describe motion adequately within a qualitative calculus, so as to obtain a tool for data and knowledge representation and for querying spatio-temporal data'. A full answer to this question needs, besides the spatio-temporal reasoning, also an exhaustive study of several database issues, increasing general performance by the use of efficient algorithms and access methods for computing intensive query operations.

Acknowledgements

This research is funded by the Research Foundation – Flanders (Project G.0344.05).

References

1. Allen, J.F., 1983, Maintaining knowledge about temporal intervals, *Comm. of the ACM*, 26(11), 832-843.
2. Freksa, C., 1992, Temporal reasoning based on semi-intervals, *Artificial Intelligence*, 54, 199-227.
3. Randell, D., Cui, Z and Cohn, A.G., 1992, A spatial logic based on regions and connection, *KR*, 165-176.
4. Egenhofer, M. and Franzosa, R., 1991, Point-set topological spatial relations, *IJGIS*, 5(2), 161-174.
5. Muller, Ph., 1998, Space-time as a primitive for space and motion, *FOIS*, 63-76.
6. Wolter, F. and Zakharyaschev, M., 2000. Spatio-temporal representation and reasoning based on RCC-8, *KR*, 3-14.
7. Hornsby, K. and Egenhofer, M., 2002. Modelling moving objects over multiple granularities, *Annals of Mathematics and Artificial Intelligence*, 36(1-2), 177-194.
8. Wolfson, O., Xu, B., Chamberlain, S., and Jiang, L., 1998, Moving object databases: issues and solutions, *SSDBM*, 111-122.
9. Erwig, M., Güting, R.H., Schneider, M., and Vazirgiannis, M., 1999, Spatio-temporal data types: an approach to modelling objects in databases, *Geoinformatica*, 3(3), 269-296.
10. Moreira, J., Ribeiro, C., and Saglio, J.-M., 1999, Representation and manipulation of moving points: an extended data model for location estimation, *Cartography and Geographic Information Systems*, 26(2), 109-123.
11. Nabil, M., Ngu A., and Shepherd A.J., 2001, Modelling and retrieval of moving objects, *Multimedia Tools and Applications*, 13(1), 35-71.
12. Van de Weghe, N., 2004, *Representing and Reasoning about Moving Objects: A Qualitative Approach*, PhD thesis, Belgium, Ghent University, 268 pp.
13. Egenhofer, M. and Al-Taha, K., 1992, Reasoning about gradual changes of topological relationships, *COSIT*, 196-219.
14. Egenhofer, M.,1997,Query processing in spatial-query-by-sketch, *JVLC*,8,403-424.
15. Freksa, C., 1992, Using orientation information for qualitative spatial reasoning, *COSIT*, 162-178.
16. Galton, A., 1995, Towards a qualitative theory of movement, *COSIT*, 377-396.
17. Galton, A., 2000, *Qualitative Spatial Change*, University Press, 409 pp.
18. Galton, A., 1995, A qualitative approach to continuity, *TSM*.
19. Galton, A., 1996, Time and continuity in philosophy, mathematics and artificial intelligence, *Kodikas/Code*, 19(1-2), 101-119.
20. Goyal, R.K., 2000, *Similarity Assessment for Cardinal Directions between Extended Spatial Objects*, PhD thesis, USA, University of Maine, 167 pp.

A Refined Line-Line Spatial Relationship Model for Spatial Conflict Detection*

Wanzeng Liu[1,2], Jun Chen[2], Renliang Zhao[2], and Tao Cheng[3]

[1] China University of Mining & Technology,XuZhou, China, 221008
luwnzg@163.com
[2] National Geomatics Center of China, 1 Baishengcun, Zizhuyuan, Beijing, China, 100044
{chenjun,zhaorenliang}@nsdi.gov.cn
[3] Department of Land Surveying and Geo-Informatics,
The Hong Kong Polytechnic University, Hung Hom, KowLoon, Hong Kong
Lstc@polyu.edu.hk

Abstract. In order to detect the spatial conflicts in the process of spatial database updating, a refined description of spatial relationships is needed. This paper presents a topological chain model for refined line-line spatial relationships. The model integrates the topological relationships and order relationships with the metric relationships, such as intersecting angles, number of intersection points, distance between two intersection points, and area of the region formed by two intersectant curves. Based upon such detailed descriptions of the line-line relationships, the spatial conflicts can be detected automatically. The validity of the model and the efficiency of the method have been approved by automatic detection of the conflicts between rivers and contours for the national spatial database at the scale of 1:50,000.

1 Introduction

In the process of spatial database updating, the spatial relationship might be inconsistent due to insimultaneous updating for the river and other topographical features, such as contours. When the updated river layer overlays the un-updated contour layer, the in-correct (illogical) relationships emerge [1], i.e. the contour falls into water, the river flows to the higher height, and the river departs from the valley, etc. These in-correct (illogical) spatial relationships are called spatial conflict, which should be detected and corrected [2].

In order to detect the spatial conflicts in the spatial database, Ubeda [3] and Servigne [4] defined the constraints of forbidden topological relationships by the 9-intersection model[5], combining with semantic information to detect the topological errors. Gadish [6] proposed a method of inconsistency detection and adjustment by using rule discovery, which is based on the 4-intersection model and valid for the inconsistency detection between area objects. The above approaches only consider the topological relations, which are suitable for checking the topological errors between

---

* The work described in this paper was supported by the Nature Science Foundation of China (under Grants No. 40337055 and No.40301042).

J. Akoka et al. (Eds.): ER Workshops 2005, LNCS 3770, pp. 239–248, 2005.

simple features, but not able to detect the complex spatial conflicts between them. The spatial relationships of rivers and other topographical features are comparatively complex, relating to not only the topological and semantic information but also the metrics, directions and attitudes. In order to detect these spatial conflicts, a refined description model called topological chain is presented. In this model, the topological relationships, order relationships and metric relationships are integrated to describe line-line spatial relationships.

This paper is organized as follows. Section 2 illustrates the line-line spatial relationships, which should be defined and described for detecting the spatial conflicts. Section 3 proposes a refined description model of line-line spatial relationships. Section 4 illustrates a method to detect and judge the spatial conflicts based on the model. Section 5 demonstrates the practical application of the proposed model for the national topographic database updating. The last section summarizes the major findings and proposes directions for further research.

2   Spatial Relationships Between Conflicting Line Objects

2.1   Topological Relationships

According to the National Topographic Map Specifications of GB/T18315-2001, a river can only intersect an identical contour once, otherwise it is considered as a spatial conflict that a river climbs up a slope. As illustrated in Figure 1, a river intersects a contour seven times, where only Point 1 is a correct intersection point. To detect this spatial conflict, the local topological relationship types, intersection numbers and the intersection sequence should be computed and described [7].

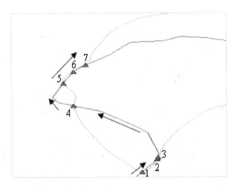

Fig. 1. A river intersects a contour seven times; it is a spatial conflict that the river climbs up the slope

2.2   Metric Relationships

The spatial relationship expresses the spatial constraints of the spatial data [8]. Among the three kinds of relationships, the metric relationship constraints are the strongest, the order relationships take second place, and the topological relationships are the weakest

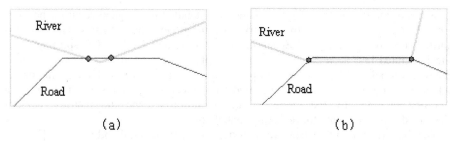

Fig. 2. The spatial conflicts between a road and a river

[9]. Generally, the metric relationships must be calculated in order to confirm the spatial conflicts. The line-line metric relationships mainly include: intersecting angle, number of intersection point, distance between two intersection points, area and perimeter of the region formed by two intersectant curves, etc. For example, Figures 2a and 2b illustrates two special cases of line-line spatial conflicts. In order to detect the spatial conflict in Figure 2a, the intersecting angle and the distance between two intersection points must be calculated and compared with the given thresholds. However, in Figure 2b, the area of the region formed by two intersectant curves is smaller than a given threshold, it still maybe an illogical relationship.

2.3 Order Relationships

Figure 3 shows a road intersecting a river several times in a small range. The spatial relationships including topological relationship, metric relationship and order relationship between them are illogical possibly. To detect and adjust this type of spatial conflict, first, each intersection must be calculated, then, the metric relationships including intersecting angle and distance between two neighboring intersection points should be evaluated, finally, left-right relation should be derived, in company with metric relationships to determine whether the relations are logical.

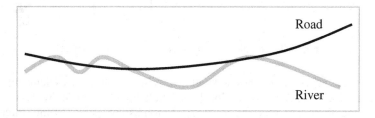

Fig. 3. A road intersects the river several times in a small range

3   A Novel Model for Describing the Line-Line Spatial Relationships

3.1 The Topology Chain for Describing the Line-Line Topological Relationships

Clementini and di Felice [7] derived a complete set of invariants for line-line relations, which includes the intersection sequence, intersection types, collinearity sense, and link orientation. By this model, the detailed topological relations between line objects are represented by a matrix of four columns and N rows. The model is further refined by Nedas and Egenhofer with metric details [10]. The splitting ratios are integrated into a compact representation of detailed topological relations, which addresses topological and metric properties of line-line relations. Though this model adopts new ideas to represent the detailed topological relationships, the description of line-line relations by the model is not intuitionistic, and the metrics relations described is not adequate for detecting spatial conflicts.

The link orientation between two lines defined by Clementini and di Felice [7] is by nature their left-right relation, as illustrated in the Figure 4. The graph in Figure 4b is derived by the graph in Figure 4a, which is overturned to 90 degrees. It is obvious that the link orientation of two scenes is inverse, but the topological relationships of two scenes are equal. Therefore, we consider the link orientation as an order relationship.

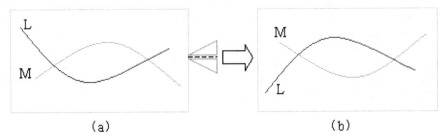

(a)   (b)

Fig. 4. The order relations of two lines

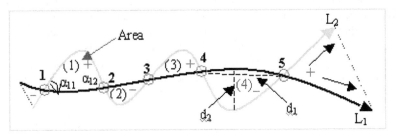

Fig. 5. The topological chain formed by two line objects

Actually, the intersection types and collinearity sense defined by Clementini can be replaced by local cross relations, local meet relations and local overlap relations. Based on the above analysis, the topology invariants are classified into three types: the local topological relationship, intersection sequence, and the number of the intersections. Thus a new description model called topological chain for line-line topological relationships is proposed.

As illustrated in Figure 5, 1, 2, 3, 4 and 5 denote intersections of two lines, called topological ties, which are divided into three types: cross, meet and overlap. As shown in Figure 6, Graph 1 is a cross topological tie, Graph 2 is an interior-meet topological tie, and Graph 3 is an end-meet topological tie. Graphs 4, 5and 6 are three kinds of overlap topological ties, with Graph 4 as a meet-overlap topological tie, Graph 5 as a cross-overlap topological tie, and Graph 6 as an end-overlap topological tie.

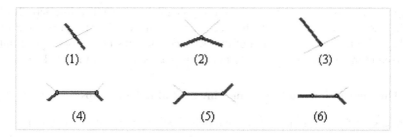

Fig. 6. The six types of the topology ties

The parts between two neighboring topological ties of two lines are defined as topological links. As illustrated in Figure 5, Graphs 1, 2, 3 and 4 denote four topological links. Suppose that each line has no self-intersections, the topological links can be divided into two types. One is the link in same directions, as illustrated in Figure 7a. The projective directions of the two lines on the straight line via the two neighboring nodes are consistent. The opposite case is the link in inverse directions, as shown in Figure 7b. In this case, the local topological relations can be defined by the topological ties, which is consistent with the intersection sequence defined by Clementini and di Felice [7]. The properties of the topological link are defined as follows:

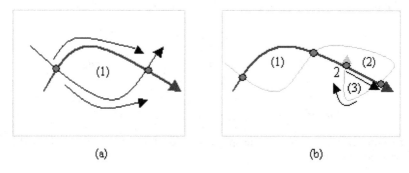

Fig. 7. The two types of the topology links, in (a), (1) denotes a positive direction link; in (b), (2) denotes a negative direction link

1) The **nodes** of topological chain are the intersection point of two lines. 2) The **direction** of topological link is defined as follows. Following the line L_1 from its first point to the end point, if line L_2 is on the left of L_1, then the direction of topological link is positive, otherwise is negative [7]. 3) The **intersecting angle on the node** is that of two intersecting line segments of the two lines. 4) The **portrait length** of the topological link is the length between two nodes, as shown in Figure 5, where d1 is a portrait length of the topological link. 5) The **lateral length** of the topological link is the maximal length of the line from the point on the line L_1 to the line L_2, which is perpendicular to the straight line via the two nodes in the topological link. As shown

in Figure 5, d_2 is a lateral length of the topological link. 6) The **line length** of the topological link is curve length between the two nodes. 7) The **perimeter** of the topological link is the summation of the two lines' lengths of the topological link. 8) The **area** of the topological link is that of the region bounded by the two lines.

3.2 The Description Model of Line-Line Spatial Relationships

It can be seen from section 3.1 that the topological chain model can represent not only the detailed line-line topological relationships, but also the detailed metric relations and the order relations by the properties of its topological links. Thus we can regard the topological chain as a carrier, translating the metric relations and order relations into the various properties of the topological links to represent the spatial relationships of line-line.

If "C" denotes local cross relation defined by the Graphs 1 in Figure 6, "M^{1}", "M^{2}" denotes local meet relation defined by the Graphs 2 and 3 in Figure 6, and "O^{1}", "O^{2}", "O^{3}" denotes local overlap relations defined by Graphs 4, 5 and 6 in Figure 6, respectively, the intersection sequence is represented by an ordinal numbers. The various properties of the topological links are expressed by a matrix of one column and six rows. The raw 1, 2, 3, 4, 5 and 6 of the matrix denotes the local order relations (direction), the topological-link's area (area), portrait length (pd), lateral length (ld), perimeter (perimeter), and the intersecting angle (a_1, a_2... a_n) at the nodes, respectively. Thus the line-line spatial relationships can be expressed by a group of ordinal numbers (k_0, k_1... k_m), the capital letters (T_0, T_1... T_m) and a matrix.

$$R(L_1, L_2) = k_0 T_0 \begin{pmatrix} direction_1 \\ \hline area_1 \\ pd_1 \\ sd_1 \\ perimeter_1 \\ \alpha_{1_1}, \alpha_{1_2} ... \alpha_{1_n} \end{pmatrix} k_1 T_1 \begin{pmatrix} direction_2 \\ \hline area_2 \\ pd_2 \\ sd_2 \\ perimeter_2 \\ \alpha_{2_1}, \alpha_{2_2} ... \alpha_{2_n} \end{pmatrix} ... k_{m-1} T_{m-1} \begin{pmatrix} direction_m \\ \hline area_m \\ pd_m \\ sd_m \\ perimeter_m \\ \alpha_{m_1}, \alpha_{m_2} ... \alpha_{m_n} \end{pmatrix} k_m T_m \quad (1)$$

For example, in the Figure 8, the spatial relationships of line objects L_1 and L_2 can be expressed as follows:

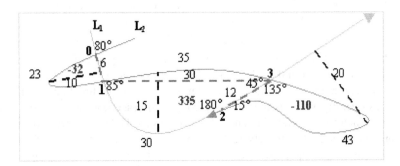

Fig. 8. The expression of line-line spatial relationships based on the topological chain model [7]

$$R(L_1, L_2) = 1C \begin{pmatrix} + \\ \overline{32} \\ 6 \\ 10 \\ 29 \\ 80°,85° \end{pmatrix} 2C \begin{pmatrix} - \\ \overline{335} \\ 30 \\ 15 \\ 77 \\ 85°,45°,180° \end{pmatrix} 4C \begin{pmatrix} - \\ \overline{110} \\ 12 \\ 20 \\ 55 \\ 135°,15° \end{pmatrix} 30^3 \qquad (2)$$

In equation 2, "1C","2C" and "4C" mean that the first, the second and the fourth cross between L_1 and L_2. "30^3" means that the third intersection between L_1 and L_2 is end-overlap. The symbol "+" in the matrix means that L_2 is on the left of L_1.

In the model, if the number (k_i) is arranged by an ascend order; it denotes that the extending directions of two lines are consistency. If two numbers are arranged by a descend order, it denotes the extending directions of two lines are inconsistency.

4 Detecting Line-Line Spatial Conflicts

The line-line spatial conflicts can be detected and judged through the following steps:

1) Detecting the forbidden topological relationships

If the forbidden topological relationship is defined as $R1$ (represented by a capital letter), search the capital letters of the whole topological chain. If R1 is detected, then identify the topological ties according to the number before R1. For example, if R1 represent that two lines cannot cross, then we can judge it by the letter "C" in the chain, and identify the topological ties determined by the number before "C".

2) Judging the more than one cross of two lines

Search the whole topological chain according to the number of the letter "C" to judge this case.

3) Judging the case that two line intersecting at a sharp angle

In the same way, search the whole topological chain, if the intersecting angle in the matrix is less than the given threshold value, then identify the nodes according to the angle sequence and the topological ties sequence.

4) Judging the inconsistent case that the two lines twist each other

This case has four characters: the first is that two lines cross more than one time continuously; the second is that portrait length of the topological link is short; the third is that the lateral length of the topological link is short, and the fourth is that the area of the topological link is small. According to the first character, we can judge whether the two lines twist each other, by the other three characters, we can further judge its rationality.

5) Judging the narrower similar polygon formed by two lines

As shown in Figure 3, first judge whether the two lines intersect at least two times and the portrait length of the topological link is more than the threshold value, then judge the case according to the value of area/pd or area/perimeter.

6) Judging the left-right relations of two lines

As illustrated in Figure 5, suppose there is no converse direction topological link within the topological chain, the left-right relations of two lines can be judged by the following method: (a) Link the two start points of two lines, and form the topological link 0, similarly, link the two end points of two lines, and form the topological link m+1;(b) Calculate the direction and area of all topological links; (c) The left-right relations of two lines can be calculated by the equation 3.

$$R_{l-r} = \frac{\sum_{i=0}^{m+1} direction_i \times area_i}{\sum_{i=0}^{m+1} area_i} \times 100\% \qquad (3)$$

If $R_{l-r} > 0$, it denotes that the L_2 is on the left of L_1, otherwise, L_2 is on the right of L_1. The larger R_{l-r} is, the more possibility the relation exists. If R_{l-r} is less than the given limits, it implies that the two lines have no obvious left-right relations.

5   Case Study

Based on the model of line-line spatial relationships proposed in the previous sections, the automatic detection of the spatial conflicts between the rivers and contours or roads has been realized by programming with VB6.0+MapObject [11].

The test data is a digital sheet of topographic map of mountainous area, which is at scale of 1:50000 and 5km*5km in size, including the updated hydrographic layer (version 2003) and un-updated contour layer. In this region, the terrain mainly is hill, the maximal value of the contour is 1530 meters, the minimal value is 1180 meters, and the contour intervals are 10 meters. Most rivers in this region are developing. There are double-line Season River, single-line Season River, double-line Perennial River and single-line Perennial River, etc, in the hydrographic layer.

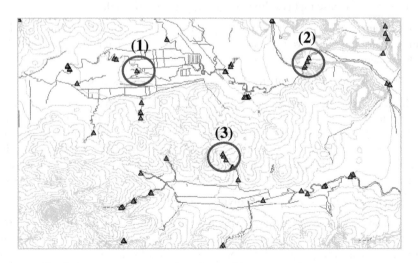

Fig. 9. The detecting result of spatial conflict between river and contours

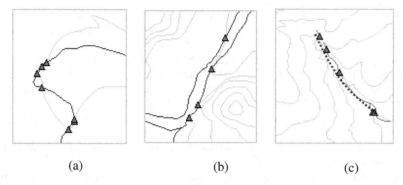

(a)   (b)   (c)

Fig. 10. Zoom in the locations of spatial conflicts shown in Fig.9. (a) zoom in location(1), (b) zoom in the location(2), (c)zoom in location(3).

The result of spatial conflict detecting is illustrated in Figure 9. The red triangle symbols denote the spatial conflict area, and it shows that 38 spatial conflicts have been detected. Zoom in on the spatial conflict area, we can discern and validate the true spatial conflict. Figure10a denotes that the river climbs up the slope. There are two types of spatial conflicts in Figure10b, one is that the contour falls into the double-line river, and the other is that the contour crosses the double-line river with a sharp angle. Figure10c represents that river departs from the valley.

The proposed method has been applied in checking the data quality of the national spatial database at the scale of 1:50000. The detecting result of 323 sheets of mountainous area shows that the detection rate of spatial conflict, which is defined by the rules, is 100%.

6   Summarizes and Further Investigations

Nowadays spatial relationships are widely used in more and more aspects in spatial information systems. It is found that the existing models of spatial relationships are inappropriate for dealing with some special problems. In order to detect the spatial conflict in the process of spatial database updating, a novel model for describing and calculating the line-line spatial relationships is presented in this paper, in which the topological relationships are integrated with order relationships and metric relationships. The method to detect and judge the spatial conflicts based on the above model is also proposed. The validity of the model and the efficiency of the method have been approved by checking the data quality of national spatial database at scale 1:50,000. The further research is how to extend this model to describe the spatial relationships between one line object and a group of line objects.

References

1. Liu, W.Z., Zhao, R.L.: A Study on the Automatic Detecting of Spatial Conflict in the Process of River Updating. The Proceeding of 8th Annual Conference of China Association for GIS, Beijing, China, (2004) 145-153 (In Chinese).
2. Chen, J., Li, Z.L., Jiang, J.: Key Issues of Continuous Updating of Geo-spatial Databases. Geomatics World. 5 (2004) 1-5 (In Chinese).

3. Ubeda, T., Egenhofer, M.J.: Topological Error Correcting in GIS. In: M. Scholl and A. Voisard (Eds): Advances in Spatial Databases SSD'97, Lecture Notes in Computer Science, Vol. 1262, Springer-Verlag (1997) 283-297.
4. Servigne, S., Ubeda, T., Puricelli, A., Laurini, R.: A Methodology for Spatial Consistency Improvement of Geographic Databases. Geoinformatica. 1 (2000) 7-34.
5. Egenhofer, M. J., Franzosa, R.D.: Point-set Topological Spatial Relations. International Journal of Geographical Information Systems 2 (1991) 161–176.
6. Gadish, D.A.: Inconsistency Detection and Adjustment of Spatial Data Using Rule Discovery. PhD Thesis, University of Guelph (2001).
7. Clementini, E. Felice, P.D.: Topological Invariants for Lines. IEEE Transactions on Knowledge and Data Engineering. 1 (1998) 38-54
8. Egenhofer, M.J.: Deriving the Composition of Binary Topological Relations. Journal of Visual Languages and Computing. 5 (1994) 133-149.
9. Schlieder, C.:Reasoning about Ordering. Lecture Notes in Computer Science, Vol. 988. Springer-Verlag, New York (1995) 341-349.
10. Nedas,k.a., Egenhofer, M.J.: Splitting Ratios: Metric Details of Topological Line-Line Relations, 17th International FLAIRS Conference, Miami Beach, FL (2004)(In Press)
11. Liu, W.Z., Chen, J., Zhao, R.L.: A Method of Spatial Conflict Detection Based on the Plane Sweep Algorithm. Greater China GIS Conference, Hong Kong (2004) (In Chinese).

Assessing Topological Consistency for Collapse Operation in Generalization of Spatial Databases

Hae-Kyong Kang[1] and Ki-Joune Li[2]

[1] Department of GIS,
[2] Department of Computer Science and Engineering,
Pusan National University, Pusan 609-735, South Korea
{hkang, lik}@pnu.edu

Abstract. Generalization of spatial databases consists of complicated operations including not only geometric transformations but also topological changes. The changes often result in an inconsistency between the original and derived databases. In order to control the quality of derived databases, we must assess the topological consistency. In this paper, we propose a set of rules to assess topological consistency for the collapse operation. The rules are based on a rigorous classification of topological properties in a collapse operation. By these rules, we can detect inconsistent topological changes in the generalization process and improve the quality of derived databases.

1 Introduction

Multi-scale database is a set of spatial databases on the same area with different scales. In general, databases with small scales can be derived from a large scale database. We call this derivation procedure *generalization of spatial database*. It contains a number of complicated operations such as aggregation, collapse, simplification, and translation, which result in a considerable transformation from the source database.

In the generalization process of spatial databases, a problem arises from the inconsistency between the source and derived databases. It results in a quality degradation of the derived database. In order to control the quality of the generalization process, the consistency between the source and derived databases must be maintained. In this paper, we propose a method to assess the topological consistency of derived database.

During the generalization process, not only geometries but also topologies in the source database are to be changed. While most of topological relations in the source database should be maintained in the derived database, a part of topologies in the source database cannot be identical with those in the source database.

For example, figure 1 shows topological changes from a source database SDB to three other derived databases $MSDB_1$, $MSDB_2$, and $MSDB_3$. Two spatial objects A and B are polygons in the source database SDB. On the other hand, in $MSDB_1$ the two objects ar a polygon and line. In $MSDB_2$, they are simply

J. Akoka et al. (Eds.): ER Workshops 2005, LNCS 3770, pp. 249–258, 2005.

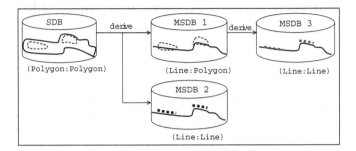

Fig. 1. Topological Changes in Generalization

lines. Figure 1 shows the change of topology from polygon-polygon topology in SDB to line-line topology in $MSDB_2$ and $MSDB_3$. This indicates that the topologies in derived databases can differ from those in the source database. In spite of the difference of topology, a set of correspondences are found between the topologies in source and derived databases. If the topologies in a derived database does not respect the correspondence, it implies that the derived database is not topologically consistent with the source database. Consequently, we can control the topological quality of derived databases by the correspondences.

Topological consistency issues are important in quality control of derived databases on multi-scale databases. Nevertheless, little attention has been paid on the issue. It is significant, however, to reconize such efforts as [6, 13, 15]. Specifically, Tryfona and Egenghofer' research[13] have made a pionering headway for the *aggregate* operation in the field. Now, we intent to continue where they left off.

The goal of this paper is to discover the correspondences between topologies of source and derived databases in case of the *collapse* operation rather than the *aggregate* operation. We propose a classification method of collapse operation by a boundary-interior model. A set of correspondence rules are proposed based on this classification. Such describe consistent topological changes to derived databases.

We discuss related work to our study in section 2. In the next section, we clarify the requirements for assessing topological consistency. Thus proposing correspondence rules to maintain topological consistency in derived databases. Finally we conclude the paper in section 5.

2   Related Work and Motivation

Topological relationships in source database are transformed to different but consistent ones on multi-scale databases. In this case, similarity or consistency between transformed relations and its original relations need to be evaluated. In [6], a boundary-boundary intersection was proposed to assess similarity of two relations on multi-scale representations. The boundary-boundary intersection is part of 9-intersection model [5]. If boundary-boundary intersections of two relations are same each other, the two relations are considered as same. The idea was developed based on the monotonicity assumption of a generalization; any topological relations between objects must stay the same through consecutive representations or continuously decrease in complexity and detail. In [13], a systematic model was proposed to keep constraints that must be held with respect to other spatial objects when two objects are aggregated. This work can be a solution when a multi-scale database is derived by aggregation. However, a solution is still lacking for multi-scale databases derived by a collapse operation from a source database.

The 9-intersection model[2, 5] is used as the topology model of this paper to represent relationships between spatial objects. According to the researches, 8 topological relations(disjoint, contains, inside, equal, meet, covers, covere-, dBy, overlap) between a polygon and a polygon, 19 topological relations($PL_1 \sim PL_{19}$) between a polygon and a line, and 33 topological relations($LL_1 \sim LL_{33}$) between a line and a line are defined. These relations will be referred to throughout this paper.

3   Classification of Collapse Conditions

In this section, we present, with an example, the requirements for assessing and maintaining topological consistency in collapse operation. The collapse operation will be classified according to the topology between the original and collapsed objects.

In figure 2, two polygonal objects A and B in the source database have MEET topological relationship according to the definitions proposed by [2]. Suppose A and B are collapsed to a and b in a derived database respectively. Accordingly, we see that the topology of derived database depends on the type of collapse operation. As depicted in figure 2(a), the topology between a and b may be MEET as the source database, while the topology between a and b in figure 2(b) is not MEET but DISJOINT. This difference comes from different topological relationships between A and a. In figure 2(a), $a \subset A^\circ$ while $a \cap A^\circ \neq \varnothing$ and $a \cap \partial A \neq \varnothing$ in figure 2(b), where ∂A and A° mean the boundary and interior of A respectively.

From this example, we observe that the topological relationship between the original object and collapsed object should be carefully examined as well as topological relationships between two objects in source database. The topology between the original object and collapsed object is determined the type of

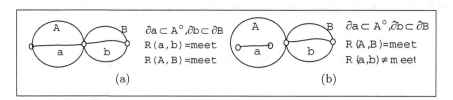

Fig. 2. Example of topological changes

collapse condition, which can be described by topology between boundary and interior of the source and collapsed objects. While we assume that the collapsed object is always contained or covered by the original object, the topology of exterior does not need to be considered.

The topologies between the original and collapsed objects are classified as table 1. Due to the paper length constraints, we assume that the geometric type of collapsed object is line. It may be important in real applications, but we can extend the method proposed in this paper to handle point objects with ease. In the table 1, although there are 9 types of collapse operations, collapse type 4 and 6 do not exist in practice, due to the continuity of line object.

Table 1. Collapse Types

Collapse-Type #	R(A,a) Derivation of A_C from A	example
type 1	$a° \subset A°$ $\partial a \subset A°$	type1 type2
type 2	$a° \subset A°$ $\partial a \subset \partial A$	
type 3	$a° \subset A°$ $\partial a \cap A° \neq \varnothing \wedge \partial a \cap \partial A \neq \varnothing$	type3 type5 (collapse-type 4 and 6 are impossible because of the continuity of line)
type 4	$a° \subset \partial A$ $\partial a \subset A°$	
type 5	$a° \subset \partial A$ $\partial a \subset \partial A$	
type 6	$a° \subset \partial A$ $\partial a \cap A° \neq \varnothing \wedge \partial a \cap \partial A \neq \varnothing$	
type 7	$a° \cap A° \neq \varnothing \wedge a° \cap \partial A \neq \varnothing$ $\partial a \subset A°$	type7 type8 type 9
type 8	$a° \cap A° \neq \varnothing \wedge a° \cap \partial A \neq \varnothing$ $\partial a \subset \partial A$	
type 9	$a° \cap A° \neq \varnothing \wedge a° \cap \partial A \neq \varnothing$ $\partial a \cap A° \neq \varnothing \wedge \partial a \cap \partial A \neq \varnothing$	

Based on the classification of collapse conditions, we can derive a set of rules to describe the consistent correspondence between topologies of the original and derived databases.

4   Rules for Accessing Topological Consistency

The goal of this study is to define a set of rules for assessing the topological consistency between the original and derived databases in case of collapse operation. In other words, we should find the correspondence rules between the

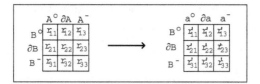

Fig. 3. Correspondence of Topological Relationship

9-IM matrix of the original and derived databases, as depicted by figure 3. If the 9-IM matrix in the derived database would differ from the matrix derived by the corresponding rules, then that the derived database is not topologically consistent with the original database.

In this section, we therefore propose a set of rules for finding thee the corresponding matrix of the derived database under given topological conditions in the source database. They are based on the classification of collapse types given in section 3. The form of rules is as $Rule : R(A, B), R(a, A) \Rightarrow R(a, B)$, where a is the collapsed object from A, $R(A, B)$, $R(a, A)$, and $R(a, B)$ are the topological relationships between A and B, a and A, and a and B. Note that the topological relationship $R(a, A)$ is determined according to collapse types defined in section 3.

The topology between two objects A and B in the source database may be described by 8-topology model, or 9-Intersection Model [2]. However, here it is classified into five cases as follows, according to point set expression. More detail classification of topology in the original database may be possible, but it leads to the same result.

- case 1 (equal set) : $A = B$
- case 2 (subset) : $A \subset B$
- case 3 (superset) : $A \supset B$
- case 4 (disjoint) : $A \cap B = \varnothing$
- case 5 (intersect) : $A \cap B \neq A, B, \ A \cap B \neq \varnothing$

Based on the five cases and the classification of collapse types defined in section 3, we derive 16 rules as summarized in table 2. For the purpose of succinctness, we omit the proof of these rules in this paper. Table 3 and 4 explain how to apply them for assessing topological consistency. Such can be clearly illustrated with an example.

Suppose $A = B$ and the collapse type is **type 1**. In order to find the topological relationship between the interior of the collapsed object a (a°) from A and B, we apply **rule 1** in table 3. Similarly, we apply **rule 3** and 6 to find the topological relationship between the boundary of a (∂a) and B for the case where $A \subset B$ and the collapse type is **type 7**.

The way to apply the rules is summarized by table 4. Note that due to the limit of paper length, it is explained only for the case where $A = B$. We can easily extend this table for the rest cases. For example, suppose that the collapse type is **type 1** and $A = B$. Then we can derive $R(a, B)$ as follows,

Table 2. Assessing Rules for Topological Consistency

Category		Assessing Rule
R(A,B)	rule #	definition
Equal-set Rule (A=B)	rule 1	$P_o^a \subset P_i^A,\ P_l^B = P_i^A\ \Rightarrow$ $P_o^a \cap P_l^B \neq \varnothing,\ P_o^a \cap P_m^B = \varnothing,\ P_o^a \cap P_n^B = \varnothing$
	rule 2	$P_o^a \cap P_i^A \neq \varnothing,\ P_o^a \cap P_j^A \neq \varnothing,\ P_l^B = P_i^A,\ P_m^B = P_j^A\ \Rightarrow$ $P_o^a \cap P_l^B \neq \varnothing,\ P_o^a \cap P_m^B \neq \varnothing,\ P_o^a \cap P_n^B = \varnothing$
Subset Rule (A ⊂ B)	rule 3	$P_o^a \subset P_i^A,\ P_i^A \subset P_l^B\ \Rightarrow$ $P_o^a \cap P_l^B \neq \varnothing,\ P_o^a \cap P_m^B \neq \varnothing,\ P_o^a \cap P_n^B = \varnothing$
	rule 4	$P_o^a \cap P_i^A \neq \varnothing,\ P_o^a \cap P_j^A \neq \varnothing,\ P_i^A \subset P_l^B\ \Rightarrow P_o^a \cap P_l^B \neq \varnothing$
	rule 5	$P_o^a \cap P_i^A \neq \varnothing,\ P_o^a \cap P_j^A \neq \varnothing,\ P_i^A \subset P_l^B,\ P_j^A \subset P_l^B\ \Rightarrow$ $P_o^a \cap P_l^B \neq \varnothing,\ P_o^a \cap P_m^B = \varnothing,\ P_o^a \cap P_n^B = \varnothing$
	rule 6	$P_o^a \subset P_i^A,\ P_i^A \cap P_l^B \neq \varnothing,\ P_i^A \cap P_m^B \neq \varnothing,$ $P_i^A \subset (P_l^B \cup P_m^B)\ \Rightarrow$ $P_o^a \cap P_l^B \neq \varnothing,\ P_o^a \cap P_m^B \neq \varnothing,\ P_o^a \cap P_n^B = \varnothing$
Super-set Rule (A ⊃ B)	rule 7	$P_o^a \cap P_i^A \neq \varnothing,\ P_o^a \cap P_j^A \neq \varnothing,\ P_i^A \cap P_l^B \neq \varnothing,\ P_i^A \cap P_m^B \neq \varnothing,$ $P_i^A \subset (P_l^B \cup P_m^B)\ \Rightarrow\ P_o^a \cap P_l^B \neq \varnothing,\ P_o^a \cap P_m^B \neq \varnothing,$
	rule 8	$P_l^B \subset P_i^A,\ P_o^a \subset P_j^A\ \Rightarrow\ P_o^a \cap P_l^B = \varnothing$
	rule 9	$P_l^B \subset P_i^A\ \wedge\ P_m^B \subset P_i^A\ \wedge\ P_o^a \subset P_i^A\ \Rightarrow\ P_n^B \cap P_p^a \neq \varnothing,$ $P_n^B \cap P_q^a \neq \varnothing,\ \text{where } P_l^B \neq B^-,\ P_m^B \neq B^-,\ P_i^A \neq A^-$
	rule 10	$P_o^a \cap P_i^A \neq \varnothing,\ P_o^a \cap P_j^A \neq \varnothing,\ P_l^B \subset P_i^A\ \Rightarrow$ $\sim (P_o^a \cap P_m^B = \varnothing \wedge P_o^a \cap P_n^B = \varnothing)$
Empty-set Rule A ∩ B = ∅	rule 11	$P_o^a \subset P_i^A,\ P_i^A \cap P_l^B = \varnothing\ \Rightarrow\ P_o^a \subset P_l^B = \varnothing$
	rule 12	$P_o^a \cap P_i^A \neq \varnothing,\ P_o^a \cap P_j^A \neq \varnothing,\ P_l^B \cap P_i^A = \varnothing, P_m^B \cap P_i^A = \varnothing$ $\Rightarrow\ P_o^a \cap P_l^B = \varnothing$
Non-empty -set Rule A ∩ B ≠ ∅, A ∩ B ≠ A, A ∩ B ≠ B	rule 13	$(P_o^a \cup P_p^a) \subset P_i^A\ \wedge\ P_j^A \cap P_l^B \neq \varnothing\ \Rightarrow\ P_q^a \cap P_l^B \neq \varnothing$
	rule 14	$P_o^a \cap P_i^A \neq \varnothing,\ P_p^a \cap P_i^A \neq \varnothing,\ P_o^a \cap P_j^A \neq \varnothing,\ P_p^a \cap P_j^A \neq \varnothing,$ $P_k^A \cap P_l^B \neq \varnothing\ \Rightarrow P_q^a \cap P_l^B \neq \varnothing$
	rule 15	$P_o^a \cap P_i^A \neq \varnothing,\ P_o^a \cap P_j^A \neq \varnothing,\ P_i^A \cap P_l^B = \varnothing,\ P_j^A \cap P_l^B = \varnothing$ $\Rightarrow\ P_o^a \cap P_l^B = \varnothing$
	rule 16	$P_o^a \subset P_i^A\ \wedge\ P_i^A \cap P_l^B = \varnothing\ \Rightarrow\ P_o^a \cap P_l^B = \varnothing$

Notaion

- $^\circ$: Interior, ∂ : Boundary, $^-$: Exterior
- $P_i^A, P_j^A, P_k^A \in \{A^\circ, \partial A, A^-\}$, where $P_i^A \neq P_j^A \neq P_k^A$, $i \neq j,\ j \neq k,\ k \neq i$.
- $P_l^B, P_m^B, P_n^B \in \{B^\circ, \partial B, B^-\}$, where $P_l^B \neq P_m^B \neq P_n^B$, $l \neq m,\ m \neq n,\ n \neq l$.
- $P_o^a, P_p^a, P_q^a \in \{a^\circ, \partial a, a^-\}$, where $P_o^a \neq P_p^a \neq P_q^a$, $o \neq p,\ p \neq q,\ q \neq o$.

Table 3. Rules for Collapse-type

Collapse-Type		A=B	$A \subset B$	$A \supset B$	$A \cap B = \varnothing$	$A \cap B \neq A, B$ $\wedge A \cap B \neq \varnothing$
type 1	a°	rule 1	rule 3 or 6	rule 8 or 9	rule 11	rule 13 or 16
	∂a	rule 1	rule 3, 6 or 7	rule 8 or 9	rule 11	rule 13 or 16
type 2	a°	rule 1	rule 3 or 6	rule 8 or 9	rule 11	rule 13 or 16
	∂a	rule 1	rule 3 or 6	rule 8 or 9	rule 11	rule 13 or 16
type 3	a°	rule 1	rule 3 or 6	rule 8 or 9	rule 11	rule 13 or 16
	∂a	rule 2	rule 4, 5 or 7	rule 10	rule 12	
type 4	a°	rule 1	rule 3 or 6	rule 8 or 9	rule 11	rule 13 or 16
	∂a	rule 1	rule 3 or 6	rule 8 or 9	rule 11	rule 13 or 16
type 5	a°	rule 1	rule 3 or 6	rule 8 or 9	rule 11	rule 13 or 16
	∂a	rule 1	rule 3 or 6	rule 8 or 9	rule 11	rule 13 or 16
type 6	a°	rule 1	rule 3 or 6	rule 8 or 9	rule 11	rule 13 or 16
	∂a	rule 2	rule 4, 5 or 7	rule 10	rule 12	rule 14 or 15
type 7	a°	rule 2	rule 4, 5 or 7	rule 10	rule 12	rule 14 or 15
	∂a	rule 1	rule 3 or 6	rule 8 or 9	rule 11	rule 13 or 16
type 8	a°	rule 2	rule 4, 5 or 7	rule 10	rule 12	rule 14 or 15
	∂a	rule 1	rule 3 or 6	rule 8 or 9	rule 11	rule 13 or 16
type 9	a°	rule 2	rule 4, 5 or 7	rule 10	rule 12	rule 14 or 15
	∂a	rule 2	rule 4, 5 or 7	rule 10	rule 12	rule 14 or 15

i. relationship between a° and B :
   According to table 4, rule 1 ($P_o^a \subset P_i^A, P_l^B = P_i^A \Rightarrow P_o^a \cap P_l^B \neq \varnothing, P_o^a \cap P_m^B = \varnothing$, and $P_o^a \cap P_n^B = \varnothing$) must be applied for this case. Let $a^{\circ} = P_o^a$ and $A^{\circ} = P_i^A$, then $P_l^B = B^{\circ}$ since $A^{\circ} = B^{\circ}$. And P_m^B and P_n^B become ∂B and B^{-}, respectively. Therefore by substituting them to **rule 1**, we obtain $a^{\circ} \cap B^{\circ} \neq \varnothing, a^{\circ} \cap \partial B = \varnothing$, and $a^{\circ} \cap B^{-} = \varnothing$.

ii. relationship between ∂a and B :
   By similar way, we obtain $\partial a \cap B^{\circ} \neq \varnothing, \partial a \cap \partial B = \varnothing$, and $\partial a \cap B^{-} = \varnothing$ by putting $\partial a = P_o^a$ and $A^{\circ} = P_i^A$ according **rule 1**.

iii. relationship between a° and B^{-} :
   If the exterior of a intersects with the interior, boundary, and exterior of B, it is evident that $a^{-} \cap B^{\circ} \neq \varnothing, a^{-} \cap \partial B \neq \varnothing$, and $a^{-} \cap B^{-} \neq \varnothing$, and $R(a^{-}, B) = (1, 1, 1)$. Thus, we exclude this case from table 2 for this reason.

Consequently we conclude that the 9-IM matrix between a and B is as follows,

$$R(a, B) = \begin{pmatrix} 1 & 0 & 0 \\ 1 & 0 & 0 \\ 1 & 1 & 1 \end{pmatrix}$$

Table 4. Equal-set Rule Example

R(A,a) Collapse-Type		R(A,B) (A = B)	Example	R(a,B) Derivation Process	R(a,B) Consistent PL
type 1	a°	rule 1	A = B; $a^\circ \subset A^\circ \wedge A^\circ = B^\circ$	$\begin{array}{c c c c}{}_a\backslash^B & o & \partial & - \\ o & 1 & 0 & 0 \\ \partial & 1 & 0 & 0 \\ - & 1 & 1 & 1\end{array}$	none
	∂a	rule 1	$\partial a \subset A^\circ \wedge A^\circ = B^\circ$		
type 2	a°	rule 1	A = B; $a^\circ \subset A^\circ \wedge A^\circ = B^\circ$	$\begin{array}{c c c c}{}_a\backslash^B & o & \partial & - \\ o & 1 & 0 & 0 \\ \partial & 0 & 1 & 0 \\ - & 1 & 1 & 1\end{array}$	PL_8
	∂a	rule 1	$\partial a \subset A^\circ \wedge \partial A = \partial B$		
type 3	a°	rule 1	A = B; $a^\circ \subset A^\circ \wedge A^\circ = B^\circ$	$\begin{array}{c c c c}{}_a\backslash^B & o & \partial & - \\ o & 1 & 0 & 0 \\ \partial & 1 & 1 & 0 \\ - & 1 & 1 & 1\end{array}$	PL_{10}
	∂a	rule 2	$\partial a \subset A^\circ \cup \partial A \wedge \partial A = \partial B$		
type 5	a°	rule 1	A = B; $a^\circ \subset \partial A \wedge \partial A = \partial B$	$\begin{array}{c c c c}{}_a\backslash^B & o & \partial & - \\ o & 0 & 1 & 0 \\ \partial & 0 & 1 & 0 \\ - & 1 & 1 & 1\end{array}$	PL_4
	∂a	rule 1	$\partial a \subset \partial A \wedge \partial A = \partial B$		
type 7	a°	rule 2	A = B; $a^\circ \subset A^\circ \cup \partial A \wedge \partial A = \partial B$	$\begin{array}{c c c c}{}_a\backslash^B & o & \partial & - \\ o & 1 & 1 & 0 \\ \partial & 1 & 0 & 0 \\ - & 1 & 1 & 1\end{array}$	PL_{12}
	∂a	rule 1	$\partial a \subset A^\circ \wedge A^\circ = B^\circ$		
type 8	a°	rule 2	A = B; $a^\circ \subset A^\circ \cup \partial A \wedge A^\circ = B^\circ$	$\begin{array}{c c c c}{}_a\backslash^B & o & \partial & - \\ o & 1 & 1 & 0 \\ \partial & 0 & 1 & 0 \\ - & 1 & 1 & 1\end{array}$	PL_{11}
	∂a	rule 1	$\partial a \subset \partial A \wedge \partial A = \partial B$		
type 9	a°	rule 2	A = B; $a^\circ \subset A^\circ \cup \partial A \wedge A^\circ = B^\circ$	$\begin{array}{c c c c}{}_a\backslash^B & o & \partial & - \\ o & 1 & 1 & 0 \\ \partial & 1 & 1 & 0 \\ - & 1 & 1 & 1\end{array}$	PL_{13}
	∂a	rule 2	$\partial a \subset A^\circ \cup \partial A \wedge \partial A = \partial B$		

Notation.

R(A,B) : relation of A and B.

PL : Polygon Line relationship in[2].

rule 1 : $P_o^a \subset P_i^A,\ P_l^B = P_i^A \Rightarrow\ P_o^a \cap P_l^B \neq \varnothing,\ P_o^a \cap P_m^B = \varnothing,\ P_o^a \cap P_n^B = \varnothing$

rule 2 : $P_o^a \cap P_i^A \neq \varnothing,\ P_o^a \cap P_j^A \neq \varnothing,\ P_l^B = P_i^A,\ P_m^B = P_j^A \Rightarrow$
$P_o^a \cap P_l^B \neq \varnothing,\ P_o^a \cap P_m^B \neq \varnothing,\ P_o^a \cap P_n^B = \varnothing$

when $A = B$ and the collapse type is type 1 according to table 2, 3, and 4. For other cases, we can derive the matrix by similar way and assess the topological consistency between the original and derived databases.

5   Conclusion

Topological changes take place during the map generalization process from a spatial database of large scale to small scaled database. These changes often result in an inconsistency between the original and derived databases. In order to control the quality of derived databases, the topological consistency must be maintained. For this, we proposed a set of rules, which describes the consistent correspondence between the topologies in the original and derived databases for the collapse operation of generalization.

These rules are based on the classification of the collapse operation by using boundary and interior topology between the original and derived spatial objects. Therefore describing the possible topological changes from the original database. Also through such findings, one can detect inconsistent topological changes in derived databases.

In this study, we have dealt with only the collapse operation from polygonal object to line object and exclude the case where the geometric type of derived object is point. Due to the fact that the polygon-point topology is much simpler than polygon-line or line-line topologies, the proposed rules in this paper can be easily extended. Such is the ground work for future research in this field. Moreover, future endeavors can also include the study on topological consistency for simplification operation in addition to collapse operation.

Acknowledgements

This research was partially supported by the Program for the Training of Graduate Students in Regional Innovation which was conducted by the Ministry of Commerce, Industry and Energy of the Korean Government and by the Internet information Retrieval Research Center(IRC) in Hankuk Aviation University. IRC is a Regional Research Center of Kyounggi Province, designated by ITEP and Ministry of Commerce, Industry and Energy.

References

1. M. J. Egenhofer and H. Herring, *Categorizing Binary Topological Relations Between Regions, Lines, and Points in Geographic Databases*, Technical Report, Department of Surveying Engineering, University of Maine, 1990.
2. M. J. Egenhofer, *Point-Set Topological Spatial Relations*, International Journal of Geographical Information Systems 5(2):161-174, 1991.
3. M. J. Egenhofer and K. K. Al-Taha, *Reasoning about Gradual Changes of Topological Relationships*, Theory and Methods of Spatio-Temporal Reasoning in Geographic Space, LNCS, VOL. 639, Springer-Verlag, 196-219, 1992.

4. M. J. Egenhofer, and J. sharma, *Assessing the Consistency of Complete and Incomplete Topological Information*, Geographical Systems, 1(1):47-68, 1993.
5. E. Clementini, J. Sharma, and M. J. Egenhofer, *Modeling Topological Spatial Relations : Strategies for Query Processing*, Computer and Graphics 18(6):815-822, 1994.
6. M. Egenhofer, *Evaluating Inconsistencies Among multiple Representations*, 6th international Symposium on Spatial Data Handling, 902-920, 1994.
7. M. J. Egenhofer, E. Clementini and P. Felice, *Topological relations between regions with holes*, International Journal of Geographical Information Systems 8(2):129-144, 1994.
8. M. J. Egenhofer, *Deriving the Composition of Binary Topological Relations*, Journal of Visual Languages and Computing, 5(2):133-149, 1994.
9. J. Sharma, D. M. Flewelling, and M. J. Egenhofer, *A Qualitative Spatial Reasoner*, 6th International Symposium on Spatial Data Handling, 665-681, 1994.
10. Muller J. C., Lagrange J. P. and Weibel R., *Data and Knowledge Modelling for Generalization in GIS and Generalization*, Taylor & Francis Inc., 73-90, 1995.
11. A.I. Abdelmoty and B. A. El-Geresy, *A General Method for Spatial Reasoning in Spatial Databases*, CIKM, 312-317, 1995.
12. M. J. Egenhofer, *Consistency Revisited*, GeoInformatica, 1(4):323-325, 1997.
13. N. Tryfona and M. J. Egenhofer, *Consistency among Parts and Aggregates: A Computational Model*, Transactions in GIS, 1(3):189-206, 1997.
14. A. I. Abdelmoty and C. B. Jones, *Toward of Consistency Maintenance of Spatial Database*, CIKM, 293-300, 1997.
15. J. A. C. Paiva, *Topological Consistency in Geographic Databases With Multiple Representations*, Ph. D. Thesis, University of Maine, 1998, http://library.umaine.edu/theses/pdf/paiva.pdf.
16. H. Kang, S. Do, and Ki-Joune Li, *Model-Oriented Generalization Rules*, Proc.(in CD) ESRI Conf. San Diego, USA, July, 2001.
17. H. Kang, T. Kim, and K. Li, *Topological Consistency for Collapse Operation in Multi-scale Databases*, 1st Workshop on Conceptual Modeling for Geographic Information Systems in Conjunction with ER2004, Lecture Notes in Computer Science 3289, Springer-Verlag, 91-102, 2004.

Spatial Relations for Semantic Similarity Measurement

Angela Schwering[1,2] and Martin Raubal[2]

[1] Ordnance Survey of Great Britain, United Kingdom
[2] Institute for Geoinformatics, University of Muenster, Germany
{angela.schwering, raubal}@uni-muenster.de

Abstract. Measuring semantic similarity among concepts is the core method for assessing the degree of semantic interoperability within and between ontologies. In this paper, we propose to extend current semantic similarity measures by accounting for the spatial relations between different geospatial concepts. Such integration of spatial relations, in particular topologic and metric relations, leads to an enhanced accuracy of semantic similarity measurements. For the formal treatment of similarity the theory of conceptual vector spaces— sets of quality dimensions with a geometric or topologic structure for one or more domains—is utilized. These spaces allow for the measurement of semantic distances between concepts. A case study from the geospatial domain using Ordnance Survey's MasterMap is used to demonstrate the usefulness and plausibility of the approach.

1 Introduction

Successful communication of concepts depends on a common understanding between human beings and computer systems exchanging such information. In order to achieve a sufficient degree of semantic interoperability it is necessary to determine the semantic similarity between these concepts. Various approaches to measure semantic similarity between concepts exist and often such calculations of semantic distances are based on taxonomic and partonomic relations. When determining semantic similarity between geospatial concepts it is important to account for their spatial relations in the calculation process. All geospatial objects have a position in space with regard to some spatial reference system and therefore a spatial relation to each other. Spatial relations are also central characteristics on the conceptual level. In this paper, we present an approach of integrating spatial relations into semantic similarity measurements between different geospatial concepts. Such integration improves the quality of the measurements by enhancing the accuracy of their results.

For the formal representation of concepts and the calculation of their semantic similarities we utilize Gärdenfors' idea of a *conceptual space*—a set of quality dimensions within a geometric structure [1]. Such a representation rests on the foundation of cognitive semantics [2], asserting that meanings are mental entities, i.e. mappings from expressions to conceptual structures, which themselves refer to the real world. They therefore allow us to account for the fact that different people have different conceptualizations of the world.

J. Akoka et al. (Eds.): ER Workshops 2005, LNCS 3770, pp. 259–269, 2005.

A case study, in which a customer specifies query concepts based on a shared vocabulary and wants to extract similar concepts from the database of a mapping agency, is used to demonstrate the importance of accounting for spatial relations in a real scenario. It is shown that without the inclusion of spatial relations, the customer is presented with answers which do not fully match her requirements.

2 Related Work

Most knowledge representations use a definitional structure of concepts to describe their semantics: Concepts are specified by necessary and sufficient conditions for something to be its extension. The nature of these conditions is distinct: Properties (features, dimensions) describe the characteristics of concepts, while semantic relations describe concepts through their relationships to other concepts.

The following section describes the formalization of natural-language spatial relations used for the description of geo-concepts. In section 2.2 we give an overview of semantic similarity measures and evaluate how they include relations. The final section describes conceptual spaces, the representational model used in this paper.

2.1 Formalization of Natural-Language Spatial Relations

Describing a concept with relations closely resembles the human way of structuring knowledge: According to the associationist theory humans memorize knowledge by building relations between concepts. The importance of spatial relations arises from the geographic reference of most of our data. All geo-objects have a position in a spatial reference system and each pair of geo-objects is spatially related. The same goes for the conceptual level: due to their functional dependence the geo-concept 'floodplain' is always situated near a water body. We consider spatial relations to be fundamental parts of the semantic description of geo-data.

While formal spatial relations—topologic [3], distance [4] and direction relations [5]—have well defined semantics, natural-language spatial relations have more complex semantics and often imply more than one type of formal spatial relation. People are more familiar with using spatial terms in their natural languages, but systems use definitions based on a computational model for spatial relations. To bridge this gap Shariff et al. developed a model defining the geometry of spatial natural-language relations following the premise *topology matters, metric refines* [6].

The *computational model for spatial relations* [7, 8] consists of two layers: first it captures the topology between lines and regions based on the 9-Intersection model. The second layer analyzes the topologic configuration according to a set of metric properties: splitting, closeness and approximate alongness.

Splitting determines the way a region is divided by a line and vice versa. The intersection of the interior, exterior or boundary of a line and a region is one- or two-dimensional. In the 1-D case the length of the intersection is measured, in the 2-D case the size of the area. To normalize length and area, they are divided either by the region's area or the length of the line or the region's boundary. *Closeness* describes the distance of a region's boundary to the disjoint parts of the line. It distinguishes between Inner/Outer Closeness and Inner/Outer Nearness. *Approximate alongness* is a combination of the closeness measures and the splitting ratios: it assesses the length of the section where the line's interior runs parallel to the region's boundary.

To capture the semantics of geo-objects with spatial relations it is important to use natural-language terms, because they are plausible for humans. Equally important is to have an unambiguous, formal interpretation of these natural language terms. The computational model by Shariff et al. provides a set of natural-language spatial relations with a formalization verified by human subject tests [8]. In our investigation we use a subset of those natural-language terms, which can be applied within the case study. People's choice of spatial relations to describe two objects differs depending on the meaning of objects, their function, shape and scale. We consider only hydrological geo-objects within a large-scale topographic map for defining spatial relations and do not take into account specific semantics of relations depending on the object's meaning, function, or shape.

2.2 Semantic Similarity Measurement

Geometric representations model objects within a multidimensional space: Objects are described on dimensions spanning a vector space [9]. Dimensions are separable into attribute-value pairs: terms that can be evaluated to a value with different, mutually exclusive levels, e.g. the flow speed of a river is either slow, middle or fast. Geometric representations evolved from multidimensional scaling (MDS) [10, 11]: while MDS starts from similarity judgments and determines the underlying dimensions, geometric models represent objects on pre-known dimensions and convert their spatial distance, interpreted as a semantic distance, to a similarity value. Similarity measures in geometric models are metric, though various extensions to account for non-metric properties exist (e.g. Distance Density Model [12], Relative Prominence Model [13]). Geometric representations use only properties for semantic description. It is not possible to describe relationships between objects or concepts.

Feature representations model concepts as sets of features. Features are unary predicates, e.g. a concept 'water body' has the feature 'flowing' or ⌐ 'flowing'. The Feature Matching Model proposed by Tversky [14] is a nonmetric similarity measure comparing two concepts as two sets of features: common features increase and distinct features decrease similarity. It was also applied in other similarity measures such as the Matching-Distance Similarity Measure [15, 16]. In feature representations the description of concepts is limited to atomic features. Relations between objects cannot be represented in a structured way: some approaches construct compound features, but compound features do not allow for structured comparison, e.g. no similarity would be detected between 'nearRiver' and 'veryNearRiver'.

Network representations describe concepts by their relation to other concepts in semantic nets. Relations are n-ary predicates with concepts as arguments. Shortest path algorithms such as Distance [17] are used as a similarity measure. The representation of relations is the strength of network models, but most similarity measures restrict the type of relations to taxonomic and partonomic relations.

Alignment models such as Goldstone's SIAM model [18] describe concepts by features and relations. For similarity measurement they additionally take into account whether features/relations describe corresponding parts: aligned matches increase the similarity more than non-aligned ones. SIAM measures similarity between spatial scenes. Applying SIAM for concept similarity is difficult: due to the different granularity of concept descriptions, an alignment of elements is often not possible. Therefore this model does not return good results when comparing geo-concepts.

This paper extends geometric models to represent spatial relations on dimensions. Geometric models are chosen, because the dimensional structure allows for modelling the degree of relations. Conceptual vector spaces provide a solid mathematical basis for representing information at the conceptual level.

2.3 Conceptual Vector Spaces as Geometric Model

The notion of a *conceptual space* was introduced by Peter Gärdenfors as a framework for representing information at the conceptual level [1]. He argued that cognitive science needs this intermediate level in addition to the symbolic and the subconceptual level. Conceptual spaces can be utilized for knowledge representation and sharing and support the paradigm that concepts are dynamical systems. A conceptual space is a set of quality dimensions with a geometric or topologic structure for one or more domains. A domain is represented through a set of integral dimensions, which are distinguishable from all other dimensions. For example, the colour domain is formed through the dimensions hue, saturation and brightness. Concepts are modelled as n-dimensional regions and every object is represented as a point in a conceptual space. This allows for expressing the similarity between two objects as their spatial distance.

In [19], a methodology to formalize conceptual spaces as vector spaces was presented. Formally, a conceptual vector space is defined as $\mathbf{C}^n = \{(c_1, c_2, ..., c_n) \mid c_i \in \mathbf{C}\}$ where the c_i are the quality dimensions. A quality dimension can also represent a whole domain, then $c_j = \mathbf{D}^n = \{(d_1, d_2, ..., d_n) \mid d_k \in \mathbf{D}\}$. The fact that vector spaces have a metric allows for the calculation of distances between points in the space. In order to calculate these so-called *semantic distances* between instances and concepts it is required that all quality dimensions are represented in the same relative unit of measurement. This is ensured by calculating the percent ranks for these values [20].

3 Case Study

A customer of the British national mapping agency Ordnance Survey, such as the Environment Agency of England and Wales, wants to set up a flood warning system [21]. An overview of existing flooding areas is needed to analyze the current flood defence situation in Great Britain.

OS MasterMap contains geographic and topographic information on every landscape feature—buildings, roads, plants, fields and water bodies. It also contains information on areas used for flooding, but these are not explicitly designated as such [22]. While labels such as 'floodplain' allude to something used for flooding, geo-objects named 'watermeadow', 'carse' or 'haugh' are identified as flooding areas by their semantic description only. The semantics of all geo-objects within OS MasterMap are described as concepts in an ontology using an accurately defined shared vocabulary. The shared vocabulary does not contain concept labels such as 'flooding area', but only terms to describe properties, e.g. 'waterlogged' and relations between concepts, e.g. 'flooding area is next to river'.

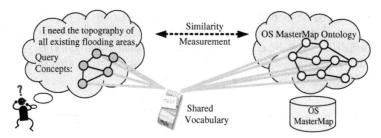

Fig. 1. Visualization of the case study

Table 1. Spatial relations in the shared vocabulary

spatial relation	examples
along	flooding areas lie along a river bank
connected to	rivers are connected to a river, a lake or the sea
in	rivers lay in a river basin
end at	rivers end at river mouths
end in	rivers end in the sea
end just inside	ship ramps end just inside rivers
end near	port feeders end near the sea
near / very near	flooding areas are near / very near a river

The customer searches for topographic information about rivers and flooding areas (figure 1). The semantics of the required information is defined within query concepts using the same shared vocabulary as in the OS MasterMap ontology[1]. To retrieve data according to their relevance, a semantic similarity measure is used to match the query concepts with OS MasterMap concepts.

The shared vocabulary was developed for this case study and contains only expressions necessary for this particular similarity measurement. It does not raise the claim of completeness nor of being a representative set of spatial relations for a geo-ontology. From the set of natural-language spatial relations formalized by Shariff et al. we identified a subset of those relations being relevant for the case study (table 1).

Many geo-concepts such as 'flooding area' and 'river' can be well described by their relation to other geo-concepts. The customer uses the spatial relations listed above and a set of dimensions to specify the query concept. The complete shared vocabulary and measurements can be found at http://ifgi.uni-muenster.de/~eidueidu/er05.zip.

4 Formalization of Measurement

For the described case study the similarity of a set of OS MasterMap concepts to the query concepts 'flooding area' and 'river' are measured. The occurring properties and spatial relations are formalized as dimensions of a conceptual vector space.

---

[1] The customer can use natural-language spatial relations, while semantics in the OS MasterMap ontology is based on the formal definition of such relations.

Concepts can be described by their properties or relations to other concepts. In a conceptual space, properties are represented by dimensions or domains. A property can be formalized by a dimension with a value— *dimension(concept) = value*. The value is defined within a specific range [1]. Figure 2 gives an example for the dimension 'waterlogged' with values for two concepts.

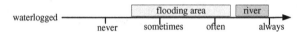

Fig. 2. Representation of dimension 'waterlogged' for concept 'flooding area' and 'river'

To model relations[2] as dimensions, the dimension need not only represent one concept and its values, but one concept and its values with regard to a second concept. We propose to represent relations between two concepts by introducing dimensions depending on the first argument of the relation. The second argument is represented with its value on the dimension in the conceptual space.

Table 2. Relations are represented on a Boolean or ordinal dimension with numerical values

relation	scale	original values	numerical values
along river	Boolean	yes	1
		not specified	0
nearness	Boolean	yes	1
		not specified	0
	ordinal	low nearness	0
		near	1
		very near	2
		not specified	-

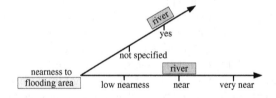

Fig. 3. Modelling relations as dimensions on Boolean and ordinal scale

We distinguish two types of relations (table 2): Boolean relations do not have any degree of existence, e.g. the relation 'along' is either applicable to two concepts or not. They are represented by one Boolean dimension. Other relations have different degrees: the 'nearness' relation can state that two objects are very near, near, or somewhere around (low nearness). These relations require a domain consisting of two dimensions—a Boolean and an ordinal: If the relation holds, the Boolean dimension has the value 'yes' and the ordinal dimension is assigned a value specifying the degree (figure 3). If the relation is not applicable the Boolean relation has the value 'not specified' and the ordinal dimension has no value.

---

[2] In this case study we limit our investigation to binary relations.

According to the rank-order rule ordinal values for the degree of relations are transformed into ordered numerical values[3] [23]. Boolean dimensions are represented by the values {1,0}. The value 0 does not state that the relation does not hold, but that it was not specified by the user or in the system[4].

For the similarity measurement we use the Euclidian and the city block metric in order to calculate distances [24]. The results of the case study demonstrate that our approach is robust and provides good results independent of the metric applied.

Here, we focus on the measurement of semantic distances on the conceptual level. Concepts are convex regions in the conceptual space. Since the Euclidian and the city block distances are between two points rather than between two regions, all concepts are approximated by their prototypes [25], i.e. representing the average value for each interval on each dimension.

5 Results of the Case Study

For the similarity calculation we compare each relation of the query concept separately to the relations of the concepts in the data source. Tables 3 and 4 show the results for the similarity measurements to the query concepts 'flooding area' and 'river' with and without spatial relations. The semantic distance values are calculated based on the differences of the standardized values for each dimension[5]. The final values are normalized by the number of dimensions used in the calculation. An Ordnance Survey expert divided the OS MasterMap concepts into three classes according to their similarities to the query concept: matching, similar (concepts must be modified to match) and non-matching.

Independent of the spatial relations the water bodies 'river', 'stream', 'channel' and 'canal' are considered as very different from the query concept 'flooding area', i.e. their semantic distances are large (table 3). The similarity measurement without spatial relations ranks 'lowland', 'meadow' and 'land' more similar to the query concept 'flooding area' than 'haugh'. Distances measured with spatial relations provide correct results. Since 'meadow' and 'land' do not necessarily lie near rivers such as flooding areas do, they are not typically used for flooding. 'Lowland' though, does not lie explicitly along water bodies, but due to the fact that it is low and rivers typically flow through lowland, it is described by being near rivers and the sea. Therefore the semantic distance does not increase much when including spatial relations.

---

[3] The different degrees of nearness result from different distances. The degrees of relation 'end' depend on the prepositions implying different distances, e.g. the distance between two concepts related via 'ends near' is greater than 'ends in'. The numerical values are applied according to the values for inner/outer closeness from the human subject test in [9].

[4] Boolean dimensions representing properties such as 'flowing' yes/no have also the values {1,0}, but here the value 0 explicitly negates the property. If a property is not applicable, this dimension of the conceptual space is not specified.

[5] Another possibility is using the z-transformation (as done in [19]), but this requires that the values for each dimension are normal [26].

Table 3. Standardized semantic distances to query concept 'flooding area'

	OS Expert	Euclidian metric		city block metric	
		without spat. rel.[6]	with spat. rel.	without spat. rel.	with spat. rel.
OSMM floodplain	match	59	40	38	20
OSMM river	no match	98	94	100	93
OSMM stream	no match	94	91	92	88
OSMM watermeadow	match	62	44	53	30
OSMM channel	no match	100	100	98	100
OSMM haugh	match	64	44	52	29
OSMM land	no match	62	79	49	69
OSMM meadow	no match	43	71	31	59
OSMM paddock	no match	66	82	54	74
OSMM lowland	similar	53	59	36	42
OSMM canal	no match	82	88	59	75
OSMM carse	match	53	38	38	22

Table 4. Standardized semantic distances to query concept 'river'

	OS Expert	Euclidian metric		city block metric	
		without spat. rel.	with spat. rel.	without spat. rel.	with spat. rel.
OSMM floodplain	no match	39	44	64	67
OSMM river	match	9	6	20	17
OSMM stream	match	7	6	21	19
OSMM watermeadow	no match	43	46	68	70
OSMM channel	similar	16	17	27	25
OSMM haugh	no match	95	85	88	87
OSMM land	no match	29	46	56	59
OSMM meadow	no match	95	96	88	88
OSMM paddock	no match	100	100	100	99
OSMM lowland	no match	95	98	97	100
OSMM canal	similar	7	11	20	19
OSMM carse	no match	32	35	52	54

The similarity measures to the query concept 'river' are shown in table 4: The similarity measure ranks with and without relations the concepts 'river', 'stream',

---

[6] The distance values are based on different numbers of dimensions. Adding new dimensions to the similarity measurement either increases the distance or it stays the same. To make the distances comparable, they are calculated relative to the number of dimensions and then scaled on a range of [0;100].

'channel' and 'canal' correctly as the most similar concepts to query concept 'river'. But the inclusion of spatial relations changes their order: 'river' and 'stream' flow towards the sea and lay in a river valley, while 'canal' and 'channel' being artificial, man-made geo-objects, do not necessarily have these relations. With spatial relations 'river' and 'stream' are classified as more similar to the query concept than 'canal' and 'channel'.

A comparison of all distance values shows that the inclusion of spatial relations leads to more sensible values for every compared concept. The results of both metrics are good, though the city block metric shows the differences between matching and non-matching concepts more explicitly. This goes along with findings that the city-block metric is more adequate with separable dimensions (e.g. [24]). To reiterate, the results of the case study demonstrate that similarity measurements are more accurate and realistic when spatial relations are included for the calculation of semantic distances between geo-concepts.

6 Discussion

In the following, the assumptions and further requirements for this similarity measurement are evaluated and discussed.

Reducing Concepts to Prototypes. To measure distances between concepts, they are represented by their prototypes in the conceptual space [27]. For some concepts this may lead to a substantial information loss, e.g. generic concepts such as 'land' with broad intervals on each dimension are semantically narrowed down to single points.

Semantic Post-Processing. All concepts are described based on a common shared vocabulary. This vocabulary does not contain concept labels, but to specify the relations other concepts such as 'river valley' are needed. Since the shared vocabulary does not define the semantics of these, they are adjusted manually, e.g. query concept 'river' is described as 'contained within river valley'. This is aligned to 'contained within river basin' of the 'OSMM river'. The concepts 'river valley' and 'river basin' are considered the same for the semantic similarity measurement. Such manual alignment could be automated by using ontologies or thesauri.

Directed Similarity. The purpose of this similarity measurement is to find the most similar concepts to the query concept. We aim at measuring directed similarity from the point of view of the customer. Therefore the similarity values are calculated based on the dimensions used to describe the query concept. Other dimensions of OS MasterMap concepts do not have any effect on the similarity.

7 Conclusions and Future Work

This paper develops a way to include spatial relations between concepts for semantic similarity measurement within conceptual spaces. We model spatial relations as dimensions and show how they can be used in similarity measurement. A case study demonstrates how conceptual spaces extended by spatial relations lead to more accurate retrieval results. Based on a shared vocabulary, a customer defines her data requirements by query concepts. Through the similarity measure the system identifies matching concepts within OS MasterMap, Britain's national topographic database.

The paper leads to different directions for future research:

1. Here we make the simplifying assumption that quality dimensions of a conceptual space are independent. This is often not true: In the case study several dimensions are used to describe the amount and time period when a concept is covered with water, e.g. 'fullOfWater' and 'waterlogged'. It will be necessary to investigate the covariances between dimensions and to account for these in the conceptual space representations. Human subject tests are a way to identify the quality dimensions for a concept and to infer their dependencies—see, for example, [24]—which would lead to non-orthogonal axes in the representation.

2. Concepts are typically convex regions in a conceptual space. As mentioned in the discussion, they are approximated by points to calculate the distances which entails information loss. To measure similarity between concepts a distance measure between regions must be developed. This can be done by calculating distances from each point of a concept to the reference concept. The resulting distance is an n-dimensional surface that can be transformed to a similarity value through its integral [28].

3. In the case study we focus on spatial relations formalized within the computational model by Shariff et al. This model is currently restricted to line-region relations. It seems possible to extend it for region-region relations and model these relations as dimensions in the same way as for the line-region relations.

Acknowledgements

We like to thank Tony Cohn and Hayley Mizen for taking part in our experiment. Moreover we thank the members of MUSIL and the 3 anonymous reviewers for their valuable comments.

References

1. Gärdenfors, P., Conceptual Spaces: The Geometry of Thought. 2000, Cambridge, MA: MIT Press. 317.
2. Lakoff, G., Cognitive Semantics, in Meaning and Mental Representations (Advances in Semiotics), U. Eco, M. Santambrogio, and P. Violi, Editors. 1988, Indiana University Press: Bloomington, Indianapolis, USA. p. 119-154.
3. Egenhofer, M.J. and R.D. Franzosa, Point-Set Topological Spatial Relations. International Journal of Geographical Information Systems, 1991. 5(2): p. 161-174.
4. Hernández, D., Relative Representation of Spatial Knowledge: The 2-C Case, in Cognitive and Linguistic Aspects of Geographic Space, D.M. Mark and A.U. Frank, Editors. 1991, Kluwer Academic Publishers: Dordrecht, The Netherlands. p. 373-385.
5. Freksa, C., Using Information Orientation for Qualitative Spatial Reasoning, in Theories and Methods of Spatio-Temporal Reasoning in Geographic Space, A.U. Frank, I. Campari, and U. Formentini, Editors. 1992, Springer: Berlin-Heidelberg-New York. p. 162-178.
6. Egenhofer, M.J. and D.M. Mark, Naive Geography, in Spatial Information Theory - A Theoretical Basis for GIS, International Conference COSIT, A.U. Frank and W. Kuhn, Editors. 1995, Springer: Berlin-Heidelberg-New York. p. 1-15.

7. Egenhofer, M.J. and A.R. Shariff, Metric Details for Natural-Language Spatial Relations. ACM Transactions on Information Systems, 1998. 16(4): p. 295-321.
8. Shariff, A.R., M.J. Egenhofer, and D.M. Mark, Natural-Language Spatial Relations Between Linear and Areal Objects: The Topology and Metric of English Language Terms. International Journal of Geographical Information Science, 1998. 12(3): p. 215-246.
9. Suppes, P., et al., Foundations of Measurement - Geometrical, Threshold, and Probabilistic Representations. Vol. 2. 1989, San Diego, California, USA: Academic Press, Inc. 493.
10. Shepard, R.N., The Analysis of Proximities: Multidimensional Scaling with an Unknown Distance Function. I. Psychometrika, 1962. 27(2): p. 125-140.
11. Nosofsky, R.M., Similarity scaling and cognitive process models. Annual Review of Psychology, 1992. 43: p. 25-53.
12. Krumhansl, C.L., Concerning the Applicability of Geometric Models to Similarity Data: The Interrelationship Between Similarity and Spatial Density. Psychological Review, 1978. 85(5): p. 445-463.
13. Johannesson, M., Modelling Asymmetric Similarity with Prominence. British Journal of Mathematical and Statistical Psychology, 2000. 53: p. 121-139.
14. Tversky, A., Features of Similarity. Psychological Review, 1977. 84(4): p. 327-352.
15. Rodríguez, A. and M. Egenhofer, Determining Semantic Similarity Among Entity Classes from Different Ontologies. IEEE Transactions on Knowledge and Data Engineering, 2003. 15(2): p. 442-456.
16. Rodríguez, A. and M.J. Egenhofer, Comparing Geospatial Entity Classes: An Asymmetric and Context-Dependent Similarity Measure. International Journal of Geographical Information Science, 2004. 18(3): p. 229-256.
17. Rada, R., et al., Development and application of a metric on semantic nets. IEEE Transactions on systems, man, and cybernetics, 1989. 19(1): p. 17-30.
18. Goldstone, R.L., Similarity, Interactive Activation, and Mapping. Journal of Experimental Psychology: Learning, Memory, and Cognition, 1994. 20(1): p. 3-28.
19. Raubal, M., Formalizing Conceptual Spaces, in Formal Ontology in Information Systems, Proceedings of the Third International Conference (FOIS 2004), A. Varzi and L. Vieu, Editors. 2004, IOS Press: Amsterdam, NL. p. 153-164.
20. Bortz, J., Statistik für Sozialwissenschaftler. Vol. 5th. 1999, Berlin, Heidelberg: Springer.
21. Schwering, A. Semantic Neighbourhoods for Spatial Relations (Extended Abstract). in Third International Conference on Geographic Information Science (GIScience). 2004. Maryland, USA: Regents of the University of California.
22. OrdnanceSurvey, OS MasterMap™ real-world object catalogue. 2001, Ordnance Survey.
23. Malczewski, J., GIS and Multicriteria Decision Analysis. 1999, New York: John Wiley. 392.
24. Johannesson, M., Geometric Models of Similarity. Lund University Cognitive Studies. Vol. 90. 2002, Lund, Sweden: Lund University. 171.
25. Rosch, E., Principles of Categorization, in Cognition and Categorization, E. Rosch and B. Lloyd, Editors. 1978, Lawrence Erlbaum Associates: Hillsdale, New Jersey. p. 27-48.
26. Devore, J. and R. Peck, Statistics - The Exploration and Analysis of Data. 4th ed. 2001, Pacific Grove, CA: Duxbury. 713.
27. Goldstone, R.L. and A. Kersten, Concepts and Categorization, in Comprehensive Handbook of Psychology, A.F. Healy and R.W. Proctor, Editors. 2003, Wiley: New Jersey. p. 599-621.
28. Schwering, A. and M. Raubal. Measuring Semantic Similarity between Geospatial Conceptual Regions. in 1st International Conference on GeoSpatial Semantics. 2005 forthcoming. Mexico City, Mexico: Springer.

Approximate Continuous K Nearest Neighbor Queries for Continuous Moving Objects with Pre-defined Paths*

Yu-Ling Hsueh, Roger Zimmermann, and Meng-Han Yang

Computer Science Department,
University of Southern California,
Los Angeles, California 90089
{hsueh, rzimmerm, menghany}@usc.edu,
http://dmrl.usc.edu

Abstract. Continuous K nearest neighbor queries (C-KNN) on moving objects retrieve the K nearest neighbors of all points along a query trajectory. In existing methods, the cost of retrieving the exact C-KNN data set is expensive, particularly in highly dynamic spatio-temporal applications. The cost includes the location updates of the moving objects when the velocities change over time and the number of continuous KNN queries posed by the moving object to the server. In some applications (e.g., finding my nearest taxies while I am moving), obtaining the perfect result set is not necessary. For such applications, we introduce a novel technique, AC-KNN, that approximates the results of the classic C-KNN algorithm, but with efficient updates and while still retaining a competitive accuracy. We evaluate the AC-KNN technique through simulations and compare it with a traditional approach. Experimental results are presented showing the utility of our new approach.

1 Introduction

Continuous K nearest neighbor queries (C-KNN) have been intensively studied in spatio-temporal databases in the recent years. C-KNN have been defined as obtaining the nearest points to all points on a given moving object's path. The query scenarios fall into two categories: (1) a dynamic query object and static data objects [2, 4] (e.g., finding a car's nearest gas station), and (2) both the query object and the data objects are dynamic (e.g., finding a car's nearest taxi [1, 3]). In recent years, there has been increasing interest in repositories of objects that are in motion due to the proliferation of e-commerce mobile services. A query example might be to "find the nearest police car while I am driving," or to "find the closest three runners around me." The dynamic nature of the data

---

* This research has been funded in part by NSF grants EEC-9529152 (IMSC ERC), MRI-0321377, and CMS-0219463, and unrestricted cash/equipment gifts from the Lord Foundation, Hewlett-Packard, Intel, Sun Microsystems and Raptor Networks Technology.

J. Akoka et al. (Eds.): ER Workshops 2005, LNCS 3770, pp. 270–279, 2005.

objects presents challenges such as frequent location updates for each moving object and expensive query processing.

In this paper, we present a new approach, termed an approximate continuous K nearest neighbor query (AC-KNN) algorithm, to maintain a KNN result set with efficient updates for moving objects. The algorithm is based on the observation that maintaining approximate continuous KNN queries can greatly reduce the computational cost. By defining *split points* on the query trajectories, moving objects can only update their locations on a segment basis regardless of the change of their velocities. In some domains, the exact continuous KNN result is not required and it is unnecessary to sacrifice disk access performance. Figure 1 shows a simple example where there are two moving objects a and b and one query object q. The initial result of q's nearest neighbor is b, which will not change until q reaches q_{s_4}. b will need to update its location only when it reaches b_{s_2}. Experimental results confirm the utility of our approach.

The rest of this paper is organized as follows. In Section 2, we outline the related work. In Section 3 we formally define the KNN problem and introduce the notation we use. In Section 4, we propose the AC-KNN algorithm in detail. In Section 5, we discuss the performance by comparing our work with an existing approach, and in Section 6 we describe our conclusions and the future work.

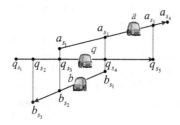

Fig. 1. Example of an AC-KNN query

2   Related Work

The problem of computing K nearest neighbors has been intensively studied. In much of the existing work, R-trees were employed to index multi-dimensional information because of its efficacy. Roussopoulos et al. [10] proposed the branch-and-bound algorithm, which traverses the R-tree in a depth-first manner to find the nearest neighbor of a query object. The primary focus of this research was on static spatial data objects (e.g, gas stations, buildings, restaurants, etc).

In recent years, the interest in databases for moving objects has been increasing. To correctly answer KNN queries, a naïve timepiece-wised approach is to update each moving object's location over time at uniform time intervals. The drawback of this approach is that it has a high overhead. Many existing algorithms assume moving objects proceed in a linear manner represented by a motion function expressed as $p(t) = \vec{x_0} + (t - t_0)\vec{v}$, where $\vec{x_0}$ is the starting point at t_0 and $\vec{v}$ is the object velocity vector. Each moving object is stored in the database in this format and updates are necessary only when the velocity vector changes, such as the direction or the speed. However, this method is not efficient when the moving object representations stored in the database are updated frequently. Many existing techniques that are based on this paradigm have been proposed [8] for handling the KNN query for moving objects.

The other type of KNN query, continuous nearest neighbor search (CNN/C-KNN), has gained similar importance due to the emergence of many e-commerce mobil services. The continuous nearest neighbor search has been defined as the set of nearest points to all points on a given moving object's path. Song et al. [9] first proposed the CNN algorithm by repeatedly performing point-NN queries at predefined sample points on the query trajectory. If the sample point rate is low, the result is likely incorrect; if the sampling rate is high, a significant overhead is incurred. Tao et al. [2] proposed a CNN algorithm to solve this problem by performing a single query to retrieve the nearest neighbors for the entire query trajectory. The query result contains a set of nearest neighbors and their corresponding split points which divide the query trajectory into segments during which the nearest neighbor results remain unchanged and the result is independent of the changing velocity of the query object. However, the algorithm focuses on static spatial data objects which are updated infrequently. Glenn et al. [3] proposed the Continuous Windowing algorithm by reducing the C-KNN query to a less expensive range query. The algorithm first invokes a range query around the query object to select at least K objects, then only those objects need to be considered when computing the C-KNN query. However, this approach requires significant rebuilds when updates are frequent or when K is large. Li et al. [1] proposed the Beach-Line algorithm (BL), which monitors only the Kth nearest neighbor to maintain the C-KNN set, instead of checking all K nearest neighbors. The idea of this algorithm is based on the observation that a necessary condition for the change in the K nearest neighbors is that the distance from the query object to the Kth nearest neighbor becomes larger than that of the $(K+1)^{th}$ one. The algorithm utilizes a delayed update approach to avoid the expensive cost of frequent disk accesses resulting from all updates. However, the cost of maintaining continuous KNN results is still significant. Arya et al. [13] proposed the sparse neighborhood graph, RNG, to process approximate nearest neighbor queries. RNG sets the error variance constant ϵ and the angular diameter δ to tune the size of divided neighboring areas around one data point. This algorithm reduces the time and space requirements for processing data objects significantly. Ferhatosmanoglu et al. [14] utilized VA$^+$-file to solve approximate nearest neighbor queries. This algorithm uses bit vectors to represent a division of the data space. In its first phase it calculates and compares distance of these spatial cells, then it computes and checks real distance between data points of nearby cells in its second phase. Berrani et al. [15] proposed another algorithm to solve approximate K nearest neighbor queries. Firstly it clusters data points in space and represents these clusters as spatial spheres. Then it sets the error variance constant ϵ to tune approximate spheres for original clusters. Data points located in the approximate clustering sphere are candidate results for the approximate KNN queries. However, all these algorithms are aimed at high dimensional and fixed data points. To our best knowledge, the problem of approximate continuous K nearest neighbor queries has not been studied before.

3   Problem Definition

We assume a set of moving objects including the query object $q = [s_q, e_q]$ and data objects $o \in O, o = [s_o, e_o]$, each moving on a pre-defined path with vector velocity $\vec{v}$. For simplicity, we assume the trajectories of moving objects are straight lines, which can be easily extended to polylines or curved lines, reflecting reality more closely. The AC-KNN algorithm computes a set of split points for each moving object in advance, and later, during the movement of the query object, the AC-KNN algorithm returns the approximate result of the KNN query when the object passes a split point on its path. The split point set for an object o is denoted $SL_o = \{o_{s_i}, o_{s_{i+1}}, ..., o_{s_{i+n}}\}$, where $o_{s_i} = s_o$ and $o_{s_{i+n}} = e_o$. In addition, since moving objects may be located in different spatial regions, we need to convert data trajectories into a transformed space, the *relative distance space*, where we compute the Euclidian distance between all points of a data trajectory and the segment of the query trajectory corresponding to the data trajectory. As an example consider Figure 2(a), with a set of data objects $\{a, b, c\}$ and q as the query object in the original space.

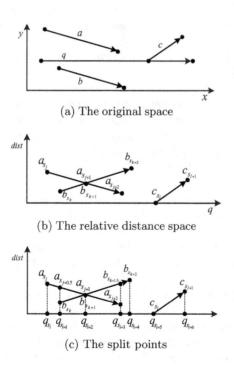

(a) The original space

(b) The relative distance space

(c) The split points

Fig. 2. The data set and the query object

Figure 2(b) shows its relative distance space and Figure 2(c) shows four sets of split points for each moving object: $SL_q = \{q_{s_i}, q_{s_{i+1}}, ..., q_{s_{i+6}}\}$ for q, $SL_a = \{a_{s_j}, a_{s_{j+0.5}}, a_{s_{j+1}}, a_{s_{j+2}}\}$ for a, $SL_b = \{b_{s_k}, b_{s_{k+1}}, b_{s_{k+1.5}}, b_{s_{k+2}}\}$ for b, and $SL_c = \{c_{s_l}, c_{s_{l+1}}\}$ for c, all of which divide the moving trajectories into segments. In case of a data object b, its split points are where there are updates on a segment basis for b. When b passes $b_{s_{k+1}}$, b is in the segment $[b_{s_{k+1}}, b_{s_{k+1.5}}]$. In case of q, the split points are not only the indicators of the location update for q but also where an AC-KNN result set needs to be updated.

At each split point, the query object receives a new set of KNN for the $[q_{s_{i+t}}, q_{s_{i+t+1}}]$ interval and the location of q is updated to $[q_{s_{i+t}}, q_{s_{i+t+1}}]$. AC-KNN algorithm computes the K nearest neighbor set based on this segment-based location information instead of specific location points in order to reduce the disk access rate. We consider a 2-D environment and Euclidian distance to measure the AC-KNN in this paper. The objective of an AC-KNN query is to retrieve approximate K nearest neighbors of q continuously during its movement to its destination. We define a *disk access* as an update request, which is triggered by one of the two following events: (1) a moving object updates its location, velocity, etc., to the server; (2) a query object requests the result updates to maintain the AC-KNN set during its movement.

In many existing systems (eg., [1, 3]), these requests take place whenever a moving object changes its velocity over time. In our AC-KNN algorithm however, a moving object only sends an update request when it reaches one of its split points. We use an R*-tree to index each moving object trajectory [11] by a single bounding rectangle for simplicity, instead of partitioning a trajectory into several bounding rectangles [6].

We observe the following lemmas when generating the split points and the AC-KNN algorithm.

Lemma 1. *All the AC-KNN results are obtained from the candidate trajectory set. The candidate trajectory set is defined as the union of trajectories of the KNN retrieved from each $r \in \{r_1 = s_q, r_2, \ldots, r_n = e_q\}$ where r_i is an increment point on the query trajectory from r_{i-1} by the length of M which is the minimum distance to cover all trajectories of KNN of r_{i-1}.*

By utilizing the *candidate trajectory set*, we are able to prune irrelevant data objects instead of considering all data sets as part of the AC-KNN result. In other words, all the trajectories in the candidate set are considered when computing the AC-KNN query. Since we use a bounding rectangle to index a trajectory, the point-NN queries invoked here can find K nearest trajectories. For instance, Figure 3, where K = 1, and the increment points are r_1, r_2, r_3, r_4 whose 1-NN are $\{a\},\{b\},\{c\},\{d\}$ respectively. Therefore, the candidate set is $\{a, b, c, d\}$.

Lemma 2. *The split points consist of (1) the starting/ending point of trajectories for moving objects, and (2) intersection points between two data trajectories or (3) intersection points between data trajectories and the query trajectory. The data trajectories are the candidate trajectories defined in Lemma 1.*

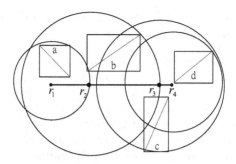

Fig. 3. The increment points and the candidate set

We have defined the split points as the points where a moving object needs to update its location on a segment basis or the points where the query object needs to send an AC-KNN update request to the server in order to maintain the AC-KNN set. Since we only consider those objects on candidate trajectories as part of the AC-KNN result set when performing the AC-KNN query, other non-candidate data objects do not need to be partitioned into segments because their updates are not relevant. A split point is used because it is where a KNN set changes, which is called an *event*. Our observation of determining a split point is based on the following: a starting point of the trajectory is where the moving object starts moving such that it should be considered as part of the AC-KNN result. Similarly, an ending point of the trajectory is where the moving object stops moving. Therefore, an ending point should be checked to see whether it is excluded from the current AC-KNN result. The intersection point of two candidate trajectories or a candidate trajectory and the query trajectory is considered since it is where an *order event* occurs; that is, one or more of the data objects in KNN set may go further or closer to the query object. The process of generating split points for the query object and data objects is divided into two parts:

1. *Relative distance space* transformation:
   Transform all the candidate trajectories into a *relative distance space* to the query trajectory. As the example in Figure 2(a), it shows the original space of the trajectories for the data set $\{a, b, c\}$ and the query trajectory q. Figure 2(b) shows its relative distance space.

2. Split point collection:
   Collect all points from all moving objects using the definition of Lemma 2. As the example shown in Figure 2(b), the points are $P = \{a_{s_j}, b_{s_k}, (a_{s_{j+1}} = b_{s_{k+1}}), a_{s_{j+2}}, b_{s_{k+2}}, c_{s_l}, c_{s_{l+1}}\}$.

3. Split point insertion:
   a. Draw a line l passing each s, $s \in P$ and perpendicular to the segment of object q.

b. Store all the intersection points between l and trajectories. The complete split point sets are as shown in Figure 2(c).

Consider Figure 2(b) and Figure 2(c) as an example, $a_{s_{j+0.5}}$ is stored because l passing the point b_{s_k} intersects the trajectory of a at $a_{s_{j+0.5}}$. In this way, while the query object is moving between $[q_{s_{i+1}}, q_{s_{i+2}}]$, it is able to know whether a is moving between $[a_{j+0.5}, a_{j+1}]$ and whether b is moving between $[b_{s_k}, b_{s_{k+1}}]$. These segment-based location information can greatly reduce the disk accesses and is relied by the AC-KNN algorithm to compute the approximate query result. The best way to represent the segment location information is using a matrix, which we call *Segment-based Location Table* (SLT).

Table 1. Segment-based Location Table

q	q_{s_i}	$q_{s_{i+1}}$	$q_{s_{i+2}}$	$q_{s_{i+3}}$	$q_{s_{i+4}}$	$q_{s_{i+5}}$	$q_{s_{i+6}}$
a	0	1	0	∞	∞	∞	∞
b	∞	1	0	0	∞	∞	∞
c	∞	∞	∞	∞	∞	1	∞

Table 1 shows an example of the SLT converted from Figure 2(c). The value of $(a, q_{s_{i+1}})$ is 1 in SLT. It represents that a is between $[a_{s_{j+0.5}}, a_{s_{j+1}}]$, but is not between $[a_{s_j}, a_{s_{j+0.5}}]$, and $[a_{s_{j+1}}, a_{s_{j+2}}]$ whose values are 0. When the value is ∞, it represents that the segment is invalid. The SLT is updated whenever a moving object moves away from its current segment. For example, while a moves from $[a_{s_{j+0.5}}, a_{s_{j+1}}]$ to $[a_{s_{j+1}}, a_{s_{j+2}}]$, $(a, q_{s_{i+1}})$ is updated to 0 and $(a, q_{s_{i+2}})$ is updated to 1.

4   AC-KNN Algorithm

The processing of AC-KNN search is to first find a candidate trajectory set where moving objects are considered when computing AC-KNN queries. Second, split points are generated at which *events/order events* take place in the *relative distance space*. The split points are converted into a SLT, storing the segment-based location information for each moving object such that the specific locations are not required in order to reduce the number of updates. The update of SLT is required only when a moving object moves away from its current segment. The query object sends an AC-KNN query each time it passes a split point on the query trajectory until it reaches its ending point. However, due to the use of the segment-based location information, the current locations of the moving objects are unknown and the Euclidian distance as the metric to find the KNN set can not be determined. We introduce a heuristic for solving this problem.

Heuristic 1. *Given two segments* $p = [s_p, e_p]$ *and* $q = [s_q, e_q]$, *an average distance between them is estimated by the following equation:*

$$\text{avedist}(p, q) = dist(midPoint(s_p, e_p), midPoint(s_q, e_q)) \qquad (1)$$

The AC-KNN algorithm adopts an incremental approach to first find the KNN from its nearest segments. If KNN is not fulfilled, the algorithm considers the nearby segments until all the KNN are found. The full AC-KNN algorithm is shown as Algorithm 1.

Algorithm 1. AC-KNN Algorithm

Input: the current split point q_{s_i} of q, a SLT, and K
Output: the AC-KNN set
1: Let C be the set of segments whose value in SLT is 1 retrieving from the elements (q_{s_i}, O) of SLT, for each $o \in O$, where O is set of all data objects.
2: Let $r = \text{maximum}(avedist = ([q_{s_i}, q_{s_{i+1}}], c))$, for each $c \in C$
3: Find the center point $q' = midPoint(q_{s_i}, q_{s_{i+1}})$ and use it to find a split point set P covered by $|q' - r|$ to $|q' + r|$, such that $P = \{q_{s_{i-n}}, q_{s_{i-n+1}}, \cdots, q_{s_{i+m}}\}$
4: Find all segments S with value 1 in SLT, for each $p \in P$
5: Find the first K objects in increasing order of $avedist([q_{s_i}, q_{s_{i+1}}], s)$ for each $s \in S$ and insert them to AC-KNN set
6: **if** (Number of AC-KNN < K) **then**
7:    Expend P by adding the next split point of $q_{s_{i+m}}$ and the previous split point of $q_{s_{i-n}}$ to P, then go to Line 4.
8: **else**
9:    Return AC-KNN set.
10: **end if**

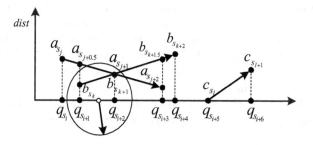

Fig. 4. $P = \{q_{s_i}, q_{s_{i+1}}, q_{s_{i+2}}\}$

Consider the example in Figure 2(c), where K = 3, while q is on $[q_{s_{i+1}}, q_{s_{i+2}}]$, initially, check with SLT (Table 1) and get $C = [a_{s_{j+0.5}}, a_{s_{j+1}}], [b_{s_k}, b_{s_{k+1}}]$. The maximum distance to the query trajectory is $r = avedist([q_{s_{i+1}}, q_{s_{i+2}}], [a_{s_{j+0.5}}, a_{s_{j+1}}])$ Therefore, as shown in Figure 4, the split point set $P = \{q_{s_i}, q_{s_{i+1}}, q_{s_{i+2}}\}$. Then check the SLT, the segments with value 1 in SLT are $[a_{s_{j+0.5}}, a_{s_{j+1}}]$ and $[b_{s_k}, b_{s_{k+1}}]$. Since $[b_{s_k}, b_{s_{k+1}}]$ has shorter $avedist$ than that of $[a_{s_{j+0.5}}, a_{s_{j+1}}]$, b is added to the AC-KNN set as the first nearest neighbor of q; a is then added as the second nearest neighbor of q. In order to find the third nearest neighbor of q, the algorithm expands P to contain more split points for retrieving more moving objects. Finally, c is added later to the AC-KNN set.

5   Experimental Evaluation

We use a timepiece-wise approach as the baseline. This is a brute-force approach which compares the current locations of all moving objects to find the K nearest neighbors of the query object at each time unit. It returns the most correct KNN set. We then compare the accuracy and disk accesses of the BL algorithm with those of the AC-KNN algorithm, respectively. Experiments are conducted with a Pentium 4, 3 GHz CPU and 1 Gbyte of memory. We set up a similar experimental environment as BL did. A 2-D 1000 × 1000 world at the scale of 1:0.02 miles is constructed, where objects are uniformly distributed and the speed vector of each object is randomly generated ranging from 0 to 5. We use the R*-tree library implemented in Java by Papadias [12]. The BL algorithm in our experiment uses four bounding rectangles to decompose each curve and the LFA is set to be two time units. The definition of the *disk access* rate has been discussed in Section 3 and for the accuracy of the KNN sets, we use a spatial accuracy metric, *Average Distance to Nearest Predicted location*(ADNP)[7], which is defined as:

$$ADNP = \frac{1}{|S|} \sum_{s \in S} dist(s, NP(s)) \qquad (2)$$

By utilizing ADNP, if the correct result KNN set is $\{a, b, c\}$ and the AC-KNN algorithm finds $\{a, c, e\}$, ADNP is calculated as $(dist(a, a) + dist(b, c) + dist(c, e))/3$. Note that the accuracy is 100% when ADNP is 0.

We use up to 50,000 moving objects and K=30 for the experiment. In Figure 5(a), we compare the ADNPs of the AC-KNN algorithm with those of the BL algorithm as a function of the number of moving objects. From the figure, we can easily observe that both the AC-KNN algorithm and the BL algorithm scale well, and that overall AC-KNN has competitive ADNP rates to the BL algorithm. From the result shown in Figure 5(b), we can see that the AC-KNN algorithm outperforms the BL algorithm in terms of disk accesses. This demonstrates that our approximate approach can greatly reduce disk accesses.

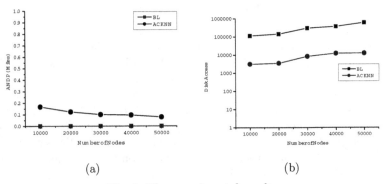

(a)   (b)

Fig. 5. The experimental results

6   Conclusions

We introduced an algorithm to perform an approximate continuous nearest neighbor search, while significantly reducing the disk access rate compared with existing C-KNN algorithms that retrieve the exact C-KNN result. By using split points on its trajectory, a moving object only needs to send a segment-based location update to the server when it reaches each split point. The server then uses the approximate segment-based location information of each moving object to compute the AC-KNN for the query trajectory. We use *avedist* defined in the paper for measuring the approximate distance between two segments. In this paper, we assume a straight line moving trajectory, which in reality might be a polyline or curve. In our future work, we intend to utilize some existing index structures such as MON trees, which we believe can be applied to handle non-linear trajectories efficiently. Our experiments show that our algorithm outperforms the Beach Line algorithm in terms of disk access rate while retaining a competitive accuracy.

References

1. Yifan Li, Jiong Yang, Jiawei Han: Continuous K-Nearest Neighbor Search for Moving Objects. SSDBM, 2004
2. Yufei Tao, Dimitris Papadias, and Qiongmao Shen: Continuous nearest neighbor search. VLDB, 2002.
3. Glenn S. Iwerks, Hanan Samet, Ken Smith: Continuous K-Nearest Neighbor Queries for Continuously Moving Points with Updates. VLDB, 2003.
4. Mohammad R. Kolahdouzan, Cyrus Shahabi: Continuous K-Nearest Neighbor Queries in Spatial Network Databases. STDBM, 2004.
5. Jun Zhang, Manli Zhu, Dimitris Papadias, Yufei Tao, Dik Lun Lee: The Location-Based Spatial Queries. SIGMOD, 2003.
6. Victor Teixeira de Almeida, Ralf Hartmut Güting: Indexing the Trajectories of Moving Objects in Networks. SSDBM, 2004.
7. Tianming Hu, Sam Yuan Sung: Spatial Similarity Measures in Location Prediction, 95-96.
8. Yufei Tao, Dimitris Papadias: Time-Parameterized Queries in Spatio-Temporal Databases. ACM SIGFMOD 2002.
9. Zhexuan Song, Nick Roussopoulos: K-Nearest Neighbor Search for Moving Query Point. SSTD, 2001.
10. N. Roussopoulos, S. Kelly, and F. Vincent: Nearest Neighbor queries. In processings of the 1995 ACM SIGMOD.
11. N. Beckmann, H.-P. Kriegel, R. Schneider, B. Seeger: The R*-Tree: An Efficient and Robust Access Method for Points and Rectangles. SIGMOD 1990. 322-331
12. Dimitris Papadias: Java implementation of R*-tree. http://www.rtreeportal.org
13. Sunil Arya, David M. Mount: Approximate Nearest Neighbor Queries in Fixed Dimensions. SODA 2005. 535-544
14. Hakan Ferhatosmanoglu, Ertem Tuncel: Vector Approximation based Indexing for Non-Uniform High Dimensional Data Sets.
15. Sid-Ahmed Berrani, Laurent Amsaleg, Patrick Gros: Approximate Searches:K-Neighbors + Procision.

Spatio-temporal Similarity Analysis Between Trajectories on Road Networks

Jung-Rae Hwang[1], Hye-Young Kang[2], and Ki-Joune Li[2]

[1] Department of Geographic Information Systems, Pusan National University, Korea
[2] Department of Computer Science, Pusan National University, Korea
{jrhwang, hykang}@isel.cs.pusan.ac.kr, lik@pnu.edu

Abstract. In order to analyze the behavior of moving objects, a measure for determining the similarity of trajectories needs to be defined. Although research has been conducted that retrieved similar trajectories of moving objects in Euclidean space, very little research has been conducted on moving objects in the space defined by road networks. In terms of real applications, most moving objects are located in road network space rather than in Euclidean space. In this paper, we investigate the properties of similar trajectories in road network space. And we propose a method to retrieve similar trajectories based on this observation and similarity measure between trajectories on road network space. Experimental results show that this method provides not only a practical method for searching for similar trajectories but also a clustering method for trajectories.

Keywords: Trajectories, Road Network Space, Similarity between Trajectories.

1 Introduction

With the spread of mobile computing, research to efficiently handle moving objects, where their movement is represented by a trajectory as a set of line segments in (x, y, t) space, has become important [1]. Since the trajectory of a moving object contains a lot of information, it is an interesting task to analyze trajectories for several application areas. One of the most important requirements for analyzing trajectories is to search for objects with similar trajectories and cluster them. For example, a query such as "Find all moving objects whose trajectories are similar to a given query trajectory" is typical.

While research has been done regarding locating similar trajectories of moving objects on Euclidean space, very little has been done regarding to moving objects in road network space. For most of real applications, we are interested in moving objects in road network space rather than in Euclidean space. In order to analyze the behavior of moving objects in road network space, a measure for determining the similarity between the trajectories of moving objects needs to be defined. This measurement of similarity allows for the retrieval of similar trajectories and the eventual discovery of their patterns and clusters.

J. Akoka et al. (Eds.): ER Workshops 2005, LNCS 3770, pp. 280–289, 2005.

Due to the properties of road network space, the methods that can be useful to search for similar trajectories differ from the methods currently in use [2][3]. The current methods have the following drawbacks. First, they assume Euclidean space, and Euclidean distance is no longer valid in road network space, where the distance is limited to the space adjacent to the roads. Since measuring similar trajectories is highly dependent on the definition of distance, the similarity measurements as defined for Euclidean space are inappropriate for road network space, and consequently the methods based on Euclidean space are not suitable for our purpose.

Second, the previously used methods do not fully exploit the spatiotemporal properties of trajectories and most of them only consider spatial similarities. For example, two trajectories passing through the same area at different times are considered similar, even though they are not similar in spatiotemporal sense. In addition two trajectories that are moving in opposite directions are considered to have similar trajectories according to those previous methods.

Our research is motivated by two requirements. First, our method should be based on the characteristics of moving objects on road network space. Second, we should simultaneously consider a spatiotemporal similarity as well as spatial similarity. Based on these ideas, we propose a search method for similar trajectories of moving objects on road networks. Our method is based on spatiotemporal properties and reflects spatial characteristics on road networks.

This paper is organized as follows. In section 2, we introduce the related work and drawbacks of the previously used methods and investigate the characteristics of similarity for the trajectory of moving objects on road networks. In section 3 we propose a method for searching for similar trajectories on road networks. Experimental results are given in section 4. Finally, we conclude and suggest future work in section 5.

2   Related Work and Motivation

In this section, we introduce related researches on moving objects on road networks. Also, we discuss the problems of existing methods by investigating the trajectory characteristics of moving objects on road networks and present the motivations of this paper.

2.1   Related Work

Searching for similar trajectories of moving objects is closely connected to two research issues: 1) representing the trajectory of moving objects and 2) defining measurements of similarity.

Concerning the first research issue, many studies have investigated ways that the trajectories of moving objects can be represented [4][5]. In particular, representation models for trajectories have been proposed based on Markovian and non-Markovian probability models in [6], which are effective in extracting useful information from trajectories. Another interesting model has been proposed in [7] that considers the geospatial lifelines of multiple granularities. These methods deal with moving objects on Euclidean space. However, most moving objects in real applications, such as vehicles or trains, are found in road network space rather than in Euclidean space. There has been some research regarding representing and handling the movement of objects in road network space [5][8][9]. A model for representing and querying moving objects on road networks is clearly presented in [5]. The representation of moving objects along a road network was also presented in [9]. A nearest neighbor search method for moving objects on road networks was introduced in [8].

Regarding the second research issue, the most important studies of search methods for similar trajectories are found in [2] and [10]. A method for finding the most similar trajectory of a given query trajectory within a database using the longest common subsequence model was proposed in [10]. However, this method has two problems when used to search for similar trajectories of moving objects on road networks. First, this method does not take temporal or spatiotemporal variables into consideration. For example, two trajectories passing through the same area at different times are considered similar. Second, since this method is based on Euclidean space, it cannot be used to search for similar trajectories on road networks as discussed in the previous section.

A method for measuring the similarity between trajectories based on shape was defined in [3]. The advantage of this definition is that spatiotemporal aspects are taken into account, unlike [2]. However, since this method assumes Euclidean space, it is difficult to apply it to road network space. A similar method was proposed in [11] but has the same problem of Euclidean distance as [2].

2.2   Similarity of Moving Object Trajectory on Road Networks

Most moving objects are in road network space rather than in Euclidean space. There are several differences between Euclidean space and road network space. First, figure 1 illustrates the different definitions of distance in Euclidean and road network space. In figure 1, the actual distance from a to b is not 4 km but 9 km. Second, different coordinate systems are employed for road network space. While the (x, y, t) coordinate system is the most popular one in Euclidean space, (Sid, d, t) is more efficient in road network space, where Sid is a road sector identifier, and d is the offset from the starting point of the road sector. Queries are given by specifying the road sector ID rather than an area in Euclidean space. It is easier to calculate distance between two points on road networks by using road network coordinate systems than Euclidean coordinate systems. Finally, road network space requires additional data to describe the connectivity between road sectors. These differences should be carefully examined and considered when analyzing trajectories in road network space.

Let's investigate trajectory properties on road networks. Figure 2 shows an example of trajectories in (x, y, t) space, where t represents the time-axis and (x, y) space, which is the projected space. TR_A, TR_B and TR_C in figure 2 are trajectories in (x, y, t) space, while TR'_A, TR'_B and TR'_C are projected trajectories onto the (x, y) plane. TR_A and TR_B pass the exact same point in the same

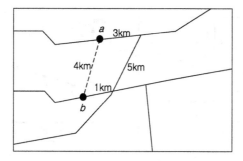

Fig. 1. Example of distances in road network space and Euclidean space

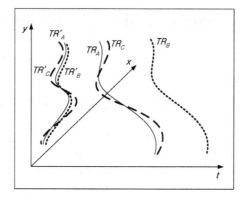

Fig. 2. Trajectories in spatiotemporal space and their projected trajectories

order but at different time intervals. On other side, TR_A and TR_C are a short distance apart from each other, but move at very similar time intervals. When we project these trajectories on the (x, y) plane, TR_A and TR_B are projected on an equal trajectory, while the projected trajectories of TR_A and TR_C are placed at different locations. This shows that if the spatiotemporal variable is considered, then TR_C would have the most similar trajectory to TR_A. This outcome differs from the similarity measures proposed by [2] and [11], which consider only the similarity between projected trajectories.

The distance between moving objects in road network space has an interesting property. Suppose that two moving object trajectories, TR_A and TR_B pass through the same points a, b and c on a road at the same time, and they take different road as depicted by figure 3(a). Then, the distance between them rapidly increases after point c according to the road network as shown by figure 3(b). In most cases, two moving objects on two different road sectors result in a relatively large distance between them and exceed the distance threshold used in similarity searches as shown by figure 3(b). This observation implies that it may be meaningless to compute distance between two moving objects if they are on different road sectors.

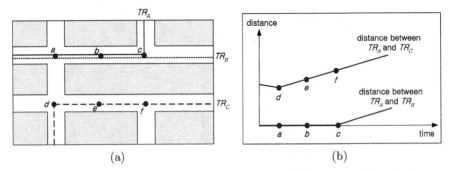

Fig. 3. Change of distance between two moving objects on road networks

3   Searching for Similar Trajectories on Road Networks

In order to retrieve similar trajectories on road networks, we could apply one of the following methods;

- Method 1 : Searching for similar trajectories based on spatiotemporal distance between trajectories.
- Method 2 : Filtering trajectories based on temporal similarity and refining similar trajectories based on spatial distance.
- Method 3 : Filtering trajectories based on spatial similarity and refining similar trajectories based on temporal distance.

We now discuss each of these methods. Method 1 looks simple and therefore attractive. In order to apply this method, we need a robust definition for measuring spatiotemporal distance, which might be the sum of spatial distance and temporal distance. However, it is impossible to define the equivalence between temporal distance and spatial distance. For example, how can the spatial distance equivalent to one minute be defined? We can define the equivalence in a specific situation by considering a parameter such that, for example, dist(1 meter) = dist(α seconds). In most cases, however, the general equivalence cannot be easily defined, and it depends on the context of the application. We leave this issue for further study.

Method 2 requires a definition for temporal similarity or distance in order to filter trajectories. For example, suppose that $[t_s(TR_A), t_e(TR_A)]$, $[t_s(TR_B), t_e(TR_B)]$ are the life spans of the two trajectories, TR_A and TR_B. In practical application, the meaning of distance between two time intervals can hardly be found. It means that the second method is not appropriate for searching for similar trajectories.

Consequently, we propose the third method for searching similar trajectories. For this method, we need a definition of spatial distance between trajectories, which are represented as curves. A widely used definition of distance between curves l and m is Hausdorff distance $dist_H(l, m)$, which is defined as:

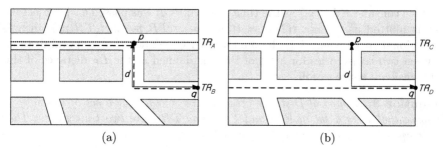

Fig. 4. Hausdorff distance on road networks

$$dist_H(l, m) = \max_{a \in l}\{\min_{b \in m} dist(a, b)\},$$

where $dist(a, b)$ is the distance between two points.

An interesting property of Hausdorff distance is found in the trajectories on road networks. In figure 4(a), the Housdorff distance between TR_A and TR_B $dist_H(TR_A, TR_B)$ is d, determined by the pair of points p and q. And we see that $dist_H(TR_C, TR_D) = d$. The distance between TR_A and TR_B is equal with that between TR_C and TR_D depending on the type of application. It means that we cannot use Hausdorff distance to measure spatial distance.

Instead of Hausdorff distance, we propose a practical method to determine the spatial similarity between trajectories based on POI(Point of Interest). For example, important intersections of roads or places can be POIs. If two trajectories pass through the same POIs, they are considered similar by the following definition.

Definition 1. *Spatial Similarity between Trajectories on road network space*
Suppose that P is a set of POIs on a given road networks. Then spatial similarity between two trajectories TR_A and TR_B is defined as

$$Sim_{POI}(TR_A, TR_B, P) = \begin{cases} 1, & if \ \forall p \in P, \ p \ is \ on \ TR_A \ and \ TR_B \\ 0, & otherwise \end{cases}$$

In order to apply Method 3 for searching for similar trajectories, in addition to a spatial similarity, we also need a measure for temporal similarity. Temporal similarity can be defined as the inverse of temporal distance. In contrast with the discussion of temporal distance in Method 2, temporal distance can be defined, when a POI is given, as the difference between the times two objects passed the same POI as follows:

Definition 2. *Temporal Distance between Trajectories for one POI*
Suppose that $p \in P$, and P is the set of POI. Then the temporal distance between two trajectories TR_A and TR_B is

$$dist_T(TR_A, TR_B, p) = |t(TR_A, p) - t(TR_B, p)|$$

If neither TR_A nor TR_B pass through p, the temporal distance is considered as infinity.

If we consider $t(TR, p_i)$ as the time the i-th POI, was passed each trajectory, TR, is plotted as a point $t(TR) = (t(TR, p_1), t(TR, p_2), ..., t(TR, p_k))$ in a k-dimensional space where k is the number of POIs. Then the temporal distance between two trajectories for a set of POIs is defined as the L_P distance of this k-dimensional space as follows:

Definition 3. *Temporal Distance between Trajectories for a set of POIs*
Suppose that P is a set of POI and TR_A and TR_B are two trajectories. Then the temporal distance between TR_A and TR_B is

$$dist_T(TR_A, TR_B, P) = L_p(TR_A, TR_B, p) = \left(\sum_{i=1}^{k} |(p_i(TR_A) - p_i(TR_B)|^p\right)^{\frac{1}{p}}$$

Algorithm 1. Searching Similar Trajectories

   Input.   input trajectories TR_{IN}, threshold δ, query trajectory tr_Q, POI set P
   Output.   similar trajectories TR_{OUT}
Begin
   $TR_{Candidate} \leftarrow \phi$
   $TR_{OUT} \leftarrow \phi$
   For each $tr \in TR_{IN}$
      If $\forall p \in P$, p is on tr
      then $TR_{Candidate} \leftarrow TR_{Candidate} \cup \{tr\}$
   For each $tr \in TR_{Candidate}$
      If $dist_T(tr_Q, tr, P) < \delta$
      then $TR_{OUT} \leftarrow TR_{OUT} \cup \{tr\}$
   return TR_{OUT}
End

Algorithm 1 summaries the search procedure explained in this section. It consists of two steps, the filtering step based on spatial similarity step and the refinement step for searching for similar trajectories based on temporal distance.

Note that we use L_2 distance for the reason of simplicity in this paper, but other types of distance can be employed. Since each trajectory is represented as a point in a multi-dimensional space, we can apply a clustering method. We call this multi-dimensional space *Temporal Trajectory Space*. A number of methods have been proposed for clustering points in this temporal trajectory space [12][13][14].

4   Experimental Results

In order to examine the feasibility of our method, we performed experiments with real trajectory data set gathered from taxis in Seoul. We clustered the trajectories based on the search method proposed in the preceding section.

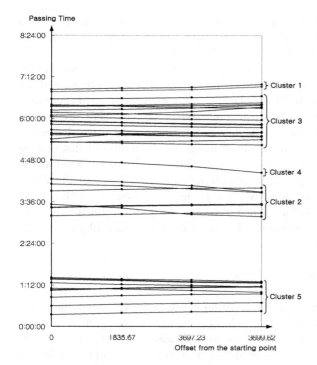

Fig. 5. Result of the first step of clustering

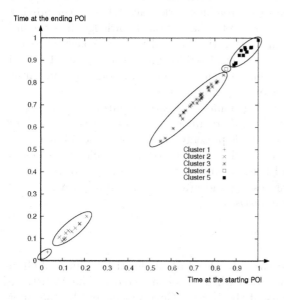

Fig. 6. Clustering of trajectories by temporal distance

For the first step, we found all of the trajectories that passed through four given POIs. For the second step we clustered the points in the temporal trajectory space by means of a shifted Hilbert curve [14].

Figure 5 shows the results of the first step, which are the trajectories passing through the given POIs, where the x-axis is the offsets of the trajectories and the y-axis is the time they passed the POIs. The descending curves represent the trajectories with opposite directions. The clusters marked on the right were obtained by the second step. We see that this result corresponds with our intuitive clustering.

These trajectories were then clustered by means of temporal distance as shown in figure 6, where only the starting and ending POIs are depicted as the x-axis and y-axis, respectively, for the reason of simplicity. Note that the x-axis and y-axis are normalized to [0,1]. An interesting fact is observed that most points in the temporal trajectory space were found around a linear graph. In fact, the slope of this line represents the speed of the moving objects, which were similar in our experiment.

5   Conclusion and Future work

Analysis of the similarity between trajectories on road networks has many potential applications. For example, it is helpful to analyze the trajectories of cars used for commercial purposes on a road for making marketing strategies. In this paper, we present the important properties of trajectories on road networks and propose a method for searching for similar trajectories on road networks.

Our method differs from the previous methods in two aspects. Firstly, our method fully exploits the properties of road network space, whereas the previous approaches assume Euclidean space. To the best of our knowledge, it is the first method for searching for similar trajectories in road network space. Secondly, spatial and temporal similarities are considered by our method, while the previous methods only took spatial similarity into account. And our method can be used to cluster trajectories as shown by experiments.

Since this work is only a staring point of research regarding searching for similar trajectories on road networks, there are a number of related issues. First of all, studies comparing other methods presented in section 3 should be carried out. And the method for selecting POIs should be studied, while we assume in this paper that they are given by users. And integration of trajectories and the attributes of drivers will be interesting and practical for real applications such as geo-marketing or insurance industries.

Acknowledgements

This research was partially supported by the Program for the Training of Graduate Students in Regional Innovation which was conducted by the Ministry of Commerce, Industry and Energy of the Korean Government and by the Internet information Retrieval Research Center(IRC) in Hankuk Aviation University. IRC is a Regional Research Center of Kyounggi Province, designated by ITEP and Ministry of Commerce, Industry and Energy.

References

1. Dieter Pfoser, Christian S. Jensen, and Yannis Theodoridis. Novel Approaches in Query Processing for Moving Object Trajectories. In *Proceedings of the 26th International Conference on Very Large Data Bases*, pages 395–406, 2000.
2. Michail Vlachos, George Kollios, and Dimitrios Gunopulos. Discovering Similar Multidimensional Trajectories. In *Proceedings of the Eighteenth International Conference on Data Engineering*, pages 673–684. IEEE Computer Society, 2002.
3. Yutaka Yanagisawa, Jun ichi Akahani, and Tetsuji Satoh. Shape-Based Similarity Query for Trajectory of Mobile Objects. In *Proceedings of the Fourth International Conference on Mobile Data Management*, pages 63–77. Springer-Verlag, 2003.
4. Laurynas Speicys, Christian S. Jensen, and Augustas Kligys. Computational Data Modeling for Network-constrained Moving Objects. In *Proceedings of the Eleventh ACM International Symposium on Advances in Geographic Information Systems*, pages 118–125, 2003.
5. Michalis Vazirgiannis and Ouri Wolfson. A Spatiotemporal Model and Language for Moving Objects on Road Networks. In *Proceedings of the Seventh International Symposium on Spatial and Temporal Databases*, pages 20–35. Springer-Verlag, 2001.
6. Ramaswamy Hariharan and Kentaro Toyama. Project Lachesis: Parsing and Modeling Location Histories. In *Third International Conference, GIScience*, pages 106–124. Springer-Verlag, 2004.
7. K. Hornsby and M. Egenhofer. Modeling Moving Objects over Multiple Granularities. *Annals of Mathematics and Artificial Intelligence*, 36:177–194, 2002.
8. Christian S. Jensen, Jan Kolárvr, Torben Bach Pedersen, and Igor Timko. Nearest Neighbor Queries in Road Networks. In *Proceedings of the Eleventh ACM International Symposium on Advances in Geographic Information Systems*, pages 1–8, 2003.
9. Nico Van de Weghe, Anthony G. Cohn, , Peter Bogaert, and Philippe De Maeyer. Representation of Moving Objects along a Road Network. In *Proceedings of the twelfth International Conference on Geoinformatics*.
10. Michail Vlachos, Dimitrios Gunopulos, and George Kollios. Robust Similarity Measures for Mobile Object Trajectories. In *Proceedings of the Thirteenth International Workshop on Database and Expert Systems Applications*, pages 721–728. IEEE Computer Society, 2002.
11. Choon-Bo Shim and Jae-Woo Chang. Similar Sub-Trajectory Retrieval for Moving Objects in Spatio-temporal Databases. In *Proceedings of the Seventh East-European Conference on Advances in Databases and Informations Systems*, pages 308–322. Springer-Verlag, 2003.
12. Tian Zhang, Raghu Ramakrishnan, and Miron Livny. BIRCH: An Efficient Data Clustering Method for Very Large Databases. In *Proceedings of the 1996 ACM SIGMOD International Conference on Management of Data*, pages 103–114. ACM Press, 1996.
13. Haixun Wang, Wei Wang, Jiong Yang, and Philip S. Yu. Clustering by Pattern Similarity in Large Data Sets. In *Proceedings of the 2002 ACM SIGMOD international conference on Management of data*, pages 394–405. ACM Press, 2002.
14. Swanwa Liao, Mario A. Lopez, and Scott T. Leutenegger. High Dimensional Similartity Search With Space Filling Curves. In *Proceedings of the seventeenth International Conference on Data Engineering*, pages 615–622. IEEE Computer Society, 2001.

Using Data Mining for Modeling Personalized Maps*

Joe Weakliam[1] and David Wilson[2]

[1] Department of Computer Science,
University College Dublin, Belfield, Dublin 4
`joe.weakliam@ucd.ie`
[2] Department of Software and Information Systems,
University of North Carolina at Charlotte,
9201 University City Blvd, Charlotte, NC 28223, USA
`davils@uncc.edu`

Abstract. There has been a vast increase in the amount of spatial data that has become accessible in recent years, mirroring the continuing explosion in information available online. People browsing the Web can download maps of almost any region when planning trips or seeking directions. However, GIS applications generating maps typically present default maps to clients without personalizing any spatial content. This gives rise to a problem whereby the most relevant map information can be obscured by extraneous spatial data, thus hindering users in achieving map interaction goals. Several applications exist that deliver personalized information, but they rely on clients providing explicit input. We describe a novel system that provides personalized map content using techniques prevalent in data mining to model spatial data interaction and to present users with automatically and implicitly personalized map content. Modeling spatial content preferences in this manner allows us to recommend spatial content to individuals whenever they request maps, without requiring the additional burden of explicit user modeling input.

1 Introduction

There has been a huge increase in the amount of spatial data that has become available online in recent years. More and more Web users are now using the Internet to download maps to locate spatial information of interest to them [1][2]. Nevertheless, many GIS applications that generate maps online typically produce default maps when providing clients with spatial information. Little attempt is made by these systems to take specific individuals' interests into consideration when generating maps, where users may have quite contrasting preferences in terms of spatial content. This is far from ideal as the users' end goals can be hindered due to the presence of irrelevant spatial content. A more suitable solution would be to provide each user with area maps containing spatial information

---

* The support of the Informatics Research Initiative of Enterprise Ireland is gratefully acknowledged.

J. Akoka et al. (Eds.): ER Workshops 2005, LNCS 3770, pp. 290–299, 2005.

tailored to their personal requirements. This allows the user to realize mapping tasks with more ease as only task-relevant detail is provided.

Data mining provides a useful means for addressing problems that influence the effectiveness of Web information search in general. These problems include: (1) the abundance problem, i.e. the phenomenon of hundreds of irrelevant documents being returned in response to a search query, (2) limited coverage of the Web, and (3) limited customization (personalization) to individual users. Many of the problems associated with presenting users with the most suitable Web information to satisfy their current requirements are mirrored by similar problems in GIS. In this paper we address the issue whereby a lack of map content personalization exists in current GIS by using data mining techniques to model users and deliver personalized spatial data.

We introduce a GIS called CoMPASS (Combining Mobile Personalized Applications with Spatial Services) [3] that generates area maps containing personalized spatial content. CoMPASS utilizes well-known data mining concepts, namely clustering and association rule mining, in order to model spatial data stored in a spatial database and for profiling user spatial preferences. All implicit map actions (panning, zooming, toggling features on/off, etc.) executed by the user are recorded along with other spatial detail related to the current map frame. User preferences regarding map features and zones of interest are inferred from the map actions executed by the user using data mining techniques. This is an attractive solution, as it requires no additional effort from the user above standard usage. Information describing user spatial preferences is then inserted into models of user interests that are updated on a regular basis to reflect the user's constantly evolving interests in various aspects of spatial information.

This paper is structured as follows. Section 2 details related work in the fields of (1) personalization in GIS and (2) data mining with emphasis on spatial data mining. In section 3 we introduce CoMPASS describing how personalized maps are generated using different data mining techniques. We provide an evaluation in section 4 with results outlining how we establish interest areas and interest map features. We conclude in section 5 and discuss future work.

2   Related Work

Many GIS applications have been developed in recent years addressing the issue of providing personalized content. In [4] Freska describes an approach for designing way-finding systems. Although Freska proposes providing map-like structures featuring only aspects of a domain relevant to a specific task whereby preventing other map features from causing a distraction, the focus is largely on spatial cognition and on how maps are interpreted by different users. The suggestion of incorporating Location-Based Services (LBS) with multi-criteria evaluation in order to provide personalized spatial decision support to users is described in [5]. The research in [5], however, is completely reliant on explicit input whereby each user must weight various non-spatial attributes used as evaluation criteria when recommending information. The goal of CRUMPET [6] is to implement and validate personalized, nomadic services for tourism using agent technology. CRUMPET is a mobile GIS that takes personal interests of users into consideration and recommends both spatial and non-spatial information based on these preferences. Although user interests are learnt over time, personalization in CRUMPET is dependent on users requesting particular services explicitly, e.g. a tourist would like detail related to Indian restaurants in Dublin. FLAME2008 [7] is a mobile GIS being developed for the Beijing Olympics in 2008 and proposes delivering personalized situation-aware information to clients using LBS. However, the architecture of FLAME2008 is based both on the user supplying explicit input and on the provision of services within a non-spatial context, e.g. the history museum is closed on Sundays.

Data mining is used by large numbers of Web applications to ascertain patterns in user behavior when users browse the Web [8]. It allows system developers to personalize Web content when recommending specific information to clients. In [9] spatial data mining is described as playing an essential role in extracting interesting spatial patterns and features, capturing intrinsic relationships between spatial and non-spatial data, and presenting data regularity concisely and at higher conceptual levels. An efficient method for mining strong spatial association rules in spatial databases is proposed in [10] where a spatial association rule is defined as a rule describing the implication of one or a set of features by an-other set of features in spatial databases, e.g. "most big cities in Canada are close to the US border". One other approach used for knowledge discovery in geographic databases is to find interesting correlations between different characteristics of various areas [11], e.g. areas with a high value for the attribute "rate of retired people" may be highly correlated with neighboring mountains and lakes. This knowledge discovery task is performed in 2 steps: (1) find areas of spatial objects, e.g. clusters or neighboring objects that are homogeneous with respect to some attribute values, and (2) find associations with other characteristics of these areas, e.g. correlate them with other attribute values.

In [12] spatial clustering is used to identify areas of similar land usage or to merge regions with similar weather patterns, etc. As a data mining function, spatial clustering can be used to gain insight into the distribution of data, to observe the characteristics of clusters, and to focus on a particular set of clusters for further analysis. According to [13] the extraction of interesting and useful patterns from spatial data sets is more difficult than extracting the corresponding patterns from traditional numeric and categorical data due to the complexity of spatial data types. A focus is drawn in [13] on the unique features that distinguish spatial data mining from classical data mining in the categories of data input, statistical foundation, output patterns, and computational process.

3   Our Approach

In this section we introduce CoMPASS as an approach to modeling personalized map content using data mining techniques. In particular we use spatial clustering and association rule mining as a means for modeling spatial content and

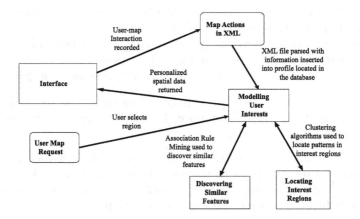

Fig. 1. CoMPASS System Architecture

generating personalized area maps. When a user requests a map all the implicit actions [14] executed by the user when browsing the map are recorded unobtrusively. Information regarding potential user spatial preferences inherent in these actions is mined from the recordings and inserted into a user model detailing user interests. Gathering data indirectly in this manner during all subsequent map requests allows us to update the user's model on a continuous basis and to reflect the user's ever-changing spatial preferences without requiring explicit input from the user. Figure 1 shows the layout of the system architecture composed of 6 main components.

3.1   User Map Request

When a user requests a map using CoMPASS, one of two things takes place. If the user has not made any previous map requests, then a map containing a default set of map features is returned, as the system has not yet established any information about possible user preferences. Once the user ceases interacting with this map, a user model is created storing spatial preferences ascertained up to the end of that first map session. Alternatively, if the user has used the system previously, then a profile of that user already exists in the database and spatial content is recommended to the user based on information in their profile.

3.2   Interface

We have developed a GUI, using non-proprietary software, allowing for standard map interactions. The GUI has been developed using OpenMap^{TM}[15], an open source java-based toolkit. We are using vector data from Tiger/Line 2000 files [16] as data source for rendering maps. Data from Tiger files has been loaded into an Oracle 9i DBMS [17] supporting Oracle Spatial. Using vector data in conjunction with Oracle Spatial has the following advantages over using raster data to represent maps:

- Vector data allows us to create distinct map layers representing different feature types in the map.
- Oracle Spatial allows us to run fast detailed spatial queries on data contained in the map as the data has been spatially indexed, e.g. highlighting all the features of a certain type falling with a tolerance distance of a specific point.

All customary map actions including panning, zooming, and toggling features on/off can be performed at the interface. The user can also execute any of the following specific spatial queries:

- Highlight aspect(s) of a particular map feature falling within a threshold distance of a selected point on the map or selected query window.
- Highlight aspect(s) of map feature intersecting a selected query window.
- Find the distance between two points on the map.

Every single map action executed by the user is captured implicitly at the interface along with additional detail related to the map. For each map action that is executed a corresponding map frame is generated as a consequence and thus each user session can be interpreted as a sequence of map frames.

3.3   Map Actions in XML

Every single map action executed by the user is recorded in log files in XML format. When users interact with maps by performing map actions, the following is recorded: (1) the action executed (pan, highlight feature, etc.); (2) the map feature(s) upon which the action was executed; (3) the number of the frame generated as a result of the action executed; (4) the time at which the action was executed; (5) the boundary of the resulting frame generated by the action executed; (6) the map feature(s) intersecting the new map frame produced by the action. The use of XML allows for fast parsing when analyzing log file content.

3.4   Modeling User Interests

Once data from the log files has been analyzed, detail from each user map session is inserted into a user model. The following outlines components of the model:

Action Sequence: This component stores the sequences of map actions executed by users during map sessions. Data mining enables us to spot patterns in sequences of map actions executed by users with similar spatial interests.

Initial Map Frames: This component stores the features present in the initial map frames of each user session. The user's response to map content presented to them in initial map frames is crucial, as spatial content in the opening frame of a map session has been recommended by the system based on mining detail from the spatial database.

Final Map Frames: This component stores the features that were present in the final map frames of each user session. Information from the final map frames is deemed significant, as the user has realized their mapping tasks for sessions at that point.

Frame Features: This component stores all the map features present in each map frame generated for every map session. The use of data mining facilitates the verification of those features most common in user sessions and the discovery of those features that users tend to group together.

3.5 Locating Interest Regions

Interest map frames are extracted from log files when inserting detail into user profiles. Interest frames are determined by the following criteria: (a) the time interval between two consecutive map frames (map actions) exceeds a threshold value, and (b) the map action resulting in the first of the two map frames being generated is performed on only a single map feature, e.g. highlighting a feature. Each interest frame recorded is assigned a relevance score dictating the significance of the frame. This score is calculated from equation (1):

$$frame\_score = num\_features * frame\_area * 1/time\_interval. \qquad (1)$$

In (1) num_features corresponds to the number of map features present in the interest frame, frame_area to the area of the interest frame, and time_interval to the time lapse before the user executes the next action. The smaller num_features, the smaller frame_area, and the larger time_interval, the lower the score and the more relevant the interest frame.

When establishing interest regions from detail in the user profile, we perform clustering on the interest frames using the k-means clustering algorithm [12]. The interest frames are clustered to locate patterns of user interest in terms of map features and map areas. When clustering interest frames, each frame is classed as a vector with 5 associated attributes: frame score, frame time, frame area, frame boundary, and a set of associated map features. Clustering the input interest frames allows us to spot trends in individual user behavior or with group behavior and the clustering process can be tested with different input attributes, e.g. frame score and frame time, or frame score and frame area. Once clusters have been created it is possible to examine them to find answers to the following questions related to user spatial preferences:

1. How many features on average are present in highest interest frames?
2. What features were present in highest interest frames?
3. Are there any interest areas of the map that the user returns to regularly?
4. How long do users spend looking at frames with varying numbers of features?
5. In frames with few features, what was the size of the area containing these features and what were these features?
6. How do users with similar mapping interests related to spatial content interact with the map differently? Do users with similar goals focus on the same aspects of the map and in what way do they alter the map content?

3.6 Discovering Similar Features

Association rule mining is the discovery of association rules in large databases. A spatial association rule is a rule indicating certain association relationships

Table 1. Map session showing feature presence in each frame

Frame	Highways	Interstates	Parks	Rivers	Lakes	Shops	Hospitals	Airports
1	1	1	0	1	0	0	0	1
2	1	1	0	1	1	0	0	1
3	1	1	1	1	1	0	0	1
4	1	1	1	0	1	0	0	1
5	0	0	1	0	1	0	0	0
6	1	1	1	0	1	0	0	1

among a set of spatial and non-spatial predicates [10]. We use the concept of association rule mining for establishing trends in how users interact with different map features and for grouping users with similar feature interests together.

Similarity measures between different pairs of map features can be calculated based on feature presence in session map frames. Map features that are highly similar are inclined to appear in frames together, i.e. feature A and feature B are "similar" if frames containing feature A also contain feature B. In (2) the similarity measure (sim_{AB}) between two features A and B is calculated as the Manhattan distance between A and B.

$$sim_{AB} = \frac{f(A \cup B)}{f(A)}. \tag{2}$$

In (2) $f(A \cup B)$ is the frequency of map frames containing both feature A and B whereas $f(A)$ is the frequency of map frames containing only feature A. We can also categorize other map features based on similarity scores calculated between pairs of features, i.e. if feature A is similar to feature B, and feature B is similar to feature C, then feature A is similar to feature C as the operation is transitive. Table 1 shows a simple map session with all the features appearing in each frame of the map session. A '1' in the map feature column indicates that the feature was present in the frame whereas a '0' indicates feature absence from the frame. Recording feature presence in this manner allows us to calculate the Manhattan distances between different pairs of features so as to determine what features different users tend to group together.

4   Evaluation and Results

We ran an experiment to evaluate the effectiveness of using data mining for modeling personalized spatial data. Six participants, with varying experience using our system, took part. They were divided into two distinct groups so that results across different users could be correlated. Mapping tasks were assigned to users for each map session where the task ultimately determined what feature(s) was the focus of the session. When assigning mapping tasks to users it was important to vary the task sequences over the course of the experiment while maintaining a degree of consistency among the features and areas at the center of tasks. Data from 80 map sessions involving both groups of users was gathered.

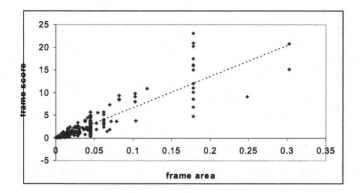

Fig. 2. Graph of clustering of interest frame area vs. interest frame score

A predetermined set of 80 mapping tasks was drafted whereby each of the 6 users was assigned the exact same task sequences for each of the first 40 sessions and for the remaining 40 sessions the task sequences were varied.

Results show that data mining is effective when locating information related to spatial content preferences, both for individual users and groups of users. Figure 2 shows a graph of a clustering of interest frame vectors based on interest frame detail gathered for 3 of the participants in the experiment with frame score on the vertical axis and frame area on the horizontal axis. This graph shows a large cluster of interest frames occurring where frame area ≤ 0.03 and frame score ≤ 2.5. Patterns in user behavior were extracted from vectors falling inside this cluster where aspects like areas of the map visited and times spent analyzing these regions were found to be similar across different users. This is what we would expect as the 3 users were assigned the same sets of task sequences over the duration of the 80 sessions. We were also able to locate map features of most significance from vectors in this cluster. As these interest frames covered only small areas of the map, fewer features were contained within these regions, and trends in feature presence were discovered. Therefore, interest frames with lowest scores are the most significant indicators of user preferences and are extremely useful for modeling and recommending personalized spatial content.

Figure 3 displays a graph of feature similarity scores vs. map session number for three participants in the experiment. The same three pairs of features were chosen for each of the three participants to show that users assigned the same mapping tasks interact with maps in the same manner. The mean Manhattan distances between each pair of features is calculated for each participant. A distance measure of 0 reveals that the two map features are exactly similar, i.e. if one feature is present in a map frame than we can be certain that the other feature is also present in that frame. A Manhattan distance of 2 between two features indicates that the two features are completely dissimilar. As can be seen from figure 3, the participants interacted with each pair of map features similarly and hence the mean feature similarity scores are alike. It should be noted that

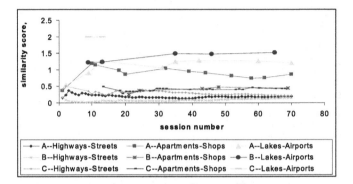

Fig. 3. Graph of session number vs. feature similarity score

some features were present in only one or two of the frames of no more than a few of the 80 sessions. Calculating the similarity scores in this manner allows us to group individual features and indeed individual users together based on their map feature preferences, e.g. all 3 users want highways and streets present in maps returned to them. As a result, we recommend these features to users every time they request maps.

5   Conclusions and Future Work

There is a distinct lack of personalization in evidence in Web-based GIS for dealing with the huge rise in the amount of spatial data available online. Several applications attempt to address this issue by providing personalization of spatial and non-spatial content. These systems, however, rely heavily on explicit input from users. We introduce the CoMPASS system that addresses the lack of spatial content personalization in GIS. CoMPASS uses well-known data mining techniques to model spatial information in a personalized format. Experiments carried out indicate that CoMPASS provides promising results when recommending spatial content in a map format.

There is much scope for future work with CoMPASS. Firstly, we intend to complete the migration of our system from its current Web-based environment to the mobile environment. Using data mining to model the map content will facilitate delivering spatial information to mobile devices as the size of data sets needed to be transmitted will be reduced significantly as only relevant content with respect to each user will be delivered to the limited mobile devices. Secondly, we are investigating the use of collaborative filtering to group users with similar spatial interests together and hence returning similar spatial content to users within the same groups. Finally, we intend to refine the system by introducing more users with more diverse goals and evaluating the system with respect to this new data, i.e. instead of assigning tasks to each user, allow them to choose their own tasks and as a result model the spatial data in a personalized format to facilitate different users with more varied sets of goals.

References

1. Mapquest. http://www.mapquest.com/.
2. Google maps. http://maps.google.com/.
3. J. Weakliam, D. Lynch, J. Doyle, M. Zhou, E. MacAoidh, M. Bertolotto, and D. Wilson. Mapping spatial knowledge for mobile personalized applications. In *Proceedings KES'05 (9th International Conference on Knowledge-Based and Intelligent Information Systems - in press)*, Melbourne, Australia, September 2005.
4. C Freska. Spatial aspects of task-specific wayfinding maps - a representation-theoretic perspective. In *Gero, John and Tversky, Barbara (eds): Visual and Spatial Reasoning in Design*, pages 15–32, 1999.
5. C Rinner and M Raubal. Personalized multi-criteria decision strategies in location-based decision support. In *Journal of Geographic Information Sciences 11*, pages 61–68 (in press), June 2005.
6. B Schmidt-Belz, S Poslad, A Nick, and A Zipf. Personalized and location-based mobile tourism services. In *Proceedings of Mobile HCI '02 with the Workshop on Mobile Tourism Support Systems*, Pisa, Italy, September 2002.
7. N Weisenberg, A Voisard, and R Gartmann. Using ontologies in personalized mobile applications. In *Proceedings of the 12th annual ACM international workshop on Geographic Information Systems (ACMGIS'04)*, pages 2–11, Washington DC, USA, November 2004.
8. M Garofalakis, R Rastogi, S Seshadri, and K Shim. Data mining and the web: past, present and future. In *Proceedings of the Workshop on Web Information and Data Management (WIDM'99)*, pages 43–47, 1999.
9. R Ng and J Han. Efficient and effective clustering methods for spatial data mining. In *Proceedings of 1994 International Conference on Very Large Data Bases (VLDB'94)*, pages 144–155, Santiago, Chile, September 1994.
10. K. Koperski and J Han. Discovery of spatial association rules in geographic information databases. In *Proceedings of the 4th International Symposium on Large Spatial Databases (SSD'95)*, pages 47–66, Portland, Maine, USA, August 1995.
11. M Ester, H Kriegel, and J Sander. Spatial data mining: A database approach. In *Proceedings of the Fifth International Symposium on Large Spatial Databases (SSD '97)*, pages 47–66, Berlin, Germany, 1997.
12. J Han, M Kamber, and AKH Tung. Spatial clustering methods in data mining: A survey. In *H. Miller and J. Han (eds.), Geographic Data Mining and Knowledge Discovery*, Taylor and Francis, 2001.
13. S Shekhar, P Zhang, Y Huang, and R Vatsavai. Trends in spatial data mining. In *H. Kargupta, A. Joshi, K. Sivakumar, and Y. Yesha (eds.), Data Mining: Next Generation: Challenges and Future Directions*, AAAI/MIT Press, 2003.
14. Y Hijikata. Implicit user profiling for on demand relevance feedback. In *Proceedings of ACM Intelligent User Interfaces Conference (IUI'04)*, pages 198–205, 2004.
15. Openmap. http://openmap.bbn.com/.
16. Tiger files. http://www.census.gov/geo/www/tiger/.
17. Oracle spatial. http://otn.oracle.com.

3D Scene Modeling for Activity Detection

Yunqian Ma[1], Mike Bazakos[1], Zheng Wang[2], and Wing Au[1]

[1] Honeywell Labs, Honeywell International Inc.,
3660 Technology Drive, Minneapolis, MN 55418, USA
{yunqian.ma, mike.bazakos, au.wing}@honeywell.com
[2] Honeywell Technology Solution Lab, No. 430 Li Bing Rd,
Zhang Jiang Hi-Tech Park, Pudong New Area, Shanghai, 201203, China
zheng.wang2@honeywell.com

Abstract. Current computer vision algorithms can process video sequences and perform key low-level functions, such as motion detection, motion tracking, and object classification. This motivates activity detection (e.g. recognizing people's behavior and intent), which is becoming increasingly important. However, they all have severe performance limitations when used over an extended range of applications. They suffer from high false detection rates and missing detection rates, or loss of track due to partial occlusions, etc. Also, activity detection is limited to 2D image domain and is confined to qualitative activities (such as a car entering a region of interest). Adding 3D information will increase the performance of all computer vision algorithms and the activity detection system. In this paper, we propose a unique approach which creates a 3D site model via sensor fusion of laser range finder and a single camera, which then can convert the symbolic features (pixel based) of each object to physical features (e.g. feet or yards). We present experimental results to demonstrate our 3D site model.

1 Introduction

The world is increasingly populated with video cameras, especially in the last few years following the terrorist actions and threats. Raw video data is becoming abundant, readily available, and in real time. In fact the volume of raw video data from hundreds or even thousands of cameras, such as at airports, seaports, casinos, etc. is overwhelming and intimidating. As a result most of it is being ignored or missed or at best is analyzed in a post-processing fashion. The key issue is that, in many applications, security people are not really interested in the "video data", but rather in the "information" contained in the video data. Every application has what is called "application specific information needs". In security applications, for example, such systems are sometimes employed to detect and track individuals or vehicles entering or leaving a building facility or security gate, or to monitor individuals within a store, office building, hospital, or other such setting where the health and/or safety of the occupants may be of concern. In the aviation industry, for example, such systems have been used to detect the presence of individuals at key locations within an airport such as at a security gate or parking garage.

Honeywell AVPS (Advanced Video Processing Solutions), or other similar COTS Automated Video Surveillance (AVS) systems, can process video sequences and

J. Akoka et al. (Eds.): ER Workshops 2005, LNCS 3770, pp. 300–309, 2005.

perform all key low-level functions, such as Motion Detection (MD), Object Tracking (OT), and even Object Classification (OC). However, while this is useful information, it may not be sufficient for a specific application, as the human has to look at all these reports and decide if the motion detects and the tracks are of interest or of threat to the application at hand. Recently, technical interest in video surveillance has moved from the low-level functions, (MD, OT, and OC) to more complex scene analysis to detect human and/or other object behaviors. i.e. patterns of activities or events [1,2]. Recognizing human behavior in real-time is becoming increasingly important, especially for high-end security applications, that video from one or many cameras must be processed in an intelligent manner which matches the needs of the application.

One way to perform intelligent video surveillance is to combine the results of an AVS system with spatio-temporal reasoning about each object relative to the key background regions and the other objects in the scene. The objective is to understand human (and other moving objects) behaviors and patterns, including the interaction of multiple people and objects. Honeywell's People Activity Detection System (PADS) can automatically detect unusual people activities and behaviors and then alert operators to potential dangers before undesirable events occur. Such behaviors could be: a person is "walking", "running", "heading in a direction", "standing at a particular location", "falling down", "loitering", "left an abandoned object", or that "crowd is forming".

Recognizing people and vehicle's behavior/activity limited to 2D image sensing is confined to qualitative events, such as a car entering a region of interest. Also activity detection based on 2D features is sensitive with respect to the viewing angle, direction of motion and distance to the tracked object. Yet, converting the 2D image coordinates of an object to 3D coordinates without additional information is not simple either. 3D sensors are expensive and stereo video is computationally intensive. Now, as for using a single traditional 2D camera, some existing methods use camera calibration [3-6], which estimates the geometrical and optical parameters of the camera and the extrinsic parameters of the camera, e.g. Heikkila [4] gives a four-step camera calibration procedure to solve the problem. [7] pointed out that existing calibration methods often favor specific camera models. Hongeng and Nevatia [2] and Bradshaw et al. [8] viewed the relation between image pixels and terrain locations as a simple 2D homography. Collins et al. [9] performed geolocation using ray intersection e.g. by a digital elevation map, given a calibrated sensor. Johnson and Bobick [10] proposed a conversion factor using known person's height (in centimeters) and measured height (in pixels), assuming the image plane is perpendicular to the ground plane.

In this paper, we propose a new approach to 3D scene modeling by sensor fusion with a laser range finder and a single camera. Also we can convert symbolic features (pixel based) of each object to physical features (e.g. feet or yards), via an operator assisted 3D site model construction. This can establish a robust representation of each object and its track in 3D space and in turn facilitates the reasoning process for detecting object behavior patterns. This paper is organized as follows. Section 2 describes related video surveillance modules. Section 3 presents proposed methods. Section 4 presents experiment results. Conclusions are given in Section 5.

2 Video Surveillance Modules

Honeywell's Digital Video Manager™ (DVM), a surveillance information manager, is a scalable, digital video management solution featuring a unique open architecture which easily integrates with existing enterprise systems.   The Advanced Video Processing Solution (AVPS) subsystem in DVM provides motion detection, motion tracking and object classification. AVPS has defined API's for all of the subsystem components. The AVPS architecture is shown in Fig. 1. We won't describe every module here. The following briefly describe the related ones.

- Video Motion Detection (VMD):   VMD detects the moving objects by separating the input image into foreground and background regions. Color and edges are used to differentiate the foreground and background pixels. The method is a modification of the Gaussian Mixture Model, which represents the characteristics of a pixel with a set of Gaussian distributions.
- Video Motion Tracking (VMT): VMT tracks the moving object from frame to frame using a set of heuristic rules and a simplified particle filter operating on a set of shape and color features.
- Object Classification: Object classification classifies the object as "human", "vehicle" or "others" using a statistical weighted average decision classifier. This classifier determines the object type based on a set of shape-, boundary-, and histogram-features, and their temporal consistency.

After motion detection, motion tracking, and object classification, we extract information of moving objects in the scene. The appearance information of detected objects is recorded. This includes not only instantaneous information of the spatial features of objects [1,11] such as width, height, and aspect ratio, but also temporal information about changes in the objects' sizes as well as motion features, such as direction of movement and speed. All these information cues, which are important features for behavioral analysis, are organized to form a composite feature vector.

Pixel-based features of an object may not be robust for behavioral analysis. The measured values in pixel may vary depending on the system, e.g., the operating characteristics, such as the location of the object within the camera's field of view (FOV), and the distance of the object to the camera. For example, if the velocity is represented by pixels/sec, the velocity of a person walking in the far end of the FOV has a lower value than that of the same person walking at the same speed in the near end of the FOV.

Representing the features in physical measurements adds more discriminatory power to the behavioral analysis. In the following section, we want to find a good correspondence solution between a group of neighboring pixels (within the same segmentation region) and the respective real world region. The corresponding 2D to 3D coordinates can be stored as a Look-up-table (LUT) along with the outline and the label of the region (other store option, e.g. XML format also applicable). This provides real world scalability and additional (and powerful) physical features for behavior reasoning, such as activities and movement patterns of the objects in the scene.

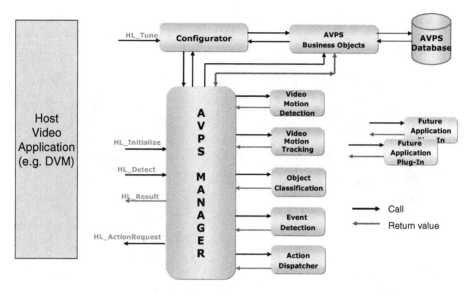

Fig. 1. Advanced Video Processing Solution (AVPS) architecture

3 Sensor Fusion and Proposed Methods

The goal of sensor fusion is to find real world 3D coordinates (x, y, and z) from 2D image coordinates (u and v) by integrating a laser range finder and a single camera. The key assumptions of our method are that each region is approximately planar and the pixels in each region belong to the same context class, e.g., "road", "buildings", "parking lot", "tree line", "sky", etc. Table 1 outlines the steps of our proposed method, see Fig. 2 for reference.

Table 1. Proposed Method

Step 1: Initialization: The operator manually outlines and labels (classifies) the key regions in the scene.
Step 2: GUI prompts the user to enter the measurements of distances and/or angles between the camera and the reference points used in defining a key region.
Step 3: Calculate the 3D coordinates for the reference points.
Step 4: Using an interpolation technique convert 2D image pixels within the region into a 3D look-up table.
Step 5: Go to Step 2, the same procedure goes to another region defined by the user in Step 1, until all the key regions are finished.

At the time of publication, the mathematical equation of the procedure couldn't be disclosed due to its proprietary nature. However, we describe the conceptual approach in the following.

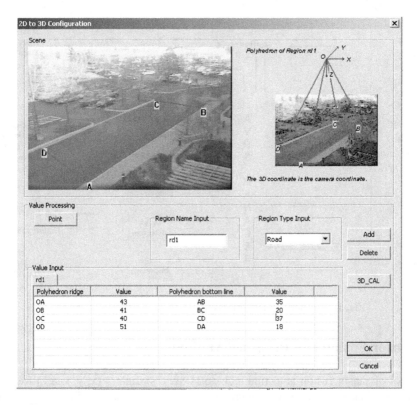

Fig. 2. Measurement input system, the upper right sub-figure shows polyhedron containing N+1 points, original point O (camera center) and N labeled points

In Step 1, manual segmentation can be accomplished using the computer mouse or other similar computer/pen device and software (as shown in Fig. 2). For example, a camera looking down into a conference room may have the following key regions that need to be outlined and labeled accordingly: floor, east wall, north wall, door, and stairs. We store the outlines of these regions along with their labels and their locations relative to one another at camera installation time.

In Step 2, after the user performs the manual segmentation, for each defined polygonal zone, the GUI will prompt the user to input the measurement of the related distance or angle of the reference points of the region and the camera. For example, a user chose a road region, as shown in Fig. 2. The region (a planar polygonal zone) is defined by four reference points: A, B, C and D. We use a camera (labeled as 'O') centered 3D coordinate system and use a laser range finder to measure the distances OA, OB, OC, OD, AB, BC, CD, DA. All these triangles form a polyhedron (the upper-right subfigure in Fig. 2) having a vertex located at the camera and a base representing the planar region outlined by the polygonal zone. For each triangle, we can measure either three sides, as illustrated above or two side lengths and the angle between them via suitable instrument.

Using the pixel features obtained from AVPS output and the calculated 2D→3D look-up table, the physical features of one or more objects are then calculated. For example, when calculating the speed of a moving object, instead of the pixel speed (*e.g.* 3 pixels/second), a more accurate physical measure of the object's speed (*e.g.* 5 miles/hour) can be obtained.

Another key element of our approach is that we can dynamically update of the 2D→3D look-up table. Initially, few accurate reference points (measured by the laser ranger finder) are used to determine the ranges to the camera. Any other points within the region are estimated based on some assumptions, such as a plain earth. Some of these assumptions may be invalid, resulting poor range estimates. Dynamic update of the 2D→3D look-up table re-estimates the ranges when new and reliable information is available. For example, one such information is the track of the same object traversing the region. Since the physical feature of the object, e.g., person's height, does not change along the track [10], it can be used to compute the ranges of the track to the camera using an inverse transformation. By dynamically updating the 2D→3D look-up table in this manner, the robustness of the surveillance system in tracking objects within more complex ROI's, scene understanding and/or behavior analysis can be improved.

4   Experimental Results

The purpose of this section is to illustrate and test the capability of our proposed algorithm.

Experiment 1: We set up a camera facing the ground, as shown in Fig. 3a. We use the 'metal grid' (labeled as object 2 in Fig. 3a) on the ground as the test region with the reference points A, B, C (red points in Fig. 3b). Fig. 3b shows the image of the metal grid in the image plane. We measure the length from the points A, B, C to the camera O, and the length between the points (as shown in Table 2) and input to our system through our software interface shown in Fig. 2.

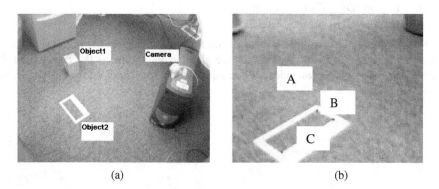

(a)   (b)

Fig. 3. (a) Experiment set-up (b) 2D image of the 'metal grid' region

Table 2. Measurement input

Line	Length	Line	Length
OC	86.9	BC	18
OB	93.1	AB	7
OA	98.9	AC	19.3

Using proposed method in Section 3, we calculate the 3D coordinates of the three points A, B, C as shown in Table 3. After that, we calculate the physical distances between the reference points, and compare results with ground truth as shown in Table 4. The errors are under 5%.

Table 3. Calculated 3D coordinates of the reference points

3D Point	X	Y	Z
A	11	10.8	97.8
B	10.9	6.8	92.3
C	18.4	20.7	82.4

Table 4. Comparison results with ground truth for experiment 1

Line	Real length	Calculated length	Error	Percentage (%)
AB	7	6.8	0.2	2.8
BC	18	18.6	0.6	3.3
CA	19.3	19.7	0.4	2.1

Experiment 2: We present a more complicated experiment here. A box is placed on the ground to simulate a 'building' in the real condition. A wedge-like object is set besides the grid to simulate a 'slope' on the ground (Fig. 4a). The operator input three regions into our system

- Region 'ABCD' of the metal grid,
- Region 'DEFG' of the wedge.
- Region 'HIJK' of the box.

As for the label points in Fig. 4b, besides the reference points of the regions, another three points (R, S, T) inside the regions are also labeled. Same as the *Experiment 1*, the length of the polyhedron sides will be input by our software interface as shown in Table 5.

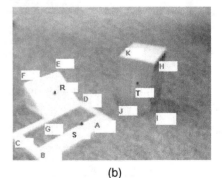

(a)   (b)

Fig. 4. (a) Experiment 2's scene (b) Label point for testing

Table 6 shows the calculated 3D coordinates of the labeled point in Fig. 4b. The three selected points R, S and T are also calculated after interpolation. After that, we calculate physical distances both inside each region and across regions (two points belong to different regions are selected and their distances are calculated). For example, Points A and S are picked within region ABCD, while Points S and T are from different region. The comparison results in Table 7 shows that the errors are almost below 5%.

Table 5. Measurement input

Polyhedron	Line	Length (cm)	Line	Length (cm)
OABCD	OA	79.8	AB	18
	OB	71.2	BC	7
	OC	76.3	CD	18
	OD	84.2	DA	7
ODEFG	OD	84.2	DE	10.9
	OE	87.8	EF	10.4
	OF	82.9	FG	10.9
	OG	79.6	GD	10.4
OHIIJK	OH	75.5	HI	14
	OI	81.5	IJ	9
	OJ	84.2	JK	14
	OK	78.9	KH	9

Table 6. Calulated 3D coordinates of labled points

Point	X	Y	Z
A	-4.57	8.29	79.24
B	-17.13	14.23	67.63
C	-21.62	11.86	72.21
D	-9.79	6.21	83.39
E	-13.83	-3.38	86.64
F	-21.46	-0.15	80.07
G	-15.88	8.84	77.50
H	10.44	-5.30	74.59
I	8.98	6.37	80.75
J	0.89	5.69	84.01
K	1.84	-6.61	78.60
R	-14.61	2.27	79.97
S	-7.21	9.48	76.68
T	6.24	-1.1	78.66

At the time of publication, the above algorithm is being tested in Honeywell parking lot, and has good performance.

Table 7. Comparison results with ground truth for experiment 2

	Line	Real distance (cm)	Calculated distances (cm)	Error	Percentage (%)
Within Region	AS	4.0	3.87	0.13	3.25
	BS	14.0	14.24	0.24	1.71
	CS	15.0	15.27	0.27	1.80
	DS	8.2	7.90	0.3	3.66
	DR	7.0	7.10	0.1	1.42
	ER	8.1	8.78	0.68	8.39
	FR	7.5	7.27	0.23	3.07
	GR	7.4	7.13	0.27	3.65
	HT	7.50	7.20	0.3	4.0
	IT	8.50	8.23	0.27	3.18
	JT	9.60	10.17	0.57	5.93
	KT	7.3	7.05	0.25	3.42
Cross Region	AF	19.3	18.90	0.4	2.07
	AR	12.1	11.73	0.37	3.06
	FS	17.1	17.53	0.43	2.51
	RS	11.3	10.84	0.46	4.07
	AH	20.1	20.77	0.67	3.33
	AT	14.6	14.33	0.27	1.85
	HS	22.6	23.16	0.56	2.48
	ST	16.7	17.23	0.53	3.17
	HR	27.2	26.72	0.48	1.76
	FH	32.0	32.78	0.78	2.44
	FT	28.5	27.75	0.75	2.63

5   Conclusion

This paper proposes a sensor fusion system for transforming two-dimensional image plane data into a 3D dense range map. It includes the steps of acquiring at least one image frame from an image sensor, manual segmentation, determining the geo-location of one or more reference points within each selected region of interest, and transforming 2D image domain data from each selected region of interest into a 3D dense range map containing physical features of one or more objects within the image frame. Using the pixel features obtained from the image frame and 2D→3D look-up table, the physical features of one or more objects can then be calculated. The physical features add robustness to the activity detection system, e.g. 5 miles/hour vs. 3 pixels/second (in symbolic features).

The current approach can easily be integrated and/or extended into an advanced, sophisticated surveillance system. We can also add camera model, registration and calibration information. The manual segmentation and range measurements of the reference points are performed once and can be part of the initialization process during the installation of the camera and AVS system.

The authors would like to thank Honeywell PADS software team: Brain VanVoorst, Ben Miller, Pradeep Buddharaju.

References

1. G. Medioni, I. Cohen, F. Bremond, S. Hongeng and R. Nevatia, "Event Detection and Analysis from Video Streams", *IEEE Trans. on Pattern Analysis and Machine Intelligence*, Vol 23. No 8. pp. 873–889, 2001.

2. S. Hongeng and R. Nevatia, "Multi-Agent Event Recognition", In Proceedings of the 8th *IEEE International Conference on Computer Vision*, Vancouver, Canada, 2001.

3. Z.Zhang, "A flexible new technique for camera calibration", *IEEE Trans. on Pattern Analysis and Machine Intelligence*, Vol. 22, No. 11, pp. 1330–1334, 2000.

4. J. Heikkila, O. Silven, "A Four-Step Camera Calibration Procedure with Implicit Image Correction". *IEEE Proc. Computer Vision and Pattern Recognition*, pp. 1106-1112, 1997.

5. O. Faugeras, *Three-Dimensional Computer Vision*. MIT Press, 1993.

6. D. Martinec and T. Pajdla, "3D Reconstruction by Fitting Low rank Matrices with Missing Data", *IEEE Proc. Computer Vision and Pattern Recognition*, pp. 198– 205, 2005

7. P. Sturm, "Multi-View Geometry for General Camera Models", *IEEE Proc. Computer Vision and Pattern Recognition*, pp. 206– 212, 2005

8. K. Bradshaw, I. Reid, and D. Murray, "The active recovery of 3d motion trajectories and their use in prediction". *IEEE Trans. on Pattern Analysis and Machine Intelligence*, 19(3):219–234, March 1997.

9. R. Collins, A. Lipton, and T. Kanade "A system for Video Surveillance and Monitoring", *8th International Topical Meeting on Robotics and Remote System*, April, 1999.

10. A. Y. Johnson and A. F. Bobick, "A Multi-view Method for Gait Recognition Using Static Body Parameters," AVBPA , pp. 301-311, 2001.

11. F. Porikli and T. Haga, "Event detection by Eigenvector Decomposition using object and feature frame", *CVPR workshop*, 2004.

SAMATS – Edge Highlighting and Intersection Rating Explained

Joe Hegarty and James D. Carswell

Digital Media Centre, Dublin Institute of Technology,
Aungier St., Dublin, Ireland.
joe@dmc.dit.ie, jcarswell@dit.ie

Abstract. The creation of detailed 3D buildings models, and to a greater extent the creation of entire city models, has become an area of considerable research over the last couple of decades. The accurate modeling of buildings has LBS (Location Based Services) applications in entertainment, planning, tourism and e-commerce to name just a few. Many modeling systems created to date require manual correspondences to be made across the image set in order to determine the models 3D structure. This paper describes SAMATS, a Semi-Automated Modeling And Texturing System, which has the capability of producing geometrically accurate and photorealistic building models without the need for manual correspondences by using a set of geo-referenced terrestrial images. This paper gives an overview of SAMATS' components, while describing the Edge Highlighting component and the Intersection Rating step from the Edge Recovery component in detail.

1 Introduction

This research investigates building reconstruction technology for creating geometrically accurate, photorealistic 3D models from terrestrial digital photography for use in LBS (Location Based Services) applications. It is envisioned that the resulting 3D model output from this work be web-enabled and made available to subsequent LBS research endeavors (e.g. for archaeologists, town planners, tourism, e-Government, etc.). Being able to produce 3D building models using terrestrial imagery allows all users to exploit the future commercialization potential of web-based LBS, as demonstrated in [1].

[10] was the first to investigate the principle of structure from motion. [9] builds on these ideas using lines instead of points, although both require correspondences to be made manually across the image set. In fact the majority of semi-automated reconstruction systems require the user to make manual correspondences across the image set in order to reconstruct a model, which is generally a very time consuming task. [3] is one of the most robust systems using this approach which allows the user to create models using a set of block primitives and by setting constraints on those primitives. A more automated modeling approach involves the modeling of roofs using aerial imagery. Models produced in this way can produce structurally accurate models but fail to capture building façades accurately, although [5], [6], and [7] have looked into the merging of façade textures with models produced from aerial imagery.

J. Akoka et al. (Eds.): ER Workshops 2005, LNCS 3770, pp. 310–319, 2005.

[2] constructs a large set of 3D building models by using spherical mosaics produced from accurately calibrated ground view cameras fitted with GPS. Although highly automated, this system was limited to modeling simple shaped buildings by simply identifying the rooflines and extruding walls downwards.

SAMATS uses a novel approach to creating building models without the need for manual correspondences to be made. [11] is an example of extracting building and window edges without the need for manual correspondence, although a rough model of the structure being modeled is required in order for this system to work. No prior building model is required by SAMATS. The ability of SAMATS to remove the manual correspondence step found in most modeling approaches is achieved by having all images geo-referenced in the same reference frame. However, the acquisition of geo-referenced terrestrial images is still a serious bottleneck that does not have a straightforward solution. Currently public GPS will give an absolute accuracy of between 1 to 10 meters using a single receiver. This resolution is not technology bound but information restriction bound, with military GPS offering centimeter accuracy. As private industries or other governments create their own satellite networks these restrictions may no longer apply - making the acquisition of accurate geo-referenced imagery as simple as regular imagery. SAMATS does not solve the difficulties in acquiring geo-referenced imagery - it only investigates the usefulness of such imagery in the overall modeling process.

This paper gives an overview of the entire SAMATS system, while focusing on the Edge Highlighting component and the Intersection Rating step of the Edge Recovery component. For a detailed description of the other components refer to [4]. Figure 1 shows a systems overview of SAMATS.

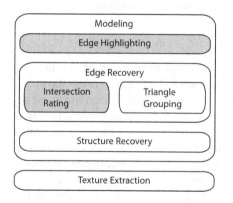
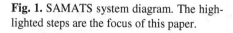

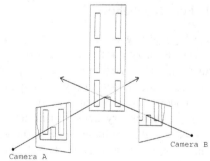

Fig. 1. SAMATS system diagram. The highlighted steps are the focus of this paper.

Fig. 2. Two point projections used to determine a point in 3-space

2   Modeling

This section describes the process used to model the geometry of a building from a set of geo-referenced images using only simple edge highlighting by the user. The basic concept behind the modeling process is as follows; if one has two images of a scene taken from different locations, and the exact position and orientation of the camera is known for each image (i.e. the exterior orientation parameters X_o Y_o Z_o Ω Φ and K) then the exact location of any point visible in both images can be determined. This is illustrated in figure 2.

The modeling process outlined in this section extends this idea by using triangle intersections to find edges rather than line intersections to find points. The modeling process can be split into three main steps; Edge Highlighting, Edge Recovery and Structure Recovery.

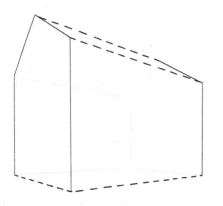

Fig. 3. House outline, primary lines solid black, secondary lines dashed black

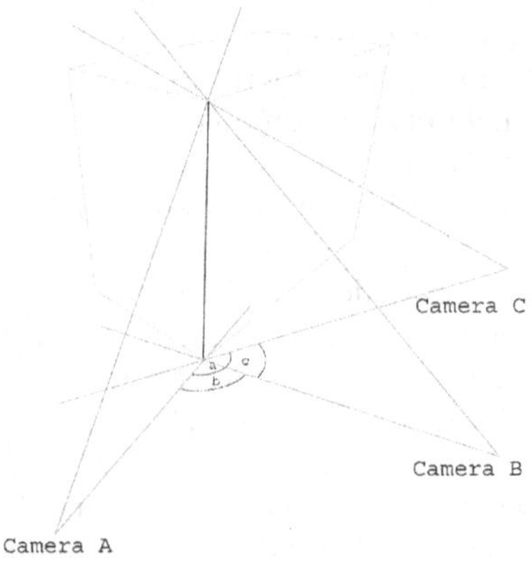

Fig. 4. For vertical edges, large disparity angles can be achieved

2.1 Edge Highlighting

Edge highlighting is the only manual step performed by the user in the modeling process. Primary lines and secondary lines are used to highlight edges in the images. Primary lines are used to recover the position of edges directly, determining the core structure of the model. They are responsible for the creation of every vertex in the final model. The endpoints of a primary line can be connected (having one or more primary or secondary lines sharing that endpoint) or unconnected (having no other lines sharing that endpoint). A secondary line is used to connect primary lines together and must have each of its endpoints connected to at least one primary line. In figure 3 the solid black lines represent primary lines while the black dashed lines represent secondary lines.

The reason the entire model is not defined by primary lines is because it is difficult to recover some edges given the input data. Primary lines are well suited to recovering the position of vertical edges because it is possible to create arbitrarily large angles of intersection about the vertical edge axis, as shown in figure 4. However, for horizontal edges near camera level it is not possible to create arbitrarily large intersection angles, making it difficult to recover the horizontal edges accurately since slight inaccuracies in the camera's intrinsic or extrinsic properties results in large errors in estimated edge location, see figure 5.

Secondary lines work by connecting primary lines, where the use of a primary line would be prohibitive, e.g. the horizontal base line of the building in figure 5. Since the primary lines will recover the vertical edges of the building, the secondary lines simply indicate to the system that these edges should be connected without trying the same recovery technique used for the primary edges.

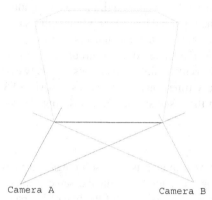

Fig. 5. For horizontal edges near camera level it is difficult to obtain arbitrarily large disparity angles

Fig. 6. Projection of primary lines. Primary edges are highlighted in white.

Primary edges should be used to recover the core structure of the building, while defining as few edges as possible. Then secondary lines should be used to define all remaining edges. A primary edge must be highlighted in at least three images, although it can be advantageous to define a primary edge in more than three images when trying to recover edges that make poor primary edge candidates. Secondary edges need only be defined in a single image.

2.2 Edge Recovery

After the edges have been highlighted, six automated steps are performed to recover the final edges; Line Projection, Triangle Intersection, Correspondence Recovery, Edge Averaging, Vertex Merging, and Secondary Edge Recovery. Each of these steps is described next.

2.2.1 Line Projection

The first step in determining the positions of the primary edges is to project the 2D primary lines to form 3D triangles. The intrinsic and extrinsic properties of the camera are used to project the primary lines from the cameras position, at the correct orientation out to infinity. This is performed for every primary line in each image, as shown in figure 6 for a scene consisting of 4 images, the final primary edges are highlighted in white.

2.2.2 Triangle Intersection

Once every 2D primary line has been transformed to a 3D triangle, the next step is to determine the intersections between the triangles. Every triangle stores a list of the triangles it intersects.

2.2.3 Correspondence Recovery

Generally each triangle intersects many other triangles even though only a small number of the triangle intersections have both their parent lines highlighting the same edge. Most systems resolve this problem by performing manual correspondences between the lines so that lines which highlight the same edge are grouped together. Once the lines are converted to triangles the only valid intersections are between members of the same group. This can be a very time consuming process. SAMATS performs this correspondence automatically in three steps; Intersection Rating, Triangle Grouping and Group Merging.

2.2.3.1 Intersection Rating

Every triangle needs to rate each of the triangles it intersects. These ratings can then be used to determine which of the intersecting triangles represent the same primary edge as itself. A naïve approach would simply use the coverage of the line of intersection as the only measure in rating each intersecting triangle, with greater coverage resulting in a better rating. This has proved to be almost completely useless because often intersecting triangles which represent a different primary edge (invalid triangles) receive better rating than those that represent the same primary edge (valid triangles).

The automated rating process does not rate an intersecting triangle on the quality of the intersection line, but on the similarity of the intersection line with other intersection lines. This is the reason for having a 3 primary line minimum when highlighting each primary edge.

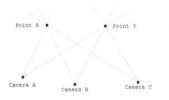

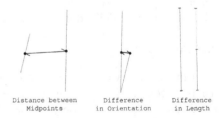

Fig. 7. 2D example of the automated correspondence determination concent

Fig. 8. The three factors considered in determining the similarities between lines of intersection

Figure 7 shows the basis of the rating algorithm in 2D. In the figure there are three cameras, **A B** and **C**, there are two points being modeled, **X** and **Y**, and there are six lines, two from each camera through the points being modeled, A_X A_Y B_X B_Y C_X and C_Y. Each line intersects every other line (although some of the intersections are off image) even though the only valid intersections are those between lines with matching subscripts. One should note that the invalid intersections are spaced quite randomly apart while the valid intersection groups have three points of intersection coincident at one location. The automated rating algorithm uses this principle of valid intersections being grouped close together when calculating the rating for each intersecting triangle.

For each triangle t_i we need to rate each of the triangles t_j in t_i's intersecting triangles set T_i. Since we know that there are at least three triangles per primary edge, we know that at least two of the intersecting triangles are valid matches. We call these valid intersecting triangles t_{j1} and t_{j2}. Note that there may be more than two valid intersecting triangles, although that fact is not important at this stage. If t_{j1} is a valid match with t_i and t_{j2} is a valid match with t_i, then t_{j1} and t_{j2} must be valid matches with each other. This implies that t_{j2} would be an intersecting triangle of t_{j1}. Therefore, when determining the rating of any t_j, we only need to consider triangles that are in both t_i's intersecting triangles set and t_j's intersecting triangles set, i.e. $T_i \cap T_j$. Note that only a sub-set of this set will contain valid intersections.

The intersecting triangle t_j can now be given a ranking based on the triangles in the set $T_i \cap T_j$. Each triangle t_k in this set intersects both t_i and t_j. Therefore, we can use the three intersecting lines, l_{ij} l_{ik} and l_{jk}, to give t_k a rating. Intersection lines are evaluated based on 3 properties; the distance between their midpoints, the relative orientations between them, and their difference in length.

The value returned for the distance between their midpoints is in the range $[0...1]$ and is described by the following equation;

$$\frac{1}{1+\left(ScalingFactor \times Distance\right)^2}$$

The *ScalingFactor* is used to set the rate at which the value declines with respect to distance. This factor is dependent on the choice of units used to model the building, e.g. if the units are meters and we only want to consider intersection lines with roughly less that 10cm spacing, then setting *ScalingFactor* to about 10 would give a good range. At 1cm the value returned by the equation would be 0.9, at 10cm the value would be 0.5, and at 100cm the value would be 0.09.

The value returned for the measure of the two lines relative orientations is calculated using the absolute value of the dot product between the lines' unit vectors and is also in the range $[0...1]$. For two lines **A** and **B** the equation is as follows;

$$\left|\hat{A} \cdot \hat{B}\right|$$

Finally, the value returned for their difference in length is in the range $[0...1]$ and is described by the following equation;

$$\frac{\max\left(\overline{A},\overline{B}\right)-\left|\overline{A}-\overline{B}\right|}{\max\left(\overline{A},\overline{B}\right)}$$

Once all these partial tests have been performed, the final rating for the lines is simply the product of the three, which is also in the range $[0...1]$. Refer to figure 8 for an illustration of each test.

Every triangle t_k in the set $T_i \cap T_j$ is given a rating based on the comparison of the three intersection lines l_{ij} l_{ik} and l_{jk}. There are three comparisons that can be made, l_{ij} with l_{ik}, l_{ij} with l_{jk}, and l_{ik} with l_{jk}. The product of these three tests is used to determine the rating of each triangle t_k in the set $T_i \cap T_j$. The product is used in favor of the sum in order to keep the ratings in the range $[0 \ldots 1]$.

Once every t_k has been given a rating there are three logical options for assigning a rating to t_j. Assign t_j the weighted sum of all the ratings in the set $T_i \cap T_j$. This has proved unfavorable since this would include triangles that are invalid. If there are a large number of low scoring invalid triangles in a particular $T_i \cap T_j$ set, the t_j will be given a poor rating even if it is a valid triangle.

A second option would be to assign t_j the weighted product of the ratings. This is a poor choice for the same reasons as assigning the weighted sum, only the problem is amplified greatly when taking the product since there are almost always a few poor scoring invalid matches, this forces the t_j rating to zero, making the rating useless.

The option that was found to work best is to assign t_j the best rated t_k in the set $T_i \cap T_j$. If t_j is a valid intersecting triangle for t_i, the best rated triangle is almost always a valid intersecting triangle for both t_i and t_j, which we'll refer to as t_k from here on. When storing the rating for each t_j, a reference to the t_k triangle responsible for this rating is also stored. This triangle is required for the triangle grouping step described briefly next.

2.2.3.2 Triangle Grouping

After the intersection rating step, for every triangle t_i, every triangle t_j in t_i's intersecting triangles set T_i will have a rating assigned to it. Also, the t_k responsible for each t_j's rating will be stored along with the rating. This information can then be used to group triangles together where each group represents a primary edge.

Essentially, the grouping process is performed in two steps. Firstly, the GSS (Group Scope Set) of each triangle is determined. The GSS for each triangle is the list of mutually high ranking intersecting triangles. Not every triangle will have the same size GSS. The size of these sets will vary depending on the number of triangles used to represent each primary edge as well as the relationship between their line intersections.

The second step in the grouping process is to use the GSSs to group the triangles into groups. The triangles are ordered based on the size of their GSS's in ascending order. Triangles with small GSSs form the initial groups. Small GSSs are more tightly coupled which is a desirable property when trying to match triangles together. After the core set of groups is created all remaining triangles are assigned a group, the vast majority being assigned to one of the existing groups with only a small minority forming their own groups.

It may not be possible to assign every triangle to a group for a number of reasons. The user may not have used three primary lines to highlight a particular primary edge or there may be too great an error to group some primary lines together either due to an error in the camera's intrinsic and/or extrinsic properties or an error in line placement. In such cases the triangles are marked as invalid. For a more detailed explanation of the Triangle Grouping step refer to [4]

2.2.3.3 Group Merging

The final step in the grouping process is group merging. If a primary edge is represented by 6 or more primary lines it may form 2 distinct groups. If the groups were left the way they were, there would be 2 primary edges representing the same building edge instead of just one. The merging step simply compares each group to each other and merges groups which are sufficiently similar.

2.2.4 Edge Averaging

Once all triangles have been assigned a group the primary edges must be determined for each group. This is simply the weighted average of all the intersection lines between all group members.

2.2.5 Vertex Merging

During the edge averaging step, each primary edge will be created totally independently from all other primary edges. In most cases this is acceptable since the majority of primary edges are not connected to any other primary edge. Sometimes however primary edges are connected. This is indicated in the edge highlighting step by having two or more primary lines share the same endpoint.

All primary edges that are connected need to have their connected endpoints coincident. This is achieved by creating a mapping between every primary line and every primary edge, and also between every primary line endpoint and every primary edge vertex. Once the mappings have been made, we can see if any of the primary lines share the same endpoints, which maps to primary edges sharing the same vertex. Once the vertices are identified they are set to the average of their positions.

2.2.6 Secondary Edge Recovery

Secondary edges are determined using the same mapping information obtained during the vertex merging step. Firstly, the secondary lines' endpoints are determined. Then the corresponding vertices are determined for these endpoints and a new group is created for each secondary line using these vertices as the secondary edge's endpoints. After all secondary edges have been highlighted the outline of the model should be complete.

2.3 Structure Recovery

Even though the outline of the model has been determined there is still no surface data associated with the model. The model is only defined in terms of vertices and lines and not in terms of surfaces and the triangles that make up each surface. Recovering this structural information is broken into three steps. The first step is to determine the models surfaces. This is achieved be treating the model like a graph, with the vertices as the graph nodes and the edges as the graph edges. Surfaces are determined by finding the shortest cycles in the graph where all the vertices are coplanar. All surface normals must then be aligned so that they all point away from the model. This is performed by aligning the normals of neighboring surfaces recursively until all normals are aligned. The final step is to triangulate all the surfaces. The algorithm used to triangulate each surface can be found in [8]. Refer to [4] for further details.

3 Texture Extraction

Coming into this section, we have an accurate model of the building, or to be exact, we have a geometrically accurate model of the building. There is still data contained in the image set that has not yet been used to increase the models realism, the buildings façades. The texture extraction process takes the façades from the images and applies them to the model. An overview of this component is presented next. For a more detailed explanation of the Texture Extraction component refer to [4].

3.1 Overview

The aim of the texture extraction process is to produce a 3D model with photorealistic textures. The texture extraction process can be broken into a number of steps. Firstly, the number of images that will contribute to each triangle is determined using back-face culling. There can be any number of contributing images, with each image's contribution first being stored in a temporary texture before they are all blended together per-pixel based on the camera-surface distance and orientation. Occlusion maps are used to prevent incorrect façade data being stored with each triangle. All triangles are then packed into a single large texture retaining the relative size of each triangle, thus creating an authalic texture map. The texture coordinates for each triangle are then set to sample the correct region of the texture map, with the texture then being assigned to the model.

4 Conclusions

This research shows that given sufficient information, user input to the modeling process can be reduced significantly. Currently user input is required for the edge highlighting step but since no correspondence is required this step could be automated using edge detection and a set of heuristics to guide the choice between using primary lines or secondary lines.

Currently SAMATS has only been used on synthetic images where the exact extrinsic and intrinsic properties of the camera are known. Achieving such precision in the real world would prove difficult without specialized equipment. New techniques will be required to facilitate the gathering of the geo-referenced images required by SAMATS in order for this system to be utilized effectively in the real world.

SAMATS has shown the ability to model rectangular and triangular roofed structures very well, however SAMATS does have trouble modeling certain structures. SAMATS has no special ability to handle curved surfaces, which makes it impossible to model such features completely accurately. Cylindrical column must be replaced by rectangular columns for instance. Another difficulty that can arise is SAMATS' inability to handle partially highlighted edges. This makes it difficult, and in some cases impossible, to model buildings in tightly confined spaces.

References

1. J.D. Carswell, A. Eustace, K. Gardiner, E. Kilfeather. An Environment for Mobile Context-Based Hypermedia Retrieval, in 13th International Conference on Database and Expert Systems Applications (DEXA2002); IEEE CS Press; Aix en Provence, France; September 2002

2. S.R. Coorg. Pose Imagery and Automated Three-Dimensional Modeling of Urban Environments. PhD thesis, MIT Ph.D. Thesis, 1998.

3. P.E. Debevec, C.J. Taylor, and J.Malik. Modeling and Rendering Architecture from Photographs: A hybrid geometry- and image-based approach. In SIGGRAPH '96 Conference Proceedings, 11-20, 1996.

4. J. Hegarty and J. Carswell. SAMATS – Semi-Automated Modeling And Texturing System. Masters Thesis, 2005.

5. S.C. Lee, S.K. Jung, and R. Nevatia. Integrating Ground and Aerial Views for Urban Site Modeling. In Proceedings of International Conference on Pattern Recognition, 2002.

6. S.C. Lee, S.K. Jung, and R. Nevatia. Automatic Integration of Façade Textures into 3D Building Models with a Projective Geometry Based Line Clustering. Computer Graphics Forum, 21(3):511-519, 2002.

7. S.C. Lee, S.K. Jung, and R. Nevatia. Automatic Pose Estimation of Complex 3D Building Models. Proceeding of the 6th IEEE Workshop on Applications of Computer Vision, 2002.

8. J. O'Rourke. Computational Geometry in C (Second Ed.). Cambridge University Press, 1998.

9. C.J. Taylor and D.J. Kriegman. Structure and Motion from Line Segments in Multiple Images. PAMI, 17(11):1021-1032, November 1995.

10. S.Ullman. The Interpretation of Structure from Motion. Proceedings of the Royal Society of London, 1976.

11. S. Zlatanova and F.A. van den Heuvel. Knowledge-based Automatic 3D Line Extraction from close range images. Web – http://www.gdmc.nl/zlatanova/thesis/html/refer/ps/SZ_FH_Corfu.pdf

Applying Semantic Web Technologies for Geodata Integration and Visualization

Zhao Xu, Huajun Chen, and Zhaohui Wu

Grid Computing Laboratory, College of Computer Science,
Zhejiang University, 310027, Hangzhou, China
{xuzhao, huajunsir, wzh}@zju.edu.cn

Abstract. Nowadays new applications ask for enriching the semantics associated to geographic information in order to support a wide variety of tasks including data integration, interoperability, knowledge reuse, spatial reasoning and many others. This paper proposes a new framework called GeoShare, which supports for semantic integration and querying over heterogeneous geospatial data sets distributed over the grid environment. In GeoShare we try to resolve semantic heterogeneity by using an RDF-View based approach for schema mediation and focus on embedding domain semantics in geographic maps. A series of grid services such as ontology service, semantic registry service, and semantic query service with a set of innovative semantic tools are brought forward to accomplish these purposes. A visualization module also presents at the upper level of GeoShare to automatically generate Scalable Vector Graphics (SVG) maps embedded with domain semantics as an ideal data view to end users.

1 Introduction

The Yangtze River Delta region, which consists of Shanghai, and parts of Jiangsu and Zhejiang provinces, is playing an increasingly vital role in China's overall economic growth, particularly in the globalization and industrialization processes. With the rapid development of economy and the explosive expansion of Chinese automobile industry around this prosperous region, more convenient, efficient and secure transportation support is urgently needed. With recent technology advances in a series of IT areas such as communication and cybernetics, Intelligent Transportation System (ITS) [9] is considered as an effective approach to solve most of the arising problems. As one of the foundational work to establish such a complicated and high-tech intelligent system, we set about trying to bring forward and implement a new framework for distributed geospatial databases integration based on semantic web technologies and standards, as well as generation of high-quality, interactive vector maps with those geodata.

As a matter of fact, the problem of semantic integration on spatial databases has been attracting extensively attention all over the world for years. Quite a number of research works have been done on finding out an ideal approach for geodata integration and sharing [6][8][16]. Most of those attempts are based on the use of ontologies for solving the problem of semantic heterogeneity. We will follow this

J. Akoka et al. (Eds.): ER Workshops 2005, LNCS 3770, pp. 320–329, 2005.

extensively accepted method and also make our own adjustment to introduce and utilize more commonly adopted technologies and standards in Semantic Web to realize the integration goal.

Based on our previous research and development work on integration of database resources with the Grid infrastructure [2][20][21], we gain lots of experience on building a database grid that supports for transparent information retrieval, data integration and knowledge sharing from geographically decentralized database resources. Some core design principles and characteristic components such as ontology-based data integration and semantic registration, in our opinion, can also be applied in the new framework to integrate heterogeneous geodata on semantic level, which will provide users a single and uniform access point to their interested information without caring the details of the distributed geodata sources. In the following sections we'll argue the major barrier to geodata integration and focus on how to utilize semantic web technologies to overcome these obstacles.

The remaining part of this paper is structured as follows. Section 2 gives a brief introduction to the related work and presents the salient features of our framework comparing with them. Section 3 gives an overview of the proposed architecture, introducing the major components and their relationship. Section 4 covers how we embed semantics in a svg map. Section 5 contains our conclusions and future work.

2   Related Work

As stated above, seamless sharing and exchange of geospatial information between systems in a distributed environment, like the Internet or database grid, is an essential requirement as well as a great challenge. The general consensus in the research field of GIS has indicated that the greatest inhibitor to continued development of GIS technology is the cacophony of terms and definitions given to abstract types, features, data models, interfaces, roles, behaviors, relationships, and so on [1]. This kind of major barrier to facilitate geodata integration and sharing is the well-known topic of semantic heterogeneity.

In order to satisfy the need for semantic integration, quite a lot of related works have come out. Within the domain of GIS research, there are many efforts concerning the integration of spatial data sources and schemas under ontology-based framework. Typical examples are SIT-SD (Semantic Integration Tool for Spatial Data) [17], ORHIDEA platform [11], and research approaches of Fonseca [5] [6] [7]. SIT-SD is a prototype of an integration tool of information sources, which is able to recognize the similarities and the differences between entities to be integrated. By materializing the models from OMT-G or OpenGIS in XMI [14], SIT-SD can help user to construct the ontologies for the objects in the model and, afterward, match the ontologies of the different schemas to integrate. At the end of the whole process, a federated schema will be constructed at the top of the framework to support for integration of heterogeneous data. ORHIDEA platform has been developed in order to perform intelligent integration of information from multiple heterogeneous GIS (spatial and geographic), and non-spatial (thematic) data sources. As a middleware, and mediator system, ORHIDEA can provide data interchange and access to data sources, distributed over the Internet, without changing how or where data is stored. Fonseca directs the research from ontologies and his solution does not create new ontologies but links between formal representations of semantics and conceptual schemas describing information stored in databases.

Besides above-mentioned works, Semantic Geospatial Web [12], a new academic area derived from the research of Semantic Web and Geographic Information Science, has recently received more and more attention. As a great complement of Semantic Web, the Geospatial Semantic Web initiative specifically looks for better support for geographic information that Semantic Web has not addressed, and wishes to enable users to retrieve more precise geodata they need, based on the semantics associated with these data.

Comparing with others' related work, our framework (temporarily called GeoShare) exhibits several technical features:

1. Firstly, it is built in a database grid environment and all core components are implemented as grid services. Since grids are persistent environments that enable software applications to integrate all kinds of resources in widespread locations, we hope such a service-oriented architecture will provide users a unified and transparent data access interface to the distributed information resources and more loosely-coupling characteristics to support additional services combination in the future.
2. Secondly, it adopted a series of Semantic Web standards such as RDF [13], SVG [3] proposed by the World Wide Web Consortium (W3C), to present the semantics embedded in the geospatial information and make geodata visible to users in an ideal, high-quality graphic format. Another major goal of utilizing these technologies is to mediate heterogeneous geospatial database schemas and enables the users to interact with the system at a semantic level.
3. Thirdly, it provides a convenient visual tool to facilitate the schema mapping from database to RDF view. The geospatial database schemas will be retrieved as views on the mediated RDF schema. With our experience, this view-based approach is more convenient and universal than those semiautomatic or automatic schema integration methods for those databases that are not normalized enough.

3   Service Architecture and Core Components

In this section, we give an overview on the core components of GeoShare. Figure 1 illustrates the relationship between them. In next paragraphs, we'll introduce the details of each component and relevant grid services in GeoShare.

3.1   Core Grid Services

A series of novel core grid services for geodata access, SVG map generation and semantic mediation has been implemented in GeoShare. They are:

(1) GeoDB Unified Access Service (GDBAS):
GDBAS brings the users a great convenience of transparently access to the underlying heterogeneous data sources. It provides a unified data access interface with which data consumers can retrieve or manipulate any kind of geodata stored in the databases without any concern about the details of the data sources such as what kind of the database is (relational or object-oriented), what product of the database is (Oracle,

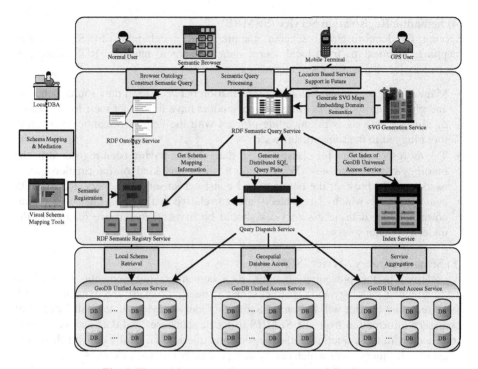

Fig. 1. The architecture and core components of GeoShare

MySQL, Informix, etc.), or what specific data access interface or SQL grammars the database supports. We make use of the factory deign pattern and thus the grid database service factory will automatically create a GDBAS instance to access the specific geospatial database that has been configured by local database administrator. The retrieved geodata will be encapsulated in XML or other formats and returned to the data consumer through the unified data access interface.

(2) Data Source Semantic Service (DSEMS):
DSEMS retrieves the semantics of the local data source and supports the vendors to publish that information. For database resource in GeoShare, the local semantic information is right the conceptual schema about the tables contained in the database. Users can inquire of this service about the local data semantics for further tasks such as semantic registration or integration.

(3) Ontology Service (ONTOS):
As we mentioned in the preceding paragraphs, ontologies can be viewed as domain semantics and mediated schema. In GeoShare, we use RDF/OWL language to define the formal ontology and publish it through Ontology Service, which will retrieve the global formal ontology from the repository and support the Semantic Browser [18] to provide users a visual navigation to the semantics and shared knowledge within GeoShare. In a word, the Ontology Service provides a unified semantic data access entry to users and supports the semantic level interactions.

(4) Semantic Registration Service (SEMRS):
Serving for local-to-global schema mapping and matching, SEMRS establishes mapping relationships from local source data schemas to mediated RDF schema. It has two major characteristics in our design:

- Maintaining the semantic mapping information between local data source schemas and global ontologies. Any data source vendors have to register their local schema to SEMRS as well as the mapping process with the help of semantic registration tool plugged in the semantic browser.
- Taking responsibility for classifying all data objects having been registered to the ontology-based taxonomy. With supports from this kind of information, when users visually browser the ontology and construct an semantic query, the SEMQS can determine which data objects are correlative within this query, and their corresponding data access services should be invoked to retrieve data from the underlying data sources.

(5) Semantic Query Service (SEMQS):
As referred in the preceding sections, data consumers use Semantic Browser to navigate global ontology and construct semantic queries visually and dynamically. All these semantic queries will be sent to Semantic Query Service, which will inquires of necessary information from the SEMRS to locate databases and data access services correlated with these queries, parse the semantic queries and finally convert them into relevant SQL query plans which can be accepted and executed by GDBAS.

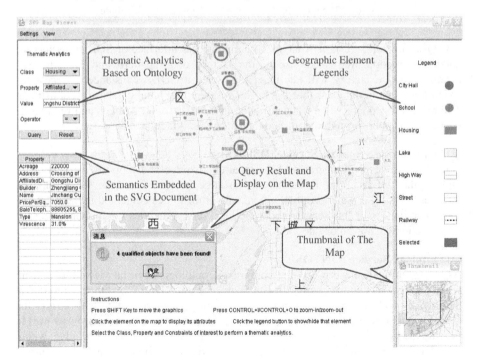

Fig. 2. The result of an SVG rendering of the generated map

(6) SVG Generation Service (SVGGS):

The results of SQL query plans returned from relevant services can be wrapped in several kinds of formats. However, as to the geospatial data sets, the best way to indicate the results is making them visualized and intuitionistic in form of graphics, which would worth a thousand words to interpret which data describes which part of the region, or something else. SVG Generation Service takes charge of converting the query results wrapped by XML format into SVG format using a pre-configured XSL-stylesheet, and then embedding the thematic information related with the region, which is also returned from the queries and wrapped in RDF format, into the SVG graphics. Figure 2 illustrates an example of the generated SVG map. In section 4 we'll discuss the detailed strategy we have employed.

3.2   Semantic Tools

Core services in GeoShare only provide underlying functions for geodata integration and visualization. Users on top of the service architecture still need a visual interface to graphically browse the RDF semantics and visually construct a semantic query for distributed geodata as well as facilitate the task of schema mapping. With such requirements we developed a general-purpose semantic toolkit, called the Semantic Browser [19], as the uniform user interface that enables users to manipulating geodata semantics in GeoShare. We investigated lots of existed and popular ontology editors and knowledge-based applications, and finally chose Protégé [15] as the main reference to our implementation. Figure 3 gives a snapshot of it. Here we summarize the technical features and major functions of the semantic browser as follow:

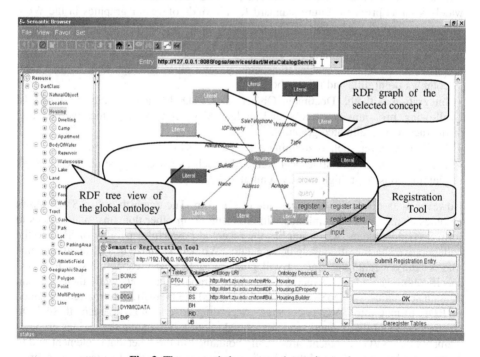

Fig. 3. The semantic browser and mapping tool

1. Improved navigation. Semantic browser allows users to connect to an ontology service from anywhere within GeoShare and visually navigation of the ontologies. With the connectivity relations among concepts of RDF graphs, the user can easily get access to relevant information from one concept to another and be aware of the relationship between the concepts.
2. Visual and Dynamic Semantic Query Generation. Users browse the information on the semantic level and by interacting with the RDF graphs of the ontology they may build customized semantic queries for interested regions during navigation.
3. Visual Semantic Registration. Local database administrator can utilize semantic browser as a semantic mapping tool to register his database and deployed data access service to the whole virtual organization. As one of the numerous plug-in to semantic browser, the mapping tool will automatically retrieve local data source schemas and help users to visually map them to global mediated RDF ontologies, which means that users can decide a specific table should be mapped onto which class and how its columns should be mapped onto corresponding properties of that class. The mapping information will be encapsulated in RDF/XML format and then submitted to the semantic registration service by users.

4   Semantic-Based Visualization

The Scalable Vector Graphics (SVG) specification [3] is an XML technology, developed by the W3C with the purpose to create a standardized vector graphics format for the Web environment. As one of W3C recommendations, SVG format is widely seen as probable future standard for all kinds of vector graphics in the Web. The major powerful features of SVG include:

- A set of graphic techniques almost as complete as the best graphics design software, such as variable transparency, anti-aliasing, fine text control, and complex point, line and fill symbols.
- Integration with the Document Object Model (DOM) and JavaScript standards, allowing programs to access and update SVG content dynamically, enabling interactive and animated graphics.
- Basic interaction, such as zooming and panning, is built into most viewers.
- Incorporation of web stylesheet standards, such as Cascading Style Sheet (CSS), yielding consistent, high-quality symbology.

However, data display is only one side of the coin. The extensibility of SVG allows graphics elements to be combined with elements from other domains/ namespaces to form SVG documents, which implies the potential powerful capability of SVG to the collaborative sharing of geographic semantics. We believe make this assumption come true is really meaningful since it will grant semantic query capabilities to the SVG specification and ideally syncretize the pure graphic data sets with their relevant domain semantics. In GeoShare, we tried to embed the thematic information presented in RDF format into SVG documents. According to common practice, these RDF statements, about the semantic and conceptual relationship among the various graphics elements in a SVG document, are usually contained in the "metadata" element and to be processed with the help of the "desc" elements present in the document to generate textual description of the SVG document for accessibility and summarization purposes. In contrast with this approach, we put the RDF contents in the "defs" element and use the XLink reference to link SVG elements with corresponding semantic fragments, the "id" attribute is used to associate the corresponding two parts. With these semantics embedded in the SVG document, we also use Jena [10] APIs to encapsulate a set of our own specific APIs for querying objects in the SVG documents. By this way only the least adjustment is needed to accommodate with the original SVG specification and conveniently reaches our goal. Figure 4 exhibits an illustration of our approach with a short fragment of the SVG document.

```
<?xml version="1.0" encoding="UTF-8"?>
<!DOCTYPE svg PUBLIC "-//W3C//DTD SVG 1.0//EN">
    http://www.w3.org/TR/2001/REC-SVG-20010904/DTD/svg10.dtd [
    ...
        <!ENTITY ns_svg "http://www.w3.org/2000/svg">
        <!ENTITY ns_xlink "http://www.w3.org/1999/xlink">
        <!ENTITY rdf "http://www.w3.org/1999/02/22-rdf-syntax-ns#">
        <!ENTITY rdfs "http://www.w3.org/2000/01/rdf-schema#">
        <!ENTITY geoshare "http://dart.zju.edu.cn/geoshare#">
]>

<svg id="root" width="708.225" height="608.3" viewBox="0 0 708.225 608.3"
    xmlns="&ns_svg;" xmlns:xlink="&ns_xlink;" xmlns:rdf="&rdf;"
    xmlns:rdfs="&rdfs;" xmlns:geoshare="&geoshare;" ...>
    <defs>
        <rdf:RDF>
            <geoshare:House id="&geoshare;house_01"
            geoshare:Acreage="220000"
            geoshare:Address="Crossing of Lishui Road and Zijing Road, Gongshu District"
            geoshare:AffiliatedDistrict="Gongshu District"
            geoshare:Builder="Zhejiang Chengjian Land Agent Construction Ltd."
            geoshare:IDProperty="house_01"
            geoshare:Name="Jinchang Culture"
            geoshare:PricePerSquareMeter="7050.0"
            geoshare:Telephone="086-571-88805255, 086-571-88805655"
            geoshare:Type="Marsion"
            geoshare:Virescence="31.0%"
            rdfs:label="Jinchang Culture"/>
            ...
        </rdf:RDF>
    </defs>

    <g id="houses">
        <rect id="house_01" x="262.1" y="133.6" width="7px" height="7px" fill="#339900"
            xlink:href="&geoshare;house_01"/>
        ...
    </g>
</svg>
```

Fig. 4. A fragment of the SVG document embedded with RDF semantics

5 Conclusion and Future Work

In this paper we have presented a framework for integration of heterogeneous geospatial databases in a grid environment by utilizing a series of latest semantic web technologies and standards such as RDF, SVG. Since W3C has released the second working draft of the SPARQL Query Language of RDF [4], we are waiting for the final results and preparing to re-implement the semantic query layer in GeoShare under the direction of this latest standard. What's more, the performance issues including response time, throughout, etc. in the grid-oriented database system will be a matter of our utmost concern. So our further investigation will focus on query processing optimization and faster dynamic generation of SVG maps. With a view to construct a more complicated intelligent transportation system in the future, dealing with data obtained in real-time, dynamic display of multiple objects, monitoring and analyzing objects' tracks also should be taken into account as well as supports for location-based services serving for cellular phone or other intelligent mobile terminal users.

References

1. C. Kottman. **White Paper on Trends in the Intersection of GIS and IT**, Open GIS Consortium, August-October, 2001.
2. Chang Huang, Zhaohui Wu, Guozhou Zheng, Xiaojun Wu. **Dart: A Framework for Grid-based Database Resource Access and Discovery**, In Proceeding of International Workshop on Grid and Cooperative Computing, GCC 2003, Shanghai, December 2003.
3. Dean Jackson, Craig Northway (Editors). **Scalable Vector Graphics (SVG) Full 1.2 Specification,** Available: http://www.w3.org/TR/SVG12/.
4. Eric Prud'hommeaux, Andy Seaborne (Editors). **SPARQL Query Language for RDF - W3C Working Draft 21 July 2005**, Available: http://www.w3.org/TR/rdf-sparql-query/.
5. F. Fonseca (2001). **Ontology-Driven Geographic Information**, PhD thesis, University of Maine, Orono, Maine 04469.
6. F. Fonseca, C. Davis, G. Camara. **Bridging Ontologies and Conceptual Schemas in Geographic Information Integration**, GeoInformatica Volume 7, No. 4, December 2003, pp. 355-378.
7. F. Fonseca, M. Egenhofer, P. Agouris and C. Camara. **Using Ontologies for Integrated Geographic Information Systems**, Transaction in GIS, 6 (3), 2002, pp. 231-257.
8. F. Hakimpour, S. Timpf. **A Step Towards Geodata Integration using Formal Ontologies**, In Proceeding of 5th AGILE Conference on Geographic Information Science, Palma April 25th-27th, 2002.
9. F. Y. Wang et al. **Integrated Intelligent Control and Management for Urban Traffic Systems**, In Proceeding of IEEE 6th International Conference on Intelligent Transportation Systems, IEEE Press, 2003, pp. 1313-1317.
10. Hewlett-Packard Labs. **Jena 2 – A Semantic Web Framework**, Available: http://www.hpl.hp.com/semweb/jena2.htm.
11. L. Stoimenov, S. Dordevic-Kajan, A. Milosavljevic. **Realization of Infrastructure for GIS Interoperability**, In Proceeding of YUINFO'02, Kopaonik, 2003 (Serbian).
12. M. Egenhofer. **Toward the Semantic Geospatial Web**, In Proceeding of the 10th ACM International Symposium on Advances in Geographic Information Systems, McLean, Virginia.
13. O. Lassila, R. Swick (Editors). **Resource Description Framework (RDF) Model and Syntax Specification**, Available: http://www.w3.org/TR/PR-rdf-syntax/.
14. OMG, O.M.G. (2002). **XML Metadata Interchange (XMI) Specification**, Available: http://www.omg.org/.
15. Protégé. Available: http://protege.stanford.edu/.

16. Villie Morocho, Felix Saltor. **Ontologies: Solving Semantic Heterogeneity in a Federated Spatial Database System**, In Proceeding of 5th International Conference on Enterprise Information System, April 2003, pp. 347-352.
17. Villie Morocho, Llius Perez-Vidal, Felix Saltor. **Semantic Integration on Spatial Database SIT-SD Prototype**, In Proceeding of VIII Jornadas de Ingenieria del Software y Bases de Datos, November 2003, pp. 603-612.
18. Yuxin Mao, Zhaohui Wu, Huajun Chen. **SkyEyes: A Semantic Browser for the KBGrid**, In Proceeding of International Workshop on Grid and Cooperative Computing (GCC 2003), Shanghai, December 2003.
19. Yuxin Mao, Zhaohui Wu, Huajun Chen. **Semantic Browser: An Intelligent Client for Dart-Grid**, Lecture Notes in Computer Science 3036: 470-473 2004.
20. Zhaohui Wu, Huajun Chen, Chang Huang, Jiefeng Xu. **Towards a Grid-based Architecture for Traditional Chinese Medicine**, In Proceeding of Challenge of Large Application in Distributed Environments International Workshop on Heterogeneous and Adaptive Computation, June 2003.
21. Zhaohui Wu, Huajun Chen, Chang Huang, Guozhou Zheng, Jiefeng Xu. **DartGrid: Semantic-based Database Grid**, In Proceeding of International Conference on Computational Science 2004, pp. 59-66.

Preface to eCOMO 2005

Willem-Jan van den Heuvel and Bernhard Thalheim

The 6th International Workshop on Conceptual Modeling Approaches for e-Business, eCOMO'05, was held in conjunction with the ER-Conference 2005.

Traditionally, modeling of business processes is achieved using different views of a single company, viz. the data or content view, functional view, process view, and the organizational view. These views should not be treated in an isolated manner. Therefore, business modeling provides an integrated view aligning these viewpoints.

With the advent of Internet technologies, business modeling has evolved to modeling of e-business processes adding several dimensions to these viewpoints, notably, another highly distributed and heterogenous communication channel, and a customer-centric dimension, increasing the need for collaboration amongst enterprises, and opening up many intricate integration challenges to business process modeling. Through the years, the eCOMO workshop series has covered these issues packaging them around contemporary research themes in the area of conceptual modelling approaches and technologies for e-business systems.

This year's theme revolves around an emerging distributed service technology named the enterprise service bus. Since two years, the enterprise service bus is touted by industry as the distributed computing technology of choice for realizing contemporary, e-business applications. In short, the enterprise service bus constitutes a distributed messaging backbone with an open and standardized architecture that allows enterprise services to be assembled, deployed and managed, enabling cross-organizational, event-driven business processes. Up till now most research efforts concentrated on developing the infrastructural backbone, focusing at technical issues such as routing, network level security, programming adapters and end-point discovery of service capabilities, while neglecting critical conceptual modeling and design issues.

This workshop targets open research problems in the latter research domain. In particular, it addresses scientific approaches and techniques to design e-business applications on top of the enterprise bus from the perspective of conceptual modeling. Topics include, but are not restricted to, cross-enterprise process modelling, semantic integration, web-service composition, and, workflow and agent modelling.

Since many e-business sytems in general, and the enterprise bus solution more in particular, are poorly equipped with collaborative modelling facilities to design shared business processes, this workshop is also directed towards presenting and discussing research results in areas such as multi-party and multiperspective process modelling, and, the automatic mapping and harmonization of the ontologies underlying enterprise models.

Out of nine submitted papers, four strong papers were selected based on a thorough, blind review process. The first paper is written by Klaus-Dieter Schewe and is entitled: "Bargaining in E-Business Systems". This paper outlines a co-design model for developing negotiation processes in business communications and is built on top of game-theoretic concepts and an existing co-design method.Secondly,

J. Akoka et al. (Eds.): ER Workshops 2005, LNCS 3770, pp. 331–332, 2005.
© Springer-Verlag Berlin Heidelberg 2005

YangPing Yang et al, have written the paper, entitled "Verifying Web Services Composition". This paper introduces a novel method for analyzing and checking web-service compositions, which is based on an extension of Colored Petri-Nets. The third paper "Process Assembler: Architecture for Business Process Reuse", authored by Sergiy Zlatkin and Roland Kaschek, presents a prototypical implementation of a tool, named the Process Assember, which helps to reuse business process specifications as process types according to a staged process. Lastly, the paper "Conceptual Content Management for EnterpriseWeb Services" by Sebastian Bossung et al., reports on the application of user-specific domain models, which are based on Conceptual Content Modeling, to facilitate the multi-party interpretation of web services.

Bargaining in E-Business Systems

Klaus-Dieter Schewe

Massey University, Department of Information Systems and
Information Science Research Centre, Private Bag 11 222,
Palmerston North, New Zealand
k.d.schewe@massey.ac.nz

Abstract. Despite the fact that bargaining plays an important role in
business communications, it is almost completely neglected in e-business
systems. In this paper we make an attempt to integrate bargaining into
web-based e-business systems. We characterise the bargaining process
and place it into the co-design approach for the design of web information
systems. Then we outline how game-theoretic concepts can be used to
develop a conceptual model for bargaining.

1 Introduction

Bargaining plays an important role in business communications. For instance, in
commerce it is common to bargain about prices, discounts, etc., and in banking
and insurance bargaining about terms and conditions applies. E-business aims
at supporting business with electronic media, in particular web-based systems.
These systems support, complement or even replace human labour that would
normally be involved in the process. In [9] it has been outlined that such systems
can only be developed successfully, if the human communication behaviour is well
understood, so that it can become part of an electronic system. Bargaining is
part of that communication behaviour.

However, bargaining is neglected almost completely in e-business. In business-
oriented literature, e.g. [6, 12] secure payments and trust are mentioned, but
negotiation latitude or bargaining do not appear. Looking at the discussion of
technology for e-business this comes as no surprise, as the emphasis is on the
sequencing of user actions and the data support, but almost never on inferences.
For instance, favourable topics in e-business modelling are business processes [1],
workflow [8], e-payment [2], trust [4], decision support [3], or web services [11].

In this paper we make an attempt to integrate bargaining into web-based
e-business systems using the co-design approach [10] to the design of web information systems (WISs). We start with a characterisation of the bargaining
process as an interaction between at least two parties. The cornerstones of this
characterisation are goals, acceptable outcomes, strategies, secrets, trust and
distrust, and preferences. We believe that before dropping into formal details of
a conceptual model for bargaining, we first need a clearer picture of what we are
aiming at. We will discuss the characteristics of bargaining in Section 2. We will
also outline the differences to auction systems.

J. Akoka et al. (Eds.): ER Workshops 2005, LNCS 3770, pp. 333–342, 2005.

In Section 3 we present the gist of the co-design approach to WIS design. We omit most of the formal details, and we also omit most of the elaborate theory underlying the method – these can be found in [10] – in order to have a very simple model of WISs, into which ideas concerning bargaining can be implanted. Basically we keep the idea of story space as a collection of abstract locations (called scenes) and transitions between them that are initiated by actions, the support of the scenes by database views, and the support of the actions by operations associated with the views.

Finally, we develop a model for bargaining based on games that are played on the conceptual model. We concentrate on bargaining involving only two parties. The moves of the players reflect offers, counteroffers, acceptance and denial. Both players aim an an optimal outcome for themselves, but success is defined as acceptable outcomes for both parties. In e-business systems the role of one player will be taken by a user, while the system plays the other role. This may be extended to a multiple-player game with not only a single human player, e.g. if bargaining becomes too critical to leave it exclusively to a system.

2   Characteristics of the Bargaining Process

Let us start looking at human bargaining processes. Examples of such processes in business are the following:

– In a typical commerce situation a customer may entering into bargaining over the total price of an order consisting of several goods, each with its particular quantity. The seller might have indicated a price, but as the order will lead to substantial turnover, he is willing to enter into bargaining. The goal of the purchaser is to reduce the total price as much as possible, i.e. to bargain a maximal discount, while the seller might want to keep the discount below a certain threshold. Both parties may be willing to accept additional items added to the order for free. This defines optimal and acceptable outcomes for both sides.

However, none of the two parties may play completely with open cards, i.e. the seller may try to hide the maximal discount he could offer, while the purchaser may hide the limit price he is willing to accept. Both parties may also try to hide their preferences, e.g. whether an add-on to the order or a discount is really the preferred option. It may even be the case that adding a presumably expensive item to the order is acceptable to the seller, while the latitude for a discount is much smaller, e.g. if the add-on item does not sell very well. So, both parties apply their own strategies to achieve the best outcome for them.

The bargaining process then consists of making offers and counteroffers. Both offers and counteroffers narrow down the possible outcomes. For instance, an offer by the seller indicating a particular discount determines already a maximal price. The purchaser may not be happy with the offer, i.e. the price is not in the set of his/her acceptable outcomes, therefore request a larger discount. Bargaining first moves into the set of mutually acceptable outcomes, finally achieves an agreement, i.e. a contract. Bargaining outside the latitude of either party may jeopardise the whole contract or require that a human agent takes over the bargaining task.

- Similar price bargaining arises in situations, when real estate, e.g. a house is sold.
- In loan applications, i.e. both personal loans and mortgages [9] the bargaining parties aim at acceptable conditions regarding disagio, interest rate, securities, duration, bail, etc. The principles are the same as for price bargaining, but the customer may bring in evidence of offers from competing financial institutions.

As a loan contract binds the parties for a longer time than a one-off sale, it becomes also important that the bargaining parties trust each other. The bank must be convinced that the customer will be able to repay the loan, and the customer must be convinced that the offer made is reasonable and not an attempt to achieve extortionate conditions. In this case the set of acceptable outcomes is also constrained by law.

In order to obtain a conceptual model from these examples let us try to extract the formal ingredients of the bargaining process. From now on we concentrate on the case that only two parties are involved in the bargaining.

1. First of all there is the object of the bargaining, which can be expressed by a parameterised view. In case of the sales situation this object is the order, which can be formalised by a set of items, each having a quantity, a price, and a discount, plus a discount for the total order. At the beginning of bargaining processes the set contains just the items selected by the customer, and all discounts are set to 0. During the bargaining process items may be added to the order, and discounts may be set. Similarly, in the loan bargaining situation the object is the loan, which is parameterised by interest rate, disagio, and duration, and the set of securities, some of which might belong to bailsmen, in which case the certification of the bailsmen becomes part of the bargaining object.

2. The set of acceptable outcomes is obtained by instantiations of the bargaining object. These instantiations are expressed by static constraints for each party. However, the constraints are not visible to the other party. They can only be inferred partially during the bargaining process. In addition to the constraints of each party there are general constraints originating from law or other agreed policies. These general constraints are visible to both parties, and they must not be violated.

   In case of the sales situation a constraint on the side of the purchaser might be a maximal acceptable price for the original order, or it might be expressed by a minimum discount in terms of any extended order. It may also be the case that the discount is expressed by a function on the set of added items, e.g. the more items are added to the order, the higher the acceptable discount must be. In case of the loan situation constraints on side of the customer can be a maximal load issued by repayments or a maximal value of securities offered. For the bank a minimum level of security and a minimum real interest rate might define their acceptable outcomes.

3. Within the set of acceptable outcomes of either party the outcomes are (partially) ordered according to preferences. For any artificial party these preferences have to be part of the system specification. For instance, in the sales situation the lower the total price, the better is the outcome for the purchaser (inverse for the seller), and an offer with more additional items is higher ranked. However, whether an offer with additional items and a lower discount is preferred over a large discount, depends on the individual customer and his/her goals.

4. An agreement is an outcome that is acceptable to both parties. Usually, bargaining terminates with an agreement, alternatively with failure.

5. The primary goal of each party is to achieve an agreement that is as close as possible to a maximum in the corresponding set of acceptable results. However, bargaining may also involve secondary goals such as binding a customer (for the seller or the bank). These secondary goals influence the bargaining strategy in a way that the opposite party considers offers made to be fair and the agreement not only acceptable, but also satisfactory. This implies that constraints are classified in a way that some stronger constraints define satisfactory outcomes. This can be extended to more than just two levels of outcomes. In general, the bargaining strategy of each party is representable as a set of rules that determine the continuation of the bargaining process in terms of the offers made by the other party.

6. The bargaining process runs as a sequence of offers and counteroffers started by one party. Thus, in principle bargaining works in the same way as a two-player game such as Chess or Go. Each offer indicates an outcome the offering party is willing to accept. Thus it can be used to reduce the set of acceptable outcomes of the other party. For instance, if the seller offers a discount, then all outcomes with a smaller discount can be neglected. Similarly, if the purchaser offers a price he is willing to pay, the seller can neglect all lower prices.

7. Furthermore, each party may indicate acceptable outcomes to the opposite party without offering them. Such playing with open cards indicates trust in the other party, and is usually used as a means for achieving secondary (non-functional) goals. In the following we will not not consider this possibility, i.e. we concentrate on bargaining with maximal hiding.

In summary, we can characterise bargaining by the bargaining object, constraints for each participating party defining acceptable outcomes, partial orders on the respective sets of possible outcomes, and rules defining the bargaining strategy of each party. In the following we will link these ingredients of a bargaining process to the conceptual model of e-business systems that is offered by the co-design method.

Note that bargaining is significantly different from auctioning system. The latter ones, e.g. the eBay system (see http://www.ebay.com) offer products, for which interested parties can put in a bid. If there is at least one acceptable bid, usually the highest bid wins. Of course, each bidder follows a particular strategy and it would be challenging to formalise them, but usually systems only play the role of the auctioneer, while the bidders are users of the system.

3   The Co-design Approach to Web Information Systems

If bargaining is to become an integral part of e-business systems, we first need a conceptual model for these systems. We follow the co-design approach [10], which we will now summarise briefly. We omit all formal details. We also omit everything that deals with quality criteria, expressiveness and complexity, personalisation, adaptivity, presentation, implementation, etc., i.e. we only look at a rough skeleton of the method. In doing so, we concentrate on the story space, the plot, the views, and the operations on the views:

1. On a high level of abstraction we may define each web information system (WIS) – thus also each e-business system – as a set of abstract locations called scenes between which users navigate. Thus, navigation amounts to transitions between scenes. Each such transition is either a simple navigation or results from the execution of an action. In this way we obtain a labelled directed graph with vertices corresponding to scenes and edges to scene transitions. The edges are labelled by action names or `skip`, the latter one indicating that there is no action, but only a simple navigation. This directed graph is called the *story space*.

2. With each action we may associate a pre- and a postcondition, both expressed in propositional logic with propositional atoms that describe conditions on the state of the system. In doing so, we may add a more detailed level to the story space describing the flow of action. This can be done using constructors for sequencing, choice, parallelism and iteration in addition to the guards (preconditions) and postguards (postconditions). Using these constructors, we obtain an algebraic expression describing the flow of action, which we call the *plot* – in [10] it has been shown that the underlying algebraic structure is the one of a Kleene algebra with tests [5], and the corresponding equational axioms can be exploited to reason about the story space and the plot on a propositional level, in particular for the purpose of personalisation.

3. On a lower level of abstraction we add data support to each scene in form of a media type, which basically is an extended view on some underlying database schema. The database schema, the view formation and the extensions (except operations) are beyond our concern here, so it is sufficient to say that there is a data type associated with each scene such that in each instance of the story space the corresponding value of this type represents the data presented to the user – this is called *media object* in [10]. In terms of the data support the conditions used in the plot are no longer propositional atoms. They can be refined by conditions that can be evaluated on the media objects.

4. Analogously, the actions of the story space are refined by *operations* on the underlying database, which by means of the views also change the media objects. For our purposes it is not so much important to see how these operations can be specified. It is sufficient to consider a predicative specification, i.e. a logical formula that is evaluated on pairs of states of the system.

Let us take a look at our sales example. Here we may have an action *select_item*, which can be executed on a scene presenting a list of goods on offer. The action may first lead to another scene, in which the item and its price are described, and on which an action *select_quantity* may become available. This action leads back to the starting scene presenting the list of goods. The whole selection process can be iterated, each time leading to a change in the database, in which an order is maintained in form of a list of selected items together with their quantity. The user may access this order at any time, change quantities or even delete items on it. This is standard followed by actions such as *enter_shipping_details* and *enter_payment_methods*, and finishing with an action *confirm_order*. Here we would assume that the total price is calculated, the order is taken to a confirmed status in the database, and the user is left to do anything else with the system, presumably leave it.

Bargaining could come in here at any time, but for simplicity let us assume that bargaining is considered to be part of the confirmation process. That is, instead of (or in addition to) the action *confirm_order* we may now have an action *bargain_price* or *bargain_discount* or even more general *bargain_order*. As before, the action may have a precondition, e.g. that the total price before bargaining is above a certain threshold, or the user belongs to a distinguished group of customers. If the bargaining action can be chosen, it will still result in a confirmed status of the order, i.e. the bargaining object, in the database. However, the way this outcome is achieved is completely different from the way other actions are executed. We will look into this execution model in the next section.

Similarly, in our loan example we find actions select_conditions_and_terms and confirm_loan. Again, if bargaining is possible, the selection of terms and conditions may become subject to a bargaining process, which will lead to an instantiated loan contract in the database – same as without bargaining. As before, the outcome of the bargaining is different from the one without bargaining, and it is obtained in a completely different way.

Therefore, in terms of the story space and the plot there is not much to change. Only some of the actions become *bargaining actions*. The major change is then the way these bargaining actions are refined by operations on the conceptual level of media types.

4   Bargaining as a Game

Let us now look at the specification of bargaining actions in view of the characteristics derived in Section 2. We already remarked that we can consider the bargaining process as a two-player game. Therefore, we want to model bargaining actions as games. There are now two questions that are related with this kind of modelling:

1. What is the ground the game is played on? That is, we merely ask how the game is played, which moves are possible, and how they are represented. This of course has to take care of the history that led to the bargaining situation, the bargaining object, and the constraints on it.

2. How will the players act? This question can only be answered for the system player, while a human player, i.e. a customer, is free in his/her decisions. Nevertheless, we should assume that both sides – if they act reasonably – base their choices on similar grounds. The way players choose their moves will be determined by the order on the set of acceptable outcomes and the bargaining strategy.

4.1 Bargaining Games

An easy answer to the first question could be to choose playing on the parameterised bargaining object, i.e. to consider instances of the corresponding data type. However, this would limit the possible moves in a way that no "reconsideration" of previous actions that led to the bargaining situation are possible. Therefore, it is better to play on the "parameterised story space".

More precisely, we consider the algebraic expression defining the plot. However, in this expression we replace the propositional conditions by predicate conditions with free variables, as they arise from the refinement of the story space by media types. Furthermore, the actions in the story space are parameterised as well, which corresponds to their refinement by operations on media types. The parameters subsume the parameters of the bargaining object. For these parameters we have deafult values that result from the navigation through the story space prior to the bargaining action.

Each player maintains a set of static constraints on the parameterised story space. These constraints subsume

– general constraints to the bargaining as defined by law and policies;
– constraints determining the acceptable outcomes of the player;
– constraints arising from offers made by the player him/herself – these offers reduce the set of acceptable outcomes;
– constraints arising from offers made by the opponent player – these offers may also reduce the set of acceptable outcomes.

With these constraints each player obtains a set of possible instantiations that are at least acceptable to him/her. The moves of the players just add constraints. If the set of instantiations reduces to a single element, we obtain an agreement. If it reduces to the empty set, the bargaining has failed.

A move by a player is done by presenting an offer. For the player him/herself this offer means to indicate that certain outcomes might be acceptable, while better outcomes are not aimed at any more. For instance, if a seller offers a discount and thus a total price, s/he gives a way all outcomes with a higher price. For the opponent player the offer means the same, but the effect on his/her set of acceptable outcomes is different.

In order to simplify the model, we may say that a move by a player consists in adding a constraint to the sets of constraints maintained by both players. The constraint can be more complicated than just a threshold on a price. It may involve enabling paths in the story space that represent additional actions, e.g. adding another item to an order. However, this constraint cannot result in an empty set of acceptable outcomes.

In addition to such "ordinary" moves we may allow moves that represent "last offers". A last offer is an offer indicating that no better one will be made. For instance, a total price offered by a seller as a last offer implies the constraint that the price can only be higher. However, it does not discard other options that may consist in additional items at a bargain price or priority treatment in the future. Thus, last offers add stronger constraints, which may even result the set of acceptable outcomes to become empty, i.e. failure of the bargaining process.

Note that this definition of "last offer" differs from tactical play, where players indicate that the offer made is final without really meaning it. Such tactics provide an open challenge for bargaining systems.

In summary, a *bargaining game* consists of a parameterised story space using the model from Section 3 plus a monotone sequence of pairs of constraints on it.

4.2 Bargaining Strategies

By making an offer or a last offer, a player chooses a constraint, the addition of which will result in an acceptable outcome satisfies all constraints arising from counteroffers. In order to make such a choice each player uses a partial order on the set of possible outcomes. Thus, we can model this by a partial order on the set of instances of the parameterised story space. We define it by a logical formula that can be evaluated on pairs of instances.

Then, whenever a player has to make a move, s/he will choose an offer that is not larger than any previous offer, and not smaller than any of the counteroffers made so far. This defines the reasonable offers a player can make. A *bargaining strategy* consists of rules determining, which offer to choose out of the set of reasonable offers. Simple ad-hoc strategies are the following:

- A *tough bargaining strategy* always chooses a maximal element in the set of reasonable offers with respect to the player's partial order. If successful, a tough strategy may end up with an agreement that is nearly optimal for the player. However, a tough strategy bears the risk of long duration bargaining and last counteroffers.
- A *soft bargaining strategy* is quite the opposite of a tough strategy choosing a minimal element in the set of reasonable offers with respect to the player's partial order. Soft strategies lead to fast agreements, but they almost jump immediately to accepting the first counteroffer.

- A *compromise bargaining strategy* aims at an agreement somewhere in the "middle" of the set of reasonable offers. Such an outcome is assumed to be mutually acceptable. The player therefore chooses an offer that lies between this compromise result and a maximal element in the set of reasonable offers, but usually more closely to the compromise than the maximum.

All these strategies are uninformed, as the only information they use are the constraints on the parameterised story space that amount to the set of reasonable offers. They do not take the counteroffers into account.

An informed bargaining strategy aims at building up a model of what is an acceptable outcome of the opponent player. For instance, if a purchaser only offers global discounts, the strategy of the seller might consist of testing, whether the purchaser would accept an increased quantity or additional items instead. If this is not the case, the seller could continue with a compromise bargaining strategy focusing exclusively on the total price. However, if the purchaser indicates that bargaining about an extended order is a possible option, the strategy might be to first increase the order volume before focussing just on the discount.

Informed bargaining strategies require to build up a model of the opponent player in terms of preference rules, thus the must be built on a heuristic inference engine.

5   Conclusion

We presented an approach to model bargaining in e-business systems on the basis of the co-design method [10]. Our conceptual model is that of a two-player game, where one part is played by a user, the one by the e-business system. The game is played on the conceptual model, and the moves of the players represent offers, counteroffers, acceptance and denial. The moves are determined by the characteristics of human bargaining processes such as goals, acceptable outcomes, strategies, secrets, trust and distrust, and preferences.

The work presented so far is only a first step toward a complete conceptual model of bargaining as part of WISs. Our future work aims at completing this model and extending the codesign method correspondingly. This includes stricter mathematical formalisation of the approach, and extensions covering multi-party bargaining, bargaining with more than one role involved, delegation, and authority seeking within bargaining. We believe it will be adavantageous to look at defeasible deontic logic [7] for these advanced goals.

References

1. Bergholtz, M., Jayaweera, P., Johannesson, P., and Wohed, P. Process models and business models - a unified framework. In *Advanced Conceptual Modeling Techniques: ER 2002 Workshops* (2002), vol. 2784 of *LNCS*, Springer-Verlag, pp. 364–377.
2. Dai, X., and Grundy, J. C. Three kinds of e-wallets for a netpay micro-payment system. In *Proceedings WISE 2004: Web Information Systems Engineering* (2004),

X. Zhou, S. Y. W. Su, M. P. Papazoglou, M. E. Orlowska, and K. G. Jeffery, Eds., vol. 3306 of *LNCS*, Springer-Verlag, pp. 66–77.

3. Hinze, A., and Junmanee, S. Providing recommendations in a mobile tourist information system. In *Information Systems Technology and its Applications - 4th International Conference ISTA 2005* (2005), R. Kaschek, H. C. Mayr, and S. Liddle, Eds., vol. P-63 of *Lecture Notes in Informatics*, GI, pp. 86–100.

4. Jøsang, A., and Pope, S. Semantic constraints for trust transitivity. In *Conceptual Modelling 2005 – Second Asia-Pacific Conference on Conceptual Modelling* (2005), S. Hartmann and M. Stumptner, Eds., vol. 43 of *CRPIT*, Australian Computer Society, pp. 59–68.

5. Kozen, D. Kleene algebra with tests. *ACM Transactions on Programming Languages and Systems 19*, 3 (1997), 427–443.

6. Norris, M., and West, S. *eBusiness Essentials*. John Wiley & Sons, Chicester, 2001.

7. Nute, D. *Defeasible Deontic Logic*. Kluwer Academic Publishers, 1997.

8. Orriëns, B., Yang, J., and Papazoglou, M. P. A framework for business rule driven web service composition. In *Conceptual Modeling for Novel Application Domains: ER 2003 Workshops* (2003), vol. 2814 of *LNCS*, Springer-Verlag, pp. 52–64.

9. Schewe, K.-D., Kaschek, R., Wallace, C., and Matthews, C. Emphasizing the communication aspects for the successful development of electronic business systems. *Information Systems and E-Business Management 3*, 1 (2005), 71–100.

10. Schewe, K.-D., and Thalheim, B. Conceptual modelling of web information systems. *Data and Knowledge Engineering 54*, 2 (2005), 147–188.

11. Simon, C., and Dehnert, J. From business process fragments to workflow definitions. In *Proceedings EMISA 2004* (2004), F. Feltz, A. Oberweis, and B. Otjacques, Eds., vol. P-56 of *Lecture Notes in Informatics*, GI, pp. 95–106.

12. Whyte, W. S. *Enabling eBusiness*. John Wiley & Sons, Chicester, 2001.

Conceptual Content Management
for Enterprise Web Services

Sebastian Bossung, Hans-Werner Sehring, and Joachim W. Schmidt

Software Technology and Systems Institute (STS),
Hamburg University of Science and Technology (TUHH)
{sebastian.bossung, hw.sehring, j.w.schmidt}@tu-harburg.de

Abstract. Web services aim at providing an interoperable framework for cross system and multiple domain communication. Current basic standards are allowing for first cases of practical use and evaluation.

Since, however, the modeling of the underlying application domains is largely an open issue, web service support for cross domain applications is rather limited. This limitation is particularly severe in the area of enterprise services which could benefit substantially from well-defined semantics of multiple domains.

This paper focuses on the representation of user-specific domain models and on the support of their coherent interpretation on both client and server side. Our approach is founded on the paradigm of Conceptual Content Management (CCM) and provides support for coherent model interpretation by automatically generating CCM system implementations from high-level domain model specifications. Our approach to CCM has been successfully applied in several application projects.

1 Introduction

In their current state of development, web services implement means for interoperable communication between computational systems and their components. Such achievements are essential to most innovative systems. In the past there have been several architectural attempts to improve component interface definition and their safe use (DCE, CORBA). Although web services have brought several improvements by essentially lifting the matter from the component level to that of reusable and interoperating services, there remain numerous open issues. In particular, web services lack any substantial support for conceptual and semantic modeling since current web service standards deal only with technical aspects such as encoding of messages or addressing of certain operations.

Following the "find and bind" metaphor for web services, one of the major visions is the automatic discovery of services given the client's requirements. For this end, conceptual and semantic models of web services are needed, and if only for semi-automatic approaches to service discovery and safe service interoperability. Since in the case of enterprise web services the involved application models usually span several domains—human resources, finances, customer relations, logistics, etc.—a shared model understanding and coherent use is particularly relevant for the web service providers as well as for their clients [3].

J. Akoka et al. (Eds.): ER Workshops 2005, LNCS 3770, pp. 343–353, 2005.

In this paper we propose to apply Conceptual Content Modeling (CCM) to the domain of web services expecting several benefits. For one, our approach allows an explicit declaration of the domain model concepts and entities (instead of declaring just technical types with little reusability, as in WSDL [17]). Secondly, it supports a coherent interpretation of such domain models by all participating parties through our generative CCM system implementation. Finally, personal extensions of domain models are possible, and support is given to integrate extended interpretations with each other. We have a prototypical implementation of web service support as part of our proven CCM technology.

The remainder of this paper is organized as follows: In section 2 we discuss the shortcomings of plain web service technology with respect to its interpretation of domain models. Section 3 gives a short summary of our CCM approach to domain modeling. Some of the requirements for enterprise services which we feel are not handled well by web service technology are discussed in section 4. In section 5 we present our support for web services by means of CCM. The paper concludes with a summary and some future research directions. Related work is discussed where appropriate.

2   Service Semantics

An important paradigm for dealing with computational artifacts is that of functional abstraction. Abstractions (also with respect to larger artifacts than functions such as classes or components) allow the creator of the artifact to pass along just the information that a user requires. Function signatures greatly enhance interoperability and are used within systems as well as for cross-system communication. Prominent examples of the latter are CORBA interface definitions and the Web Service Description Language (WSDL), but interface models achieve little in providing semantics for a computational entity.

In order to bring together objects in a computer system and their domain semantics, one needs to attach meaning to the former. One way to do this is through ontologies (e.g. [4] with regard to enterprises). These are essentially a vocabulary of terms with some specification of the meaning of each. Such terms have to be combined with the objects in a computer system they describe. Another possibility to describe the relationship of objects in a computer system to their application domain is the use of a conceptual modeling (see e.g. [2, 6]). We will introduce our conceptual modeling approach in the next section.

The web service standards that are in widespread use today (WSDL, UDDI, XML Schema, etc.) mainly deal with the syntax of service communications. Ontologies can be used to define the concepts a web service deals with. OWL-S[1] is an approach to modeling web services by means of an ontology. Modeling frameworks like WSMF (Web Service Modeling Framework, [5]) can then make use of this ontology to connect it to existing web service technology. While we basically agree with this approach, there are several problems with it, mainly

---

[1] http://www.daml.org/services/owl-s/1.1/

centered around the way it causes semantics to be modeled. In particular we would like to emphasize that:

- The domain model is specified from the point of view of the web service. That is, input and output parameters of (possibly several) web services are given, causing a partial domain model to be implicitly defined by the union of all these parameters. Aspects that happen not to be needed for the collection of web services at hand are not modeled, resulting in an incomplete and difficult to understand "domain model".
- Reuse of such definitions is difficult [17], especially with regard to their incompleteness.
- Agreed-upon ontologies seem unlikely in the heterogeneous environment of web services, but subjective views of different communication parties will coexist (also see [9]). In the enterprise environment, this makes costly integration projects necessary [7].
- The proper implementation of the definitions (be they in ontologies or a commonplace web service standard) is still left entirely to the developers at both ends of the communication channel. Code generators for web services can do very little to enforce correct use of the domain model if all they know about this model are some fragments of XML Schema.
- In general, integration of ontologies into the rest of the system is not trivial [10], with regard to the amount of legacy systems found in usual enterprise environments this is even more evident.

In the next sections we detail how we deal with these problems in our Conceptual Content Management Systems (CCMSs).

3   Conceptual Content Management

Entities from the real world—concrete as well as abstract ones—are often described using content of various forms. However, these descriptions can only

```
model BigNewsAgency                     class Person {
from Taxonomy import Classifier           content photo: Image
class Photograph {                        concept
  content reproduction: Image               characteristic name: String
  concept                                   relationship children: Person* ; ...
    relationship depicted: Person* ; ...  constraint
}                                             children.dateOfBirth > dateOfBirth}
class News {
  content text: String                  model EditorSmith
  concept                               import News from BigNewsAgency
    relationship relevantPrs: Person*   class PersonalNews refines News {
    relationship classification:          concept
      Classifier* ; ...                     characteristic author: Person
}                                       }
```

Fig. 1. Simple example of an asset model

be interpreted correctly with a conceptual understanding of the entity's nature. Therefore, entities must be described by a close coupling of content and concept, which we call an *asset*. In this section we will briefly introduce the key concepts of our asset language. More details can be found in [12, 13, 14].

3.1 Assets

Asset class definitions are organized in a model under the keyword model (fig. 1). They describe entities in a dualistic way by means of content-concept pairs. As an example, consider the definition of the class Person. In the content compartment a list of (multimedial) content objects is given. Possible handle types depend on the base language, which is currently Java. Content is treated as opaque by an asset-based system. It could for example be a photographic reproduction for a piece of art or an informal description (i.e., natural language text) in the case of more abstract entities.

The concept compartment provides a conceptual description of the entity. Characteristic attributes are those which are inherent to an entity (such as the person's name in the example), relationships are established between autonomous entities (e.g., the person's children). Note that attributes can be single or multi-valued, as indicated by the asterisk (*). Finally, constraints can be imposed on an asset class which are checked for all its instances. Constraints are defined over Java expressions in the case of characteristics, by relationship navigation, or by asset queries. Standard comparators (equal, less-than, etc.) are interpreted based on a type dependent meaning. Constraint expressions can formed by conjunction, disjunction, and negation.

Definitions from other models can be reused by importing them into a model (from ... import ...). They can then also be used as basis of asset class definitions, which is signified by the refines keyword.

3.2 Openness and Dynamics

Often—if not most of the time—different users will not be able to agree on an asset class definition to model entities from the real world. CCMSs acknowledge this fact by *openness* and *dynamics*. A system is open if its users are able to change the asset definitions in case they do not agree with what they find. A system is dynamic if it can react to such (re)definitions without requiring human (i.e. programmer) intervention.

CCMSs are both open and dynamic, which allows users to have their own subjective view of the world. Nevertheless, a CCMS also supports users in interacting with other users who might have a different model of the same entities. This is done with the help of mappings implemented in modules (section 3.3).

In the example of fig. 1 consider Mrs. Smith who is an editor at BigNews-Agency. She largely agrees with the definition of News by her employer, but would like to add a bit of information in her subjective view. To this end, she creates her own PersonalNews which also stores the author. By means of the

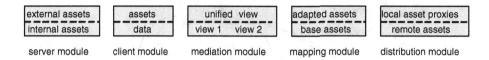

external assets	assets	unified view	adapted assets	local asset proxies
internal assets	data	view 1 view 2	base assets	remote assets
server module	client module	mediation module	mapping module	distribution module

Fig. 2. CCMSs are built from different kinds of modules

CCMS she is able to do so without the help of a developer. The system reacts to her subjective view and incorporates it into its architecture (see next section).

3.3   Architecture

We have developed an architecture similar to the mediator architecture described in [16] which our Conceptual Content Management Systems use. This architecture makes use of five different kinds of modules arranged in layers and combined into components. We refer to such a combination as a system *configuration*. Fig. 2 gives an overview of the modules along with their interfaces. For more details please refer to [13, 14].

Such a system is generated by means of a compiler which is designed as a framework. The framework incorporates a number of generators that create the appropriate modules from asset definitions given by the user. The generators exchange information on their output by an extended kind of symbol table, and the framework synchronizes them according to their dependencies. Being available at runtime, the compiler enables a CCMS to react to user changes of asset definitions and thus plays a crucial role in implementing dynamics. Section 5 gives details on a compiler configuration for creating web-service-enabled systems (with an example in fig. 3). Sample system configurations for web services are also given in that section (specifically fig. 5).

4   Enterprise Conceptual Modeling

Enterprise applications work on large amounts of data which are crucial to their proper operation. These data need to be understood and modeled properly in order to enable the enterprise to do business. This section looks at three aspects of enterprise-wide models: cross-domain cooperation, personalization of domain models, and domain model evolution.

4.1   Cross-Domain Cooperation

Developing enterprise applications consistently requires cross-domain cooperation. Each participating enterprise will typically have at least one model of its application domain [15]. This problem is reflected in the extensive literature on integration problems in enterprise applications (see e.g., [8]). To enable communication, it is beneficiary to have a conceptual domain model for each of the partners. This greatly facilitates integration along the conceptual route and can even be combined with reasoning on ontologies, see [10].

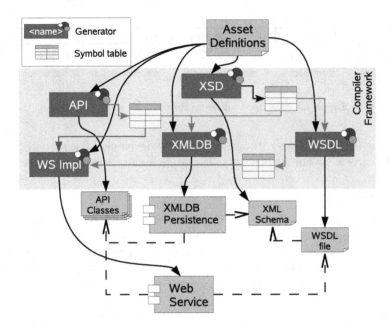

Fig. 3. Compiler Framework

4.2   Different Understanding of Domain Concepts

Even if two enterprises work in the same general area, they normally have slightly
different understandings of their domain, yielding distinct but related models. In
order to enable these enterprises to do business with each other, it is necessary to
put their models into relation. There usually is a provider-requestor relationship
between the business partners (e.g., a seller and a buyer). This aspect makes is
feasible to describe one model as a personalization [11] of the other. A model of a
company might have different personalization relationships with several models
(importing from some, being the basis for others), which is not a problem when
models can be composed of several sub-models.

4.3   Evolution with Respect to Domain Models

Most implementation technologies assume the domain model to be fixed, any
modifications lead to a vast number of changes in almost all parts of the application. Nevertheless, one of the most common things to do for a domain model is
to change. Reasons for this are manifold: Business circumstances might change,
users might not agree to the opinion of others, or new use cases may arise. This
calls for systems which support evolution of domain models in order to (a) allow changes in the first place (preferably in a way where most of the work is
done automatically), and (b) provide clear semantics for such changes to enable
backwards compatibility to the old system.

5   CCM for Enterprise Services

The requirements from the previous section are met by CCMSs whose modules are generated and configured accordingly. In our approach the basis for this is not a computational model of the web services' implementation, but a common domain model which describes the entities in the enterprise. The invocational details are left to standard web service technology.

Users wishing to set up a component participating in a CCMS use such a domain model or a personalized variant of it. The properties discussed in the previous section are available to asset definitions in form of the asset language. In this section we illustrate this by giving a small sample asset model. From such definitions CCMSs with the desired properties are generated. We will discuss some basic configurations of such CCMSs and argue their benefits.

5.1   A Sample Asset Model

Fig. 4 shows a model that is defined on top of the model BigNewsAgency introduced in fig. 1. In that model classes News and Person have been defined as part of the model of a news agency. From some other given model Taxonomy a class Classifier has been made available to classify instances of News.

```
model TabloidPress
import News from BigNewsAgency
class News {
  constraint text.length () <= 5000 and
    classification <= lookfor Classifier { name = "Triviata" }
}
```

Fig. 4. Personalized view of a client of BigNewsAgency

A user working for Tabloid Press might use the model BigNewsAgency in conjunction with a new model TabloidPress shown in fig. 4. This model gives a new definition of the class News. For such news there now is a constraint which requires news to be at most 5000 characters long (using a method of the standard Java class java.lang.String) and to be classified under the Classifier "Triviata" (with the operator "<=" interpreted as "subset of", its default interpretation for multi-valued relationships).

When starting the asset model compiler it receives both the models BigNewsAgency and TabloidPress. It will generate modules for a configuration where the new system holds instances of News, Person, and Classifier. Having access to the original model the CCMS might be configured to interoperate with other systems which are based on the model BigNewsAgency, e.g., to take over existing News and to check whether they match the new definition. Likewise, Classifiers might be retrieved from a remote system.

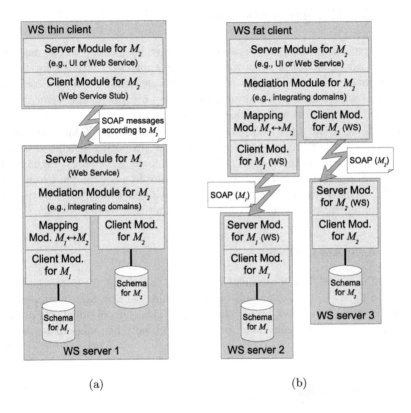

(a)   (b)

Fig. 5. CCMS configurations for enterprise services. (a) Three-tier configuration with application server and thin client, (b) Two-tier configuration with fat client.

5.2   Sample CCMS Configurations

Fig. 5 shows two abstract sample configurations of CCMSs which meet the three requirements outlined in section 3—cross-domain cooperation, personalization of domain models, and their evolution. The asset models M_1 and M_2 shown in the examples represent two domain models. Depending on the demands to be met they can fill various roles as will be explained below.

Both systems outlined in fig. 5 show a client-server scenario incorporating two components—one offering a service, and the other using it. Both client components include a topmost server module by which the system is accessed. In the case of an interactive client this server module would be a graphical user interface, a web server, or the like. If the component is itself offering a web service the server module will implement a web service interface. The client components are based on a web service client module which sends request to components providing a web service.

The lower part of each of the systems shown in fig. 5 constitutes such service providers (WS server 1...3). Both include a server module to accept web service requests and a client module to access a third-party component. Instead of the latter, web service client modules could be used if other services are used by the service at hand.

The system shown in fig. 5(a) shows a typical three-tier scenario in which a client component accesses a service on request. On the server side such a request is analyzed by a mediation module as part of the application logic and is delegated to a subsystem which handles either assets of a domain model M_1 or a model M_2.

Fig. 5(b) shows a configuration where the logic of using several servers and combining assets resides on the client ("fat client"). Depending on the assets involved, a web service request is sent to one of the server components WS server 2 or WS server 3 which host assets of models M_1 or M_2 respectively.

The asset models M_1 and M_2 shown in the examples represent two domain models. Depending on their relationship different scenarios are realized. Examples of models are:

1. M_1 and M_2 represent two domains and M_1 is integrated into M_2; assets from M_1 are possibly adapted before they can be integrated into M_2
2. M_2 is a personalized variant of M_1; assets in the public view represented by M_1 are lifted to M_2 by a mapping module
3. M_1 and M_2 are revisions of a model, where M_2 is the successor revision of M_1; by means of adaptation the assets from the outdated model are integrated into the newer version

Thus, the examples shown in fig. 5 exemplify CCMSs with the contributions outlined in section 4.

Other configurations of CCMSs than the ones indicated by fig. 5 cover different scenarios. E.g., assets from that part of the component WS server 1 that handles assets of M_2 could be adapted by an additional mapping module before being integrated into one model through the mediation module.

5.3   Benefits of Systems Generation

More complete systems than those in the examples from the previous section incorporate more layers in which mediation and mapping modules define the application logic. In enterprise applications there will be a number of components which use each other's services. The service to use is chosen by a mediation module as shown in the client component of fig. 5(b).

The key to the realization of the discussed system features—domain interoperation, personalization, and evolution—are the properties of openness and dynamics. These features allow servers and clients to share models of the domain at hand while still being able to change the definitions. This is especially important in enterprise systems as integration is expensive [7].

As pointed out in section 3 dynamics is achieved by generating all parts of a CCMS from asset definitions using the asset model compiler. As all components are generated on the basis of the same domain model, they interpret requests coherently. This ensures that modules which form components of a system are created in a fashion that allows them to interoperate: Either they share the same domain model, or there are differences which some user stated explicitly. In the latter case mapping modules which allow the interoperation of components based on variants of a common domain model are derived from constraints describing model interrelations.

In the case of web services, generation covers interface definitions in WSDL as well as web service implementations in the form of server and client modules. The WSDL declarations contain XML schema information for the types involved which match the asset definitions. The web service operations are determined by the generators and implemented by corresponding configurations.

The asset model compiler framework has been introduced in fig. 3. This figure also shows several generators for the parts of a CCMS related to web services. A WSDL specification matching the asset class definitions is produced by the WSDL generator, while the server and client modules to provide and access services are created by the generator named WS Impl. The WSDL generator in turn uses XML Schema definitions created by the XSD generator which is also used, e.g., to define the schema of an XML database. As every module conforms to the uniform module API of CCMSs the generator for the web service server and client modules also uses the API definitions given by the API generator.

6   Summary and Outlook

To achieve interoperability through web services, semantic and conceptual models are highly important. We have shown that in this respect much can be gained by a coherent domain model which should be implemented in an open and dynamic way. Specifically, domain model entities should not just be modeled with respect to the service operations in which they are used. Supporting an application-wide approach through Conceptual Content Management has substantial benefits, which we have outlined in this paper.

In the future we plan to enhance web service support for CCMS in several ways. We will improve the coupling of web services (and possibly other technologies) with CCMS, for example in order to integrate legacy systems. Furthermore, the integration of standard web services (i.e., web services that are not implemented in a CCM enabled system) needs to be addressed, especially with regard to interoperability in an enterprise services environment. In a different area, we are also working on bringing together ontology-based semantic models, description logics and CCMS. We see two approaches to be pursued: firstly, description logics might be used to express the constraints in asset classes. The implications of changing semantics from closed-world to open-world will have to be explored. Secondly, with a mapping from asset class definitions to terms in ontologies [1] it would be possible to reason on the level of classes in contrast to performing model checking for given instances. This will require a mapping from asset classes to terms in ontologies. We aim to employ description logics to then perform reasoning in these systems.

References

1. Alex Borgida and Ronald J. Brachman. *The Description Logic Handbook: Theory, Implementation, and Applications*, chapter Conceptual Modeling with Description Logics, pages 349–372. Cambridge University Press, 2003.
2. Michael L. Brodie, John Mylopoulos, and Joachim W. Schmidt, editors. *On Conceptual Modelling: Perspectives from Artificial Intelligence, Databases, and Programming Languages*. Topics in Information Systems. Springer-Verlag, 1984.
3. Dov Dori. The Visual Semantic Web: Unifying Human and Machine Knowledge Representations with Object-Process Methodology. In *Proc. 1st Int. Workshop on Semantic Web and Databases*, 2003.
4. Fadi George Fadel, Mark S. Fox, and Michael Gruninger. A Generic Enterprise Resource Ontology. In *Proc. 3rd Workshop on Enabling Technologies: Infrastructure for Collaborative Enterprises*, 1994.
5. Dieter Fensel and Christoph Bußler. The Web service modeling framework (WSMF). In *Database and Information Research for Semantic Web and Enterprises*, 2002.
6. Richard Hull and Roger King. Semantic Database Modeling: Survey, Applications, and Research Issues. *ACM Computing Surveys*, 19(3), 1987.
7. Jinyoul Lee, Keng Siau, and Soongoo Hong. Enterprise Integration with ERP and EAI. *CACM*, 46(2), 2003.
8. David S. Linthicum. *Next Generation Application Integration: From Simple Information to Web Services*. Addison-Wesley, 2004.
9. Peter Mika, Daniel Oberle, Aldo Gangemi, and Marta Sabou. Foundations for Service Ontologies: Aligning OWL-S to DOLCE. In *Proc. of the 13th Int. WWW conf. 04*. ACM, 2004.
10. Boris Motik, Alexander Maedche, and Raphael Volz. A Conceptual Modeling Approach for Semantics-Driven Enterprise Applications. In *Proc. of DOA/CoopIS/ODBASE 2002*, pages 1082–1099, London, 2002. Springer-Verlag.
11. Gustavo Rossi, Daniel Schwabe, and Robson Guimares. Designing personalized web applications. In *Proc. WWW 01*, pages 275–284, New York, NY, USA, 2001. ACM Press.
12. Joachim W. Schmidt and Hans-Werner Sehring. Conceptual Content Modeling and Management: The Rationale of an Asset Language. In *Proc. PSI 03*, volume 2890 of *LNCS*, pages 469–493. Springer, 2003.
13. Hans-Werner Sehring. *Konzeptorientiertes Content Management: Modell, Systemarchitektur und Prototypen*. PhD thesis, Hamburg University of Science and Technology (TUHH), 2004.
14. Hans-Werner Sehring and Joachim W. Schmidt. Beyond Databases: An Asset Language for Conceptual Content Management. In *Proceedings of the 8th ADBIS*, volume 3255 of *LNCS*, pages 99–112. Springer-Verlag, 2004.
15. René van Buuren, Henk Jonkers, Maria-Eugenia Iacob, and Patrick Strating. Composition of Relations in Enterprise Architecture Models. In *LNCS*, volume 3256, pages 39 – 53, 2004.
16. G. Wiederhold. Mediators in the Architecture of Future Information Systems. *IEEE Computer*, 25:38–49, 1992.
17. Jian Yang and Mike. P. Papazoglou. Web Component: A Substrate for Web Service Reuse and Composition. In *LNCS*, volume 2348, page 21, 2002.

Verifying Web Services Composition[*]

YanPing Yang[1], QingPing Tan[1], and Yong Xiao[2]

[1] Computer College of National University of Defense Technology, China
yanpingyang@nudt.edu.cn
[2] National Lab of Parallel Distributed Process, China
yongxiao@nudt.edu.cn

Abstract. Current Web services composition proposals, such as BPEL, BPSS or WSCI, provide notations for describing the control and message flows in service collaborations. However, they remain at the descriptive level, without providing any kind of mechanisms or tool support for verifying the composition specified in them. In this paper, we present an approach based on CP-net formalism to analyze and verify Web services composition. We provide translation scheme from composition language into CP-nets and the techniques to analyze and verify effectively the CP-nets to investigate several behavioral properties. Our approach is essentially independent of the language describing composition. As an example, to show the effectiveness of our technique, in this paper, we present the transformation of WSCI to CP-nets, which can be analyzed, verified and simulated as prototypes of WSCI models by the CP-net tools.

1 Introduction

A Web service is a software system designed to support interoperable machine-to-machine interaction over a network. There might be frequently the case that a Web service does not provide a requested service on its own, but delegates parts of the execution to other Web services and receives the results from them to perform the whole service. In this case, the involved Web services together can be considered as Web services composition.

There is a need for a language to describe how various Web services are composed. A landscape of languages (BPEL, BPSS, WSFL and WSCI) has been proposed and no one stands out as yet. By using one of the languages, we can express composition descriptions as programs. However, the descriptions may contain errors, because we probably write such programs in an ad-hoc manner. Executing a buggy program, consumes tremendous amount of network resources which are shared publicly. Thus, verifying the Web service composition prior to its execution in the Internet is mandatory.

In this paper, we're interested in how much the Colored Petri Nets (CP-nets) [2] analysis and verification techniques can be used as a basis for raising reliability of Web services composition.

---

[*] The paper is supported by the National Grand Fundamental Research 863 Program of China under Grant No.2003AA001023.

J. Akoka et al. (Eds.): ER Workshops 2005, LNCS 3770, pp. 354–363, 2005.
© Springer-Verlag Berlin Heidelberg 2005

Most of existing approaches to verify business process are based on model checking techniques. Using Petri nets to model and verify business processes is another choice, and the discussion of the advantages and the disadvantages of the two approaches are out of scope.

For an overview of modeling business processes by means of Petri nets, we refer the reader to, for example, the work of Van der Aalst [6], Martens [5], Stahl [7] and Narayanan [4]. As we say above, because CP-nets combine the strengths of Petri nets with the expressive power of high-level programming languages, we claim that using the colored token of CP-nets to model different message and event type of business process are more natural.

We have analyzed and verify the composition written in Business Process Execution Language (BPEL) (*http://www-128.ibm.com/developerworks/library/ws-bpel/*) in [1]. In this paper, we want to extend our work to Web services composition written in Web service Choreography Interface (WSCI) (*http://ifr.sap.com/wsci/speci fication/wsci-spec-10.htm*). We present the transformation rules from WSCI to CP-nets in a constructive way. Therefore we can translate WSCI composition into CP-nets, which can be analyzed and verified by the existing specialized CP-net tools such as Design/CPN (*http://www.daimi.au.dk/designCPN/*) and CPN tools (*http://www.da imi.au.dk/CPNtools/*). So we make tool support available for and analyzing and verifying WSCI composition.

2   Backgrounds

WSCI addresses Web services composition from two primary levels. At the first level, WSCI builds up on the WSDL *portType* capabilities to describe the flow of messages exchanged by a Web service. The *interface* construct introduced by WSCI permits the description of the externally observable behavior of a Web service, facilitating the expression of sequential and logical dependencies of exchanges at different operations in WSDL *portType* element. At the second level, WSCI defines the *model* construct, which allows composition of two or more WSCI *interface* definitions (of the respective Web services) into a collaborative process involving the participants represented by the Web services. WSCI calls this *global model*, which provides a global, message-oriented view of the overall process.

CP-nets were formulated by Jensen [2] as a formally founded graphically oriented modeling language. CP-nets are useful for specifying, designing, and analyzing concurrent systems. In contrast to ordinary Petri nets, CP-nets provide a very compact way of modeling complex systems, which makes CP-nets a powerful language for modeling and analyzing industrial-sized systems. This is achieved by combining the strengths of Petri nets with the expressive power of high-level programming languages. Petri nets provide the constructions for specifying synchronization of concurrent processes, and the programming language provides the constructions for specifying and manipulating data values.

Practical use of CP-nets has been facilitated by tools to support construction and analysis of systems by means of CP-nets. Design/CPN and CPN tools are two outstanding of them. In this paper, we use CPN tools to illustrate our work.

Table 1. Behavior Properties of CP-nets

Property	Original Meaning	Meaning in Verification
Reachability	The possibility of reaching a given state	Whether it is possible for a process to achieve the desired result.
Boundness	The maximal and minimal number of tokens which may be located on the individual places in the markings.	If a place is CP, then the number of tokens it contains is either o or 1, otherwise this indicates errors. If a place is a MP, then boundedness can be used to check whether the buffer overflows or not.
Dead Transitions	The transitions which will never be enabled.	There are no activities in the process that cannot be realized. If initially dead transitions exist, then the composition process was bad designed.
Dead Marking	Markings having no enabled binding element.	The final state of process instance is one of dead marking. If the number of dead markings reported by state space analysis tool is more than expected, then there must be errors in the design.
Liveness	A set of binding elements remaining active.	It is always possible to return to a activity if we wish. For instance, this might allow us to rectify previous mistakes.
Home	About markings to which CP-net is always possible to return	It is always possible to return to a state before. For instance, to compare the results of applying different strategies to solve the same problem.
Fairness	How often the individual transitions occur.	Fairness properties can be used to show the execution numbers in each process. We can find the dead activity that will never be executed.
Conservation	Tokens never destroyed	Certain tokens are never destroyed. Hence, resources are maintained in the system.

The properties of CP-nets to be checked include boundness, deadlock-freedom, liveness, fairness, home, and application specific properties. The application specific properties are expressed as reachability of CP-nets. All the properties have their specific meaning in verifying Web services composition (cf. Table 1).

CP-net models can be structured hierarchically. This is particularly important when dealing with CP-net models of large systems. The basic idea underlying hierarchical CP-nets is to allow the modeler to construct a large model from a number of smaller CP-nets called pages. These pages are then related to each other in a well-defined way.

In a hierarchical CP-net, it is possible to relate a so-called substitution transition (and its surrounding places) to a separate CP-net called a subpage. A subpage provides a more precise and detailed description of the activity represented by the transition. Each subpage has a number of port places and they constitute the interface through which the subpage communicates with its surroundings. To specify the relationship between a substitution transition and its subpage, we must describe how the port places of the subpage are related to so-called socket places of the substitution transition. This is achieved by providing a port assignment. When a port place is assigned to a socket place, the two places become identical.

3 Transforming WSCI to CP-Nets

The aim of this section is to provide a transformation from WSCI specification to CP-nets in a constructive way. The overview of the transformation idea can be concluded as follows:

✓ The *interface* element describes the WSCI view of a Web service participating in a choreographed, long-lasting and stateful message exchange with other services. Each *interface* is represented by a CP-net called Interface Net (I-Net).

✓ Messages are represented by tokens. Different message type can be represented by the products type of the component part type of messages.

✓ A WSCI *activity* is usually mapped to a CP-net transition. We do so for several reasons. First, mapping subactivities into places poses the following problem: if a place represents a subactivity state, when the actor returns the subactivity will start again. Thus, to represent the leaving point where a subactivity continues would be impossible. Secondly, the hierarchical modeling technique is by means of substitution transitions, and therefore if a transition represents a subactivity, there always remains the possibility of decomposing it into various actions (other transitions) and resting points (places) that enable interruptions and returns. Thirdly, modeling subactivities by transitions allows us to model data flow in the places of the subactivity flow more clearly. The detailed transformations see section 3.2.

✓ The control flow relations between activities specified by WSCI semantics are captured with CP-nets token firing rules and the arc inscriptions and transition guard expressions.

✓ Each WSCI *model* is represented by a hierarchical CP-net called Composition Net (C-Net).

An example used for further discussion is a benchmark Ticket Reservation process within a virtual enterprise comprising a Traveler, a Travel Agent, and an Airline in WSCI specification (*http://ifr.sap.com /wsci/specification /wsci-spec-10.htm*).

The Traveler planning on taking a trip decides the destination and calendars and submits her choice to a Travel Agent. The Agent finds the best travel plan and asks the Airline to verify the availability of seats. The Airline provides information about the availability of seats. The Traveler provides the Agent with her Credit Card information to properly book the seats. She also provides her contact information for the Airline to receive an e-Ticket. The Airline books the seats and issues e-tickets to the Traveler. Finally, the Agent charges the Traveler's Credit Card and sends the notification of the charge.

3.1 Interface Net

WSCI aims to describe how Web services participate in choreographed, long-lasting and stateful message exchanges. The focus of the described behavior is on the temporal and logical dependencies among the messages the Web service exchanges with one or more other services in the context of a given scenario. WSCI maps this description to the notion of an interface. We transform an interface into an I-Net.

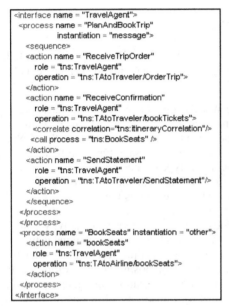
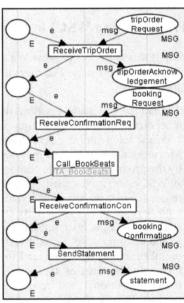

Fig. 1. We give a transformation example: (a) gives a simplified WSCI *interface* definitions of TravelAgent Web service and it can be transformed into the I-Net as (b). In (b), a circle place is the graphical notation for a control place, representing a state of the web service with respect to possible message exchanges it can be involved in; an ellipse place is the graphical notation for a message place. A token available on a message place represents the reception/production of a message.

I-Net is a hierarchical CP-net where:

✓ *Places are of three different types, specifically control places, referred to as CP, input message places, referred to as IMP, and output message places, referred to as OMP; let us define MP = IMP $\cup$ OMP and P = CP $\cup$ MP;*

✓ *Transitions are of three different types. The fist type is auxiliary transition, referred to as AUT, which is used to implement composite construct such as loop or conditional fork. The second type is substitution transition, referred to ST, which is abstract representation of subpages modeling sub processes. The third type is activity transition, referred to as ACT; let us define T= AUT $\cup$ ST $\cup$ ACT;*

✓   *Tokens placed on control (input/output message) places are referred to as control (message, respectively) tokens;*
✓   *Arc connected with control (input/output message) places are referred to as control (message, respectively) arcs.*
✓   *Each place∈MP is labeled with a message, i.e., a function mess: MP→ε is defined (ε is the set of messages specified for the Web service, according to the formalization provided in [3])*

The interface in Fig.1(a) involves two top-level processes PlanAndBookTrip and BookSeats. The former is the main process and the latter will be called by the former as a reuse unit. The PlanAndBookTrip process starts by receiving a TripOrder message sent by the traveler. Then, the traveler sends a BookRequest to the Travel Agent containing her credit card information to finalize the reservation. Next, the Travel Agent invokes the bookSeats operation of the Airline service to finally book the seats. The Travel Agent sends a ReserveResult message to the traveler as an acknowledgement. Last, the TravelAgent send traveler the payment information.

This scenario can be transformed into the I-Net as in Fig.1(b). Taking transition ReceiveOrder as an example, upon receiving the message tripOrderRequest, service does some work and sends the message tripOrderAckowledgement as output; at this point, service is ready to accept new input messages. When a token is available on the upper circle place and another one is available on the upper square place, the transition can fire, thus moving a token on the lower circle place and another one on the lower ellipse place.

3.2 Activity Transformations

WSCI describes the behavior of a Web service in terms of choreographed activities. Activities may be atomic or complex, i.e. recursively composed of other activities.

Atomic Activity Transformation. Atomic activities represent the basic unit of behavior of a Web service. The most important atomic activities include actions dealing with messages corresponding to the execution of operations defined in static service definition languages such as WSDL. They can be associated with one of the following types of WSDL operations:

✓   One-way: The action performed by the service receives a message.
✓   Request-response: The action performed by the service receives a message and sends a response back to the sender.
✓   Notification: The action performed by the service sends a message to another service.
✓   Solicit-response: The action performed by the service sends a message to another service and waits for a response.

We list the atomic activity transformations in Table 2.

Table 2. Atomic Activity Transformation

Activity	Sample	Transformation

One-way	`<action name = "ReceiveReservation"`
	`  role = "tns:travelAgent"`
	`  operation = "tns:TAtoTraveler/ReserveTickets">`
	`<operation name="CancelReservation">`
	`<input message`
	`="defs:reservationCancellationRequest"/>`
	`</operation>`	

Request-response	`<action name = "ReserveTickets"`
	`  role = "tns:Traveler"`
	`  operation                =`
	`"tns:TravelerToTA/ReserveTickets">`
	`<operation name = "ReserveTickets">`
	`  <input message = "defs:reservationRequest"/>`
	`  <output message =`
	`    "defs:reservationConfirmation"/>`
	`</operation>`	

Notification	`<action name = "SendStatement"`
	`  role = "tns:TravelAgent"`
	`  operation = "tns:TAtoTraveler/SendStatement"/>`
	`<operation name = "SendStatement">`
	`  <output message = "tns:statement"/>`
	`</operation>`	

Solicit-response	`<action name = "OrderTrip"`
	`  role = "tns:Traveler"`
	`  operation = "tns:TravelerToTA/OrderTrip"/>`
	`<operation name = "OrderTrip">`
	`  <output message="defs:tripOrderRequest"/>`
	`  <input message =`
	`    efs:tripOrderAcknowledgement"/>`
	`</operation>`	

Complex Activities. Complex activities are recursively composed of other activities, and WSCI supports the definition of the following kinds of composition:

✓ The complex activity *sequence* operates on a single activity set and performs the activity set exactly once in sequential order.

✓ The complex activity *all* operates on a single activity set and performs the activity set exactly once in non-sequential order, possibly parallel.

✓ The complex activity *choice* and *switch* selects one activity set based on a triggered event and a condition respectively, and performs that activity set exactly once in sequential order.

✓ The complex activity *foreach, until and while* operates on a single activity set and performs the activity set repeatedly.

The transformations of most frequently used complex activity are shown in Table 3.

Table 3. Complex Activities Transformation

Activity	Sample	Transformation

Sequence	`<sequence>`
	`<action name = "CancelReservation"`
	`  role = "tns:TravelAgent"`
	`  operation =`
	`ns:TAtoAirline/RequestCancellation"/>`
	`<action name = "ReceiveCancellationNotification"`
	`  operation = tns:TAtoAirline/AcceptCancellation">`
	`</action>`
	`<action name = "NotifyOfCancellation"`
	`  operation=`
	`  "tns:TAtoTraveler/NotifyOfCancellation"/>`
	`</sequence>`	

All	`<all><action name = "ReceiveTickets"`
	
	`operation=tns:TravelerToAirline/ReceiveTickets">`
	`</action>`
	`<action name = "ReceiveStatement"`
	`  operation= tns:TravelerToTA/ReceiveStatement">`
	`</action></all>`	

Choice	`<choice> <onMessage>`
	`<action name = "ReceiveCancellationRequest"`
	`  operation="tns:AirlineToTA/CancelReservation">`
	`</action> </onMessage>`
	`<onMessage>`
	`<action name="PerformBooking"`
	`  role ="tns:Airline"`
	`  operation="tns:AirlineToTA/BookSeats">`
	`</action>`
	`<action name="SendTickets" role ="tns:Airline"`
	`  operation="tns:AirlineToTraveler/SendTickets"/>`
	`</onMessage>`
	`</choice>`	

switch	`<switch> <case>`
	`<condition>tns:cancelItinerary</condition>`
	`<action name = "CancelItinerary"`
	`  operation = "tns:TravelerToTA/CancelItinerary"/>`
	`</case>`
	`<default>`
	`<action name = "ReserveTickets"`
	`  operation =`
	`"tns:TravelerToTA/ReserveTickets"/>`
	`</default>`
	`</switch>`
		=

| while | ```
<while name="ReserveSeats">
  <condition>defs:notLastSeat</condition>
  <action name="ReserveNextSeat"
    operation="tns:AirlineToTA/ReserveSeat">
  </action>
</while>
``` | 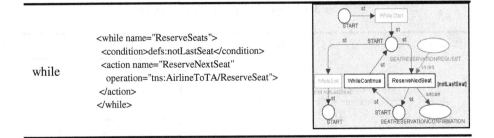 |

3.3 Composition Net

WSCI describes the coordination by means of the WSCI *Model*. The global *Model* is described by a collection of interfaces of the participating services, and a collection of links between the operations of communicating services. Links between operations indicate that the respective services will exchange messages across those links.

Each *Model* can be represented as an Composition Net, which is a specific net connecting at least two I-Nets, and specifying the routing of messages and the act of passing the task of the orchestration from an organization to another one.

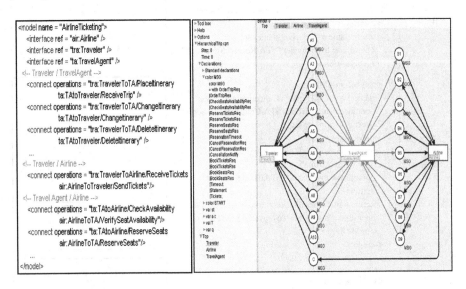

Fig. 2. We give a global model transformation example. The WSCI global *Model* presented in (a) contains references to the interfaces of the three participants in the Ticket Reservation Process. The global model in (a) can be transformed into the C-Net as (b).

Composition Net (C-Net) is a hierarchical CP-net where:

✓ *Places are of one type, specifically message places. Each place will be assigned to input/output message places belong to different I-Nets.*

✓ *Transitions are also of one type, specifically organization transitions. Each transition is an abstract representation of I-Net in the superpage .They are labeled with an organization; the availability of a token in an place connecting the organization transition means that the task of the composition of the overall process is currently assigned to the organization labeling the transition.*

✓ *For each place p, there exist omp∈ OMP and imp∈ IMP such that p, omp and imp are members of one fusion set, meanwhile omp and imp necessarily belong to different I- Nets.*

Fig.2 shows a global model transformation example. The Global Model describes how the PlaceItinerary operation defined within the TravelerToTA port type and the ReceiveTrip operation defined within the TAtoTraveler port type are connected and, together, describe how the trip order is sent by the Traveler and received by the Travel Agent. The port assignment between C-Net and its subpages I-Nets are implemented by the connect specification in the WSCI global *Mode*.

4 Conclusions

In this paper, we introduce an approach to verify and analyze Web services composition. We pick up WSCI and present the transformation algorithms from WSCI to CP-nets. These generated CP-net models can be analyzed, verified and simulated as prototypes of the former by many existing and specialized analysis and verification tools.

References

1. Y.P. Yang, Q.P. Tan, J.S. Yu, Feng Liu, Transformation BPEL to CP-Nets for Verifying Web services Composition, the International Conference on Next generation Web services Practices (NWeSP'05), IEEE Computer Society, Seoul, Korea, August 2005.
2. K. Jensen, "Colored Petri Nets Basic Concepts, Analysis Methods and Practical Use", Volume 1, 2 and 3, second edition, 1996.
3. M. Mecella, B. Pernici, and P. Craca, "Compatibility of Web services in a Cooperative Multi-Platform Environment", in Proceedings of VLDB-TES 2001, Rome, Italy, 2001.
4. S. Narayanan and S.A. McIlraith. Simulation, verification and automated composition of Web services. In Proceedings of the 11th International World Wide Web Conference, ACM, Honolulu, May 2002, pages 77–88.
5. A. Martens. Distributed Business Processes - modeling and Verification by help of Web services, PhD thesis, Humboldt-University at zu Berlin, July 2003, available at http://www. informatik.hu-berlin.de/ top/download/documents/pdf/Mar03.pdf.
6. W.M.P.van der Aalst and K.M.van Hee, Workflow Management: Models, Methods, and Systems. The MIT Press, 2002.
7. Christian Stahl. Transformation von BPEL4WS in Petrinetze. Diplomarbeit, Humboldt-UniversitÄat zu Berlin, April 2004.

Towards Amplifying Business Process Reuse

Sergiy Zlatkin and Roland Kaschek

Department of Information Systems,
Massey University, New Zealand
{S.Zlatkin, R.H.Kaschek}@massey.ac.nz

Abstract. This paper proposes an approach to the reuse of business
processes. We consider business process reuse as a problem of software
economy. An infrastructure component that we call Process Assembler is
supposed to aid in business process reuse. That component together with
a number of other components could realize an environment for assessing,
pricing and trading business processes. We work out the key functionality
that needs to be provided by a software component that aids in reusing
business processes. Furthermore we discuss a high level architecture for
the Process Assembler and argue that this software component can aid
users in performing the key reuse functionality.

1 Introduction

Many organizations, be they companies or non-profit organizations, require key
composite activities be modeled as so-called business process, i.e. ([32]), "a set
of one or more linked procedures or activities which collectively realize a business objective or policy goal, normally within the context of an organizational
structure defining functional roles and relationships". Such a process describes
for example the production of goods or the way services are provided. A process consequently contains specifications of the material- or human resources,
the know-how, and the know-what that is needed for good creation or service
provisioning.

We presuppose a broad view of business process reuse as the use of any kind of
development artifacts for a non-intended purpose. We identify two main areas of
business process reuse as particularly important. Firstly, using existing artifacts
for creating new ones. Secondly, using the existing artifacts for educating staff.
We concede that the impact of both of these points is limited (compare for
example [10]). However, we expect the development time to decreased and the
process quality to increased if business processes could be reused easily. Business
process reuse could be significantly simplified if an easy to use and extensible
reuse- environment would be available.

Effectively reusing software artifacts is to a large extent a problem of the
economy of creation, use, and maintenance of such artifacts. We therefore think
the existence of a market for business processes would increase the production
of reusable business processes as well as the use of such processes. Currently no
market seems to exists for trading business processes. Such a market requires the

J. Akoka et al. (Eds.): ER Workshops 2005, LNCS 3770, pp. 364–374, 2005.
© Springer-Verlag Berlin Heidelberg 2005

existence of an environment in which the process quality can be assessed, business deals negotiated and prices determined. Our Process Assembler is supposed to be a software component that provides key functionality for business process reuse. Its model-based architecture enables for simple extensions by new languages for modeling business processes or tools for process analysis. A business process bazaar, i.e., a portal in which business processes actually can be traded is a much more complex environment than the Process Assembler. We restrict this paper to discussing the Process Assembler.

Paper structure: in the next section we discuss related work. In section 3 we discuss reuse of business processes. We work out and introduce an architecture for the Process Assembler in section 4. We conclude the paper with an outlook on our future work in section 5 and list then our references.

2 Related Work

There exists a number of catalogs of business processes such as the one[1] of OA-SIS (Organization for the Advancement of Structured Information Standards) and UN/CEFACT. We ignore most of these here, as many of them appear to be paper-ware only. However, the MIT Process Handbook Project (see, e.g. [21], [23]) has implemented a process repository that deserves particular attention since it comes with thousands of process types that could help staff in organizations to create new ideas for business processes, goods, services or business models. The repository implements a simple and unsatisfactory search function, i.e., only search for names is supported. The repository additionally implements an interesting navigation function. Navigation takes place in two dimensions. Firstly, a level of abstraction may be choosen from which a business process may be viewed. Secondly, links may be followed that connect business processes.

Van der Aalst et al (see, e.g. [1], [2] and [3]) have studied so-called workflow patterns, i.e., frequently occurring operators for combining new workflows from given ones. The patterns they analyzed were categorized as: Basic Control Patterns, Advanced Branching and Synchronization Patterns, Structural Patterns, Patterns Involving Multiple Instances, State-based Patterns, Cancellation Patterns. Those patterns may be applied for analyzing, understanding process types. Workflow patterns may be used for modifying process types (especially from control perspective) and assembling new process types as a set of process type constructors.

The Workflow Management Coalition (WFMC) is an international non-profit organization committed to the development of standards regarding workflow technology. In its workflow reference model (for details see e.g. [1], [32]) interfaces for data exchange between different components of workflow management systems are contained. The interfaces 1 and 4 are related to our work. Interface 1 connects a workflow enactment service, i.e. workflow engine, with a modeling tool. This interface is relevant if processes from a repository are shipped to a

---

[1] See http://www.ebxml.org/specs/bpProc.pdf

workflow engine in its internal format. Interface 4 is relevant if prior to shipping a translation into the internal format of the target workflow engine has to be performed. The former sometimes is called design-time interoperability and the latter run-time interoperability. The WFMC did not develop a special translator; rather they proposed several standard languages (Meta-types) such as WPDL (WFMCs process definition language, see e.g. [33]) for interface 1 and Wf-XML (see e.g. [31]) for interface 4 respectively. It would be useful to incorporate those Meta-types to Process Assembler in future as they are supported by many systems and tools.

3   Reusing Business Processes

Our starting point are the "facts" 15 to 20 of [10] "Reuse-in-the-small (libraries of subroutines) began nearly 50 years ago and is a well-solved problem. ... Reuse-in-the-large (components) remains a mostly unsolved problem, even though everyone agrees that it is important and desirable. ... Reuse-in-the-large works best in families of related systems and thus is domain dependent. This narrows the potential applicability of reuse-in-the-large. ... There are two 'rules of three' in reuse (a) it is three times as difficult to build reusable components as single use components, and (b) a reusable component should be tried out in three different applications before it will be sufficiently general to accept into a reuse library. ... Modification of reused code is particularly error-prone. If more than 20 to 25 percent of a component is to be revised, it is more efficient and effective to rewrite it from scratch. ... Design pattern reuse is one solution to the problems inherent in code reuse." Obviously creating reusable components requires as surplus charge being paid. Without solving the problem to adequately allocating that charge to the users of reusable components such components only in exceptional cases will be created.

We do not comment on the accuracy of these "facts". We think, however, that they address the key problem of reuse, i.e., the economy of software creation, use, and maintenance. Our initial heuristic for business process reuse is: (1) build-up domain specific process libraries, (2) limit modifications of the process structure to a sensible degree, (3) take into account modifying the resources involved in a business process, (4) use an effective and efficient business process specification and access facility, (5) implement a reuse infrastructure that is based on a reuse-organization model such as the two-library model for software reuse (see, for example [11]), and (6) put in place workable models of process pricing and usage. In this paper we mainly discuss the basic aspects of the Process Assembler, i.e. we focus on item (4). We consider it as one of the key in the mentioned bazaar.

3.1   Effectively and Efficiently Accessing Business Processes

We approach the problem of retrieving a business process that likely is of some use from the angle of understanding processes. We therefore first look at aided knowledge acquisition regarding a business process. We base that aid on explanations of the business process. The Oxford English Dictionary Online defines explanation as "That which explains, makes clear, or accounts for; ...; a statement that makes things intelligible." Philosophers according to [28] distinguish several kinds of explanations, in particular causality explanation (an explanation in terms of cause and effect), teleological explanation (an explanation of a course of action taken in terms of purposes and intentions), deductive-nomological explanation (an explanation in terms of general laws, rules of deduction, and initial conditions), and statistical explanation (an explanation in terms of probabilities). A widely used schema for deductive-nomological explanation is the Hempel-Oppenheim schema, see, for example, [24]. In that source also the limitations of that schema are discussed indicating that a restriction to deductive-nomological explanations would be too limiting. We think that deductive-nomological explanations and statistical explanations are not really helpful fur our purposes. We therefore do not consider them further. We, however, include similarity explanations (explanations in terms of similarity or dissimilarity) as a further kind of explanation and include explanations of concepts into what we are going to consider for aiding understanding.

We intend to use the specified kinds of explanations in the following way. An electronic dictionary (such as WordNet, see [8], or similar) is supposed to be accessible to the Process Assembler such that on the one hand lexical relations (such as synonymy, hyponymy, homonymy, and collocation, see, e.g. [34] for more detail) can be exploiting for better understanding the terms that occur in the process (concept explanation). On the other hand an electronic thesaurus can be used to introduce a qualitative similarity concept between business processes. That relation can be defined modeling notions that occur in business processes and that are related to each other via lexical relations in the electronic thesaurus. Another similarity-based approach to process retrieval works via a set of similarity measures implemented on top of business processes. Another approach to similarity is structure oriented. It focuses on refinement or specialization of processes as well as on aggregation.

One can benefit from the idea of teleological explanations by capability to retrieve those business processes that contain specified parts in a sequence or alternative. The relation of that to teleological explanations becomes particularly apparent if one presupposes a process modeling concept such as "goal" or "purpose" being available and thus particular ways of achieving given goals or meeting given purposes can be associated with sequences of those process parts.

3.2   Key Functionality for Business Process Reuse

Following Langefors' (see, e.g. [12–p. 11]) we define information system as a technically implemented medium for recording, storing, and disseminating linguistic expressions as well as for the purpose of drawing inferences from such expressions. As the Process Assembler is supposed to be an information system for business process reuse its key functionalities are:

- **Adapt** a business process to the case at hand.
- **Assess** a business process as to whether or not it should be reused in the case at hand.
- **Cast** a business process, i.e., translate it from a language $\mathcal{L}$ into a language $\mathcal{L}'$.
- **Display** to a user the available business processes in a suitable form that can be parameterized by the user.
- **Navigate** the displayed business processes.
- **Retrieve** the business processes that best match a given specification.
- **Specify** a set of business processes. We assume currently that respective specifications will be obtained in terms of an SQL-like dedicated query language.

These key functionalities of business process reuse cover the stages of workflow type reuse as given in [20]. The navigation function, as was the case with the MIT process catalog, can follow the abstraction dimension (more or less detail of processes will be displayed to users), the direct links between business processes. It can, however, also follow indirect links established via the thesaurus' lexical relations.

4   An Architecture of a Reuse Environment

Our Process Assembler is a Web information system and supposed to aid in executing the functionalities identified in the previous section. The figure 1 shows its high-level layers and blocks architecture.

The key components in that diagram are briefly explained below. They are in an obvious relation to the key functionalities identified in the previous section.

- **Assessor**, for aiding users in assessing the suitability of a business process for reuse in the case at hand. Here we intend using ripple down rules (these are essentially if-then-else rules that can be refined by exceptions and counter conditions, see, for example [25]) for representing and acquiring the assessment knowledge.
- **Exchanger**, for importing and exporting business processes.
- **Mapper**, for casting business processes.
- **Modeler**, for creating and adapting business processes.
- **Registry**, for registering business process modeling languages with the Process Assembler. Obviously the registry is a key component of the model based architecture and it needs to provide to the other components the functionality required for handling the processes described in a particular modeling language.
- **Repository** for storing business processes.
- **Retriever**, for aiding users in obtaining business process specifications.
- **Viewer**, for aiding users in inspecting the process repository and navigating through it.

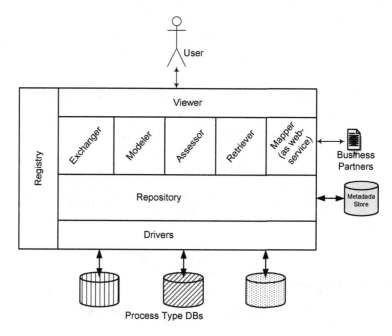

Fig. 1. Process Assembler Architecture

We find our architecture complying with the repository system architecture requirements of [15]. In that source the authors say that "... the major modules of the architecture are the repository management system and the data store". The components Assessor, Retriever, Modeler, and Repository, together implement the functions of the repository management system. The architecture above allows for post-implementation addition of new components and interfaces. Potential future extensions of the Process Assembler could include an analyzer (for assessment of structural quality aspects of business processes), and a simulator or animator (for assessing dynamic quality aspects of business processes).

4.1   Metadata Store

The Metadata store contains meta information regarding the business process modeling languages that have been registered at the Process Assembler. The main concepts regarding which information needs to be stored in the Metadata store concerns the concepts in Fig. 2). These concepts are briefly explained below:

- **Business Area**, i.e., the kind of business done such as car manufacturing, education, software development, or an individual organization in such an area such as BMW AG, Massey University, Microsoft Inc. etc.
- **Functional Area**, i.e., a class of operations to which the process belongs, such as manufacturing, operations, marketing, sales accounting, human resources, etc.

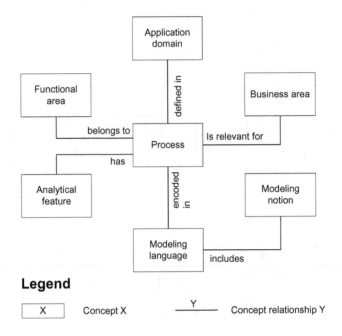

Fig. 2. Meta information

- **Application Domain**, i.e., the domain of applicability of the process such as production workflow, collaborative design, web-services, etc.
- **Analytical Features**, i.e., additional types of information chunks that users might need such as the frequency of reuse, reuse scenarios, the process' author, the process complexity, etc.

4.2   An Example Business Process Modeling Language

To illustrate the flavor of the modeling languages for business processes with which we currently experiment we include the diagram in Fig. 3 that shows an extension of CoCA as a high level class diagram. For precursors of it see, e.g. [16], [26], [18]). We are interested in CoCA as it supports all perspectives of the ARIS "house" (see [13] and [29] for details), i.e., the functional-, operational-, behavioral-, informational-, and organizational perspective. We expect that processes encoded in CoCA will be highly reusable.

Unlike many other process modeling languages CoCA offers a natural nesting of modeling notions (process, phase, cooperation, and contribution) that allows a drill-up and -down in processes that are encoded in CoCA. Additionally a navigation via the generalization of processes may be performed, see [17] for the respective detail. Finally, the association of contribution and process in Figure 3 allows for a navigation along process links. As described above, the association

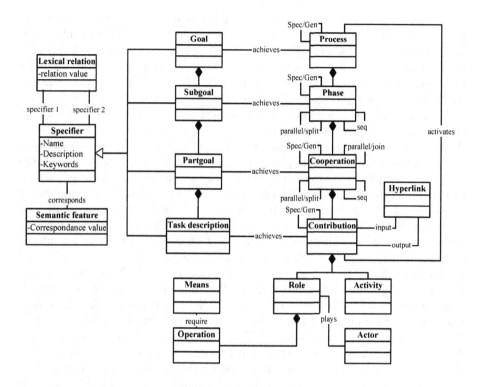

Fig. 3. CoCA Meta-type

between process parts (such as process, phase, cooperation, and contribution) to words in the thesaurus that is established via the nested goal structure enables the use of general language structures and semantics for navigating and accessing business processes.

5   Outlook

The exposition above, certainly, is not at the level of formality required for implementation of the Process Assembler. We have started to formalize things. We define the basic process related terminology as follows: A **grammar** G (see chapter 6 of [30]) is a four-tuple (Σ, N, S, R), such that Σ, and N are finite non-empty sets, the **alphabet** (or terminal symbols) and the **non-terminal symbols** respectively. Furthermore $S \in N$ is the **start symbol** and R is a set of **rules** of the form $L \rightarrow R$, where L is a string of terminal- and non-terminal symbols that contains at least one non-terminal and R is a sequence of terminals and non-terminals. The **language** $L(G)$ generated by G is the set of all strings that can be derived by finitely often applying rules in R when starting with the start symbol S. A **process type system** (PTS) is a four-tuple $P = (\mathcal{B}, \mathcal{C}, \mathcal{G}, \Phi)$ such that $\mathcal{B}$ is a set of **basic process types** such as, phase, task, role, etc.; $\mathcal{C}$ is a set of **process type constructors** such as sequence, if-then-else, while-loop, chose-m-out-of-n, etc.; $\mathcal{G} = (\Sigma, N, S, R)$ is a context free grammar such that $\mathcal{C} \subseteq N$, and $\mathcal{B} \subseteq \Sigma$. The language $L(G)$ is denoted as $\mathcal{T}(P)$. Each element of that set is called a **type** of P; Φ is a mapping defined on $\mathcal{T}(P)$ such that $\Phi(B)$ is a set E_B, the **extent** of B, $\forall B \in \mathcal{B}$, and such that $\Phi(T) = \Phi(C)(\Phi(t_1), \ldots, \Phi(t_m))$ is a set E_T, called the extent of T, $\forall T = C(t_1, \ldots, t_m) \in \mathcal{T}(P)$. Each element $x \in E_T$ is called a **process**. With our concept of PTS we aim at considering workflows more as type expressions in programming languages than as processes. That in fact means that we distinguish a requirements or usage metaphor (process) from an implementation or semantics metaphor (tuple). On the latter metaphor one wouldn't say that tasks are carried out one after another or in parallel. Rather one would say that users have a shared and cooperative access to data. In that metaphor it is clear that theoretical concepts of process similarity such as **bisimulation** not necessarily are adequate for introducing a semantic equivalence of process types. It is not likely that bisimulation would be of any help here, as (1) bisimulation is undecidable for Petri Nets (see [14]), and (2) We can define PTS that generate as their language Petri Nets, Event Driven Process Chains, and CoCA respectively. We thus expect that bisimulation is undecidable for PTS. We intend to extend the list of process modeling languages that can be covered by our formalization. We conceptualize the drill-down from a PTS such that the language of that PTS is a business process modeling language. A type is then a model encoded in that language and a process is an instance of a type. A process differs from that type by the individual modeling notions occurring in the type being given a particular interpretation (which is achieved by adequate labeling and using an "a priori semantics").

We have defined the concept of type cast. Let $P = (\mathcal{B}_P, \mathcal{C}_P, \mathcal{G}_P, \Phi_P)$ and $Q = (\mathcal{B}_Q, \mathcal{C}_Q, \mathcal{G}_Q, \Phi_Q)$ be PTS. A mapping $\tau : L(\mathcal{G}_P) \rightarrow L(\mathcal{G}_Q)$ is called **type cast** from P to Q if $\tau(C(t_1, \ldots, t_m)) = N^Q_{C, t_1, \ldots, t_m}(\tau(t_1), \ldots, \tau(t_m))$, holds for all $C(t_1, \ldots, t_m) \in L(\mathcal{G}_P)$, where $N^Q_{C, t_1, \ldots, t_m}$ is a possibly derived type constructor of $\mathcal{G}_Q$ of arity m. We have constructed an type cast that translates Petri nets into event driven process chains. We intend to increase the number of type casts known to us and implement them in a Process Assembler prototype. Type casting involves of course semantic issues. It cannot be guaranteed that lossless casts are possible, since the expressivity of process modeling languages may be quite different. Occasionally such equivalences could be established, as it is the case with Petri Nets and ACP_δ terms, that are known to be of equal expressivity in the sense of the existence of bijective functors of the respective categories, see [27] for more detail. We currently do not address in our research issues of semantics preservation. We more rely on post-cast modifications being made if that should be required. Note that workflow definition languages of workflow management system are candidates for PTS. Availability of the required type casts therefore would even help enacting workflows by workflow engines for which they were not defined.

The currently existing Process Assembler prototype can store CoCA processes. We are working on an SQL like business process query language that will be integrated into the Retriever. We expect that language to work such that in the "FROM" clause terminals or non-terminals of the grammars may occur that are part of PTS that are registered at the Process Assembler. We expect the "WHERE" clause such that additionally to the comparison operators known from SQL the lexical relations maintained by the electronic thesaurus can be used.

References

1. Van der Aalst, W.M.P.; van Hee, K.: Workflow management: Models, Methods, and Systems, The MIT Press, 2002.
2. Van der Aalst, W.M.P.; ter Hofstede, A.H.M.; Kiepuszewski, B.; Barros, A.P.: Workflow Patterns. BETA Working Paper Series, WP 47, Eindhoven University of Technology, Eindhoven, 2000.
3. Van der Aalst, W.M.P.; ter Hofstede, A.H.M.: Workflow Patterns: On the Expressive Power of (Petri- net-based) Workflow Languages, Proc. of the Fourth International Workshop on Practical Use of Colored Petri Nets and the CPN Tools, Aarhus, Denmark, August 28-30, 2002.
4. Arkin, As.; Intalio: Business Process Modeling Language, BPMI, URL: http://www.BPMI.org., 2002.
5. Batini, C.; Ceri, S.; Navathe, S. B.: Conceptual database design: an Entity-Relationship approach, The Benjamin/Cummings Publishing Company, 1992.
6. Business Process Execution Language for Web Services, Specification, Version 1.1, 5 May, 2003.
7. Chen, P.: The Entity-Relationship Model: Toward a unified View of Data, ACM TODS, 1976.
8. Fellbaum, C.: WordNet - An Electronic Lexical Database, The MIT Press, 1998.
9. Fowler, M.: Patterns of Enterprise Application Architecture, Addison-Wesley, 2003.
10. Glass, R. L.: Facts and Fallacies of Software Engineering, Addison-Wesley, 2003.
11. Graham, I.: Requirements Engineering and Rapid Development: A Rigorous Object-Oriented Approach, Addison-Wesley, 1998.
12. Hirschheim, R.; Klein, H. K.; Lyytinen, K.: Information Systems Development and Data Modeling, Conceptual and Philosophical Foundations. Cambridge, UK: Cambridge University Press, 1995.
13. Jablonski, S.; Bussler, C.: Workflow Management: Modeling Concepts, Architecture and Implementation, International Thompson Computer Press, 1996.
14. Jancar, P.: Decidability Questions for Bisimilarity of Petri Nets and Some Related Problems, In: Enjalbertand, P.; Mayr, E. W.; Wagner, K. W. (Eds.): Proceedings of STACS'94, Springer Verlag, 1994, pp. 581 - 592.
15. Jablonski, S.; Petrov, I.; Meiler, C.; Mayer, U.: Guide to Web Application and Platform Architectures, Springer, 2004.
16. Kaschek, R.; Kohl, C.; Mayr, H.: Cooperations - An Abstraction Concepts suitable for Business Process Reengineering, ReTIS'95 Proceedings, OCG, 1995.
17. Kaschek, R.: Process Ontology as Factor of Business Process Modeling, (In German), In: Pohl, K.; Schürr, A.; Vossen, G. (Eds.): Modellierung '98, Technical Report 6/98-I, University of Münster.
18. Kaschek, R.; Wiltsche, M.; Rinderer, Th.: CoCA - Towards Reuse Support at the Business Process Meta Level, In: Rozman, I. (Ed.): Proceedings of IS'98, International Conference on Development and Reengineering of Information Systems, Ljubljana, Slovenija, October 1998; pp. 5 - 8.

19. Keller, G.; Nüttgens, M.; Scheer, A.-W.: Semantische Prozeßmodellierung auf der Grundlage "Ereignisgesteuerter Prozeßketten (EPK)", in: Scheer, A.-W. (Hrsg.): Veröffentlichungen des Instituts für Wirtschaftsinformatik, Heft 89, Saarbrücken 1992.
20. Kradolfer, M.: A Workflow Metamodel Supporting Dynamic, Reuse-based Model Evolution, PhD Thesis, University of Zurich, 2000. URL: http://www.ifi.unizh.ch/ifiadmin/staff/rofrei/Dissertationen/Jahr_2000/ thesis_kradolfer.pdf
21. Malone, T.W.; Crowston, K.; Herman, G.A.: Organizing Business Knowledge, The MIT Process Handbook. The MIT Press, Cambridge, Massachusetts, 2003.
22. Miheev, A.; Orlov, M.: Perspective of Workflow Management Systems, PC Week, 2004. URL: http://www.pcweek.ru, (from Russian).
23. Pentland, B.T.; Osborn, C.S.; Wyner, G.; Luconi, F.: Useful Descriptions of Organizational Processes: Collecting Data for the Process Handbook, Massachusetts Institute of Technology Sloan School of Management, 1999.
24. Poser, H.; Philosophie of Science: A Philosophical Introduction, (In German), Philipp Reclam jun. GmbH & Co, Stuttgart, 2001.
25. Pham, S. B.; Hoffmann, A.: Intelligent Support for Building Knowledge Bases for Natural Language Processing, In Kaschek, R. (Ed.): Intelligent Assistant Systems, Idea Group Inc., 2006.
26. Rinderer, T.: CoCA Approaches and Means for object-oriented Business Process Modeling, (In German), Master Thesis, University of Klagenfurt, Austria, 1998.
27. Rittgen, P.: Prozesstheorie der Ablaufplanung, B. G. Teubner, 1998.
28. Regenbogen, A.; Meyer, U.: Dictionary of Philosphical Concepts, (In German), Felix Meiner Verlag, Hamburg, 1998.
29. Scheer, A.-W.: ARIS - Business Process Frameworks, Springer, 2000.
30. Tucker, A. B., (Ed. in chief): Computer Science Handbook, Chapman & Hall/CRC, 2004.
31. Wf-XML 2.0: XML Based Protocol for Run-Time Integration of Process Engines, URL: http://www.wfmc.org, October 2003.
32. Workflow Management Coalition Terminology and Glossary, Document Number WFMC-TC- 1011, February 1999, URL: .
33. Workflow Process Definition Interface – XML Process Definition Language, Document Number WFMC-TC-1025, URL: http://www.wfmc.org.
34. Yule, G.: The Study of Language, Cambridge University Press, 1996.

Preface to QoIS 2005

Isabelle Comyn-Wattiau and Samira Si-Saïd Cherfi

Quality is emerging as a key issue for information systems researchers and practitioners. The information system is often defined as "a system, whether automated or manual, that comprises people, machines, and/or methods organized to collect, process, transmit, and disseminate data that represent user information". Information system quality aims at the evaluation of its main components, i.e. system quality, data quality, information quality as well as model quality and method quality. Ongoing research encompasses theoretical aspects including quality definition and/or quality models. These theoretical contributions lead to methods, approaches and tools for quality measurement and/or improvement. Most approaches focus on specific environments, such as web site quality, data warehouse quality, ontologies quality, etc. This workshop is devoted to present and discuss papers related both to theoretical and practical aspects of information systems quality. Two sessions are devoted respectively to :

- Model quality: frameworks, criteria, and experiments are proposed to facilitate the evaluation of model and language quality.
- Quality driven process: this session will demonstrate how quality can be used to improve methods, tool selection or information retrieval.

We received 17 papers from over 10 countries and the program committee finally selected seven papers. We would like to thank the organizing committee of the 24th International Conference on Conceptual Modeling (ER2005) for recognizing the relevance of this workshop and ER2005 workshop chairs for their helpful attitude. We are very indebted to all program committee members and additional reviewers who have very carefully and timely worked. We would also like to thank all the authors who submitted papers to our workshop.

We hope this workshop will be an opportunity for stimulating exchange between researchers and practitioners about information system quality.

J. Akoka et al. (Eds.): ER Workshops 2005, LNCS 3770, p. 375, 2005.
© Springer-Verlag Berlin Heidelberg 2005

Measuring the Perceived Semantic Quality of Information Models

Geert Poels, Ann Maes, Frederik Gailly, and Roland Paemeleire

Management Informatics Research Unit,
Faculty of Economics and Business Administration, Ghent University – Ugent,
Hoveniersberg 24, B-9000 Gent, Belgium
{Geert.Poels, A.Maes, Frederik.Gailly,
Roland.Paemeleire}@UGent.be

Abstract. Semantic quality expresses the degree of correspondence between the information conveyed by a model and the domain that is modelled. As an early quality indicator of the system that implements the model, semantic quality must be evaluated before proceeding to implementation. Current evaluation approaches are based on ontological or meta-model analysis and/or use objective metrics. They ignore the model user's perception of semantic quality, which also determines whether the benefits of using a faithful model will be achieved. The paper presents the development of a perceived semantic quality measure. It presents a measure pre-test, i.e. a study aimed at refining and validating a new measure before its use in research and practice. The results of the pre-test show that our measure is reliable and that it is sufficiently differentiated from other perception-based measures of information model use like ease of use, usefulness, and user information satisfaction.

1 Motivation

A key quality of a modeling grammar is semantic expressiveness. This property is the ability "to express everything that needs to be modeled without much effort from the modeler" [9, p. 43]. The expressiveness of a grammar can be evaluated by comparing its constructs against that of a benchmark, e.g. a meta-model or an ontology [9], [21]. Alternatively, it can be assessed via scripts produced using the grammar, in which case it must be measured.

Objective measures (also called *metrics*) for the 'semantic quality' [11] of a script have been proposed in [1], [7], [16]. Using metrics or ontological analysis assumes that each script user will interpret the underlying reality and its representation in exactly the same manner [6]. It is our position that also the user's perception of semantic quality, rather than just a theoretically verified semantic quality, determines whether benefits result from using a 'faithful' script (e.g. increased information retrieval performance and user satisfaction [6], better comprehension of the domain semantics conveyed by the script [15], and ease of understanding [1]). Therefore, an empirical approach that recognizes possible differences in user perception of semantic quality is needed to complement more theoretically-oriented evaluations.

J. Akoka et al. (Eds.): ER Workshops 2005, LNCS 3770, pp. 376–385, 2005.
© Springer-Verlag Berlin Heidelberg 2005

The only literature reference to a measure for semantic quality, as perceived by script users, is [6], where a one-item seven-point Likert scale instrument with the assertion "The documentation I received provided me with a realistic representation of the accounting information flows of the business" and anchored at agree = 1 and disagree = 7 was proposed. This instrument was used in an empirical study to assess the relative perceived semantic expressiveness of alternative accounting information models. Without rewording the item statement, it cannot be used in other information modeling research contexts. Furthermore, compared to a single-item instrument, the use of a multi-item instrument is preferred because of its ability to diminish measurement error resulting from the specificity of individual items [3].

The goal of our research is to develop a more general and multi-item measure for perceived semantic quality (PSQ). The paper reviews the first steps taken in the measure development process and presents a *measure pre-test*, i.e. a study aimed at refining and validating a proposed measure before its use in research and practice. This study was conducted in the context of an experiment investigating user comprehension effects of pattern recognition in enterprise information models. The results show that after some refinement, a valid and reliable measure is obtained that is sufficiently differentiated from other perception-based measures for ease of use, usefulness, and user information satisfaction.

Section 2 briefly describes our previous work on the creation of the measure as well as a former pre-test. In section 3, first the measure is revised based on the preliminary empirical results obtained and next the design and operation of a new pre-test and resultant validity and reliability analysis are presented. Finally, section 4 discusses the pre-test findings and their implications for the further validation and use of the refined PSQ measure.

2 Measure Development

Following the approach of [8], the first steps in the measure development process were to generate relevant items for the measurement instrument (subsection 2.1) and conduct a pilot study to test the initial measure (subsection 2.2). Further details of these first steps can be found in [17].

2.1 Item Generation

In [11] a literature survey was conducted to identify the quality properties for conceptual schemas that are related to semantic quality. Assuming that these properties cover the content domain of semantic quality, we used them to generate items for our measurement instrument. Table 1 shows the quality properties considered, their definition, and the corresponding item statement in our initially proposed instrument. The wording of the item statements was adapted to a pilot study (confer infra). The style of the item statements was varied in order to avoid monotonous responses.

Table 1. Items of initially proposed measure

Item name (Item code)	Quality property	Definition [11]	Item statement
Correct (PSQ1)	Correctness	All statements in the schema are *correct* and relevant to the problem domain	The conceptual schema represents the business process correctly
Relevant (PSQ2)	Correctness	All statements in the schema are correct and *relevant* to the problem domain	The conceptual schema shows only relevant entities, relationships and structural constraints
Complete (PSQ3)	Completeness	The schema contains all statements about the problem domain that are correct and relevant	The conceptual schema gives a complete representation of the business process
Adequate (PSQ4)	Completeness	The schema contains all statements about the problem domain that are correct and relevant	Entities, relationships or structural constraints must be added to adequately represent the business process
Minimal (PSQ5)	Minimality	The schema does not contain statements that overconstrain the problem domain	None of the entities, relationships and structural constraints in the conceptual schema can be removed
Consistent (PSQ6)	Consistency	The schema does not contain contradictory statements	The conceptual schema contains inconsistencies
Realistic (PSQ7)	*Based on single-item measurement instrument in [6]*		The conceptual schema is not a realistic representation of the business process

The correctness property was split into two items (PSQ1 and PSQ2) as it also incorporates the notion of relevance. The dual concept of correctness is completeness, which is equally important in determining the correspondence between a script and the reality that is modeled [11]. Hence, we decided to create also two items for the completeness property (i.e. the 'complete' (PSQ3) and 'adequate' (PSQ4) items).

The property of minimality is subsumed by correctness [11]. As it might be a sub-dimension of correctness, we decided to retain it in the measurement instrument, but create only one item for it (PSQ5). In [11] it is further shown that the property of consistency is subsumed by both correctness and completeness. As it is, however, one of the most commonly mentioned quality properties and it is recommended to have an overinclusive item pool when developing new measurement instruments [12], we decided to create also a separate item for this property (PSQ6).

Finally, the 'realistic' item (PSQ7) is based on the single-item measurement instrument used in [6]. The realism property seems to capture the essence of semantic quality, meaning a good fit between script and reality modeled. As it is common to include an "overall" item for the construct measured [5], we decided to add the 'realistic' item to the instrument.

2.2 First Pre-test

The initial measure was tested on a convenience sample of 42 students, after they performed a comprehension task on an entity-relationship diagram that served as a structural model of a business process. The diagram showed the business concepts involved, their interrelationships, and the business rules that govern the process. For each PSQ item a 7-point Likert scale with response options ranging from 'strongly disagree' to 'strongly agree' was offered. The measurement instrument also contained the items of two other perception-based measures frequently used in human factors related conceptual modeling research: perceived ease of use (PEOU), adapted from [5] and user information satisfaction (UIS), adapted from [19] (Table 2). It is important to include marker variables or concepts to which the construct under study is expected to relate and from which it must be differentiated [4]. Since all items in the instrument capture user perceptions about the diagram, we wished to verify whether the PSQ items measure another construct than what is measured by the PEOU and UIS items. If in later research the impact of PSQ on UIS and PEOU is investigated, then it is essential to have different measures for these constructs.

Table 2. PEOU and UIS items included in the first pre-test measurement instrument

Item code	Item statement/question
PEOU1	I found the conceptual schema cumbersome (confusing) to use
PEOU2	Using the conceptual schema required a lot of mental effort
PEOU3	The conceptual schema was clear and understandable to me
PEOU4	Overall, I found the conceptual schema easy to use
PEOU5	Using the conceptual schema was frustrating
UIS1	How adequately do you believe the conceptual schema meets the information needs that you were asked to support?
UIS2	How efficient is the conceptual schema for providing the information you needed?
UIS3	How effective is the conceptual schema for providing the information you needed?
UIS4	Overall how satisfied are you with the conceptual schema for providing the information you needed?

The reliability of the PSQ measure, calculated using Cronbach's alpha, was only 0.71, which is barely above the value of 0.70 that is usually required for measurement instruments to be deemed reliable [14]. An inter-item correlation analysis showed that the 'minimal' item (PSQ5) was not significantly correlated (at the 0.05 level) to any of the other instrument's items. Removing this item from the PSQ measure would increase its reliability to 0.74. Further problems were noted with the 'consistent' (PSQ6), 'relevant' (PSQ2), and 'adequate' (PSQ4) items, although their removal did not increase the measure's reliability. The 'consistent' item was significantly correlated to only one other PSQ item, the 'correct' (PSQ1) item, suggesting that the pilot study participants found it difficult to distinguish between the correctness and consistency concepts. The 'relevant' item was significantly correlated only with the 'realistic' (PSQ7) item, but with several of the PEOU and UIS items, indicating that it cannot discriminate between the constructs measured. The 'relevant' item statement might have been misunderstood by the pilot study participants. The entity-relationship diagram they were given also showed attributes, whereas the item statement only mentions entities, relationships and structural constraints. It is possible that the participants therefore focused on the word "only", and not on "relevant". Although the 'adequate' item (PSQ4) correlates well with the 'correct' (PSQ1), 'complete' (PSQ3), and 'realistic' (PSQ7) items, it also showed a significant correlation with UIS1, probably because of the common use of the word 'adequately'.

3 Measure Refinement and Validation

3.1 Item Refinement

We reckoned that the problems encountered were due to lack of content validity (i.e. participants did not associate them with the semantic quality concept) or confusing phrasing. To increase validity, a better conceptual definition of semantic quality was needed. Next, item statements were derived from this conceptual definition.

A comprehensive definition of semantic quality was found in the four quality properties for conceptual schemas proposed in [20]. This proposal maps the four kinds of ontological deficiencies[1] of modeling grammars onto script properties. Hence, from the perspective of ontology, a faithful representation of a domain is obtained if a script adheres to these four prescriptions.

Table 3 compares the quality properties of [20] to the quality properties underlying our initial PSQ measure. The theoretical foundation (in ontology) of the quality properties in [20] provides a suitable conceptual basis for a refined PSQ measure. Not only is it an assurance for content validity; it also combines a theoretical, ontological approach with an empirical, perception-based approach.

Table 3. Comparison of quality properties

Quality properties and definitions in [20]		Properties underlying PSQ measure
Accuracy	The schema should accurately represent the semantics of the problem domain as perceived by the focal stakeholders.	Correctness
Completeness	The schema should completely represent the semantics of the problem domain as perceived by the focal stakeholders.	Completeness
Conflict-free	The semantics represented in different parts of the schema should not contradict one another.	Consistency
No-redundancy	The schema should not contain redundant semantics.	Minimality

---

[1] These are *construct excess* (i.e. a grammatical construct might not map to any ontological construct), *construct deficit* (i.e. an ontological construct might not map to any grammatical construct), *construct overload* (i.e. several ontological constructs map to one grammatical construct), and *construct redundancy* (i.e. several grammatical constructs map to one ontological construct) [21].

Table 4 shows how some of the problematic initial item statements were refined using the property definitions in [20]. At the same time, the wording of item statements was made more general, no longer referring to entity-relationship modeling. The term 'business process' was retained as a reference to the focal domain, but can be replaced by a more appropriate description of the focal domain if necessary.

Table 4. Items of refined PSQ measure (changed item statements indicated in *italic*)

Item name (Item code)	PSQ property	Corresponding property in [20]	Item statement
Correct (PSQ1)	Correctness	Accuracy	The conceptual schema represents the business process correctly
Relevant (PSQ2)	Correctness	Accuracy	*All the elements in the conceptual schema are relevant for the representation of the business process*
Complete (PSQ3)	Completeness	Completeness	The conceptual schema gives a complete representation of the business process
Adequate (PSQ4)	Completeness	Completeness	*Elements must be added to faithfully represent the business process*
Minimal (PSQ5)	Minimality	No-redundancy	*The conceptual schema contains redundant elements*
Consistent (PSQ6)	Consistency	Conflict-free	*The conceptual schema contains contradicting elements*
Realistic (PSQ7)	Realism	-	The conceptual schema is a realistic representation of the business process

The changes in item statements can be summarized as follows:

- The more general formulation of PSQ2 made the use of "only" superfluous;
- "adequately" in the PSQ4 statement was replaced by "faithfully";
- The PSQ5 statement now restates the no-redundancy property of [20];
- In the PSQ6 statement "inconsistencies" was replaced by "contradicting elements", as in the conflict-free property of [20].

3.2 Second Pre-test

Because of the changes made to four of the item statements, a new measure pre-test was needed. In a pre-test no hypotheses are formulated that involve the construct for which a new measure is developed. However, the construct should be measured as one of the study's variables, in conditions that are realistic for the subsequent use of the measure. It was thus required that the study participants could develop a belief about the semantic quality of the conceptual schema they had to rate using the refined PSQ items. Further, it was required that participants could also develop beliefs about the schema and attitudes towards the use of the schema that relate to other constructs than semantic quality, like ease of use and user information satisfaction.

These requirements were met in an experiment that investigated the impact of pattern recognition on the user comprehension of conceptual data models in UML class diagram notation. Using a within-subjects experimental design, 17 graduate-level business students had to perform an experimental task on a class diagram that showed occurrences of modeling patterns previously learned and a similar task on a diagram that did not show such pattern occurrences. The diagrams used represented two domains. For each domain there was one class diagram with pattern occurrences and one without. Counterbalancing was used to alleviate order effects. The experimental task to be performed using a diagram required checking a diagram's conformity with textual scenarios describing the domain.[2]

User comprehension was measured with performance-based measures (task completion time, task accuracy) and perception-based measures (perceived ease of interpretation, perceived usefulness). The perceived ease of interpretation (PEOI) measure of [10] is similar to the PEOU measure used in the first pre-test (confer section 2.2), but more specific to schema comprehension tasks. The perceived usefulness (PU) measure was taken from [13]. Also user information satisfaction (UIS) was measured. The PEOI and PU items are shown in Table 5. The UIS items were the same as in the first pre-test (confer Table 2). A questionnaire with the PEOI, PU, UIS, and refined PSQ items (intermingled) was presented to each participant after completing the experimental task with a diagram. So, each participant filled the questionnaire twice.

Table 5. PEOI and PU items included in the second pre-test measurement instrument

Item code	Item statement
PEOI1	It was easy for me to understand what the conceptual schema was trying to model
PEOI2	Using the conceptual schema was often frustrating
PEOI3	Overall, the conceptual schema was easy to use
PEOI4	Learning how to read the conceptual schema was easy
PU1	Overall, I think the conceptual schema would be an improvement to a textual description of business process
PU2	Overall, I found the conceptual schema useful for understanding the process modelled
PU3	Overall, I think the conceptual schema improves my performance when understanding the process modelled

After collecting the data, an inter-item correlation matrix was calculated.[3] The rewording of the 'minimal' item (PSQ5), now based on the no-redundant property [20], was not successful. As in the first pre-test, PSQ5 is not significantly correlated to any of the other PSQ items. As a consequence, the reliability of the PSQ measure is relatively low (Cronbach's alpha = 0.73). Removing PSQ5 increases the reliability to 0.81, which is well above the 0.70 cut-off [14].

---

[2] Page limits prohibit the inclusion of the experimental materials, which can be found in [18].
[3] Not shown because of page limits. A copy can be sent upon request (contact the first author).

The other problematic item is PSQ4 ('adequate'). Though it correlates well with PSQ6 ('consistent') and PSQ7 ('realistic'), it is not correlated with the other PSQ items, not even with PSQ3, the other completeness item. Removing both PSQ4 and PSQ5 increases the reliability of the PSQ measure to 0.84.

The inter-item correlations for all other pairs of PSQ items are significant at the 0.05 level. Removing any of these items decreases the reliability of the PSQ measure. Also, no problems are noted for the PEOI, UIS, and PU items; the inter-item correlations for all pairs of items purported to measure a same construct are significant. The reliability of the PEOI, UIS, and PU measures is respectively 0.91, 0.91, and 0.82.

The significant correlations between items of different measures indicate the existence of relationships between the constructs measured. Only PSQ and PEOI do not seem to be related. To assure that items belonging to different measures measure different constructs (and not just one or a few latent constructs), a formal validity test was undertaken. The Campbell and Fiske test [2] for an item i calculates the average of the correlation coefficients for i and the other items of the measure that i belongs to (*CV value*); calculates the average of the correlation coefficients for i and the items belonging to other measures (*DV value*); and verifies whether i's *CV value* is larger than i's *DV value*. If this test fails for i, then i is more closely associated with the items of other measures than with the items of the measure that i belongs to. In that case, item i should be rejected. Our calculations show that, after removing PSQ4 and PSQ5, all remaining items pass the Campbell and Fiske test.

4 Discussion

Comparing the first and second pre-test (on respectively the initial and refined PSQ measure), reliability and validity were higher in the second pre-test. Excluding PSQ4 ('adequate') and PSQ5 ('minimal'), a reliability of 0.84 is obtained versus a reliability of 0.74 in the first pre-test (after removing the initial PSQ5). The refinement of PSQ2 ('relevant') and PSQ6 ('consistent'), using more general terms and based on the quality definitions in [20], was successful. The hypothesized improvement in content validity of these items was reflected in their convergent validity as they were significantly correlated to the other PSQ items.

The PSQ4 item did not correlate well to some other PSQ items. Perhaps the pre-test participants did not understand the meaning of "faithfully" in the item statement (being non-native English speakers)? As PSQ4 was intended to capture the same underlying quality property as PSQ3, i.e. completeness, it can be removed from the PSQ measure without loss of information.

Neither pre-test supports the inclusion of PSQ5 in the measure. A plausible reason is that the diagrams used in the pre-tests did not show obvious redundancies. Hence, participants were not motivated to express a well-articulated opinion about this item. Without further empirical evidence we advise not to include PSQ5 in the measure. In future confirmatory analyses, redundant elements could be injected in the experimental materials to see if this semantic quality problem gets noticed by users and influences their perception.

A threat to the validity of the second pre-test results is the low number of participants. Because of the within-subjects design, each of the seventeen participants rated the PSQ measure twice, so a total of 34 data points were available for analysis. Although this sample size is common for measure pre-tests (for which convenience samples are acceptable [8]), confirmatory analyses should have bigger sample sizes (such that also the factorial validity of the measure can be tested).

Concluding (and acknowledging the need for further measure validation), a purified measure (with the PSQ1, PSQ2, PSQ3, PSQ6, and PSQ7 items), seems to reliably capture the semantic quality of a schema as perceived by schema users. This measure enhances objective measures of semantic quality that employ a theoretical evaluation approach based on meta-models, reference models or ontologies. An *objective* semantic quality measure forces each and every schema user to refer to a same quality benchmark. The use of a metric assumes that the user agrees on the underlying assumption of 'perfect' quality of the meta-model, ontology or reference model that is used as the benchmark. On the contrary, a *perceived* semantic quality measure allows individual users to express an opinion about the correctness, completeness, consistency, and realism of the schema, without having to agree on a quality benchmark (or even having to explicitly formulate such a benchmark). The use of a perception-based measure also recognizes that users may only partly understand the reality that is modeled or may possibly not reach an agreement on a common understanding of this reality.

Measuring perceptions of semantic quality is important, as they might determine whether benefits will result from using a 'faithful' schema. For instance, [6] demonstrated that greater perceived semantic expressiveness is associated with greater task accuracy (in information retrieval). The implication for research is that studies investigating means to assure or improve the semantic quality of information models should also look at the user perception of semantic quality. Theoretically-based quality improvements that go unnoticed are less likely to achieve the benefits (e.g. increased user comprehension) they aim for. The proposed PSQ measure offers a research instrument to verify whether quality improvements are also perceived by users. Likewise, in business practice, the measure might be used to assess how end users react to alternative models (which can have, objectively, the same semantic quality).

In our future work, we will use the PSQ measure to investigate the relationship between perceived semantic quality and other perception-based variables of conceptual schema use, such as ease of use, usefulness and user information satisfaction. The goal of this work is a comprehensive user beliefs and attitudes model towards the use of information models. The end-result we aim for is an evaluation model for information models, similar to the Technology Acceptance Model [5].

References

[1] Akoka, J., Comyn-Wattiau, I., 2004. Les systèmes d'information comptables multidimensionnels: Comparaison de deux modèles. Working Paper.

[2] Campbell, D.T., Fiske, D.W., 1959. Convergent and Discriminant Validation by the Multitrait-MultiMethod Matrix. Psychological Bulletin, 81-105.

[3] Churchill, G.A., 1979. A Paradigm for Developing Better Measures of Marketing Constructs. J. Marketing Research 16, 64-73.

[4] Clark, L.A., Watson, D., 1995. Constructing Validity: Basic Issues in Objective Scale Development. Psychological Assessment 7 (3), 309-319.

[5] Davis, F., 1989. Perceived Usefulness, Perceived Ease of Use, and User Acceptance of Information Technology. MIS Quarterly 13(3), 319-339.

[6] Dunn, C.L., Grabski, S.V., 2000. Perceived semantic expressiveness of accounting systems and task accuracy effects. Int'l J. Accounting Information Systems 1(2), 79-87.

[7] Etien, A., Rolland, C., 2005. A Process for Generating Fitness Measures. Presented at the 17th Conf. Advanced Information Systems Engineering.

[8] Froehle, C.M., Roth, A.V., 2004. New measurement scales for evaluating perceptions of the technology-mediated customer service experience. J. Operations Management 22(1), 1-21.

[9] Gemino, A., Wand, Y., 2003. Evaluating modeling techniques based on models of learning. Communications of the ACM 46(10), 79-84.

[10] Gemino, A., Wand, Y., 2005. Complexity and Clarity in Conceptual Modeling: Comparison of Mandatory and Optional Properties. Data & Knowledge Engineering.

[11] Lindland, O.I., Sindre, G., Sølvberg, A., 1994. Understanding Quality in Conceptual Modeling. IEEE Software 11(2), 42-49.

[12] Loevinger, J., 1957. Objective tests as instruments of psychological theory. Psychological Reports 3, 635-694.

[13] Moody, D.L., 2001. Dealing with Complexity: A Practical Method for Representing Large Entity Relationship Models. PhD Dissertation. University of Melbourne.

[14] Nunally, J., 1978. Psychometric Theory. McGraw-Hill.

[15] Parsons, J., Cole, L., 2005. What Do the Pictures Mean? Guidelines for Experimental Evaluation of Representation Fidelity in Diagrammatic Conceptual Modeling Techniques. Data & Knowledge Engineering.

[16] Patig, S., 2004. Zur Ausdrucksstärke der Stammdaten des Advanced Planning and Scheduling. Wirtschaftsinformatik 46(2), 97-106.

[17] Poels, G., Maes, A., Gailly, F., Paemeleire, R., 2004. A Measure for the Perceived Semantic Expressiveness of Accounting Information Models. Presented at the 5th Int'l Research Symposium on Accounting Information Systems.

[18] Poels, G., Maes, A., Gailly, F., Paemeleire, R., 2005. User Attitudes towards Pattern-Based Enterprise Information Models: A Replicated Experiment with REA Diagrams. Presented at the 2nd Int'l Conf. on Enterprise Systems and Accounting.

[19] Seddon, P., Yip, S.-K., 1992. An Empirical Evaluation of User Information Satisfaction (UIS) Measures for Use with General Ledger Accounting Software. J. Information Systems 6(1), 75-92.

[20] Shanks, G., Tansley, E., Weber, R., 2003. Using ontology to validate conceptual models. Communications of the ACM 46(10), 85-89.

[21] Wand, Y., Weber, R., 2002. Information Systems and Conceptual Modeling – A Research Agenda. Information Systems Research 13(4), 363-376.

Situated Support for Choice of Representation for a Semantic Web Application

Sari E. Hakkarainen, Anders Kofod-Petersen, and Carlos Buil Aranda

Department of Computer and Information Science,
Norwegian University of Science and Technology,
7491 Trondheim, Norway
{sari, anderpe}@idi.ntnu.no, builaran@stud.ntnu.no
http://www.idi.ntnu.no

Abstract. As more and more companies are augmenting their data to include semantics, it is imperative that the choices made when choosing the modelling language are well founded in knowledge about the language and the domain in question. This work extends the Semiotic Quality Framework with computational and situated instruments. Furthermore, it demonstrates how the extended Semiotic Quality Framework can facilitate the choice of the most suited language for a real world application. The application is a directory services system, which currently is being moved into the realms of the Semantic Web.

1 Introduction

The IT industry is currently changing focus from providing storage, processing and network services to provide knowledge intensive information and services to large numbers of customers. The diversity and multitude of resources and applications on the Web places elaborate requirements on methods and tools for efficient generation, manipulation and compositional usage of information and services. Metadata, ontology/domain model and semantic enrichment can bridge the heterogeneity and facilitate the efficient usage of information assets on the Semantic Web [1]. However, a formal, standardised representation of signs and meaning is required [2] for supporting ontologies, i.e. explicit and shared conceptualisations [3] of the domain.

Several general-purpose models for description of Web-resources have emerged, where the intention is to facilitate the search, aggregation, filtering, selection, reasoning and presentation of information assets on, and for the (semantic) Web. However, the number of languages and models is large, as is the number of types of prospective applications. Applications can be categorised according to the kind of domain they address (medical, commerce, education, library, oil drilling, etc), the kind of application they target (knowledge management, process monitoring, archival, etc.) or the kind of modelling environment they are supposed to fit in (taxonomies, data flows, data models, process models, etc.). The span for each of these categories is seemingly endless.

In conceptual modelling there are a number of frameworks suggested for evaluating modelling approaches in general. For instance, the Bunge-Wand-Weber ontology [4] has been used on several occasions as a basis for evaluating modelling techniques,

J. Akoka et al. (Eds.): ER Workshops 2005, LNCS 3770, pp. 386–397, 2005.
© Springer-Verlag Berlin Heidelberg 2005

e.g. NIAM [5] and UML [6], as well as ontology languages in [7]. The semiotic quality framework first proposed in [8] for the evaluation of conceptual models has later been extended for evaluation of modelling approaches [9] and used for evaluating UML and RUP [10], This framework was also used in evaluation of ontology languages and tools in [11]. Similarly, [7, 12, 13] evaluate various ontology languages. These studies concentrate on evaluating the technical features of representation languages and partly tools, independent of situational factors of particular development projects. Such studies target audience of highly skilled modelling experts rather than the wide spectrum of potential developers of Semantic Web applications. [14] extends and evaluates methods rather than languages. The framework suggested by [15] is meant for requirements specifications, but is still fairly general. There are also more specific quality evaluation frameworks, e.g. [16] for process models, and [17, 13] data/information models.

The objective here is to develop support for the choice of appropriate Web-based knowledge representation formalism. The way-of-working is to 1) evaluate existing representations in general, using an existing semiotic quality framework for conceptual models, 2) to extend the quality framework with computational and situated features, 3) to develop trial ontologies using a common ontology creation tool and the language specifications, and 4) to evaluate the existing representations in an industrial case study. In the case study, the aim is to support the development of an integrated knowledge-based system for directory services by moving from traditional relational data models to semantically richer representations.

The paper is organised as follows: First, an overview of the Semiotic Quality Framework is given. Secondly, the case study is described. This is followed by a discussion of the results obtained. Finally, a conclusion and an outlook on future work is given.

2   The Semiotic Quality Framework

In order to evaluate the Web representation languages, the *Semiotic Quality Framework* (SQF) [9, 11], a model quality framework consisting of five semiotic factors of quality modelling languages, is chosen. The framework has three main characteristics that make it well-suited as an evaluation instrument: 1) it distinguish between goals and means separating what to achieve from how, 2) it is closely related to linguistics and semiotic concepts, and 3) it is based on a constructivist world-view, the framework recognises that models are build from interaction between the designer and the user. The main model of the semiotic quality framework is as follows.

A – **Audience** refer to the individual, A_i, organisational, A_s, and technical actors, A_t who relate to the model. This includes both human participants and artificial actors.

K – **Participant knowledge** is the explicit knowledge that is relevant for the audience A. It is the combined knowledge of all participants in the project.

L – **Language extension** is what can be represented according to the graphical symbols, vocabulary and syntax of the language; the set of all statements that may be informal L_i, semi-formal L_s, or formal L_f.

M – **Model externalization** is the set of all statements in an actor's model of a part of a perceived reality written in a language L.

I – **Social actor interpretation** is the set of all statements which the externalised model consists of, as perceived by the social audience A_i and A_s.

T – **Technical actor interpretation** is all the statements in the conceptual model L as they are interpreted by the technical audience A_t.

D – **Modelling domain** is the set of all statements that can be stated about a particular situation.

The framework evaluates the physical, empirical, syntactic, semantic, pragmatic, perceived semantics, social and knowledge quality; it evaluates the quality of conceptual models, modelling environments, and modelling languages. This work focuses on the evaluation of the Web representations as modelling languages.

2.1 Adapted appropriateness of languages

The Semiotic Quality Framework consists of five quality factors, called appropriateness, namely: Domain Appropriateness (DA), Participant Knowledge Appropriateness (PAK), Knowledge Externalizability Appropriateness (KEA), Comprehensibility Appropriateness (CA), and Technical Actor Interpretation Appropriateness (TAIA). Here we modify the DA as in [18], as follows.

DA covers seven perspectives for languages: 1) Structural Perspective refers to the static structure, classes and properties, 2) Functional Perspective refers to the processes, activities, and transformations, 3) Behavioural Perspective refers to the states and transitions between them, 4) Rule Perspective refers to the rules for certain processes, activities, and entities, 5) Object Perspective refers to the resources, processes and classes, 6) Communication Perspective refers to the language actions, meaning and agreements, and 7) Actor and Role Perspective refers to the actor, role, society and organisation.

With the modification of the DA we acquire the elements needed to analyse the most practical features of the languages. With the PKA we measure the knowledge of the user. With the KEA we analyse if the language provides enough elements to represent the domain model specified. With CA we analyse if the language is consistent enough and provides clear elements for modelling the domain, and with TAIA we analyse if the language provides enough features for allowing automatic reasoning, the key concept in our investigation. The quality factors are further developed in the sequel.

2.2 Selection Criteria for Quality Factors

For the quality of conceptual modelling languages, Sindre [19] identifies criteria for the constructs of the language and how these constructs are presented visually. Four main groups of sub-criteria are identified: perceptibility, expressive power, expressive economy, method tools and potential. Seltveit [20] adds the criteria of reducibility, meaning the features provided by the model to handle large and complex models.

Let $\mathcal{CF}$ be an evaluation framework such that $\mathcal{CF}$ has a fixed set $\mathcal{A}$ of appropriateness categories a, where $\mathcal{A} = \{a_1, a_2, a_3, a_4, a_5\}$ and $a_i \in \mathcal{A}$. Each a is a quadruple $< id, \ descriptor, \ C, \ cw >$, where id is the name of the category, $descriptor$ is a natural language description, C is a set of selection criteria ac, and cw defines a function of S that return -1, 1 or 2 as coverage weight, where S is a set of satisfied elements ac in the selection criteria C of each appropriateness category in $\mathcal{A}$. Intuitively, we define a number of selection criteria alongside an associated coverage weight function for each category in the evaluation framework. The appropriateness categories with attached descriptors, selection criteria and coverage weight functions are as follows.

a_1 – *Domain appropriateness (DA)* indicates whether the method guidelines address the problems of eliciting/representing relevant facts of the problem domain. Ideally, $D\backslash L = \emptyset$, i.e. there are no statements in the expected application domain that cannot be expressed in the target language, and one should not be guided to express things that are not in the domain (limited number of constructs). The former criterion means that a_1c_1 - the developer is guided to make use of high expressive power whereas the latter means that a_1c_2 - there is a limited number of modelling constructs that are generic, composable and flexible in precision. The equation 1 holds for each modelling perspective of a_1p_1 - structural (SP), a_1p_2 - functional (FP), a_1p_3 - behavioural (BP), a_1p_4 - object (OP), a_1p_5 - communication (CP), and a_1p_6 - actor-role (AP) perspective.

a_2 – *Participant knowledge appropriateness (PKA)* indicates whether the method corresponds to what participant in the modelling activity perceive as a natural way of working. Ideally, $K \cap L\backslash L = \emptyset$, that all the statements in the models of the languages used by the participants are part of their explicit knowledge. Hence a method guideline a_2c_1 - should not promote usage of statements not in a participant's knowledge, a_2c_2 - external representation should be intuitive, and a_2c_3 - non-intuitive representations should be introduced carefully.

$$cw_1(S_1) = \begin{cases} 2 & if \quad a_1c_1 \wedge a_1c_2 \in S_1 \\ 1 & if \quad a_1c_1 \vee a_1c_2 \in S_1 \quad (1) \\ -1 & if \quad S_1 = \emptyset \end{cases} \qquad cw_2(S_2) = \begin{cases} -1 & if \quad |S_2| = 0 \\ 1 & if \quad 0 < |S_2| \leq 1 \quad (2) \\ 2 & if \quad 2 < |S_2| \leq 3 \end{cases}$$

a_3 – *Knowledge externalization appropriateness (KEA)* indicates whether the method assists the participants in externalising their knowledge. $K \cap L\backslash K = \emptyset$, i.e. there are no statements in the explicit knowledge of the participant in the modelling activity that cannot be expressed in the target language. This appropriateness focuses on how relevant knowledge may be articulated in the language rather than what knowledge is expressed. This implies the partial quality goals of generality, a_3c_1 – the guidance to use the language should be as domain independent as possible, and completeness a_3c_2 – there is guidance for all possible usages of the language.

a_4 – *Comprehensibility appropriateness (CA)* indicates whether the participants are able to comprehend the method guidelines. Ideally, $L\backslash I = \emptyset$, i.e. all the possible statements of the language are understood by the participants in the modelling effort using the method guidelines. Thus, a_4c_1 - the described modelling constructs are easily distinguished from each other, a_4c_2 - the number of constructs is reasonable or organised in a natural hierarchy, a_4c_3 - proposed use of modelling constructs is uniform for all the statements expressed in the target language, a_4c_4 - the guidance is flexible in the level of detail in the target language, and a_4c_5 - separation of concerns and multiple views is supported.

$$cw_3(S_3) = \begin{cases} 2 & if \quad a_3c_1 \wedge a_1c_2 \in S_3 \\ 1 & if \quad a_3c_1 \vee a_1c_2 \in S_3 \\ -1 & if \quad S_3 = \emptyset \end{cases} \quad (3) \qquad cw_4(S_4) = \begin{cases} -1 & if \quad 0 < |S_4| \leq 1 \\ 1 & if \quad 1 < |S_4| \leq 3 \\ 2 & if \quad 3 < |S_4| \leq 5 \end{cases} \quad (4)$$

a_5 – *Technical actor interpretation appropriateness (TAIA)* indicates whether the method guidelines lend themselves to automated tool support or assist in support for reasoning. Ideally, $T \backslash L = \emptyset$, all possible mechanisms in the technical participants interpretation are supported by the target language. This implies the partial quality goals for automatic reasoning support in the instructions provided for the target language, i.e. a_5c_1 - both formal syntax and semantics are operational and/or logical, a_5c_2 - efficient reasoning support is provided by executability, a_5c_3 - natural language reasoning is supported, and a_5c_4 - information hiding constructs are provided enabling encapsulation and independent components.

$$cw_5(S_5) = \begin{cases} 2 & if \quad a_5c_1 \wedge (a_5c_2 \vee a_5c_3 \vee a_5c_4) \in S_5 \\ 1 & if \quad a_5c_1 \vee a_5c_2 \vee a_5c_3 \vee a_5c_4 \in S_5 \\ -1 & if \quad S_5 = \emptyset \end{cases} \quad (5)$$

The selection criteria for the appropriateness categories above are exhaustive in the categories a_2, and a_4, whereas the set of satisfied criteria S of the remaining categories may also be the empty list. None of the criteria are mutually exclusive. The coverage weight cw is independent of any category-wise prioritisation. Since the intervals are decisive for the coverage weight they can be adjusted depending on preferences of the evaluator. However, when analysing different evaluation occurrences the intervals need to be fixed in comparison, but may be used as dependent variable.

2.3   Weighted Quality Requirements

Here, we adopt the PORE methodology [21] to prioritise the classification criteria based on company's requirements in order to evaluate the ontology building guidelines in this particular situation. The method has been applied successfully on SQF in [14, 22] for method guideline evaluation and classification, respectively. Hence, the importance weights for each appropriateness category are calculated as follows.

Let $R(CF)$ be a set of weighted requirements such that R has a fixed set $R\mathcal{A}$ of categories ra, where categories in $R\mathcal{A}$ correspond with categories $\mathcal{A}$ of an evaluation framework EF, i.e. $R\mathcal{A} = \mathcal{A}$, and $a \in \mathcal{A}$, $ra \in R\mathcal{A}$. ra is a triple $< id, \quad descriptor, \quad iw >$, where id is the name of the appropriateness requirement category, *descriptor* is a natural language description of the appropriateness requirement, and iw_{ra} defines a function of I that returns 0, 3, or 5 as importance weight based on priorities and policy of the company, where I is a set of importance judged elements ra in the selection criteria C of each category in $R\mathcal{A}$.

$$iw_{ra}(I) = \begin{cases} 1 & if \quad ra \quad is \quad optional \\ 3 & if \quad ra \quad is \quad recommended \\ 5 & if \quad ra \quad is \quad essential \end{cases} \quad (6)$$

Table 1. Requirements

Appropriateness	Importance *iw*	
Domain Appropriateness	High	5
Participant Knowledge Appropriateness	Medium	3
Knowledge Externalizability Appropriateness	Low	1
Comprehensibility Appropriatenss	Low	1
Technical Actor Interpretation Appropriateness	high	5

3   An Industrial Semantic Web Application

Our industry partner proposed an investigation of directory services. The problem consist of two databases containing the directory service data of Sweden and Norway. This means millions of records with information about people, streets, companies, and the respective country. Further, it is difficult to obtain accurate reasoning mechanisms or results from queries based on the databases solely.

The company wants to move into the emerging Semantic Web technologies. The investigation consists of creating ontologies that represent Norwegian and Swedish databases based on the databases schemata. The ontologies are created in three different ontology languages. RDF(S) [23] for analysing if the expressiveness of this language is enough for our case and OWL [24] and Topic Maps [25], two more complex ontology languages which offer more facilities for representing data in a proper way. The SQF representation requirements of the company are summarised in Table 1.

The databases were in different formats. The one containing the Swedish data was schematised in XML and the DTD was given, making it easier to approach the creation of the ontologies. The other, containing the Norwegian data, was schematised in a text document without a formal schema. Here, a reverse engineering process was required.

In order to obtain more objective basis for the analysis, the ontologies were 1) created using the [26] ontology building method, 2) given scope according to a common UML representation as a control ontology, and 3) implemented using Protégé 2000 as the editing tool.

3.1   UML Models Based on the Databases

A consolidated UML model for the two national databases is depicted in Figure 1. The UML model of the Swedish database is extracted from the DTD and the interpretations of each class are tailored while creating the model. The classes are used for representing the information of persons in the database. There is no distinction between persons and companies in the Swedish data.

The Norwegian data was provided in a text format, without a structured specification of classes, relations or attributes. As a re-engineering effort, we extracted a basic model from the file.

The Norwegian database model is an approximation to the original database. The original data was not considered definitive and was not adequate to build an ontology from. Thus, some attributes have been added and the structure of the database has been re-designed in order to be similar to the Swedish database. Finally, the consolidated

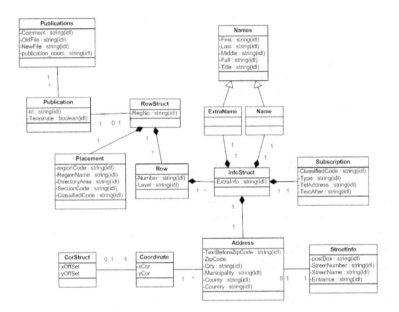

Fig. 1. UML model of the Swedish data

UML model consists of 14 classes, 14 relations, and 38 attributes. This model is used as control model for creating ontologies in the below representations.

RDF(S) ontology. Almost all the classes in the RDF(S) ontology are the same as in the UML model in Figure 1. The classes are: Publications, Publication, Row, Address, Subscription, Placement, StreetInfo, Coordinate and the two kinds of names with the superclass Names. The only classes missing are RowStruct and InfoStruct. These two classes are not necessary in RDF(S) because we are able to specify the domain and the "Allowed classes" for the instances properties. All the properties assigned to this slot will be applied to all its elements. The disadvantage of this method is that we have to create a class for the attribute RegNo since we want to apply the property to the attribute.

OWL ontology. The UML model represented in Figure 1 can be fully designed within OWL. All the classes and properties in the OWL ontology have a corresponding class or relationship in the OML model. OWL offers enough facilities to represent this model easily. Furthermore, we can, unlike within RDF(S), add more relations and restrictions to these relations. The model created is the same as the model created with RDF(S) and the model specified in UML.

Topic Maps ontology. The hierarchy of classes created in Topic Maps is almost the same as the one created within OWL and RDF(S). In the Topic Maps ontology there are the Name superclass with the children Name and ExtraName, Publications, Placement, Row and the other relevant classes extracted from the UML model.

The main difference is the parent class of every class. This parent is the Topic superclass. All classes of our model inherit the properties of Topic, and we have a better organisation of our model. The attributes (slots or facets), which can be just a typical slot typed as Integer or String or an instance of any class, can be defined in two different ways. Just like a simple slot or like a Topic Name, this allows us to create more specific relations. To create associations we follow the same way as the one used to create the classes or the attributes. First creating an instance of a Topic, and then referencing the association to the Topic or creating attributes like instances, as with OWL or RDF(S).

4   Comparative Evaluation of Representations

The adapted SQF evaluation model provides an instrument to evaluate quality of Web representation languages. Below, we summarise the evaluations of RDF(S), OWL and Topic Maps in the general terms, independent of target application. In the evaluation table (Table 2), the columns are the various appropriateness levels from section 2.2, NL description of the coverage, and the assigned criteria weights.

RDF(S). The table shows that RDF(S) provides the basic elements in order to satisfy the Domain Appropriateness requirements, because it covers the Structural Per spective fully and partially the Object Perspective and gets a weight of 1. The other appropriateness are covered by RDF(S) only providing the basic requirements of these appropriateness and therefore have the weight of 1.

OWL. The table shows that OWL provides the basic elements in order to satisfy the Domain Appropriateness requirements, because it covers the Structural Perspective and Rule Perspective fully and partially the Object Perspective and gets a weight of 2. The Technical Actor Appropriateness is fully covered because OWL is designed to support this appropriateness and gets a weight of 2. The other appropriateness are covered by OWL only providing the basic requirements of these appropriateness, thus they receive a weight of 1.

Topic Maps. The table shows that Topic Maps provides the basic elements to satisfy the Domain Appropriateness requirements, because it covers the Structural Per spective fully and partially the Object Perspective and gets a weight of 1. Topic Maps also covers Technical Actor Appropriateness because it provides more el ements to detailed data and facilitate the automatic reasoning by the information agents and gets a weight of 2. The other appropriateness are covered by Topic Maps only providing the basic requirements of these appropriateness and receives a weight of 1.

4.1   Comparison of the Three Languages

After this analysis of the languages in Tabel 2, it is possible to compare with the quality requirements of the cooperating company. First, the total coverage weights Tw_i for each representations i are calculated. In Table 3, we have summarised which of the studied ontology building methods that meet the situated, quality-based requirements specified by our industry partner. Here, the importance weights from Table 1 are multiplied by the coverage weights from Table 2 and total weights calculated using Equation 7 are used as overall feasibility rate for supporting the choice of ontology building guidelines.

$$Tw_i = \sum_{ra \in \mathcal{A}} (cw_{ra} \times iw_{ra}) \tag{7}$$

The weights assigned in Table 1 are based on the requirements specified in Section 3 , in Table 1. The requirements are specified in natural language. We have translated these requirements to a more proper representation in order to make a comparison with results obtained from the language analysis. The situated comparison of the languages with the requirements are shown in the Table 3.

In Table 3 we have added a new column where it is specified the final weight of the languages. We have multiplied the results obtained for each language by the weight of the appropriateness given in the Table 1. With this new column and the new weights

Table 2. Analysis of the three languages

Appropriateness	Description	cw
RDF(S)		
DA	RDF(S) covers structural perspective. Objective perspective is partially covered. Does not cover rule, behavioural, functional, actor or role perspective	1
PKA	Dependent on the desinger's experience	-1
KEA	Posibility to model determinated situations	1
CA	RDF(S) elements are easily distinguished. The number of phenomena are reasonable. The structure of RDF(S) is partially consistent.	1
TAIA	RDF(S) partially covers this appropriateness. Provides basic elements for automatic reasoning.	1
OWL		
DA	OWL covers structural and rule perspective. Objective perspective is partially covered. Does not cover behavioural, functional, communication, actor, or role perspective.	2
PKA	Dependent on the desinger's experience	-1
KEA	Domain dependent. Possibility for modelling main database concepts.	-1
CA	OWL elements are easily distinguished. The number of phenomena is reasonable. Symbol discrimination is not fully covered. The structure of OWL is consistent.	1
TAIA	OWL covers this appropriateness.	2
Topic Maps		
DA	Topic Maps covers structural perspective and partially object perspective. Does not cover functional, behavioural, rule, object, communication, actor or role perspective.	1
PKA	Dependent on the desinger's experience	-1
KEA	Domain dependent	-1
CA	Topic maps diferentiates between symbols. The number of phenomena are reasonable, but less reasonable than RDF(S) or OWL. The structure is partially consistent. Not enough expressive economy.	1
TAIA	Topic Maps covers this appropiateness.	2

Table 3. Situated comparison of languages

Appropriateness	RDF(S)	OWL	Topic Maps
DA	$1x5$	$2x5$	$1x5$
PKA	$-1x3$	$-1x3$	$-1x3$
KEA	$-1x1$	$-1x1$	$-1x1$
CA	$1x1$	$1x1$	$-1x1$
TAIA	$1x5$	$2x5$	$2x5$
Total weight	7	17	10

added we can distinguish more easily the most adequate language and eventually elaborate with slightly modified importance weights.

Thus, the most appropriate language in our case, is the Web Ontology Language, OWL. This is due to the elements that it provides for creating first order predicates. The other languages do not offer this feature because RDF(S) is not designed to offer these facilities and Topic Maps is orientated to create a better quality relationships between its elements but does not provide enough elements for constraining these relations.

5   Conclusions and Future Work

An evaluation of representations for Semantic Web applications was conducted adapting the extended [14, 22] Semiotic Quality Framework (SQF) [9, 11]. The comparative analysis was performed in two steps, one general evaluation, i.e. their applicability for Semantic Web applications in general, and one contextual, i.e. how appropriate they are for ontology development in a real world project in particular. The applicability of situated SQF was tested in a case study. The main results are as follows.

- SQF is suited for evaluating semantic web representations. Use of the numerical values for the weights and adoption of the PORE methodology [21] produce explicit evaluation results.
- In both steps, the general classification and the evaluation against the situated requirements, OWL [24] came out on top, meeting many of the evaluation criteria. This is also the only representation which support the rule perspective.
- Following OWL, Topic Maps [25] proved to be slightly better than RDF(S) [23].

The contribution of this paper is two-fold: First, an existing evaluation framework was tried out with other evaluation-objects than it has been used for previously; Second, numerical values and metrics were incorporated to the quality factors of SQF and thus, supporting qualification of weighted selection. The case study suggests that, given the small adjustments, the framework originally intended for evaluation of conceptual modelling languages, is applicable in evaluation of semantic Web representations, regardless if the analysis is used for their selection, quality assurance, or engineering

The concrete ranking of methods may be of limited use, as new ontology languages and methodologies and associated tools are developed, and as the existing languages evolve. Nevertheless, it can be useful in terms of guiding the current and future creators of such languages and modelling environments. Drawing attention to the weakness of current proposals, these can be mended in future proposals, so that there will be several high quality representations to choose from in the future. The underlying assumption for our work is that the option to choose appropriate representation suited for any application at hand may increase and widen the range and scalability of the Semantic Web ontologies and applications.

There are several interesting topics for future work, such as supplementing the theoretical evaluations with empirical ones as larger scale semantic Web applications arise utilising the empirical nature of [9], as well as evaluating more representations as they emerge, e.g. OWL-S and SWRL. Further possibilities are in investigating the appropriateness of the formalisation quality criteria in the Unified methodology [2] as a complement to the semiotic quality framework in order to conduct evaluation of the process oriented methodological frameworks that were out of scope of this study.

Acknowledgement

We would like to thank the company Invenio for their cooperation and invaluable support in conducting this research.

References

1. Berners-Lee, T., Hendler, J., Lassila, O.: The semantic web. Scientific American (2001)
2. Uschold, M., Grüninger, M.: Ontologies: Priciples, methods, and applications. Knowledge Engineering Review **11** (1996) 93–155
3. Gruber, T.R.: A translation approach to portable ontology specifications. Knowledge Acquisition **5** (1993) 199–220
4. Wand, Y., Weber, R.: Mario Bunge's Ontology as a formal foundation for information systems concepts. In Weingartner, P., Dorn, G., eds.: Studies on Mario Bunge's Treatise. Rodopi (1990)
5. Weber, R., Zhang, Y.: An analytical evaluation of NIAM's grammar for conceptual schema diagrams. Information Systems Journal **6** (1996) 147–170
6. Opdahl, A.L., Henderson-Sellers, B.: Ontological evaluation of the UML using the Bunge-Wand-Weber model. Software and Systems Modelling (SoSyM) **1** (2002) 43–67
7. Davies, I., Green, P., Milton, S., Rosemann, M.: Using Meta-Models for the Comparison of Ontologies. In: Proceedings of the Eighth CAiSE/IFIP8.1 International Workshop on Evaluation of Modeling Methods in Systems Analysis and Design (EMMSAD'03). (2003)
8. Lindland, O.I., Sindre, G., Sølvberg, A.: Understanding Quality in Conceptual Modeling. IEEE Software **11** (1994) 42–49
9. Krogstie, J., Sølvberg, A.: Information systems engineering - Conceptual modeling in a quality perspective. Kompendiumforlaget, Trondheim, Norway (2003)
10. Krogstie, J.: Using a Semiotic Framework to Evaluate UML for the Development of Models of High Quality. In Siau, K., Halpin, T., eds.: Unified Modeling Language: System analysis, design, and development issues. IDEA Group Publishing (2001)
11. Su, X., Ilebrekke, L.: A comparative study of ontology languages and tools. In Pidduck, A.B., Mylopoulos, J., Woo, C.C., Ozsu, M.T., eds.: Advanced Information System Engineering, 14th International Conference, CAiSE 2002, Toronto, Canada, May 27-31, 2002, Proceedings. Volume 2348 of Lecture Notes in Computer Science., Springer Verlag (2002) 761–765

12. Gómez-Péres, A., Corcho, O.: Ontology Languages for the Semantic Web. IEEE Intelligent Systems (2002) 54–60

13. Schuette, R.: Architectures for evaluating the quality of information models – a meta and an object level comparison. In Akoka, J., Bouzeghoub, M., Comyn-Wattiau, I., Métais, E., eds.: Proceedings of 18th International Conference on Conceptual Modeling, Paris, France. Volume 1728 of Lecture Notes in Computer Science., Springer Verlag (1999)

14. Hakkarainen, S., Hella, L., Strasunskas, D., Tuxen, S.: Choosing Appropriate Metod Guidelines for Web-Ontology Building. In: Proceedings of the 24th Internation Conference on Cenceptual Modeling (ER 2005). (2005) To appear.

15. Pohl, K.: Three dimensions of requirements engineering: a framework and its applications. Information Systems **19** (1994) 243–258

16. Becker, J., Rosemann, M., von Uthmann, C.: Guidelines of Business Process Modeling. In Aalst, W., Desel, J., Oberweis, A., eds.: Business Process Management, Models, Techniques, and Empirical Studies. Volume 1806 of Lecture Notes In Computer Science., Springer Verlag (2000) 30–49

17. Moody, D.L., Shanks, G.G., Drake, P.: Evaluating and Improving the Quality of Entity Relationship Models: Experiences in Research and Practice. In Ling, T.W., Ram, S., Lee, M.L., eds.: Proceedings of the 17th International Conference on Conceptual Modeling, Singapore (ER '99). Volume 1507 of Lecture Notes in Computer Science., Springer Verlag (1998)

18. Su, X., Ilebrekke, L.: 14. In: Using a Semitic Framework for a Comparative Study of Ontology Languages and Tools. IDEA Group Publishing Group (2004) 278–299

19. Sindre, G.: Hicons: a general diagrammatic framework for hierarchical modelling. PhD thesis, Norwegian Institure of Technology (1990)

20. Seltveit, A.H.: Complexity reduction in information systems modelling. PhD thesis, Norwegian Institute of Technology (1994)

21. Maiden, N.A., Ncube, C.: Acquiring cots software selection requiremtents. IEEE Software **15** (1998) 46–56

22. Hakkarainen, S.E., Strasunskas, D., Hella, L., Tuxen, S.M.: Weighted evaluation of ontology building methods. In: CAiSE 2005 Forum. (2005) To Appear.

23. Brickley, D., Guha, R.V., eds.: RDF Vocabulary Description Language 1.0: RDF Schema. W3C (2004)

24. McGuinness, D.L., van Harmelen, F., eds.: OWL Web Ontology Language Overview. W3C (2004)

25. Biezunski, M., Bryan, M., Newcomb, S.R., eds.: ISO/IEC 13250:2000 Topic Maps: Information Technology – Document Description and Markup Language. International Organization for Standardization (1999)

26. Noy, N.F., McGuinness, D.L.: Ontology development 101: A guide to creating your first ontology. Technical Report KSL-01-05, Stanford Knowledge Systems Laboratory (2001)

Towards Systematic Model Assessment

Ruth Breu and Joanna Chimiak-Opoka

University of Innsbruck, Institute of Computer Science,
Techniker Str. 21a, Innsbruck, Austria
{Ruth.Breu, Joanna.Opoka}@uibk.ac.at
http://qe-informatik.uibk.ac.at/

Abstract. In this paper a novel approach for the tool–based quality assurance of models is presented. The approach provides a meta model framework for domain specific and tool–independent quality assessment in heterogeneous model landscapes. In our framework we provide the concepts of queries, checks and views defined on meta model level and interpreted over the whole model landscape. Queries, checks and views are described in a predicative language based on the structures of the meta model.

1 Introduction

After years of intensive standardisation activities in the context of UML (Unified Modelling Language) and the development of methods and tools, models have found their way to real software development. More and more companies and organisations use models to fill the gap between informal textual descriptions of requirements and the realizing code. Usage scenarios of models range from the analysis of business processes and system requirements to the documentation of software architectures and model driven software development.

Not surprisingly the use of models in real applications reveals new requirements and challenges. One of these challenges is the **quality assurance** of the models developed. Complex model landscapes (sets of related models) as e.g. developed within software architecture documentation in general contain inconsistencies and gaps. Quality assurance of these model landscapes cannot be done by pure manual inspection or review but requires tool assistance.

Drawing an analogy to quality assurance of code one can identify at least two important sub–disciplines of model quality assurance: model testing and static analysis of models.

Model testing can be applied in cases where the models are attached with a kind of executability—like in model driven software development or model simulation. We will not deal with this aspect in this paper and refer to [1] for a testing approach in the context of workflow models.

In this paper we address the **static analysis aspect of models**. Modern static code analysis deals with the quantitative analysis of dependencies within the code and is intimately connected with the notion of code metrics [2]. Analogous approaches to static model analysis can be found in the context of UML

J. Akoka et al. (Eds.): ER Workshops 2005, LNCS 3770, pp. 398–409, 2005.

diagrams [3]. However, the proposed indicators of high quality models, such as for example general diagramming metrics (e.g. number of elements of diagrams, number of stereotypes of diagrams) or diagram–specific metrics (e.g. for class/package diagrams: depth of inheritance hierarchy, number of child classes, number of parameters), are far less accepted than indicators of high quality code.

In this paper we present a novel approach for model assessment which goes into a different direction. Our first observation is that the quality of a model landscape is mainly influenced by the **interplay of model elements and diagram types**. For example, consider a requirements specification consisting of the following set of models: use case model, scenarios referring to the use cases, class model, state diagrams and schematic use case descriptions. All these models are based on a set of *model elements* like use case, actor and business class. The use of these model elements in different models creates **complex interrelationships** and multiple sources of inconsistency. In the example, the initiating actor in the use case description has to be an actor related with the use case in the use case model. The model elements and their interrelationships cannot be defined in a general way for all kinds of UML diagram types but depend on the underlying method and application context.

A second observation is that in many cases the quality checks of a model landscape cannot be done in an automatic way but the quality manager can be supported by *views* that provide **aggregated information** about the model landscape. For instance, a view on an enterprise model may list all information objects and applications together with the information which information object is related with which application in the model. This information can be generated from the business process model and enables the quality manager to perform cross–checks on the model landscape. In general we distinguish *queries, checks* and *views*. Queries are functions on the current status of the model landscape returning some value or model element, checks are queries with Boolean result. Queries may be embedded in views that represent queries for multiple input.

The applications of our approach are many–fold contexts in which complex model landscapes are developed. This ranges from requirements specification to the documentation of software architectures and enterprise models. In particular, the concepts presented in the sequel are aligned with the requirements of three of our cooperation projects with industrial partners (MedFlow—Quality Assessment of Business Processes in Health Care, ProSecO—IT Security Assessment in Enterprises, Pro2SA—Model–Based Strategic Alignment).

An important further aspect coming out of these applications is the **heterogeneity of the model landscape**. We want to reason about model elements which are defined in (UML–)tools, semi–formal or formal text documents or even code. As a common syntactic framework we make therefore use of XML (Extensible Markup Language).

The sequel of the paper is structured as follows. In section 2 we present related work and give more information about the application context in our projects. Section 3 introduces the basic framework and concepts and in section 4 we introduce the concepts of queries, checks and views. Section 5 draws a conclusion.

2   Related Work

There are several approaches, mostly in industrial contexts that deal with quality checks of model landscapes. In the ARCUS project [4] in cooperation of the Bayerische Landesbank and sd&m a meta model for describing IT landscapes has been defined. An important element of this meta model are relationships between the model elements that define a notion of traceability. A plug–in for an UML tool supports a fixed set of queries over the model landscape. In a study at BMW [5] a set of quality checks for model landscapes that are input to an in–house MDA (Model Driven Architecture) tool has been developed. As further example Siemens Princeton has developed a tool for checking requirements specifications [6]. For the implementation these approaches use the scripting facilities or programming interfaces of UML tools or graphic programs such as Rational Rose, Together, Visio or Adonis [7, 8]. These three examples are indicators of the practical relevance of the topic. Our approach goes one level beyond in the respect that it provides a generic platform for describing checks and views over a given set of interrelated model elements.

Concerning the syntactic and semantic framework we use parts of the OCL (Object Constraint Language, v. 2.0) for describing navigations over model elements. Existing methods and tools are rather dedicated to homogeneous environments. For example the Executable UML [9] can be used to execute and test models defined in UML. This is enabled by a formal action semantics for models (in an action language) and specification of constraints tags (in OCL). And with the OCLE tool [10] it is possible to check UML models against well–formedness rules, methodological, profile or target implementation language rules expressed in OCL. It is also possible to obtain metric information about UML models. This tool also uses XML, but only as a data tier, while we use XML also for modelling purposes. Our method provides mechanisms to carry out checks for domain specific models defined in heterogeneous environment.

In the subsequent chapters we present the basic concepts of QUARC (QUAlity Requirement Checks). Currently we work on a tool–based realisation which will be presented in accompanying papers. QUARC is aligned with the requirements of three of our cooperation projects: MedFlow, ProSecO, and Pro2SA. In context of our projects the focus of QUARC is the automatic check of model consistency and the generation of views for supporting manual checks, e.g. concerning media or applications ruptures and appropriate tool support of the actors (MedFlow); the generation of aggregated views in a highly linked heterogeneous model landscape (ProSecO); information aggregation in model landscapes and evaluation of quantitative data associated with model elements (Pro2SA).

Although the case study presented in the subsequent sections has been taken out of the MedFlow project, the method is not specific for clinical process and it can be also applied to models from other domains (e.g. industrial, commercial, enterprise). In MedFlow we develop an approach for the systematic quality check of models describing business processes in health care. The background of this project is the task of targeting standard hospital information systems towards the needs of complex organisational processes in hospitals. More about this project can be also found in [11].

3   Basic Concepts

In this section we introduce the basic concepts and model packages our approach is based on. Because the concepts are tightly correlated, we used arrow symbols for cross–definitions ($\rightarrow$ *definition*).

Model. A model (at instance level) is a structured document that is subject of the quality assessment. We consider any type of UML models like class diagrams and sequence diagrams, but also text documents or code. Models at instance level are based on $\rightarrow$ *meta model elements* to describe properties of systems. Each model at instance level has an associated $\rightarrow$ *model type*.

Meta model. The meta model defines the universe of discourse for the quality assessment. The meta model is a class diagram modelling the $\rightarrow$ *model elements* and their relationships. This class diagram is contained in the $\rightarrow$ *meta model package*. A meta model may be associated with a specific method (e.g. use–case based requirements specification), a particular application domain (enterprise models, embedded systems) or with a particular development environment (e.g. a meta model associated with an MDA–tool).

Meta model element. A meta model element is a class in the $\rightarrow$ *meta model* describing a basic concept used in the $\rightarrow$ *models* at instance level. Examples for meta model elements are *actor, information resource, business process, action* and *logical tool*. Meta model elements are the basic units over which queries and views can be formulated. Meta model elements may have attributes (e.g. the medium of an information resource) and may be linked with other meta model elements. These links have to be directed (in either or both directions) indicating where the $\rightarrow$ *model type package* link is maintained.

Meta model package. The meta model package contains the $\rightarrow$ *meta model*, the $\rightarrow$ *EM–mapping*, the $\rightarrow$ *model type package*, and the queries, checks and views.

Model Type. A model type groups models in a category. A model type (at instance level) is subject of the quality assessment. The model type characterises the role of the model within the underlying method. For instance, we may have model types *business class diagram* and *technical class diagram* in a context where we assess the documentation of software architectures. The interdependencies of the model types are described in the $\rightarrow$ *model type package*.

Model type package. The model type package contains a class diagram describing the interrelationships between the $\rightarrow$ *model types*. We call an instance of this class diagram a **model landscape**.

EM–mapping (Meta Model Element–Model Type Mapping). The EM–mapping maps the $\rightarrow$ *meta model elements* to the $\rightarrow$ *model types* defining in which model type the meta model elements are defined and used. As an example, the model element *information* is defined in the *Information Model* and used in the *Business Process Model*.

In the sequel we provide an example together with more detailed information about the basic concepts. The description of queries, checks and views is introduced in section 4.

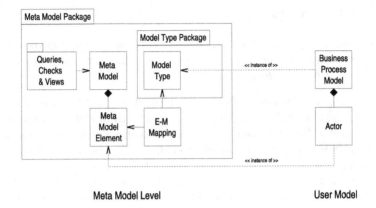

Fig. 1. The notions on meta model level and user model

Figure 1 summarises the notions at meta model level and user model level together with their interrelationships. The two levels correspond to the M2 layer models (Meta Model Level) and M1 layer models (User Models) defined in the MOF (Meta–Object Facility) Metadata Architecture.

3.1 Meta Model

Example 1. Figure 2 depicts (a portion of) the meta model for the quality assessment of clinical process descriptions. Processes are defined in a hierarchical way based on the notion of sub–processes and actions. Each action is associated with the executing actor, input and output information and logical tools supporting the executing actor.

3.2 Model Type Package

The model type package describes the model landscape that is subject of the quality assessment. The model types are related with «uses» relationships which means that model elements of one model type are used in the other model type. Aggregation is used for a hierarchical structuring of model types.

Example 2. Figure 2 depicts the set of model types used in our case study.

Each model type is associated with a type which is either an UML diagram type or XML, i.e. we assume non–UML models to be interconnected via XML structures.

Example 3. The following models, from Example 2, are defined in UML: *Business Process Model* as activity diagrams, *Organisation Model* and *Information Model* as class diagrams, *Logical Tool Model* as component diagrams and finally *Physical Tool Model* as deployment diagrams. The other models are defined in XML: *Description of Actions* and *Permission Model*.

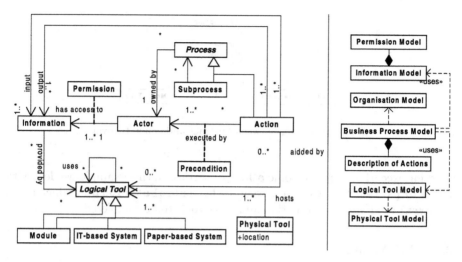

Fig. 2. A meta model (on the left) and model types (on the right) for clinical process

For each model type we assume an XML–representation. We have chosen XML because it is a standard notation and already supported in the UML–context by XMI (XML Metadata Interchange) as a standard format for model interchange. The most important features of XMI are its built–in nesting mechanism and the possibility of transformation from OCL–navigation–expressions to XPath–expressions.

In our case study the information model is an UML class diagram interconnected via its XMI representation, whereas the description of actions is a text document with an XML interface. The action description contains an informal description and information about the executing actor, input and output information and the logical and physical tool used in this action.

Example 4. Figure 3 depicts an action description at instance level. The XML description is based on our XML Schema for action description (due to limited space we do not present its full syntax here), in which the following attributes of an action are defined: **name** of the action, **role** involved in execution of it, **input** and **output** information needed or produced by it, **tool**, both logical and physical,

```
1   <action name="check of patient's data">
2     <role name="control station"/>
3     <input>
4       <information name="referral">
5     </input>
6     <tool logical="MEDAS (KIS)" physical="PC-LST-1"/>
7     <tool logical="Power Chart (KIS)" physical="PC-LST-1"/>
8     <tool logical="RIS"  physical="PC-LST-2"/>
9   </action>
```

Fig. 3. Sample Ation description

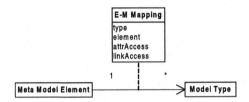

Fig. 4. Mapping between abstract and concrete level

used for execution of it and some other properties. In this example we deal with `check of patient's data` action executed by `control station` role, which needs `referral` as an input information and three tools to be completed.

3.3 EM–Mapping

The EM–mapping provides the interconnection between meta model elements and the model types. More precisely we define for each meta model element,

- in which model type it is defined,
- in which model types it is used, and
- how attributes and outgoing associations can be retrieved.

Figure 4 illustrates the kind of information that is provided for each meta model element. The type is either *defined* or *used*, the element attribute maps the meta model element to some model element of the target model type. For UML diagram types this means that the meta model elements are mapped to elements of the UML meta model (in some cases the meta model may itself contain elements of the UML meta model). The attributes attrAccess and linkAccess define the access to the attributes and outgoing links of the meta model element at XML level, an aspect that is not treated in more detail in this paper.

Example 5. Table 1 depicts a part of the mapping for the meta model elements in our case study.

Table 1. Sample EM–Mapping (Schematic)

Meta Element	Type	Model Type	Element
Action	def	*Description of Actions*	XML::action.xsd::action
Action	use	*Business Process Model*	UML::Activity Diagram::Action
Information	def	*Information Model*	UML::Class Diagram::Class::Information
Information	use	*Description of Actions*	XML::action.xsd::action::input::information
Information	use	*Description of Actions*	XML::action.xsd::action::output::information
Logical Tool	def	*Logical Tool Model*	UML::Component Diagram::Component
Logical Tool	use	*Description of Actions*	XML::action.xsd::action::tool.logical
...	...	...	...

4   Queries, Checks and Views

Based on the structure of meta model elements in the next step a set of queries, checks and views can be defined. These are defined by method responsibles whose task is to assist developer teams in the systematic quality assessment of the models developed.

Query. A query is a function over meta model element instances returning a value (e.g. a Boolean value or an integer) or meta model element instance(s). The result of a query may depend on the input parameters and the network of currently existing instances of meta model elements in the model landscape. The goal of a query is to provide the modeller with information about the model landscape.

Example 6. Examples of queries in our case study may be the following: Amount of actors in the model landscape; The set of logical tools an actor is related with in the business process model (via the actor–action–logical_tool relationships).

Check. A check is a query with Boolean value as result. The goal of a check is to assess a model landscape based on a given constraint. Moreover, we associate each model landscape with a set of predefined (well–formedness) checks that are related with the model type package and the EM–mapping. The user models have to conform to the model structure described in the model type package. For instance, each model element that is *used* in the model landscape should also be *defined* in some model (checking the consistency of *used* and *defined* relations in the EM–mapping).

Example 7. An example of a check is the following: There exists at least one actor and one information class in the model landscape.
An example of a predefined check: Each action used in *Business Process Model* has to have a textual description in *Description of Actions*.

View. A view is a query whose result is represented for all (or a restricted set of) input elements and may be equipped with further information regarding the quality assessment of the result. The goal of a view is to present aggregated information over a model landscape and to support the modeller in model inspection. The benefit of the view is to support the modeller in a cross–check of the business process model.

Example 8. Table 2 depicts the example view `InformationInLogicalTool` listing all information types and logical tools defined in user models and indicating if the given information is related with the given logical tool. Here *Referral*, *Diagnostic Findings* and *Image* are classes in the *Information Model* (class diagram) and *KIS*, *PACS*, *PaterNoster* are components in *Logical Tool Model* (component diagram). The result is defined by the OCL–like expression as given in Example 9 and is true for a given information and a given logical tool, if the information is saved in the tool.

Queries, checks and views in our approach are described by OCL based predicative language expressions that are constructed over the class diagram of the

Table 2. Sample View InformationInLogicalTool

Information	Logical Tool	Result
Referral	KIS	true
Referral	PACS	false
Referral	PaterNoster	true
Referral	...	...
Diagnostic Findings	KIS	true
Diagnostic Findings	PACS	false
Diagnostic Findings	PaterNoster	true
Diagnostic Findings	...	...
Image	KIS	false
Image	PACS	true
Image	PaterNoster	true
Image	...	...
...	...	...

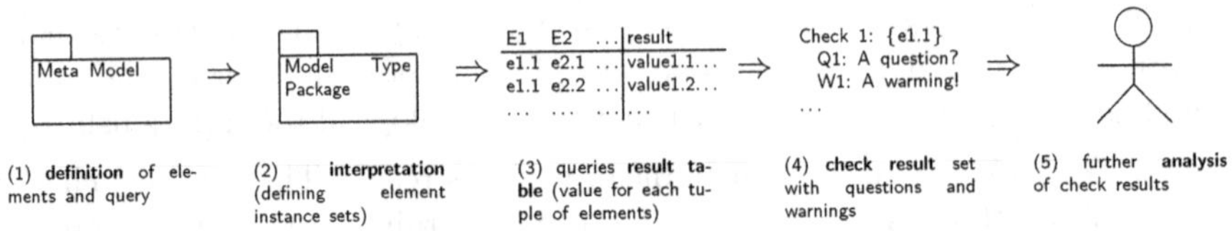

Fig. 5. The process of defining, interpreting and executing a view

meta model. These expressions are embedded into XML structures that provide the syntactical framework.

In Fig. 5 the process of defining, interpreting and executing a view is depicted. In the first step the view is defined over the meta model elements. Then in the second step the sets of instances of meta elements are collected from the corresponding user models (via the model database). The result values are calculated for each combination of elements of the given sets and in the third step the result table is presented. In the fourth step the result table is subject of further analysis and automatically executed checks. The result of the fourth step are sets of elements fulfilling the condition in the given check. For non–empty check sets questions and warnings may be shown to the user. The questions are used in checks, for which additional analysis of the result is needed. The warnings could be used in checks for well–formedness rules. In the fifth step the analysis of the questions and warnings is made by user.

In the sequel we will present in more detail the structure of queries, checks and views together with sample expressions.

4.1   Queries and Checks

Checks and queries are defined as functions over individual or aggregated elements. The formal definition of such a function is expressed as follows:

$$\mathcal{Q}(\mathbf{p}_1 : \mathcal{T}_1, \ldots, \mathbf{p}_n : \mathcal{T}_n) \quad \mathcal{T} : \mathcal{E} \tag{1}$$

where $\mathcal{Q}$ is a query or check name; $\mathbf{p}_i$ is an instance of meta model element $\mathcal{T}_i$; a result is an expression $\mathcal{E}$ of a given type $\mathcal{T}$. If type $\mathcal{T}$ is Boolean we call the function $\mathcal{Q}(\cdot)$ a **check**.

Example 9. A simple check could answer the question if a given information is saved in a given logical tool:

```
InformationInLogicalTool(i:Information, lt:LogicalTool) Boolean
    : i.logicaltool.select(name = lt.name).notEmpty().
```

4.2   Views

The views are built according to information from concrete models and formally we define them as follows:

$$\mathcal{V}(\mathcal{T}_1 \; [\mathcal{F}_1], \ldots, \mathcal{T}_n \; [\mathcal{F}_n]) \quad \mathcal{T} \tag{2}$$

where $\mathcal{V}$ is a view name; $\mathcal{T}_i$ is one of n types (meta elements), which could be optionally filtered with a given filter $\mathcal{F}_i$.

Let $\mathcal{P}_i$ denote the set of instances of the given meta element $\mathcal{T}_i$ occurring in the user models. To complete the view we have to consider all tuples from the Cartesian product $\mathcal{P}_1 \times \cdots \times \mathcal{P}_n$. For each tuple $(\mathbf{p}_1 : \mathcal{T}_1, \ldots, \mathbf{p}_n : \mathcal{T}_n)$ we calculate the result $(\mathbf{r})$ using a query defined in the view $(\mathcal{Q}(\cdot) : \mathcal{T})$, thus $\mathbf{r} = \mathcal{Q}(\mathbf{p}_1, \ldots, \mathbf{p}_n)$. We extend the tuple adding the result and we therefore obtain extended tuples in form $(\mathbf{p}_1 : \mathcal{T}_1, \ldots, \mathbf{p}_n : \mathcal{T}_n, \mathbf{r} : \mathcal{T})$. The result of the view is a set of extended tuples.

Example 10. We define a view for informations and logical tools:

$$\mathcal{V}_{InformationInLogicalTool}(Information, LogicalToool) \; Boolean.$$

In this view we use the query `InformationInLogicalTool(·)`, as defined in Example 9. The result is a set of tuples:

```
Tuple(i : Information, lt : LogicalTool, r : Boolean),
     where r = InformationInLogicalTool(i,lt).
```

An example result of an evaluation of the view is shown in Table 2.

If we would like to consider only a subset of the input set of instances we have to apply a filter. The filter $\mathcal{F}_i$ defines a constraint for the set $\mathcal{P}_i$. A set with filter is defined as $\mathcal{P}_i = \{\mathbf{p}_i : \mathcal{F}_i(\mathbf{p}_i)\}$.

Example 11. If we would like to consider only physical tools (**pt**) located in the radiology ward then we use the following filter.

$$\text{pt.location='Radiology'}$$

Complementary Checks. Additionally we support the definition of complementary checks, questions and warnings within the view. Complementary checks are defined as queries over the set of tuples (result table). As a result of a complementary check we obtain the set of elements or tuples fulfilling the query.

Example 12. Let's say we would like to find unsaved information. We use the result of the view defined in Example 10 ($\mathcal{V}_{InformationInLogicalTool}$), and make a complementary check over this result. If we denote the result by `view:Set(Tuple)` then we can find all unsaved information using the following expression:

```
context view def:
collect (info : Information |
        self.select(i = info and r = 'true').size() = 0).
```

If we obtain a non–empty set as a result of complementary check, the warnings (see Example 13) or the questions (see Example 14) are shown. **Warnings** and **questions** are described in natural language and are not processed automatically. Warnings could be defined for checks over well–formedness rules.

Example 13. If the result of the complementary check defined in Example 12 is a non–empty set then the set will be listed and the warning *Each Information should have a medium!* will be shown.

Example 14. Let's say we would like to find redundant information, i.e. information saved in many logical tools. We use the result of the view defined in Example 10 ($\mathcal{V}_{InformationInLogicalTool}$), and make a complementary check over this result using the following expression:

```
context view def:
collect (info : Information |
      self.select(i = info and r = 'true').size() > 1).
```

If the result of the complementary check is a non–empty set then the set will be listed and the question *Is consistency of the redundant information warranted?* will be shown.

5   Conclusion

In the preceding sections we have presented a novel approach for the tool–based quality assurance of models. A main idea of this approach is to provide a meta model framework supporting application–specific quality assessment, tool–independent expression of quality assessment criteria and quality assessment in heterogeneous model landscapes both comprising (UML) models and textual models.

In our approach we provide the concepts of queries, checks and views. Queries are model retrievals, checks support automatic check of model constraints. Views support the modeller with aggregated information about the model landscape and may be associated with informal checks and heuristic quality indicators.

The approach presented is work in progress. Currently we both work on the final definition of an OCL based predicative language to describe queries, checks and views and on the software architecture of the related tool. Our work is driven by practical requirements of cooperation projects with industrial partners.

References

1. Breu, R., Breu, M., Hafner, M., Nowak, A.: Web service engineering—advancing a new software engineering discipline. In: Proc. of 5th International Conference on Web Engineering. (2005) (accepted).
2. Fenton, N.E., Pfleeger, S.L.: Software Metrics — A Rigorous and Practical Approach. Thomson, London (1997)
3. Gronback, R.: Model validation: Applying audits and metrics to uml models. In: Proc. of Borland Conference. (2004) (available on http://bdn.borland.com/borcon2004/).
4. Heberling, M., Maier, C., Tensi, T.: Visual Modeling and Managing the Software Architecture Landscape in a Large Enterprise by an Extension of the UML. In: Second Workshop on Domain-Specific Visual Languages. An OOPSLA Workshop, Seattle, WA (2002)
5. Jug, F.: Methods and techniques for quality assurance in software development process in bmw group (in german). Master's thesis, Technical University of Munich, Dep. of Computer Science (2004)
6. Berenbach, B.: Evaluating the quality of a uml business model. In: Proc. of 11 IEEE International Requirements Engineering Conference, Monterey, CA, USA (2003)
7. Junginger, S., Kuehn, H., Strobl, R., Karagiannis, D.: The next generation business process management toolkit ADONIS (in German). In: Wirtschaftsinformatik. Volume 42. University of Trier (2000) 392–401
8. BOC: Adonis. http://www.boc-eu.com/advisor/adonis.html (2000) access 2005-04-24.
9. Mellor, S.J., Balcer, M.J.: Executable UML. A Foundation for Model-Driven Architecture. Addison-Wesley (2002)
10. LCI team: Object constraint language environment (2005) Computer Science Research Laboratory, "BABES-BOLYAI" University, Romania.
11. Saboor, S., Ammenwerth, E., Wurz, M., Chimiak-Opoka, J.: Medflow—improving modelling and assessment of clinical processes. In: Proc. of 19th Medical Informatics Europe, MIE 2005. (2005) (accepted for oral presentation).

A Fuzzy Based Approach to Measure Completeness of an Entity-Relationship Model

Tauqeer Hussain, Mian M. Awais, and Shafay Shamail

Department of Computer Science, Lahore University of Management Sciences,
DHA, Lahore, Pakistan 54792
{tauqeer, awais, sshamail}@lums.edu.pk

Abstract. Completeness is one of the important measures for semantic quality of a conceptual model, an ER model in our case. In this paper, a complete methodology is presented to measure *completeness* quantitatively. This methodology identifies existence of functional dependencies in the given conceptual model and transforms it into a multi-graph using the transformation rules proposed in this paper. This conversion can be helpful in implementing and automating computation of quality metrics for a given conceptual model. The new Fuzzy Completeness Index (FCI) introduced in this paper adopts an improved approach over Completeness Index proposed by authors in the previous research. FCI takes into account the extent a functional dependency has its representation in the conceptual model even when it is not fully represented. This partial representation of a functional dependency is measured using the fuzzy membership values and fuzzy hedges. The value of FCI varies between 0 and 1, where 1 represents a model that incorporates all the functional dependencies associated with it. Computation of FCI is demonstrated for a number of conceptual models. It is illustrated that the quality in terms of completeness can effectively be measured and compared through the FCI based approach.

1 Introduction

Conceptual modeling is one of the most demanding and challenging steps in the database design methodology. Various parameters, frameworks and methodologies [1-5] have been proposed in the literature to define and evaluate quality of a conceptual model. Among these parameters, semantic quality is of major interest in the context of conceptual modeling. Assenova and Johannesson [6] define semantic quality as the degree of correspondence between the schema and the problem domain. Semantic quality can further be classified as *validity* and *completeness* of a conceptual model where validity suggests that the model should only contain true statements about the domain and completeness means that it should contain all true statements about the domain. Thalheim has also suggested design quality parameters of a conceptual model [7] which include Completeness, Naturalness, Minimality and Flexibility. Thalheim describes completeness as the representation of all relevant features of the application domain.

Although these parameters contribute towards a better understanding of quality aspects of a conceptual model, there are only a few measures and metrics [4, 9, 10, 11]

J. Akoka et al. (Eds.): ER Workshops 2005, LNCS 3770, pp. 410–422, 2005.

which are quantitatively defined in the literature. Therefore, the need of defining appropriate quantitative measures has been emphasized [1,8] in order to reduce subjectivity in evaluation of quality of conceptual models. For developing a conceptual model, the entity-relationship model or entity-relationship diagram (ERD) has established its wide acceptance in the database community. In this paper, therefore, we are primarily concerned with the semantic quality measuring completeness of an Entity-Relationship (ER) model. Our motivation is to define an appropriate metric which can easily be used in practice to measure completeness of an ERD. In this regard, the authors have already proposed a metric named Completeness Index (CI) in their previous research [12]. This metric views functional dependencies identified in a problem domain as important features, statements or requirements of the problem which should be 'completely' present in a conceptual model. Thus, this view supports definitions of completeness given by Assenova and Johannesson, and Thalheim. It is worth mentioning here that, in practice, functional dependencies (FDs) belonging to a problem domain are identified during the conceptual modeling phase as their identification is critical to know the key attributes of every entity type in the ERD and also due to the fact that it is a lot easier to define a set of functional dependencies while concentrating on attributes of just individual entity types (in the conceptual modeling phase) as compared to selecting from all attributes of a database schema (in the logical modeling phase).

This paper outlines an attempt improving upon the previously suggested metric. Previously, Completeness Index was defined as the ratio between number of functional dependencies represented in an ERD and total number of FDs identified in the problem domain, that is, $CI = n(f) / n(\Im)$. The definition uses a projection set f of $\Im$ on ξ defined as $f = \pi_\Im (\xi)$ representing the set of functional dependencies represented in the ERD and where $\Im$ is the set of all functional dependencies identified from a problem domain, and ξ denotes the ERD. The definition also uses a count operator 'n' such that $n(\Im)$ represents total number of functional dependencies in $\Im$ and $n(f)$ represents total number of functional dependencies in f. Formally, $n(f) = \sum count(FD_i)$ where $count(FD_i) \in \{0,1\}$; 0 if i^{th} functional dependency in $\Im$ is not represented in the ERD; otherwise 1. The use of CI has been shown to be effective [12] in knowing the fraction of the total functional dependencies that is represented in an ERD. The motivation of this paper is to measure the extent a functional dependency is represented or not represented in an ERD rather than having just binary values for $count(FD_i)$. That is, $0 \le count(FD_i) \le 1$. This suggests the need for introducing a fuzzy measure of the conceptual model for measuring completeness that indicates the degree to what extent a functional dependency is represented in an ERD. This approach is especially helpful in comparing the quality of two given conceptual models in terms of completeness if both have the same CI which is less than 1, as elaborated in the following example:

Consider the two ERDs given in Fig. 1 and Fig. 2 developed for the same problem domain. Let $\Im$ be defined as $\{A \rightarrow B, B \rightarrow C, W \rightarrow Y, Y \rightarrow Z, X \rightarrow W\}$ for this problem. Now, for Fig. 1, $f = \{A \rightarrow B, X \rightarrow W\}$, so CI = 2/5 = 0.4. Similarly, for Fig. 2, $f = \{A \rightarrow B, W \rightarrow Y\}$ and hence CI = 2/5 = 0.4. Therefore, we conclude that based upon CI, Fig. 1 and Fig. 2 are equally good for completeness; although they are not, intuitively, at least.

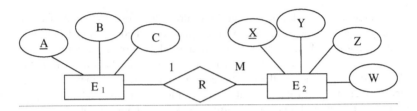

Fig. 1. A sample ERD for computing CI

In this paper, we define the concept of Fuzzy Completeness Index (FCI) and propose rules to fuzzify [13] the measure of completeness index. We also elaborate, with the help of an example:

- the computation of FCI, and
- that, the quality in terms of FCI of a conceptual model gets improved following the schema transformation rules already given by authors in [12].

The paper is arranged in the following way: In section 2, a set of rules for transforming an ERD to a graph is presented. Our approach requires for ease of understanding and implementation that a given ERD be first transformed to a graph representation having nodes and edges. Section 3 summarizes schema transformation rules which can be used to improve a given ERD. Section 4 introduces the methodology for calculating the FCI; section 5 discusses the application of FCI to three different ERDs which are actually schema transformations of an ERD. The final section concludes the findings of this research.

2 Transforming ERD to TAS Graph

In this section, we propose a technique which can be applied to transform any ERD to a graph. This transformation is required to understand the computation of FCI as well as to automate its computation. The proposed technique uses a colored multi-graph representation [14] to define a Tauqeer-Awais-Shamail (TAS) graph which then can be searched to compute FCI.

2.1 Transformation Rules

Following are the rules proposed for the transformation of a given ERD:

1. For every entity type E_i, draw a colored node e_i on the graph. Such a node is called an E-node.
2. For every non-recursive relationship type R_k between entity types E_i and E_j, draw a node r_k on the graph. Such a node is called an R-node. Also draw edges connecting r_k with E_i and E_j.
3. For every attribute A_j, which is not composite or part of a composite attribute, draw a node a_j. Such a node is called an A-node.
    a. If A_j exists on an entity type E_i, then draw an edge connecting a_j and e_i.
    b. If A_j exists on a relationship type R_i, draw an edge connecting a_j and r_i.

4. For every composite attribute A_j on the entity type E_i which is broken into attributes $B_1, B_2, ..., B_k$, draw nodes $b_1, b_2, ..., b_k$ respectively. Such nodes are called SA-nodes. Also draw edges connecting each SA-node with e_i.
5. For every recursive relationship type R_j with entity type E_i, draw a node r_j and a bi-directional edge between r_j and E_i.

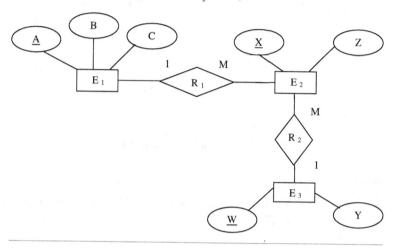

Fig. 2. An alternate ERD for the ERD in Fig. 1

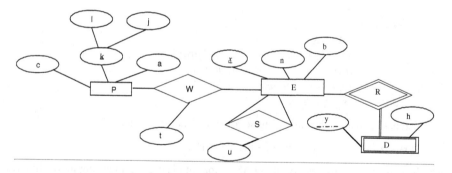

Fig. 3. A Sample ERD for Transformation to TAS Graph

For demonstration, the above-mentioned rules are applied to an ERD given in Fig. 3 and a TAS graph is generated as shown in Fig. 4.

2.2 Mapping Functional Dependencies on TAS Graph

When an ERD has been transformed to its TAS graph, the next step is to mark FDs on the graph as per our proposed algorithm 1.

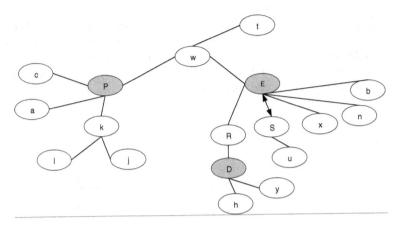

Fig. 4. TAS Graph for the ERD in Fig. 3

Algorithm 1

Let 𝔍 be the set of functional dependencies identified from the problem domain.
For every FD X → Y in 𝔍:
If X is a singleton set,
          search A-node X on the TAS graph,
          search A-node Y on the TAS graph,
          draw a directed edge from X to Y
Else
          together := null
          For every A_i ∈ X
                    search A-node A_i on the TAS graph,
                    draw a directed edge from A_i to Y,
                    together := together + edge (A_i, Y)
          and_edges (together)

For instance, consider the following set of functional dependencies:

$$\{x \rightarrow b,\ x \rightarrow n,\ x\ y \rightarrow h,\ l\,j \rightarrow c,\ l \rightarrow a\}$$

which is used to map these on the TAS graph, applying the mapping algorithm 1, as shown in Fig. 5.

2.3 Searching TAS Graph

Searching the TAS graph for any E-node, A-node or R-node is quite straight forward, so functions like *findEnode( )*, *is_Enode (Enode)*, *findAnode(Enode)*, *is_Anode (Anode)*, *findRnode( )* and *is_Rnode (Rnode)* can be defined. However, identifying key(s) for every entity type involves application of functional dependencies marked on the TAS graph. For this purpose, an algorithm for a function *findKey (Enode)* can be developed which searches an A-node from where uni-directed edges can span all A-nodes connected with the respective E-node.

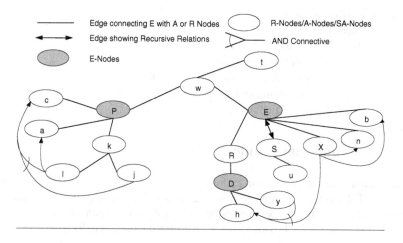

Fig. 5. TAS Graph with Functional Dependencies

3   Schema Transformation Rules

For ease of reference, rules for schema transformations and the related definitions as given in [12] are presented below:

Definition 1 - Functional Dependency for an Entity Type: *There exists a functional dependency X → Y for an entity type E if for every entity in E the set of attributes, X, uniquely determines the attribute Y. Formally, let e_i represent an ith entity in E where i=1,2,3, …, n and $e_i[X]$ represent the value of attribute X for e_i, then there exists a FD X → Y if for any two entities e_i and e_j in E:*
$$e_i[X] = e_j[X] \text{ implies } e_i[Y] = e_j[Y]$$

Definition 2 – Key of an Entity Type: *Let Attr(E) represent the set of all attributes of an entity type E and there exists a set of FDs $X → Y_i$ where $Y_i \subseteq$ Attr(E) such that $X \cup (Y_i) = Attr(E)$ and there does not exist any attribute Z of E such that $Z \in X$ and X-{Z} → Attr(E) then X is a key of E.* It can be denoted by Key(E).

Definition 3 – Key Attribute of an Entity Type: *Let K be any Key(E) and there exists an attribute A such that $A \in K$ then X is a key attribute of E.*

Definition 4 – Non-Key Attribute of an Entity Type: *Let K be any Key(E) and there exists an attribute $A \in Attr(E)$ such that $A \notin K$ then X is a non-key attribute (NK) of E.*

Schema Transformation Rule 1:

> For every FD X → Y of an entity type E in a given ERD,
>> where $X \subset$ Key (E), and $Y \subset$ [Attr(E)-Key (E)]
>>> i) create an entity type E' such that:
>>>> Attr(E') := X ∪ Y, and Key(E') := X
>>> ii) mark E as a weak entity type of E'
>>> iii) Attr(E) := Attr(E) – [X ∪ Y]

Schema Transformation Rule 2:
    For every FD X → Y:
        where X,Y ⊂ Attr(E)-Key(E)
            i) create an entity type E' such that:
                Attr(E') := X ∪ Y, and Key(E') := X
            ii) create a relationship type R between E and E'
            iii) Attr(E) := Attr(E) – [X ∪ Y]

4   Fuzzy Completeness Index (FCI)

In this section, we introduce a fuzzy logic [13] based approach to measure complete-ness of an ER model and propose a metric called Fuzzy Completeness Index (FCI). The fuzzy concept of membership values and hedges has been incorporated, in this research, to define the FCI of an Entity-Relationship Diagram (ERD).

4.1   Proposed Fuzzy Completeness Index (FCI)

The following expression represents the Fuzzy Completeness Index (FCI):

$$FCI_{overall} = \frac{\sum_{i=1}^{i=n} FCI_i}{n(\Im)}, \tag{1}$$

where FCI_i is the FCI value of ith functional dependency.

For a given functional dependency FD_i: $X \rightarrow Y$, the FCI value can be calculated using the formula given below.

$$FCI_i = 1 - \frac{TAS(FD_i)}{A_{Total}}, \tag{2}$$

where TAS is a fuzzy operator applied on a functional dependency for a given TAS graph, and A_{Total} represents the total number of attributes present in the ERD under consideration. The TAS operator works on the basis of following rules:

Rule F1: The value for $TAS(FD) \rightarrow$ ZERO, if $\{X,Y\} \subseteq$ Attr(E_i) ¦ X is a Key con-nected to some i^{th} E-Node (found using $findKey(E_i)$) in a TAS Graph. This suggests that FCI, in this case, approaches 1 using formula 2.

Rule F2: The value for $TAS(FD) \rightarrow$ ONE and consequently FCI approaches 0, if

$\{X,Y\} \not\subset \cup$ (Attr(E_i)) , that is, either X or Y (A-Node or SA-Node) or both do not be-long to any E-Node in a TAS graph.

Rule F3: The value of $TAS(FD) \rightarrow \gamma$, if

(a) $\{X\}$ = Key(E_i) and Y ∈ Attr(E_j) for some E-Nodes E_i, and $E_i \neq E_j$
(b) $\{X\}$ = Key(E_i) and $\{Y\} \subseteq$ Attr(CA_j); for some E_i and CA_j is the jth A-Node on E_i.

(c) $Y \in Attr(E_i)$ and $\{X\} \subseteq Attr(CA_j)$ and $\{X\} = Key(E_i)$; where E_i is some E-Node in the TAS graph and CA_j is the jth A-Node with SA nodes as its children on E-Node E_i.

(d) $\{X\} \neq Key(E_i)$

For Rule F3(a), the value of γ represents the effort required to bring X and Y on the same E-Node. This effort is calculated by applying the counting operator on the number of R-Nodes that separate E_i and E_j.

For Rule F3(b) and F3(c), the concept of fuzzy hedge is used to find out the FCI. In this case the formula for calculating FCI is modified as below:

$$FCI_i = \left[1 - \frac{TAS(FD_i)}{A_{Total}} \right]^m .$$
(3)

Here γ is found by applying the counting operator on the number of composite attributes nodes present between X and Y and the number of R-Nodes that separate E_i and E_j when $E_i \neq E_j$, and $m = \gamma + 1$.

For Rule F3(d), The TAS operator approaches γ that is obtained by applying rule F3(a), F3(b) or F3(c) plus the effort to mark X as the key attribute on E_i. This effort can be envisaged through hypothetically applying the schema transformation rules given in section 3, according to which a new E-Node E_p is introduced which connects to E_i through a new R-Node. The calculation of the TAS operator can be illustrated with the help of an example. Consider a TAS graph as given in Fig. 6(a). The count for the TAS operator in this case will be $(j=1)+(k=2x1)=3$. Here j represents the number of new R-Nodes that need to add in the TAS graph , whereas k is the number of R-Nodes between E_i and E_p for each A-Node that has to be placed on E_p (2 in this case). This count can easily be determined after the schema transformation rules proposed in section 3 are applied to Fig. 6(a), that results in a modified TAS graph shown in Fig. 6(b). Now, we define algorithm 2 for finding the FCI using the TAS graph process.

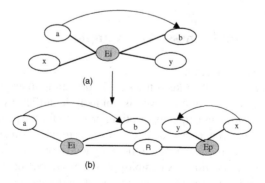

Fig. 6. Schema Transformation based upon $X \rightarrow Y$

Algorithm 2
  Convert an ERD into a graph
      For a given Functional Dependency $X \rightarrow Y$
          Find attribute nodes X and Y in the graph
          IF X or Y not found THEN
                  $FCI_i = 0$

      ELSE

                  Counter = 0, $\gamma = 0$
                  IF X is not a key attribute THEN
                      Mark X as a key attribute
                      Increment Counter by 1
                  IF Y belongs to the set of all attributes on which directed-edges emerging from X incident THEN
                              $TAS(FD_i) = 0$, and $FCI_i = 1$ from Eq. 2

              ELSE

                      Count the number of R-Nodes (stored in CountR) and A-Nodes (stored in CountA) that separate X and Y
                      Count the nodes separating X from its nearest E-Node (stored in CountX) and the nodes separating Y from its nearest E-Node (stored in CountY)
                      Counter = Counter + CountR + CountA + CountX + CountY
                      $\gamma = Max (CountX, CountY)$
                      $TAS(FD_i) = Counter$

                      $m = \gamma + 1$
                      Calculate FCI_i from Eq. 3

      REPEAT above steps for all FDs
  Calculate $FCI_{overall}$ from Eq. 1

5 Evaluating Completeness: An Example

Consider the following scenario:

"In a company, we want to keep track of information about its employees, projects, and departments. An employee is uniquely identified by an ID. His name, department#, department name, its location, project# for the project he is presently working on, its location, cost and supervisor are stored. A supervisor supervises a project location. Same project #s can be assigned to two different projects if they are at two different locations. An employee belongs to one department only and is not allowed to work on more than one project. Queries can refer to any of the data specified here."

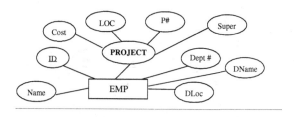

Fig. 7. An ERD For A Company Database

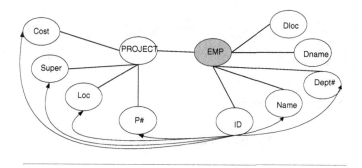

Fig. 8. TAS Graph for ERD in Fig. 7

Now, for this description, we may get an ERD given in Fig. 7. The corresponding TAS graph is shown in Fig. 8. In order to compute CI for Fig. 8 let us first identify $\mathfrak{S}$ as consisting of the following set of FDs:

FD1: P#, LOC → Cost
FD2: LOC → Super
FD3: ID → Name
FD4: ID → Dept#
FD5: Dept# → DName
FD6: Dept# → DLoc
FD7: ID → P#
FD8: ID → LOC

It can be observed that f consists of only FD3, FD4, FD7 and FD8. So, CI = 4/8 = 0.5. Values of FCIs for FD1 – FD8 as per rules described in the previous section are 0.308, 0.444, 1, 1, 0.875, 0.875, 0.79, and 0.79 respectively.

Overall FCI can be computed as:

$$FCI_{overall} = \frac{\sum_{i=1}^{i=n} FCI_i}{n(\mathfrak{S})} = \frac{6.082}{8} = 0.7603 \cdot \tag{4}$$

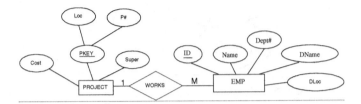

Fig. 9. Improved Conceptual Schema after First Transformation

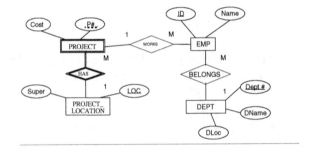

Fig. 10. TAS Graph for ERD in Fig. 9

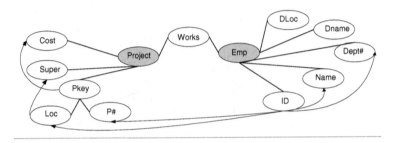

Fig. 11. Improved Conceptual Schema after Second Transformation

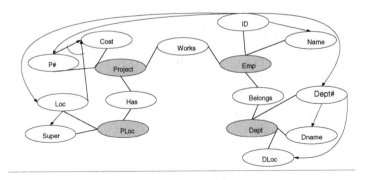

Fig. 12. TAS Graph for ERD in Fig. 11

Another ERD for the same problem domain is represented in Fig. 9, which is an improved version of Fig. 7. The corresponding TAS graph for Fig. 9 is shown in Fig. 10. The CI value for the new TAS graph is 0.7 as FD1, FD3, FD4, FD7 and FD8, whereas the overall FCI value comes out to be 0.972.

Fig. 12 represents yet another TAS graph for the conceptual schema shown in Fig. 11. For TAS graph in Fig. 12 the CI value is 1 and the overall FCI is 1.000.

6 Conclusion and Future Work

In this paper a metric, Fuzzy Completeness Index to measure completeness of an ER model is proposed. The proposed measure provides a more realistic method of quantifying the completeness index of an ERD. The measure indicates the effort required to make the model more complete, simply because the metric varies between 0 and 1 and has been fuzzified. The research uses fuzzy membership values and fuzzy hedges to calculate the proposed FCI. The proposed measure is applied to a scenario for real life situation and it is demonstrated how quality is improved in terms of completeness for various schema transformations using rules already proposed by authors in their earlier research. In our future work, we want to extend this approach to involve higher degree relationships, hierarchies and other constructs of the Extended ER model.

References

1. Lindland, O., Sindre, G. and Solvberg, A. (1994) Understanding Quality in Conceptual Modeling. IEEE Software, 11(2), pp.42-49
2. Moody, D. L. and Shanks, G. G. (1994) What Makes a Good Data Model? Evaluating the Quality of Entity Relationship Models. Proceedings of the 13th Int'l Conf. on the Entity Relationship Approach 1994, pp. 94-111
3. Moody, D. L., Shanks, G. G. and Darke, P. (1998) Improving the Quality of Entity Relationship Models – Experience in Research and Practice. Proceedings of the 17th Int'l Conf. on Conceptual Modeling 1998, pp. 255-276
4. Kesh, S. (1995) Evaluating the Quality of Entity Relationship Models. Information and Software Technology, 37(12), pp.681-689
5. Schuette, R. and Rotthowe (1998) The Guidelines of Modeling – An Approach to Enhance the Quality in Information Models. Proceedings of the 17th Int'l Conf. on Conceptual Modeling 1998, pp. 255-276
6. Assenova, P., Johannesson, P. (1996) Improving Quality in Conceptual Modelling by the Use of Schema Transformations. ACM SIGMOD 15th Int'l Conf. On Conceptual Modeling 1996, pp. 277-291
7. Thalheim, B. (2000) Entity-Relationship Modeling: Foundations of Database Technology. Springer-Verlag
8. Krogstie, J., Lindland, O. and Sindre, G. (1995) Towards a Deeper Understanding of Quality in Requirements Engineering. Proceedings of the 17th Int'l Conf. on Advanced Information Systems Engineering (CAISE) 1995, pp. 82-95
9. Gray, R., Carey, B., McGlynn, N. and Pengelly, A. (1991) Design metrics for database systems, BT Technology, 9(4), pp. 69-76

10. Moody, D. L. (1998) Metrics for Evaluating the Quality of Entity Relationship Models. Proceedings of the 17th Int'l Conf. on Conceptual Modeling 1998, pp. 213-225
11. Piattini, M., Genero, M. and Jimenez (2001) A Metric-Based Approach For Predicting Conceptual Data Models Maintainability. Int'l Journal of Software Engineering & Knowledge Engineering, 11(6), pp. 703-729
12. Hussain, T., Shamail, S. and Awais, M. (2004) Schema Transformations – A Quality Perspective. Proceedings of 8th Int'l Multi-Topic Conf., IEEE INMIC 2004, pp. 645-649
13. Ross,T.J. (1995) Fuzzy Logic with Engineering Applications. McGraw-Hill.
14. Cormen, T., Leiserson, C., Rivest, R., Stein, C. (2001) Introduction to Algorithms, 2nd Ed. MIT Press.

Managing Information Quality in e-Science: A Case Study in Proteomics

Paolo Missier[1], Alun Preece[2], Suzanne Embury[1], Binling Jin[2], Mark Greenwood[1], David Stead[3], and Al Brown[3]

[1] University of Manchester, School of Computer Science, Manchester, UK
[2] University of Aberdeen, Computing Science, Aberdeen, UK
[3] University of Aberdeen, Molecular and Cell Biology, Aberdeen, UK
info@qurator.org
http://www.qurator.org

Abstract. We describe a new approach to managing information quality (IQ) in an e-Science context, by allowing scientists to define the quality characteristics that are of importance in their particular domain. These preferences are specified and classified in relation to a formal IQ ontology, intended to support the discovery and reuse of scientists' quality descriptors and metrics. In this paper, we present a motivating scenario from the biological sub-domain of proteomics, and use it to illustrate how the generic quality model we have developed can be expanded incrementally without making unreasonable demands on the domain expert who maintains it.

1 Introduction

A key element of e-Science is the development of a stable environment for the conduct of information-intensive forms of science. Increasingly, scientists expect to make use of information produced by other labs and projects in validating and interpreting their own data, while funding bodies expect the results generated by their projects to have greater longevity and wider usefulness. In this context, information does not merely document the state of the art in a domain, it also becomes a fundamental resource in the discovery of new knowledge. Hence the increasingly stringent requirements by funding bodies and publishers that scientists place their experimental data in the public domain in forms that are amenable to analysis by software tools as well as by humans.

At present, a variety of obstacles prevent the full realisation of this e-Science vision, not least of which are those caused by the inevitable variations in the quality of the information being shared [3]. It is tempting to view this as a problem for data producers, and to concentrate on defining standards and procedures for data capture (such as those defined by the MGED consortium for the capture and recording of information about microarray experiments [4]). While such standards are important and worthwhile, they cannot provide a complete solution. They can do little to address the quality of the volumes of legacy data that have so far been amassed. Moreover, like other forms of quality, information quality (IQ) is typically a function of the requirements of the information consumer rather than its producer. A scientist searching for information relating

J. Akoka et al. (Eds.): ER Workshops 2005, LNCS 3770, pp. 423–432, 2005.

to a drug that is about to be used in patient trials will have more stringent requirements than one searching for examples to be used in a textbook, for example. Similarly,
one scientist might think of "accuracy" in terms of some calculated experimental error,
while another might define it as a function of the equipment that captured the data.

What is required, therefore, is some means by which we can determine the quality of
a specific data set relative to the needs of a specific user. For example, data sets that are
incomplete or inaccurate can still be used to good effect by those who are aware of these
deficiencies and can work around them. The viability of this approach depends on the
ability to elicit and manage detailed specifications of the IQ requirements of individual
users (or, at best, communities of like-minded individuals). The task of specifying new
forms of quality preference should not be too onerous on users or those managing the
information environment. IQ preferences should ideally be expressed in a formal language so that the definitions are machine-manipulable, both to allow (semi-)automatic
determination of the quality of a data set, and to facilitate browsing and searching of
the quality model.

The Qurator project[1] aims to provide the software infrastructure needed to support
this form of domain-specific IQ management, focussing specifically on two domains of
post-genomic biology: proteomics and transcriptomics [5]. We envision an e-Science
environment in which a new user (scientist) can use IQ tools to discover potentially-
useful IQ preferences for adaptation and reuse, and which allows new customised preferences to be defined without involving an expensive knowledge capture exercise.

The existing IQ literature offers useful starting points to meet these goals, by providing a common terminology for describing quality properties, or *dimensions* [11, 9].
However, it falls short of providing principled solutions to the problem of expressing
quality requirements in a formal way, let alone to the problem of expressing complex
quality-oriented views of data. In this paper, we describe a knowledge-intensive approach to modelling both the quality and application domains, which may serve a foundation to address these problems. We present the ontological model of IQ that forms the
heart of our approach (Section 3), and show how the ability to reason over the model
allows it to be self-managing under the addition of new quality preferences (Section 4).

The ontology is implemented in OWL and makes use of OWL-DL reasoning features[2]. Although we do not add any new theoretical elements to the Semantic Web
framework, its application to this problem is, to the best of our knowledge, novel. This
project is still in its early stages, and validation of the ideas presented here is in progress
with the collaboration of the Aberdeen Proteomics Facility, at the University of Aberdeen, UK. Tool support for exploiting the ontology is in the planning stage.

2   Background on Protein Identification

To motivate the ontology presented in this paper, we present a scenario from the area of
proteomics that illustrates the kinds of quality preference which arise from the domain-
specific approach we are investigating. Proteomics is the study of the set of proteins
that are expressed under particular conditions within organisms, tissues or cells. One

---

[1] Qurator is funded by the EPSRC Fundamental Computer Science for e-Science Programme.
[2] http::/www.w3.org/TR/owl-guide/

experimental approach that is widely used to gain information about the large-scale expression of proteins involves extracting the proteins from a biological sample, then separating them by a technique known as 2-dimensional gel electrophoresis (2DE). With this technique, the proteins are separated into a 2D matrix, where they are distinguished by net charge and molecular size. These two separating factors are typically enough to differentiate each protein in the sample, so that each spot on the gel contains just one kind of protein. The spots can be examined individually and the amount of protein in each can be estimated after staining and densitometric scanning.

In a typical proteomic experiment, several different samples are subjected to the procedure outlined above and the resulting 2DE maps are compared. This allows the biologist to compare the expression rates of various proteins under contrasting conditions, for example to examine the different expression rates between a healthy tissue sample and a diseased one. By comparing the gel images that are produced from the samples, the biologist can hypothesise that the changes in protein expression thus highlighted may be a significant cause or result of the biological phenomenon under study.

Before such a hypothesis can be fully stated, it is necessary to identify the proteins that are present in the spots that indicate varied expression levels. This task is routinely performed using the technique of peptide mass fingerprinting (PMF). In PMF, the protein within the gel spot is first digested with an enzyme that cleaves the chain of the protein at certain predictable sites. The fragments of protein that result (called *peptides*) are extracted and their masses are measured using mass spectrometry. The list of peptide masses is then compared against theoretical peptide mass lists, derived by simulating the process of digestion on protein sequences extracted from a protein database (e.g. NCBInr[3]). Since, for various reasons, it is unlikely that an exact match will be found, the protein identification search engines typically return a list of potential protein matches, ranked in order of search score. Different search engines calculate these scores in different ways, so their results are not directly comparable. Furthermore, although some search engines (e.g. Mascot[4]) attempt to estimate the probability that a match is valid, others (e.g. MS-Fit[5]) do not, and it may be difficult for the experimenter to decide whether a particular protein identification is acceptable or not.

It would be useful for biologists seeking to interpret the results of proteomic experiments to be able to assess the credibility of a protein identification result by comparing readily accessible metrics for a list of protein matches. There are three metrics that can be used for this purpose, and which are independent of the search engine used:

– Hit ratio: the number of peptide masses matched, divided by the total number of peptide masses submitted to the search. Ideally, most of the masses should be accounted for by the protein identified, but because of additional peaks in the mass spectrum (originating from the presence of other proteins, for example) the hit ratio is unlikely to reach unity.

---

[3] http://www.ncbi.nih.gov/BLAST/

[4] http://www.matrixscience.com/

[5] http://prospector.ucsf.edu/

- Excess of limit-digested peptides: calculated by subtracting the number of matched peptides containing a missed cleavage site from the number of peptides with no missed cleavages. Ideally, a complete (limit) digest will have been achieved during PMF, in which case the number of missed cleavage sites would be zero. However, in practice a small number of missed cleavages are to be expected.
- Sequence coverage: the number of amino acids contained within the set of matched peptides, expressed as a percentage of the total number of amino acids recorded for the protein in the database. The higher the coverage, the greater the confidence in the match, but limitations of the experimental technique mean that full coverage is never achieved. It is also necessary to consider the size of the protein. A lower coverage of, say, 15% may be satisfactory for a large protein where many other peptides have been successfully matched. A similar coverage for a smaller protein, on the other hand, would be indicative of a poor match. Therefore, care must be exercised when interpreting the value of this metric.

These three metrics can be combined in a logical expression that allows us to classify protein matches as being either acceptable or unacceptable. A software tool can then be envisaged that allows the user to set acceptance criteria for each metric independently and to see the effect in real time of altering any or all of the threshold values on the acceptability of the data set.

While we use this as a simple example, a more general approach for the creation of quality preferences is described in the next section. The choice of these metrics also represents a simplification. In protemics, many more variables can be used to formulate statements regarding the quality of experimental results. For in-depth reviews of the field and of the variables involved, please see [6, 7, 1].

3   Ontology-Based Modelling of Information Quality

Although we are focussing on two specific application domains within the Qurator project (i.e. proteomics and transcriptomics), our ultimate goal is to produce a model of IQ that can be instantiated to produce domain-specific IQ models for a wide variety of application areas. In order to achieve this, it was necessary to find some over-arching organisational structure that would give meaning to the domain-specific terms and allow for comparison and analysis of quality preferences provided by multiple users. For this purpose, we have adapted generic IQ concepts that have been in use within the IQ community for a long time, and which are grounded in the wealth of existing literature on this topic [3, 8, 10, 9]. These concepts, such as accuracy, completeness and currency, give useful placeholders for common IQ concerns but they are not sufficiently well defined to be directly applicable to real applications. Instead, the user (or group of users) will wish to talk about specific properties relating to the domain of interest. Rather than speaking of accuracy, she will talk of equipment tolerances or scores resulting from error models, for example.

The Qurator quality ontology must therefore bridge the gap between these generic quality concepts (i.e. quality properties) and concepts from the users' own domain. Figure 1 shows a fragment of the OWL ontology we have created for the proteomics

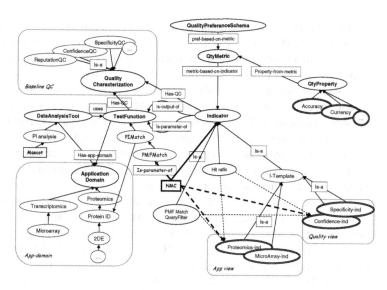

Fig. 1. Fragment of the IQ ontology

scenario described in Section 2, and which we will use to illustrate how this bridging is achieved[6]. Here, specific domain knowledge coming from the biologists and bioinformaticians involved in the project has been imported into the Qurator ontology from the *my*Grid project data ontology [12], which describes basic biological concepts as well as a number of biological databases and data analysis tools. In Figure 1, the ApplicationDomain classes represent domain data concepts that have either been lifted from *my*Grid, or added to it. We have then extended the model with terms of specific interest to proteomics specialists.

We will use the following simple but realistic proteomics scenario to explain how the ontology supports the definition of new domain-specific quality preferences with the help of the generic quality terms. Suppose that a biologist wishes to rank a set of protein identification experiments performed using 2DE technology. The ranking is based on the scores obtained by matching PMFs against the NCBInr database, using the Mascot analysis tool. This score (the QualityMetric) is itself based on some function of the hit ratio and the number of possible missed cleavages (NMC) found during the match.

The user's first task is to create a QualityPreferenceSchema, which will be used to rank the experimental results. Although the ontology fragment shows only a single generic concept for preference schemas, in practice one would expect a range of more specific schemas to be defined (to partition experiments into classes, for example, or to filter them based on a threshold for the associated quality metric).

---

[6] No standard notation currently exists for expressing the logical features of the OWL language. We use a graphical notation in which ovals represent classes, rectangles represent individuals and lines represent object properties [2]. We show user-defined individuals and properties using thick lines, and represent subsumption properties or individuals' classifications obtained through reasoning using dotted lines.

The next step is to define the `QualityMetric` itself, which must be based on one or more `Indicators`. An `Indicator` is some value that can be provided by the environment, either directly by retrieval from some persistent store or metadata repository, or by computation. In our example, the two indicators are `HitRatio` and NMC. We will assume that the former is already present in the domain-specific part of the ontology, but that the user must add the NMC indicator to the model. In practice, it is common in e-Science for indicators to be associated with particular data analysis tools, as modelled by the `TestFunction` class. For example, `HitRatio` is part of the output produced by an analysis program which performs protein identification matches against a protein database. There may be many such programs, and the class `PIMatch` represents their general form in the ontology.

Next, we associate the indicators with the generic quality concepts. In the ontology, the quality domain is divided into two levels, which together support self-management of the ontology. The lower level is represented by the root class `QualityCharacterization`, or "QC" for short, which is extended with a small and fairly stable collection of key concepts describing information quality, such as `ReputationQC`, `ConfidenceQC` and `CurrencyQC`. These classes are used to characterize the quality of domain concepts that are part of the ontology. In particular, an `Indicator` is defined as any data entity that can be quality-characterized (through the `has-QC` property). Some indicators are domain-independent. For example, time stamps on data are commonly used to assess currency of information, and would therefore be associated to `CurrencyQC`. Others, such as `HitRatio`, are completely domain-specific.

The default associations shown in the model reflect the intended meaning of the quality terms, and have been introduced based on domain experts' recommendations. For instance, a `ConfidenceQC` indicator provides users with a level of confidence in the outcome of an experiment (as is the case for `HitRatio`, for instance) while a `ReputationQC` can be associated to indicators that scientists use to assess the overall reputability of an experiment outcome – these may include the laboratory that performed the experiment, and the standing of the associated journal publications. As the QC concepts form a primitive collection of ontology terms, they also act as the axiomatic base for the quality domain. Qurator makes an effort to ensure that domain experts use these terms consistently when introducing new domain-specific concepts.

The second method of encoding quality into the Qurator ontology is a higher level model of the more traditional data quality terminology found in the literature. These concepts are rooted at the `QtyProperty` class. The mapping between the two levels of quality concepts represents additional knowledge regarding quality, and provides for added flexibility in the specification of the semantics of the terms. For instance, we might define *Accuracy* (i.e. the property that describes how closely a data entity reflects the actual state of the real world entity that it stands for) in terms of *confidence* and *specificity*. The way to read this association is as follows:

"a quality metric that is based on confidence or on specificity indicators, expresses the intention of the expert to capture accuracy properties of the underlying data."

Of course, there are many who would disagree with this as a definition of accuracy. The key point here is not the exact form of the definition, but the fact that there is a principled way to establish the logical associations between the users' operational definitions of quality, implemented using indicators and metrics, and a shared conceptualization of data quality.

4 A Self-managing Quality Model Through DL Reasoning

The combination of domain knowledge and generic quality concepts has proved to be surprisingly powerful when combined with the kinds of reasoning facility offered by standard description logic ontologies. As the ontology is implemented in OWL-DL, we have benefited from its ability to provide consistency checks and entailment in supporting additions to and modifications of the ontology. In particular, the inferencing capabilities provided by description logics allow newly introduced concepts, such as indicators, to be classified against the quality model automatically, and therefore to become available to other scientists for reuse.

The principal mechanism that underlies this self-classifying ability is the set of QCs. We have already discussed how indicators can be classified relative to the current set of QCs. Test functions may also be treated in this way, and QCs can be automatically propagated from them to the indicators that act as their parameters or results. For instance, we can formalize the (sufficient) condition that an indicator is a `Confidence-ind` if it is a parameter of any `TestFunction` whose QC includes `ConfidenceQC`. In OWL DL, this can be written as:

(1) `Confidence-ind` $\equiv \exists$ `is-parameter-of` .

$$(\exists \text{ has-QC ConfidenceQC})$$

Note that the `has-QC` property is many-to-many, so the existential quantifier indicates that *at least* one of the indicator's QCs must be a `ConfidenceQC`.

To see this in action, suppose that when the new user-defined indicator `NMC` is introduced, the only information the user can provide about it is that it is a parameter to a matching algorithm for PMFs. This algorithm is already known to the ontology as the concept called `PMFMatch`, which has a `ConfidenceQC`:

(2) `NMC` $\in$ `Indicator` $\sqcap$ ($\exists$ `is-parameter-of` . `PMFMatch`)

(3) `PMFMatch` $\sqsubseteq \exists$ `has-QC` . `ConfidenceQC`

Assertion (2), here, defines `NMC` as an individual whose class is the domain of the `is-parameter-of` property, with a range of the class `PMFMatch` (this is called an *anonymous class*). Assertion (3) states that `PMFMatch` is a sub-class of any anonymous class that is a domain of the `has-QC` property, with range `ConfidenceQC` (a necessary condition). By applying standard DL reasoning[7] to these three assertions, we infer that `NMC` is a member of the `Confidence-ind` class (shown as a thick dashed line in Figure 1):

---

[7] Our   implementation   makes   use   of   the   RACER   DL   reasoner (http://www.racer-systems.com/).

(4) NMC ∈ Confidence-ind.

Note that similar entailments are performed on the built-in indicators (thin dashed lines in the figure). Following this approach, it is possible to classify indicators with respect to both the quality and the application domains; the application domain hierarchy is rooted at ApplicationDomain in Figure 1. The role of the classes under the I-template class is to provide both a *quality view* and an *application view* of the indicators. Once they have been populated through reasoning, such views provide a basis for a variety of user queries regarding available indicators, including the user-defined ones. For example, they make it possible to query the ontology to discover what indicators exist that are suitable for a specific quality purpose (e.g. "give me all confidence indicators") so that users can browse the currently available resources before they go to the trouble of creating their own from scratch.

4.1 A Detailed Scenario

We can now illustrate how the ontology supports the user scenario introduced earlier. The following assertions, stated informally, capture the user's intuitions regarding quality preferences for his data collection:

1. a new indicator, NMC, must be introduced;
2. NMC is used as a parameter of any PMF matching algorithm (i.e. it does not need to be associated with a specific algorithm instance at this stage);
3. a ranking criterion aRC is introduced, as a function of NMC and the pre-existing indicator HitRatio;
4. a particular quality preference schema, aDPS, is defined;

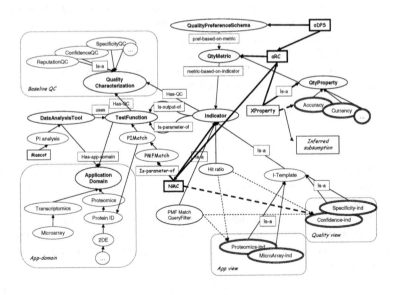

Fig. 2. Entailments for the user scenario (dotted lines)

5. it is stated that aDPS uses the aRC metric;
6. it is stated that aDPS applies to any proteomic data collection.

The effects of these assertions are shown in Figure 2 by rectangles (individuals) and thick solid lines (properties).

In addition to this domain-specific information supplied by the user, we also assume the existence of assertions within the ontology stating that a Confidence-ind is either an indicator whose set of QCs includes a ConfidenceQC, or that it is either the parameter or the output of a test function whose QC includes ConfidenceQC. Note that this definition generalizes assertion (1) to capture the case in which a direct assertion regarding the QC properties of an indicator is made. Finally, Accuracy is assumed to be defined in terms of quality metrics and QC-indicators as follows:

> Accuracy is the quality property for which there exists a metric qm that is based on a set of indicators, at least one of which is either a Confidence-ind or a Specificity-ind.

The DL expressions corresponding to these informal assertions are presented in the Appendix. Based on these expressions, Qurator introduces an additional individual, XProperty ∈ QtyProperty, and establishes an association between the user annotations and the shared quality terminology, by asserting that XProperty is related to metric aRC.

With these definitions, the reasoner infers that (a) NMC ∈ Confidence-ind, and (b) XProperty ∈ Accuracy. In practice, we have used the reasoning capabilities associated with OWL DL to classify elements from the user input, in a way that is consistent with a predefined ontology. The resulting quality annotations are consistent with the model and can be stored for future querying.

5 Conclusion

By bridging the gap between the quality domain and application domains, the Qurator ontology allows scientists and bioinformaticians to describe their personal perceptions of data quality in e-science in a natural yet formal way, while relationships with a shared quality model are automatically established. This allows us to provide a controlled environment for managing user extensions to the ontology, which in turn facilitates incremental development of domain-specific quality models. As our scenario illustrates, users are only expected to provide information about their own tools and indicators. Qurator then searches for relationships between the new domain-specific concepts and the existing quality concepts. This combination of extensibility and reusability has the potential to produce rich, community-supported bases of shared knowledge that eases the navigation and exploitation of the information resources provided by e-Science. It is expected to have particular value for scientists who are not experts in the domain of the data (e.g. proteomics scientists wishing to make use of transcriptomics data).

In addition to the entailment patterns described in this paper, a number of other patterns can potentially be supported by the Qurator model. We are currently exploring these, as well as expanding our understanding of the uses of information quality in our two application areas. We need to learn more about the ways in which quality annotations of the kind provided by our model can be used in practical applications, and how the model can be better embedded within upcoming e-Science tools.

A   DL Assertions for the User Scenario

1. aDPS ∈ QualityPreferenceSchema
2. aRC ∈ QualityMetric
3. aDPS pref-based-on-metric aRC
4. metric-based-on-indicator HitRatio
5. aRC metric-based-on-indicator NMC
6. NMC ∈ Indicator ⊓ (∃ is-parameter-of . PMFMatch)
7. PMFMatch ⊑ ∃ has-QC . ConfidenceQC
8. Confidence-ind ≡ (∃is-parameter-of
                     (∃has-QC ConfidenceQC))
                   ⊔ (∃is-output-of(∃has-QC ConfidenceQC))
                   ⊔ (∃has-QC ConfidenceQC)
9. Accuracy ≡ ∃ QtyProperty-from-metric
             (∃ metric-based-on-indicator
             (Specificity-ind ⊔ Confidence-ind))
10. XProperty ∈ QualityProperty
11. XProperty property-from-metric aRC

References

1. R. Aebersold and M. Mann. Mass spectrometry-based proteomics. *Nature*, 422:198–207, March 2003.
2. F. Baader, I. Horrocks, and U. Sattler. Description Logics. In S. Staab and R. Studer, editors, *Handbook on Ontologies*, pages 3–28. Springer-Verlag, 2004.
3. L. English. *Improving Data Warehouse and Business Information Quality*. Wiley, 1999.
4. A. Brazma *et al.* Minimum Information about a Microarray Experiment (MIAME) — Toward Standards for Microarray Data. *Nature Genetics*, 29:365–371, December 2001.
5. P. Missier, S. Embury, M. Greenwood, A. Preece, and B. Jin. An ontology-based approach to handling information quality in e-science. In *Proc 4th e-Science All Hands Meeting*, 2005.
6. A. Pandey and M. Mann. Proteomics to study genes and genomes. *Nature*, 405:837–846, June 2000.
7. S. D. Patterson and R. H. Aebersold. Proteomics: the first decade and beyond. *Nature Genetics*, 33(Supplement):311–323, March 2003.
8. T.C. Redman. *Data Quality for the Information Age*. Artech House, 1996.
9. R.Y.Wang, M.Ziad, and Y.W.Lee. *Data Quality*. Advances in Database Systems. Kluwer Academic Publishers, 2001.
10. Y. Wand and R.Wang. Anchoring data quality dimensions in ontological foundations. *Communications of the ACM*, 39(11), 1996.
11. R. Y. Wang and D. M. Strong. Beyond accuracy: What data quality means to data consumers. *Journal of Management Information System*, 12(4), 1996.
12. C. Wroe, R. Stevens, C. Goble, A. Roberts, and M. Greenwood. A suite of DAML+OIL ontologies to describe bioinformatics web services and data. *International Journal of Cooperative Information Systems*, 12(2):197–224, 2003.

Tool Support and Specification Quality: Experimental Validation of an RE-Tool Evaluation Framework

Raimundas Matulevičius and Guttorm Sindre

Dept. of Computer and Information Science,
Norwegian Univ. of Science and Technology,
Sem Sælands vei 7-9, NO-7491 Trondheim, Norway
{raimunda, guttors}@idi.ntnu.no

Abstract. Automated support for the requirements engineering (RE) process is a recognised research area. However, the practice still relies on office tools rather than RE-tools. Reasons include financial causes and difficulty to evaluate the available RE-tools. This work reports on an experiment trying to validate a previously proposed framework for evaluating RE-tools. The experiment participants used several alternative tools for making requirements specifications, and then evaluated the tools by means of the framework. This enables us to look at the participants' performance with the various tools, evaluation approaches, and their perceptions about the same tools. The findings indicate advantages of using the evaluation framework, and of combining several evaluation techniques. The experiment indicates that RE-tools provide better support than office tools, leading to higher quality specifications.

1 Introduction

Requirements engineering (RE) tools are software tools which provide automated assistance during the RE process [14]. The need for automated support may vary in different projects; if a company does not have a mature RE process, automation won't necessarily help. But if the company deals with system requirements specifications (SRS) which need to evolve over time, RE-tool support could clearly be useful. But the mainstream RE practice relies on office tools (editors and drawing tools) rather than targeted RE-tools (CaliberRM, RequisitePro, and DOORS). Reasons for not using RE-tools include financial causes, like high price and low return on investment.

Because of their limited use in practice it is difficult to evaluate RE-tools in terms of their impact on an organisation's processes. It is also difficult to examine tools in an experiment, as it is difficult to control for the developers' capabilities. Moreover, RE-tools provide the benefit for large projects; while an experiment requires prescribed tasks of a fairly limited size. There is a need for a cheaper evaluation that can be done analytically rather than empirically. A problem of such evaluations is that they easily become subjective. Hence, to support the evaluations' completeness, they should be grounded in a sound framework providing methodological guidance.

A framework to evaluate RE-tools according to the functional requirements is presented in [14]. The evaluation framework is based on analytical arguments, but not on an empirical investigation, except for limited trials of some parts of it. Generally, validation could be performed through perception, performance and correctness tests.

J. Akoka et al. (Eds.): ER Workshops 2005, LNCS 3770, pp. 433–443, 2005.
© Springer-Verlag Berlin Heidelberg 2005

Perception involves investigation of framework usability, ease to use and user satisfaction [6]. This paper reports on an experiment where the framework's performance and correctness are tested. Three research questions are formulated:

RQ1: *Does the evaluation framework select tools which yield a high-quality SRS?*
RQ2: *Do evaluation scenarios help to test RE-tools better than RE-tool tutorials?*
RQ3: *Do RE-tools provide better support for the RE process and better maintenance of the SRS quality than office tools?*

The first question analyses the performance and correctness of the framework [14]. The second question considers evaluation techniques. The third question investigates tool support to maintain the SRS quality. The second and third questions target validity of the RE-tool evaluation approach [16], which guides the framework application.

The paper is structured as follows: section 2 analyses related work. Section 3 describes the research method. Section 4 considers the experiment results. Finally, section 5 concludes the paper and discusses the lessons learned.

2   Related Work

The literature suggests several approaches to select commercial off-the-shelf (COTS) products. Procurement Oriented Requirements Engineering (PORE) integrates requirements acquisition and tool selection using templates [13]. Off-The-Shelf Option (OTSO) describes how to incorporate tools into the systems used in an organisation [9]. COTS Acquisition Process (CAP) describes three-process software acquisition [17]. The *scenario-based* selection [2] proposes a comparison between baseline scenarios and tool scenarios which maps a baseline scenario into a future where a tool is applied. Social-technical approach to COTS evaluation (STACE) comprises requirements elicitation, social-technical criteria definition, alternatives identification, and evaluation [11]. The *quality-based* approach [3] constructs an ISO/IEC 9126-based model for a tool domain. Approaches are criticised for labour-intensive activities, their application is time consuming and domain knowledge demanding.

RE-tool frameworks [5, 7] specify the RE-tool requirements. But the frameworks lack application guidelines. The approach and framework compositions are proposed in [1, 7]; however empirical findings from their application have not been reported.

The framework (Fig. 2) for evaluation of RE-tool functional requirements [14] consists of three dimensions, inspired by the NATURE project [18]. The *representation* dimension deals with the degree of formality, where requirements are represented using informal, semiformal and formal languages. The *agreement* dimension deals with the degree of agreement among participants by collaboration means and rationale maintenance. The *specification* dimension deals with the degree of requirements understanding at a given time moment and completeness of specification. The framework features are reconsidered according to the Lang and Duggan [12] requirements. The framework application follows the RE-tool evaluation approach [16]. In comparison to the COTS selection approaches, the RE-tool evaluation approach guides the user of the framework [14] through the RE-tool selection and assessment domain.

This work reports on an experiment, where the framework [14, 16] is applied to assess the RE-tools. The purpose of this work is to investigate what influence assessment of the tools and if the selected tool(s) yield the high-quality SRS'es.

3 Research Method

The research method is shown in Fig. 1. The experiment was executed at the Norwegian University of Science and Technology. 44 students of the 4th year participated in the experiment which was a part of the exercises in the course TDT4250 Modelling of Information Systems. The students were divided into ten groups of 4-5 persons. The treatment involved the course material and theoretical lectures given to the students. But the attendance of the lectures was not compulsory, so participants had different knowledge of the experiment settings.

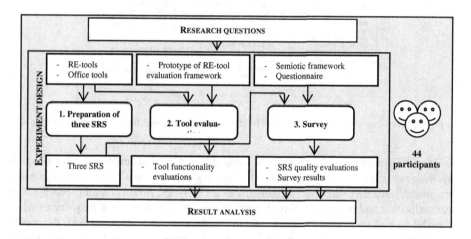

Fig. 1. Research method

The **problem** used in the experiment is a natural language case description of the information system dealing with network fault handling. The problem statement also included a list of requirements which should be maintained in SRS'es.

Tools. Four RE-tools were downloaded from the Internet according to tool surveys [12, 14]. They are: CaliberRM, CORE, RDT, and RequisitePro. The RE-tools were fully functional versions limited by the entry number or evaluation period.

Office tools included text editors (e.g., MS Word™ and Latex), and drawing tools (e.g., Visio). The participants chose them according to their own preferences.

Both **evaluation scenarios** and **RE-tool tutorials** describe the RE-tool functionality. Tutorials are downloaded from the Internet and focus on teaching tool functions. Scenarios describe steps to prepare an SRS for *the same* problem using an RE-tool.

The **framework prototype** [15] shown in Fig. 2 is implemented to support the framework [14]. The RE-tool evaluation is guided through five steps: *framework feature selection (FS)* and *prioritisation (FP)*, *RE-tool selection (RS)* and *evaluation*

(EV), and *result analysis (RA)*. The final score for each tool is calculated as a sum of multiplications between framework activity priorities and RE-tool evaluation score.

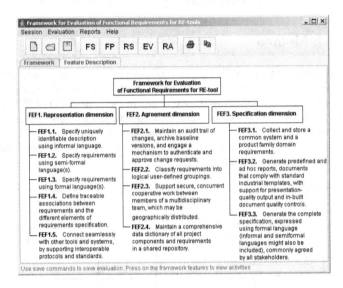

Fig. 2. Prototype of the Framework for Evaluation of Functional Requirements for RE-tool

The **SRS quality** could be measured using the goals and means of the semiotic quality framework [10]. which separates the quality into physical, empirical, syntactic, pragmatic, semantic, perceived semantic, and social types. *Physical quality* has goals of externalisation and internalisability. *Empirical quality* deals with error frequencies when a model is read or written. *Syntactic quality* is the correspondence between the model and the language in which the model is written. *Semantic quality* is the correspondence between the model and the domain. *Pragmatic quality* is the correspondence between the model and audience's interpretation. *Perceived semantic quality* is the correspondence between the model interpretation and participant knowledge. *Social quality* has the goal of participant agreement. In this work, the semiotic quality framework is used to compare the quality of SRS'es produced in the experiment.

Tasks. In the first task participant groups used RE-tools and office tools to prepare SRS'es. In order to prepare the first SRS, the group used *RE-tool* #1 and the *evaluation scenario*. The second SRS was prepared for the same problem, but the group got familiar with *RE-tool* #2 functionality according to the *RE-tool* #2 *tutorial*. The third SRS for the same problem was prepared using *office tools* chosen by the group.

The second task was performed individually. The participants used the framework prototype [15] to evaluate the tool functionality. The prototype aimed to provide the rationale, and participants could communicate by commenting the evaluation issues.

The third task comprised the survey. First, the semiotic quality framework [10] was applied to evaluate the quality of SRS'es. Next, the participants compared the evaluation techniques and how RE-tools and office tools support the RE process.

Post-evaluation was performed to reduce the subjectivity of the SRS quality evaluation. Four teaching assistants (TA) evaluated the performance of each group and selected the "best-quality" SRS'es after individual inspection and discussion.

4 Results

This section presents the findings. First the result analysis method is described; next the research questions are discussed. The section concludes with validity issues.

4.1 Analysis Method

The analysis method (Fig. 3) comprises correlation analysis, descriptive statistics and hypothesis testing [19]. Table 2 shows the correlation between the tool functionality and SRS quality. The *tool functionality* is a sum of framework activity evaluations:

$$Tool\ functionality = \sum_{i}^{n}\sum_{j}^{m} p_{i,j}e_{i,j}, \tag{1}$$

where n – group size, m - number of framework activities, $p_{i,j}$ – the priority of activity j, and $e_{i,j}$ – evaluation of activity j, evaluated by group member i.

The *SRS quality* evaluation is calculated as a sum of all quality type evaluations:

$$SRS\ quality = \sum_{i}^{n}\sum_{j}^{t} q_{i,j}, \tag{2}$$

where t – the number of quality types, and $q_{i,j}$ – the evaluation of the quality type j, evaluated by group member i. *Correlation coefficients* are calculated between tool functionality and SRS quality.

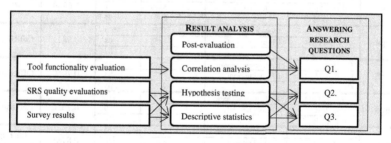

Fig. 3. Result analysis method

Hypothesis testing. Three null hypotheses are formulated:

H$_{10}$: *Evaluation scenarios and RE-tool tutorials contributes to the same RE-tool functionality evaluation.*

H$_{20}$: *Evaluation scenarios and RE-tool tutorials contributes to the same SRS quality.*

H$_{30}$: *The SRS'es prepared with the RE-tools and office tools are of the same quality.*

The hypothesis follows the results in Table 2. The *pair t-test* is applied. H_{10} and H_{20} compare performance of the evaluation techniques to assess the tools and the SRS quality (Tables 3 and 4). H_{30} compares the SRS quality, where SRS'es are prepared using different tools (Tables 7). If the null hypothesis is not rejected, nothing can be said about the outcome [19]. Therefore, the survey performed in task 3, is helpful.

Descriptive statistics. The survey defines a number of subjective measures (Table 1) which are used to gather participants' opinion about evaluation techniques and tool support for the RE process. The central tendency measures are calculated in Tables 5 and 8. The *correlation* between measures is calculated in Tables 6 and 9.

Table 1. Measures of evaluation techniques and of the tool support for the RE process

Evaluation technique measure	Measures of tool support for the RE process
X1 – understandability of the RE-tool functionality; X2 – learning of the RE-tool functionality; X3 – understandability of the problem for which SRS'es were prepared; X4 – preparation of the SRS; X5 – satisfaction of the RE-tool usage; X6 – quality of the SRS; X7 – evaluation of the RE-tools.	Y1 – understandability of the RE process; Y2 – support for RE activities; Y3 – functionality to prepare the SRS; Y4 – means to ensure the quality of the SRS; Y5 – satisfaction of the tool usage during RE; Y6 – usability to execute different RE activities; Y7 – changeability and traceability of the requirements and SRS.

Table 2. Evaluation of tool functionality and SRS quality

Groups	Tools	Evaluation technique	Tool func-tionality	SRS quality	Correla-tion coefficient	Post-evaluation (SRS'es preferred by TA)
Group1	RDT	Evaluation scenario	3603	119	0,9674	RequisitePRO
	RequisitePRO	RE-tool tutorial	3672	128		
	Office tools	-	2370	98		
Group2	CORE	Evaluation scenario	4844	101	-0,7450	CORE and MS Office
	CaliberRM	RE-tool tutorial	5583	102		
	Office tools	-	4448	127		
Group3	CORE	Evaluation scenario	4155	96	-0,6549	CORE
	CaliberRM	RE-tool tutorial	5378	81		
	Office tools	-	2933	91		
Group4	CORE	Evaluation scenario	1396	105	0,8811	CORE
	RequisitePRO	RE-tool tutorial	1466	98		
	Office tools	-	805	87		
Group5	CORE	Evaluation scenario	2165	85	0,7060	Office tools
	CaliberRM	RE-tool tutorial	1850	83		
	Office tools	-	2238	105		
Group6	RequisitePRO	Evaluation scenario	3571	110	-0,7626	RequisitePRO
	CaliberRM	RE-tool tutorial	4536	109		
	Office tools	-	3167	126		
Group7	RDT	Evaluation scenario	1338	99	0,8765	CORE and Office tools
	CORE	RE-tool tutorial	1496	136		
	Office tools	-	1620	134		
Group8	RDT	Evaluation scenario	1626	70	0,2849	RDT, CORE and Office tools
	CORE	RE-tool tutorial	1656	87		
	Office tools	-	1173	75		
Group9	RDT	Evaluation scenario	3168	114	0,4881	RequisitePRO
	RequisitePRO	RE-tool tutorial	4255	148		
	Office tools	-	2460	135		
Group10	RDT	Evaluation scenario	4208	90	0,9926	CaliberRM and RDT
	CaliberRM	RE-tool tutorial	5108	93		
	Office tools	-	2133	77		

4.2 RQ1: Selected Tools Yield Better SRS'es

In Table 2 the correlation between the tool functionality and SRS quality shows a direct dependency in seven groups (1, 4, 5, 7, 8, 9, and 10). In the post-evaluation the TA selected the same 7 SRS'es prepared with high-evaluated tools. But in groups 2, 3, and 6 the correlation is negative. Negative or low correlation does not mean that there is no dependency. It rather points that there are some other influential factors.

The findings in Table 2 show tendency that the framework [14] helps to select the tools which yield a high-quality SRS. But it is also important to evaluate tool non-functional characteristics, like usability, performance and evaluator's experience.

4.3 RQ2: Scenarios and Tutorials Provide Evaluation on the Same Level

Hypothesis Testing. Table 3 shows summary statistics for the RE-tool functionality assessment using evaluation techniques. The t-test is found higher than critical value of t-test (2.2621). This means that H_{10} have to be rejected ($\alpha=0.05$).

Table 3. t-test of the RE-tool functionality assessment using evaluation techniques

	Mean	Standard deviation	t-test, $|t_0|$
Evaluation scenario	3007.4	1283.706	2.858
RE-tool tutorial	3500	1712.709	

Table 4 shows summary statistics for the SRS quality using evaluation techniques. The t-test is found lower than critical value (2.2621). H_{20} cannot be rejected ($\alpha=0.05$).

Table 4. t-test of the SRS quality evaluation using evaluation techniques

	Mean	Standard deviation	t-test, $|t_0|$
Evaluation scenario	98.9	14.579	1.051
RE-tool tutorial	106.5	23.377	

Descriptive Statistics. Table 5 highlights better understandability of the problem to which SRS'es were prepared (X3), preparation of the SRS (X4), satisfaction of the RE-tool usage (X5), quality of the SRS (X6) and the evaluation of the RE-tools (X7) while using evaluation scenarios. The understandability of the RE-tool functionality (X1) and learning of tool functionality (X2) is higher using tool tutorials. The results are confirmed by H_{10}. The groups commented that, first, they studied RE-tool tutorials to learn functionality; later on, they used evaluation scenarios to prepare SRS'es.

In Table 6, correlation coefficients show a relatively strong dependency between RE-tool functionality (X1) and learning of functionality (X2), satisfaction of RE-tool usage (X5), and quality of the SRS (X6). The coefficient also indicates, that in order to evaluate an RE-tool (X7), it is important to learn tool functional features (X2).

The experiment findings indicate that the evaluation scenarios provide better or at least equal means to evaluate and compare RE-tools as the RE-tool tutorials.

Table 5. Descriptive analysis of evaluation techniques (evaluation scale 1-to-6: 1 – evaluation scenario is better; 6 – tool tutorial is better)

Measure	Mode	Median	Mean	Variance	Standard deviation
X1	5	4	3.66	2.46	1.57
X2	4	4	3.50	1.84	1.36
X3	3	3	2.89	1.56	1.26
X4	2	3	2.91	1.53	1.24
X5	2	3	3.20	1.93	1.39
X6	2	3	3.11	2.09	1.43
X7	2	3	3.43	2.02	1.42

Table 6. Correlations coefficient between evaluation technique measures

	X1	X2	X3	X4	X5	X6	X7
X1	1,00						
X2	0,49	1,00					
X3	0,17	0,06	1,00				
X4	0,38	0,40	0,31	1,00			
X5	0,45	0,41	0,09	0,24	1,00		
X6	0,53	0,01	0,35	0,28	0,30	1,00	
X7	0,34	0,60	0,07	0,21	0,35	0,25	1,00

4.4 RQ3: RE-Tools Provide Better Support for RE Process

Hypothesis Testing. Table 7 shows summary statistics of the SRS quality evaluation using different tools. The t-test is found lower than critical value of t-test (2.2621) in both cases with different evaluation techniques. H_{30} cannot be rejected (α=0.05).

Table 7. t-test of the SRS quality, where SRS'es are prepared with RE-tools and office tools

	Mean	Standard deviation	t-test, $|t_0|$
RE-tools and evaluation scenario	98.9	14.579	1.051
Office tools	105.5	23.372	
RE-tools and tool tutorials	106.5	23.377	0.171
Office tools	105.5	23.372	

Descriptive Statistics. In Table 8 only the mean of Y5 shows better satisfaction of the office tools. Other measures indicate higher performance of the RE-tools.

Table 8. Descriptive analysis of tool usage (evaluation scale 1-to-6: 1 – RE-tools are better, 6– office tools are better)

Measure	Mode	Median	Mean	Variance	Standard deviation
Y1	2	2	2.91	2.5	1.58
Y2	2	2	2.50	1.65	1.28
Y3	2	2	2.80	1.47	1.21
Y4	2	2	2.43	1.23	1.11
Y5	3	3	3.53	1.92	1.39
Y6	2	3	3.00	1.63	1.28
Y7	1	2	2.25	1.73	1.31

In Table 9 correlation coefficients suggest that satisfaction of tool usage (Y5) depends on understandability of the RE process (Y1), functionality to prepare SRS (Y3), usability to execute different RE activities (Y6) and the changeability and traceability of the requirements and SRS (Y7). Functionality to prepare the SRS (Y3) depends on usability to execute different RE activities (Y6).

The experiment results suggests to use the RE-tool to support the RE process and to maintain the SRS rather than office tools.

Table 9. Correlation coefficient between tool measures

	Y1	Y2	Y3	Y4	Y5	Y6	Y7
Y1	1,00						
Y2	0,35	1,00					
Y3	0,32	0,13	1,00				
Y4	0,13	0,48	0,00	1,00			
Y5	0,40	0,26	0,48	0,01	1,00		
Y6	0,15	0,34	0,47	0,16	0,50	1,00	
Y7	0,45	0,14	0,19	0,05	0,41	0,10	1,00

4.5 Threats to Validity

The experiment involves 44 students rather than practitioners who would be more relevant since the goal is to help industry to evaluate tools. The participants had basic knowledge but limited experience in RE practice. But they all were following the same study program for 3,5 years, i.e., they were quite homogeneous regarding age and background. The use of students is a common research approach in software engineering [8]. Since the participants were in their 4th year and had only 1 study year left, their knowledge were quite close at least to practitioners who just graduated.

The quality of the evaluation scenarios which are prepared by one of the TA, could influence both the RE-tool assessment and quality of the SRS'es. To mitigate this threat the scenarios were executed by other two TA before the experiment.

Time limits and fatigue may influence the results. After preparing two SRS'es the participants were tired. The learning effect could also be noticed here: the participants already knew the problem, so the better SRS could be prepared with office tools. To investigate these threats the future work might involve different research designs.

Validity is influenced by the requirements number which was relatively small. Therefore many participants preferred office tools instead of RE-tools. The situation could be different if dealing with a large number of requirements changing over time, where the usefulness of advanced tools would be more evident.

5 Conclusions and Lessons Learnt

This paper reports on the experiment which considers the performance and correctness of the RE-tool evaluation approach [14, 16]. The experiment also compares two evaluation techniques and tool support for RE. The findings indicate:

- The framework [14] guides to the selection of the tools which yield a high-quality SRS. But the evaluation is subjective and much affected by the user experience.
- Combination of evaluation techniques allows better RE-tool assessment than using techniques separately. The findings recommend to study the RE-tool tutorials first, and to learn the tool functionality. Next, the evaluation scenarios help

to analyse the RE-tool performance in the applied study according to the same problem.
– RE-tools do provide better support for the RE process and better maintenance of the SRS quality. When using office tools more manual work is needed.
– Tool selection much depends on tool usability [4]. The experiment indicated a poor RE-tool usability, although the functionality is higher than office tools. The RE-tools are designed for the specialists [12, 14] proficient both in engineering methods and tool functionality. But this is not the case when evaluating RE-tools as the users could have different experience and understanding of the RE process.

The experiment contributes to the validity of the framework [14, 16]. But the experiment does not consider framework usefulness compared to other frameworks [5, 7]. Therefore the validation is still limited. The validation would be stronger if it was performed in industry, but not academia. The future work involves comparing of various frameworks and performing similar experiments involving practitioners.

References

1. Carvallo, J.P., Franch, X., Quer, C.: A Quality Model for Requirements Management Tools, Requirements Engineering for Sociotechnical Systems, Idea Group Publishing (2004).
2. Feblowitz, M.D., Greenspan S.J., Scenario-Based Analysis of COTS Acquisition Impacts, Requirements Engineering, 3 (1998) 182-201.
3. Franch, X., Carvallo, I.: A Quality-Model-Based Approach for Describing and Evaluating Software Packages, Proceedings of the IEEE Joint International Conference on Requirements Engineering (RE'02), Germany (2002) 104-111.
4. Goowin, N.C.: Functionality and Usability, Communications of the ACM, 30 (3) (1987) 229-233.
5. Haywood, E., Dart, P.: Towards Requirements for Requirements Modelling Tools, Proceedings of the 2nd Australian Workshop on Requirements Engineering, Australia (1997) 61-69.
6. Hands, K., Peiris, D.R., Gregor, P.: Development of a Computer-Based Interviewing Tool to Enhance the Requirements Gathering Process, Requirements Engineering, 9 (2004) 204-216.
7. Hoffmann, M., Kuhn, N., Weber, M., Bittner, M.: Requirements for Requirements Management Tools, Proceedings of the IEEE Joint International Conference on Requirements Engineering (RE'04), Japan (1997) 301-308.
8. Karlsson, L., Berander, P., Regnell, B., Wohlin, C.: Requirements Prioritisation: An Experiment on Exhaustive Pair-Wise Comparison versus Planning Game Partitioning, Proceedings of the Empirical Assessment in Software Engineering, Scotland (2004).
9. Kontio, J.: A Case Study in Applying a Systematic Method for COTS Selection, Proceedings of the 18th International Conference on Software Engineering (ICSE'96) (1996).
10. Krogstie, J., A Semiotic Approach to Quality in Requirements Specifications, Proceedings IFIP 8.1 Working Conference on Organisational Semiotics (2001).
11. Kunda, D.: STACE: Social Technical Approach to COTS Software Selection. In: Cechich, A., Piattini, M., Vallecillo A. (ed.): Component-Based Software Quality - Methods and Techniques. Lecture Notes in Computer Science, Springer-Verlag (2003) 85-98.

12. Lang, M., Duggan, J.: A Tool to Support Collaborative Software Requirements Management, Requirement Engineering, 6 (3) (2001) 161-172.
13. Maiden, N.A., Ncube, C.: Acquiring COTS Software Selection Requirements, IEEE Software (1998) 46-56.
14. Matulevičius, R.: Validating an Evaluation Framework for Requirement Engineering Tools, In: Krogstie, J., Halpin, T., Siau K. (ed.): Information Modeling Methods and Methodologies (Adv. Topics of Database Research), Idea Group Publishing (2005) 148-174.
15. Matulevičius R., Prototype of the Evaluation Framework for Functional Requirements of RE-tools. Accepted for the Proceedings of the 13th IEEE International Requirements Engineering Conference (RE'05), Paris, France, August-September (2005).
16. Matulevičius R., Sindre G., Requirements Engineering Tool Evaluation Approach. Accepted for the Proceedings of the 14th International Conference on Information Systems Development (ISD 2005), Karlstad, Sweden, August, 2005.
17. Ochs, M.A., Pfahl, D., Chrobok-Diening, G., Nothhelfer-Kolb, B.: A COTS Acquisition Process: Definition and Application Experience, Proceedings of the 11th ESCOM Conference (2000) 335-343.
18. Pohl, K.: The Three Dimensions of Requirements Engineering: a Framework and its Applications, Information Systems, 19 (3) (1994) 243-258.
19. Wohlin, C., Runeson, P., Høst, M., Ohlsson, M.C., Regnell, B., Wesslen A.: Experimentation in Software Engineering, Kluwer Academic Publishers (2002).

Improving Object-Oriented Micro Architectural Design Through Knowledge Systematization

Javier Garzás[1] and Mario Piattini[2]

[1] OCU, Oficina de Cooperación Universitaria,
Chief Technology Officer, C / Arequipa, I, Bloque I,
Planta 5, Madrid, Spain
javierg@ocu.es, jgarzas@gmail.com
[2] Alarcos Research Group. University of Castilla-La Mancha,
Ronda de Calatrava, s/n. 13071, Ciudad Real, Spain
Mario.Piattini@uclm.es

Abstract. Designers have accumulated much knowledge referring to OO systems design and construction, but this large body of knowledge is neither organized nor unified yet. In order to improve OO micro architectures, using the accumulated knowledge in a more systematic and effective way, we have defined a rules catalog (that unifies knowledge such as heuristics, principles, bad smells, etc.), the relationships between rules and patterns and an improvement method based on these subjects. We have carried out a controlled experiment which shows us that the usage of a rules catalog and its relationship with patterns really improves OO micro architectures.

1 Introduction

According to the [10], design is both "the process of defining the architecture, components, interfaces, and other characteristics of a system or components" and "the results of (that) process". A design must describe the architecture of a system, how the system is decomposed and organized into components. Generally, in software engineering the design is performed at two abstraction levels: macro architectural (high level) and micro architectural (low level). Object Oriented (OO) Micro architectural design is an old and well-known area within software engineering. Designers have accumulated a large body of knowledge regarding OO micro architectural design. Nevertheless, it is neither organized nor unified yet, and this area is still suffering from a lack of structured and classified knowledge.

In OO Micro-Architectural Design, *patterns* are the most popular and most refined example of accumulated knowledge; [4], [9], [3], [5], [6], etc, are popular references in this field. There are many and popular examples of design patterns, as *Observer, Decorator, State* or *Command* [9]. Nevertheless, even now, patterns application implicates several types of problems: difficult application, difficult learning, temptation to recast everything as a pattern, pattern overload, deficiencies in catalogs, and so on. Therefore, nowadays, patterns application is a real and important problem, and this fact has been brought up at several major congresses, for example OOPSLA 2001 - Workshop *"Beyond Design: Patterns (mis)used"*, where authors such as [20]

J. Akoka et al. (Eds.): ER Workshops 2005, LNCS 3770, pp. 444–453, 2005.

say *"We got more and more aware that a good description of the proposed solution is necessary, but useless for the reader if the problem and the forces that drive the relationship between problem and solution are not properly covered"*. Furthermore, several important books have dealt with this problem [8]. In many cases, the reason of these important problems is that the OO micro architectural design knowledge is associated exclusively with the pattern concept without taking into consideration that other elements of knowledge exist, such as ***principles, heuristics, best practices, bad smells***, etc.

According to ***principles,*** the main contributions are [14], [9] and [13]; and examples of principles are the *Dependency Inversion Principle, Don't Concrete Super class Principle*, etc. With regard to ***heuristics,*** we can refer mainly to [19] or [2], for example: *"if two or more classes only share a common interface (i.e. messages, not methods), then they should inherit from a common base class only if they will be used polymorphically."* [19]. *Concerning* best practices, *we can highlight the Venners's work, for example:* "see objects as bundles of behaviour, not bundles of data." [22]. Regarding ***bad smells,*** the main work in this field is that of [7] in which several bad smells such as *"Refused request, Subclasses that do not use what they inherit"* are *enumerated*. But, again, the application of these bad smells and the differences between them are not clear: many of them concern a single concept with different names, while others sometimes do not contain knowledge gained from experience, and still others are simply vague concepts. This confusion leads to a less efficient use of knowledge.

In order to improve OO micro-architectural designs, using all the accumulated knowledge in a more systematic and effective way, we have defined a rules catalog (which unifies knowledge such as heuristics, principles, bad smells, etc.), and the relationship between rules and patterns. We have also created an improvement method for knowledge application which will be stated in the next section. In section three, we will summarize a controlled experiment aimed at demonstrating that the improvement method could really help us improve OO micro architectures. In section four, we will carry out a short presentation about the related work. Finally, the conclusion will point out the most important arguments of our work as well as the further research.

2   A Method to Use OOD Knowledge to Improve Micro Architectural Quality

We have observed that principles, heuristics, best practices, bad smells, etc., have an analogous structure, since all of them can be expressed as a Rule – they posit a condition and offer a recommendation. It should be stressed that the "recommendation" is not a solution like that of a pattern. Patterns are more formalized than Rules and pattern descriptions are always broader. They propose solutions to problems, while rules are recommendations which a design should fulfill. Unlike patterns, rules are highly based on using natural language, which can be more ambiguous [15].

In this regard, we have developed a unified rules catalog (some examples are shown in table 1), which are named and unified according to their condition in order to improve their detection. To describe these rules, we have used the sections used in [9] catalog to describe a design pattern (See table 2): Name, Intent, Also known as, Motivation, Structure, Applicability, Participants, Collaborations, Consequences, and Known Uses. With the exception of Implementation and Sample code (renamed and unified as Sample Design) and Related Patterns (renamed as "Implies the use of Patterns")

Table 1. Some OOD Rules

Rule of IF There are dependencies on concrete classes
Rule of IF an object has a different behavior according to its state
Rule of IF a class hierarchy is composed of many classes
Rule of IF anything is used a bit or never
Rule of IF a super class knows to any sub class
Rule of IF a class collaborates with many others
Rule of IF an interface changes to many clients
Rule of IF between an interface and its implementation there is not an abstract class
Rule of IF a super class is concrete
Rule of IF in a service there are many parameters
Rule of IF a class is very large
Rule of IF users interface elements are on domain class
Rule of IF a class use more external things than those of its own
Rule of IF a class refuses any delegate
Rule of IF the attributes classes are public or protected

The last section of the rule description reflects that rules could imply the use of patterns. Often, when we introduce a rule, we obtain a new design, which needs a pattern. One example of this situation is the application of *if there are dependencies on concrete classes* rule, which introduces an abstract class or an interface, which in turn needs a creational pattern [9] to create instances and associate objects in the new situation. We can observe that this does not always happen (cardinality 0 to n), not all the rules imply the introduction of a pattern; a clear example of this can be seen when we apply rules which work only inside a module, for example the "Long Method" Rule, a bad smell according to [7]. Therefore, for each rule in the catalog, if apply, we have enumerated the patterns implied by it, and these patterns are listed in the "Imply the use of (patterns)" section.

Based on this rule catalog, we have defined an OO micro architecture improvement method, which helps designers in their improvement activities (figure 1). In a first step, violated rules are detected within a micro architecture. This step is performed using the rules catalog. A second step consists of identifying the related patterns implicated by rules. With all this knowledge, we can improve the quality of OO micro architecture in a rational and systematized way.

Table 2. Detail of a Rule

NAME: "IF THERE ARE DEPENDENCIES ON CONCRETE CLASSES"
Intent
Strategy to depend on interfaces or abstract classes rather than on concrete elements.
Also known as
Dependency Inversion Principle [13] or Programming to an Interface, not an Implementation [9].
Motivation
The structural design shows a particular type of dependency where high level modules depend on low-level ones. So, why do high-level modules depend directly on implementation modules? OO architecture shows a dependency mostly on abstractions and the modules containing implementation details also depend on these abstractions, not vice versa. The dependency has been inverted. This rule implies that each dependency within the design must have as its objective an interface or an abstract class; the dependencies must not have concrete classes as objectives. Concrete things are much more likely to change than abstract ones.
Applicability
Use this rule when: *you find dependencies on, or associations with concrete classes which may change.* Do not use this rule: *If dependency exists on a concrete class which is not likely to change (for example, a library class such as String).*
Recommendation
IF there are dependencies on concrete classes, **THEN** these dependencies should be on abstractions.

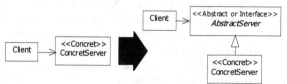

Participants
Client, Concrete Server (implementation) and Abstract Server Class (Concrete Server Interface).
Collaborations
Client communicates with Abstract Server Class and Concrete Server implements Abstract Server.
Consequences
Among others, this rule has the following benefits and liabilities: To introduce abstractions with which the design can be extended without being modified, to limit the impact of the variations in design, all the subclasses can respond to requests of the interface, and the subtypes of the abstract class and the clients will not be aware of the specific types of the objects being used.
Known Uses
This rule is used in many design patterns, frameworks, and components models.
Implies the use of (Patterns) One of the most common places where design depends on concrete classes is when instances are created. Concrete classes have to be instanced and the creational patterns (Abstract Factory, Builder, Factory Method, Prototype and Singleton) allow this instantiation.

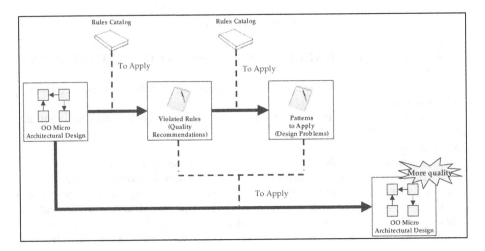

Fig. 1. A method to improve OO micro architectural quality

3   Empirical Validation

In this section, we will present a description of the main steps of the process followed to carry out the empirical validation, which is based on [23] [16] [17] [18] and [11] [12]. The main objective of this controlled experiment was to compare the effectiveness and efficiency of "traditional" OO design improvement methods with this new approach. Moreover, we aimed at analyzing if disposing of a rules catalog that unifies design knowledge as principles, best practices, heuristics, etc., and their relations with patterns has influence on the effectiveness and efficiency in the improving of the quality of OO micro architectures. Based on the GQM (Goal Question Metrics) template, the goal definition of our experiment can be summarized as follows:

Analyze	The improvement method based on the rules catalog
for the purpose of	evaluating
with respect to	effectiveness and efficiency
form the point of view of	software engineers
in the context of	software companies in Spain

3.1   Planning

The experiment is specific since it is focused on one technique applied to one domain; the ability to generalize from this specific context is further elaborated below when discussing threats to the experiment. The experiment addresses a real problem, i.e., if the method presented is more effective and efficient to be used in OO micro architectural quality improvement. Eighteen professionals of two companies carried out the experiment. The selected subjects were professionals having extensive experience in OO Design. We classified the subjects into two groups according to their professional experience. The subjects were asked to fill a questionnaire out about their expertise, and taking into consideration the collected responses, we formed two groups of subjects, trying to have the same number of subjects with good marks and bad marks in each group. Both groups had [9] patterns catalog, but only one of them had the rules catalog. In addition to this, in a previous 30 minutes session, we explained to this group some notions about rules and their relationships with patterns; and how to apply the rules catalog. For each subject, we had prepared a folder with the experimental material. Each folder contained one micro architectural diagram and a questionnaire for answers.

We had to consider what independent variables or factors were likely to have an impact on the results. These are OO Micro Architecture. We considered two dependent variables [21]:

- **Effectiveness**: Number of Defects Found / Total Number of Defects. This is the percentage of the true improvements found by a designer with respect to the total number of defects.
- **Efficiency:** Number of Defects found / Inspection Time. Where Inspection Time is related to the time that subjects spent on inspecting the micro architecture; it is measured in minutes.

3.2 Hypotheses Formulation

Our purpose was to test two groups of hypotheses, one for each dependent variable.

Effectiveness Hypotheses

- H0,1. There is no difference regarding effectiveness of subjects in detecting the violation of rules using a rules catalog and their relationship with patterns as compared to subjects without using the rules catalog. // H1,1 : ¬ H0,1
- H0,2. There is no difference regarding effectiveness of subjects in detecting the application of patterns implicated by rules using a rules catalog s and their relationship with patterns as compared to subjects without using the rules catalog. . // H1,2 : ¬ H0,2
- $H_{0,3}$. There is no difference regarding effectiveness of subjects in detecting the application of patterns not implicated by rules using a rules catalog and their relationship with patterns as compared to subject without using the rules catalog // $H_{1,3} : \neg H_{0,3}$

Efficiency Hypotheses

- H0,4. There is no difference regarding efficiency of subjects in detecting the violation of rules using a rules catalog and their relationship with patterns as compared to subjects without using the rules catalog. . // H1,4 : ¬ H0,4
- H0,5. There is no difference regarding efficiency of subjects in detecting the application of patterns implicated by rules using a rules catalog and their relationship with patterns as compared to subjects without using the rules catalog. . // H1,5 : ¬ H0,5
- H0,6. There is no difference regarding efficiency of subjects in detecting the application of patterns not implicated by rules using a rules catalog and their

relationship with patterns as compared to subjects without using the rules catalog. // H1,6 : ¬ H0,6

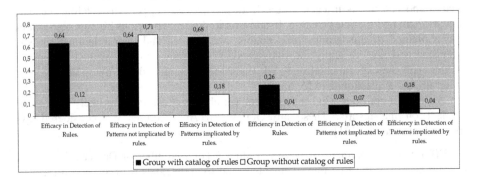

Fig. 2. Averages obtained from the experiment

3.3 Operation

In this section, we will describe the preparation, execution, and data validation of the experiment. Before the day of the experiment execution, we gave a seminar to the subjects of the group which would use the rules catalog. In this seminar, we explained to the subjects how to apply the rules catalog. The subjects had to fulfill manually their proposed solution, writing down the start and end time of the activity. We collected the forms filled out by the subjects, checking if they were complete.

3.4 Analysis and Interpretation

Figure 2 shows the averages obtained from the experiment. Outliers have not been identified. In order to decide how to test the validity of the hypotheses, we evaluated if the data followed a normal distribution, the result was normal, we decided to perform a t-Student test. In table 3, the results obtained by means of t-Student are shown. The first column represents the t-stat and the second column shows the t critical two – tail.

Table 3. Results obtained by means of t-Student

	t stat	t Critical two-tail
Efficacy in Detection of Rules.	5.38887	2.26215
Efficacy in Detection of Patterns not implicated by rules.	-0.22360	2.20098
Efficacy in Detection of Patterns implicated by rules.	3.36269	2.20098
Efficiency in Detection of Rules.	7.03868	2.26215
Efficiency in Detection of Patterns not implicated by rules	0.22269	2.26215
Efficiency in Detection of Patterns implicated by rules	4.35678	2.17881

We have obtained the following results. Firstly, it was confirmed by the t-Student test that the group with the rules catalog obtained better results in "Efficacy and Efficiency in Detection of Rules" and "Efficacy and Efficiency in Detection of Patterns implicated by rules". In the second place, the t-Student test could not confirm that the group with the rules catalog obtained better results in "Efficiency in Detection of Patterns not implicated by rules". However, this group obtained better averages; we have to highlight that "Efficiency in Detection n of Patterns not implicated by rules" is not influenced by rules catalog, since these patterns are not in catalog because they are not implicated by rules, and the application of these patterns will result in the detection of design problems more than design recommendations. Lastly, in a similar way, we could not confirm by using the t-Student test that the group without the rules catalog obtained better results in "Efficacy in Detection of Patterns not implicated by rules"; however, again, this result is not influenced by rules catalog.

3.5 Threats to Validity

A list of issues that threats the validity of the empirical study is identified below:

Conclusion Validity

The results confirmed by means of the t-Student test that there was a significant difference between the two groups, and that the new approach seems to be more effective and efficient for carrying out the OO micro architectural quality improvement. The statistical assumptions of each test were verified, so that, the conclusion validity was fulfilled.

Internal Validity

- Differences among subjects. We formed two groups, and the subjects were apportioned between these two groups according to their expertise and skills. For this reason, the subjects were asked to fill out a questionnaire about their expertise, and taking into account the collected responses, we formed the two groups of subjects.
- Differences among OOD diagrams. We used only one OOD diagram.
- Precision in the time values. The subjects were responsible for recording the start and finish times of each test, so they could introduce some imprecision but we think it is not very significant.
- Fatigue effects. The average time for carrying out the experiment was 20 minutes, so fatigue effects were minor.
- Persistence effects. Persistence effects are not present because the subjects had never participated in a similar experiment.
- Subject motivation. The subjects were very motivated.
- Other factors, such as plagiarism and influence among subjects were controlled.

External Validity

- Subjects. We are aware that more experiments with professionals must be carried out in order to be able to generalize these results. However, the subjects could be considered "common" OO designers at least in the context of Spanish software companies.

• Material used. We believe that the documents used might not be representative of an industrial problem, so more experiments with larger diagrams are needed.

5 Conclusions and Future Work

There has not been much effort made on empirical studies about OO design knowledge, and the few works we have found are mainly focused on design patterns. We should consider that (a) OO micro architectural design knowledge is associated with the pattern concept, but other elements exist, such as *principles, heuristics, best practices, bad smells*, etc. (b) these other elements show a confused description, unification, definition, etc. (c) rules and patterns are related elements (d) with a unified rules catalog and related patterns is easier to detect quality recommendations and to apply patterns.

We are conscious that, in this experiment we have chosen to investigate with an individual technique that could interact with many other development techniques and procedures ("the life-cycle issue", [12]), the life-cycle model (light or weight) [1], the modeling language used to express the design artifacts, etc. So, large-scale empirical studies are needed to obtain conclusions about the effect of the knowledge systematization in the OOD improvement.

References

1. Boehm B. and Turner R. (2003). Using Risk to Balance Agile and Plan-Driven Methods. Computer. Vol. 36, No. 6, pp. 57-66.
2. Booch G. (1996). Managing the Object-Oriented project. Addison-Wesley
3. Buschmann F., Meunier R., Rohnert H., Sommerlad P. and Stal M. (1996). A System of Patterns: Pattern-Oriented Software Architecture, Addison-Wesley.
4. Coad P. (1992). Object-Oriented Patterns. *Comunications ACM, v*ol. 35, n 9, pp. 152-159.
5. Coplien J. and Schmidt. (1996). Patterns Languages of Program Design. Addison Wesley.
6. Tichy W. (1997). A catalogue of general-purpose software design patterns. TOOLS '97.
7. Fowler M. (2000). Refactoring improving the design of existing code. Addison Wesley
8. Fowler M. (2002). Patterns of Enterprise Application Architecture. Addison Wesley.
9. Gamma E., Helm R., Johnson R. and Vlissides J. (1995). *Design patterns: Elements of Reusable Object Oriented Software*. Addison-Wesley.
10. IEEE (1990). STD 610.12-1990. IEEE Standard Glossary of Software Engineering Terminology.
11. Kitchenham B. A., Pfleeger S., Pickard L., Jones P., Hoaglin D., El Emam K., and Rosenberg J. (2002). Preliminary Guidelines for Empirical Research in Software Engineering. IEEE Transactions on Software Engineering, IEEE Computer Society, 28 (8).
12. Kitchenham, B.A., Dybå, T. and Jorgensen, M. (2004). Evidence-based Software Engineering. Procedigns of the 26th International Conference on Software Engineering (ICSE´04), IEEE.
13. Martin 1996. The Dependency Inversion Principle. C++ Report.
14. Meyer B. (1988). *Object Oriented Software Construction*. Prentice Hall.

15. Pescio C. (1997). Principles Versus Patterns. IEEE Computer, 30 (9): 130-131.
16. Prechelt L., Unger B., Philippsen M., and Tichy W. (1997). Two Controlled Experiments Assessing the Usefulness of Design Pattern Information During Program Maintenance. IEEE Transactions on Software Engineering, 28(6), pp. 595-606.
17. Prechelt L. and Unger B. (1998). A Series of Controlled Experiments on Design Patterns: Methodology and Results. Proc. Softwaretechnik '98 GI Conference (Softwaretechnik-Trends 18(3)), pp. 53-60, Paderborn, September 7-9, 1998.
18. Prechelt L., Unger B., Tichy W., Bössler P. and Votta G. (2001). A Controlled Experiments in Maintenance Comparing Design Patterns to Simpler Solutions. IEEE Transactions on Software Engineering. 27(12), pp. 1134-1144.
19. Riel A. J. (1996). *Object-Oriented Design Heuristics*. Addison-Wesley.
20. Schwanninger C. (2001). "Patterns as problem indicators" Workshop on Beyond Design Patterns (mis)Used, OOPSLA 2001.
21. Thelin T., Runeson P., Wholin C., Olsson T. and Andersson C. (2004). Evaluation of Usage – Based Reading – Conclusions after three Experiments. Empirical Software Engineering, 9 (1-2): 77 - 110.
22. Venners B. (2004). *Interface Design Best Practices in Object-Oriented API Design in Java*. Retrieved February 2004 from http://www.artima.com/interfacedesign/contents.html
23. Wohlin C., Runeson P., Höst M., C. Ohlsson M., Regnell B., and Wesslén A. (2000). Experimentation in Software Engineering – An Introduction. Kluwer Academic Publishers.

Tutorials

Tutorial 1:
eduWeaver – The Courseware Modeling Tool

Judit Bajnai[1], Dimitris Karagiannis[1], and Claudia Steinberger[2]

[1]University of Vienna, Faculty of Computer Science,
Institute of Business and Knowledge Engineering,
Bruenner Str. 72, A-1210 Vienna, Austria
[2]Alpen-Adria University Klagenfurt,
Department of Business Informatics and Application Systems,
Universitätsstrasse 65-67, A-9020 Klagenfurt, Austria

1 Motivation

Technology has changed the way we live, think and work. Technology has revolutionized business and now it must revolutionize teaching and learning. Live classroom based training is becoming too costly and cumbersome. 80 % of teachers and students already use computers. So e-learning can be seen as a means supporting life long learning with a lot of benefits. The worldwide web, high-capacity networks and high speed computers make learning available to people 24 hours a day in their office, at home or also in hotel rooms during a business trip around the globe. E-learning enables the access to learning when it is convenient.

Although there are a lot of web based e-learning solutions on the market, using these tools to conceptually model courses still stays a very difficult task for most teachers. They are supposed to digitize and multimedialize their course contents and to organize virtual courseware themselves. However mostly passing the first euphoria courseware creation mostly turns out to be incredible costly and time consuming [1].

A platform-independent modeling tool for the creation of courseware is introduced in this tutorial.

2 The Conceptual Modeling Tool - eduWeaver

In 2001 an Austrian project called eduBITE (Educating Business and Information Technologies) [2], [3] started. The project was funded the Austrian Federal Ministry for Education, Science and Culture (bm:bwk) within the initiative "Neue Medien in der Lehre" (new media in education). The main focus of this project was the development of an instructional design method called eduWeaver at the University of Vienna. EduWeaver is based on the meta-modeling platform ADVISOR® [4], [5]. Within this meta-modeling platform the e-learning specific courseware modeling method eduWeaver was implemented [6]. eduWeaver supports teachers by creating new e-learning courses reusing existing multimedia learning objects created and provided by different higher educational institutes in Austria [7]. eduWeaver provides teachers with a so called "content pool" for content management of existing teaching

J. Akoka et al. (Eds.): ER Workshops 2005, LNCS 3770, pp. 457–458, 2005.
© Springer-Verlag Berlin Heidelberg 2005

materials. This content pool is linked with the modeling core of eduWeaver, offering a graphical tool in order to do instruction design work. eduWeaver also offers a standardized interface for SCORM [7] based content export in order to provide e-learning courses within a learning management system.

The modeling core of eduWeaver consists of four modeling levels. Each level contains learning construct instances that correspond to the model types Course-Overview, Course, Module and Lesson. These model types are hierarchically linked to each other by internal references. Within each modeling level sequences of instruction can be graphically modeled by using according object and relation classes representing different granularities of the process level [8].

Within this tutorial the Meta-Model of eduWeaver will be introduced, further a practical introduction into modeling with eduWeaver will be given. Past experiences and future development plans are discussed with interested researchers.

References

1. Cisco Systems, Inc., Reusable Learning Object Strategy, Definition, Creation Process and Guidelines for Building, Version 3.1, April 22, 2000
2. The project eduBITE, http://www.edubite.ac.at
3. Karagiannis, Dimitris; Bajnai, Judit; Mayr, Heinrich C.; Steinberger, Claudia: eduBITE - Educating Business and Information Technologies, for Dagstuhl-Seminar 03191, Conceptual, Technological and Organisational Aspects of Electronic Learning, May 04-09, 2003, Dagstuhl, Germany
4. ADVISOR http://www.boc-eu.com/advisor/start.html
5. BOC Information Technologies Ltd. http://www.boc-eu.com
6. Bajnai, Judit; Karagiannis, Dimitris: ADVISOR®– Meta-Modeling Tool for Individual Instructional Design, published in the Proceedings of ICL2004, September 2004, Villach, Austria
7. Sharable Content Object Reference Model http://www.adlnet.org/
8. Bajnai, Judit; Steinberger, Claudia: eduWeaver – The Web-Based Courseware Design Tool, IADIS International Conference, Proceedings of WWW/Internet 2003, 5. – 8. November 2003, Algarve, Portugal

Tutorial 2:
FOOM - Functional and Object Oriented Methodology: An Integrated Approach

Peretz Shoval

Ben-Gurion University, Department of Information Systems Engineering, Israel

1 Aims and Objectives

The tutorial will provide an overview of FOOM methodology, which integrates the functional and object-oriented approaches. It will start with a rationale and need for the integrated approach, in light of the traditional methodologies on one hand, and the object-oriented and UML-based methodologies on the other hand. The main part of the tutorial will be devoted to a description of the analysis and design stages and sub-stages of FOOM, and their products. The description will be accompanied by a running example.

2 Summary of the Tutorial

2. 1 Background on Development Methodologies

This part of the tutorial will provide a brief introduction on IS development methodologies, making a distinction between traditional (process-oriented) and object-oriented (UML-based) methodologies. The objective is highlight the strengths and weaknesses of each approach, and provide a rationale for a methodology which combines the two approaches.

2.2 Overview of FOOM Methodology

This is the main part of the tutorial. It will describe the activities and products of the analysis and design phases according to FOOM, along with a running example. Here they are described in brief:

The Analysis Phase
The analysis phase consists of two main activities: data analysis and functional analysis. The main products of this phase are:

a) a data model in the form of an **initial class diagram**, consisting of data classes, i.e. classes derived from the application/user requirements.
b) a functional model in the form of **hierarchical OO-DFDs**, consisting of functions, external entities (user, time and real-time entities), data classes (instead of the traditional data stores) and data flows among them.

J. Akoka et al. (Eds.): ER Workshops 2005, LNCS 3770, pp. 459–460, 2005.

The two product of the analysis phase are synchronizes so that all class appearing the class diagram appear also in the OO-DFDs, and vise versa.

The Design Phase

This phase consists of the following stages:

Top-level design of transactions: The transactions of the system are derived from the OODFDs; eventually they will become the class methods. The products of this stage include transactions diagrams and their top-level descriptions, expressed in pseudo-code.

Design of the user-interface: The methodology provides a semi algorithmic method to design a menus-tree interface from the hierarchy of OO-DFDs. The resulting menus become objects of the Menus class.

Design of the inputs and outputs: The data flows from external entities to functions and from functions to external entities are used to design the input and output screens/reports. These become objects of the Forms and Reports classes, respectively.

Design of the system behavior: The methodology provides guidelines for converting the toplevel descriptions of the transactions into detailed descriptions, and then for mapping each transaction description to class methods. A distinction is made between "basic" (CRUD) methods, "application-specific" methods, and "transaction" methods. The methodology provides two alternative techniques for describing each method in detail: pseudo-code and message-chart. The products of the design phase will be used (at the implementation stage) to create the software using any programming environment.

2.3 Evaluation of the Methodology

This part of the tutorial will present some research results in which FOOM methodology was evaluated and compared to another development methodology that combines the functional and OO approaches.

Tutorial 3:
Domain Engineering – Using Domain Concepts to Guide Software Design

Iris Reinhartz-Berger[1], Arnon Sturm[2], and Yair Wand[1]

[1] Department of Management Information Systems,
University of Haifa, Haifa 31905, Israel
`iris@mis.hevra.haifa.ac.il`
[2] Department of Information System Engineering,
Ben-Gurion University of the Negev, Beer Sheva 84105, Israel
`sturm@bgumail.bgu.ac.il`

1 Introduction

As the variability of information and software systems has increased, the need for an engineering discipline concerned with building reusable assets (such as specification sets, patterns and components) on one hand and representing and managing knowledge in specific domains on the other hand has become crucial. This discipline, called domain engineering, supports the notion of a domain, defined as a set of applications that use a set of common concepts for describing requirements, problems and capabilities. The purpose of domain engineering is to identify, model, construct, catalog, and disseminate a set of software artifacts that can be applied to existing and future software in a particular application domain. As such, it can support the effective and efficient management and development of software assets. Hence, it is important to introduce this discipline among software engineering practitioners and researchers.

As an evolutionary approach, domain engineering relates to many different areas, such as conceptualization, ontologies and ontology deployment, metadata and metamodeling, knowledge base integration, reuse, patterns, model driven approaches, and reference modeling.

The purpose of this tutorial is to present and discuss domain engineering concepts, methods, problems, and solutions. We will focus on modeling domains and applications and validating application models against their domain models. We will review several domain engineering methods, explaining their basics, rationale, advantages, and shortcomings. We will compare and discuss these methods using a case study, emphasizing how they support the reuse and knowledge management activities.

2 What Is Domain Engineering?

Domain engineering is a discipline that supports the notion of a domain. A domain can be defined as a set of applications that use common concepts for describing

J. Akoka et al. (Eds.): ER Workshops 2005, LNCS 3770, pp. 461–463, 2005.
© Springer-Verlag Berlin Heidelberg 2005

requirements, problems and capabilities. The purpose of domain engineering is to identify, model, construct, catalog, and disseminate a set of software artifacts that can be applied to existing and future software systems in a particular application domain. Similarly to software engineering, domain engineering includes three main activities: domain analysis, domain design, and domain implementation. Domain analysis identifies a domain and captures its ontology. It should specify the basic elements of the domain, identify the relationships among these elements, and represent this understanding in a useful way. Domain design and domain implementation are concerned with mechanisms for translating requirements into systems that are made up of components with the intent of reusing them to the highest extent possible.

Domain modelling is especially important because of two main reasons. First, modelling, especially visual modelling, can help understanding complex system specifications and support communications among the various stakeholders engaged in the development process. Secondly, the core elements of a domain and the relationships among them usually remain unchanged, while the technologies and implementation environments are in progressive improvement. Hence, domain models usually remain valid for long periods.

Many methods and techniques have been developed to support domain modelling. FODA and PLUS, for example, are feature-oriented approaches that emphasize the common and different features of applications in a specific domain. GME, metaEdit+, and DOME are examples of metamodeling environments which enable definition of domain specific languages (DSL) and support using them to describe particular applications.

Nowadays, domain engineering and domain specific languages receive special attention from communities which deal with conceptualization, ontologies and ontology deployment, metadata and metamodeling, knowledge base integration, reuse, patterns, model driven approaches, and reference modelling. However, domain engineering has been criticized as dealing with too broad areas (domains) which are usually understood only during the development process. Most of these problems can be solved by taking a special care of the way domain engineering is woven into software engineering.

In this tutorial, we will present and discuss domain engineering concepts, problems, and solutions through the Application-based Domain Modelling (ADOM) approach. ADOM binds domain and application models into a general framework and enables the definition of mutual constraints between these types of models. An example of a domain can be process control systems which monitor and control the values of certain variables through a set of components that work together to achieve a common objective or purpose. Application areas within this domain include engineering and industrial control systems, control systems in the human body, and financial derivation-tracking products. All of these application types share concepts such as controller, controlled value, executer, sensor, etc. In addition, they should obey constraints, such as "keep the system within the boundaries of the controlled value." Being influenced by Meta Object Facility (MOF), the architecture of ADOM consists of three layers: the *application layer*, the *domain layer*, and the (modeling) *language layer*. The application layer consists of models of particular applications, including their structure (scheme) and behaviour, e.g., a home climate control system and a financial derivation-tracking product. The language layer comprises metamodels of modeling languages, e.g., UML. The intermediate domain layer consists of specifications of various domains, e.g., the process control systems domain. The ADOM architecture also enables enforcing constraints among the different layers; in particular the domain layer enforces constraints on the application layer. The fulfilment of these constraints can be checked due to the ability to classify application elements according to the domain terminology.

3   Tutorial Objectives

The main purpose of the tutorial is to present and discuss domain engineering concepts and methods, focusing on domain modelling activities. This can be divided into the following objectives:

- Introducing basic concepts of domain engineering.
- Reviewing and comparing common domain modeling methods.
- Discussing domain engineering problems and solutions to these problems.
- Exemplifying domain engineering concepts and their integration into software engineering processes through the Application-based Domain Modeling (ADOM) approach.

Tutorial 4:
Reasoning About Web Information Systems

Klaus-Dieter Schewe[1] and Bernhard Thalheim[2]

[1] Massey University, Department of Information Systems and
Information Science Research Centre,
Private Bag 11 222, Palmerston North, New Zealand
k.d.schewe@massey.ac.nz
[2] Christian Albrechts University Kiel,
Department of Computer Science and Applied Mathematics,
Olshausenstr. 40, D-24098 Kiel, Germany
thalheim@is.informatik.uni-kiel.de

1 Content Outline

We will start with a brief introduction describing the various aspects of web information systems (WISs) such as purpose, usage, content, functionality, context, presentation. Following this we plan to present three major blocks (of more or less the same size) dealing with an overview of the co-design approach to WIS design, propositional reasoning about WISs, and consistency of WISs.

2 The Co-design Approach to WIS Design

The co-design approach to WIS design [18] specifies systems on different levels of abstraction. In the tutorial we will focus on two of these levels. On a high-level of abstraction co-design emphasises *storyboarding*, which concerns the usage of the intended system. It consists of three interconnected parts: the modelling of the story space, the modelling of the actors, i.e. clases of users, and the modelling of tasks.

For the story space the obvious idea is to regard a web information system as a collection of abstract locations (called scenes), between which users navigate. While navigating through the system a user will execute certain actions. Thus, we first obtain a rough story space language, which consists simply in modelling labelled directed graphs. In a second step we take a closer look into the sequencing of actions executed by a user. Using sequential and parallel composition, a choice operator and pre- and postconditions the micro-level of the story space is given by an assignment-free process algebra. In a third step we will also show that the story algebra can be represented in a graphical way using StateCharts. This enables animation and simulation of a story space design.

For the actors we first address the modelling of roles, then the profiling of users. The used model combines characteristics and then defines important user types. For each user type preference rules describing the user behaviour can be formulated. These preference rules will form the basis for the personalisation of the story space.

J. Akoka et al. (Eds.): ER Workshops 2005, LNCS 3770, pp. 464–467, 2005.
© Springer-Verlag Berlin Heidelberg 2005

The tasks link the actors with the story space. A task will consist of a goal, involved actors and required actions. Reasoning about tasks will permit to set up task execution plans. The goals will also link the tasks to the personalisation of the story space.

On a lower level of abstraction the co-design approach emphasises content and functionality modelling. This level takes a deeper look into the scenes of the story space and links the abstract storyboard layer with the necessary database support. Obviously, each scene will be supported by a view on some underlying database that will result in the content to be presented. we will formally define such views. However, co-design takes the point of view that the content of the whole WIS including the navigation structure also defines a view. That is, the navigation structure will become part of the view construction mechanism.

With respect to functionality operations are added to the views. These operations have to be understood as detailed specifications of the actions that appear in the story space. We will present an abstract language for modelling these operations. In a last step we introduce two more extensions to the views. The one is hierarchies, which enable more coarse or more detailed presentation of information. we will indicate how to define hierarchies and which operations have to be made available to switch presentation along the hierarchy. A third extension addresses adaptivity to technical restrictions such as channel bandwidth or enddevices. This leads to cohesion, for which co-design employs preorders and proximity values. We will show that with such a model there is an algorithmic solution of the adaptivity problem, while the model only requires the specification of cohesion.

3 Propositional Reasoning about WISs

Starting from the algebra that was used to specify story spaces, we will now show that this algebra can in fact be represented as a Kleene algebra with tests (KATs) [16]. This enables a simple form of system personalisation. Preferences and goals of users that are modelled by equations on the KAT can be exploited for simple, but effective term rewriting for the purpose of simplifying the story space according to a user's needs. We will demonstrate this approach to personalisation by examples. In a second step we will then briefly discuss decidability and complexity of this reasoning process [10].

As this form of personalisation by propositional reasoning depends on the user types, the question arises, whether the classification of users can be dispensed with. In doing so, user characteristics, preferences and goals just become part of the equations used in the reasoning with KATs. However, the implications on decidability and complexity are still not known, so we will only indicate this idea and point to open research problems.

The roles of actors are associated with rights and obligations. This leads to a deontic logic. We will formally describe the usage of the logic for expressing the requirements, then outline how the logic can be used to reason about tasks. This requires setting up proof obligations that state, whether a task can be satisfied. Reasoning about tasks in this way will permit to set up task execution plans.

4 Consistency of WISs

The added details on the lower level of abstraction enable a more detailed level of reasoning. First we will link the operations back to the story space showing that now on a more detailed level we can express desirable properties of the system by formulae of a higherorder dynamic logic [9]. In doing so, we obtain various proof obligations for consistency, personalisation and satisfiability of goals.

For the first of these proof obligations we will introduce static and dynamic constraints, both formalised in the logic. We will then argue how consistency can be expressed by other formulae. How to prove these formulae will only be sketched briefly. For personalisation we will outline that the approach that was used on the higher level of abstraction still applies to this level as well. Translated into the higher-order dynamic logic we obtain a more complicated formula to prove. Finally, satisfiability of goals considers postconditions leading to the aim to determine preconditions that would be sufficient for reaching the goals.

In a second step we will reconsider the deontic logic used to formalise rights and obligations. Similar to the first part the deontic logic now becomes a high-order one. It enables formalising proof obligations for task satisfiability depending on the rights. In particular, users are modelled by agents.

5 Literature

The work presented in the tutorial is partly subject to the book [19], which will subsume our journal publications [18, 12, 20, 13, 7] on the subject, a previous tutorial [14], and various conference publications [3, 15, 2, 9, 11, 1, 8, 5, 4, 16] and book chapters [6, 17]. Major publications dealing with logical aspects of WISs are [18, 16, 9, 10]. A draft version of parts of the book may be available at the time of the conference.

References

1. Binemann-Zdanowicz, A., Kaschek, R., Schewe, K.-D., and Thalheim, B. Contextaware web information systems. In Conceptual Modelling 2004 – First Asia-Pacific Conference on Conceptual Modelling (Dunedin, New Zealand, 2004), S. Hartmann and J. Roddick, Eds., vol. 31 of CRPIT, Australian Computer Society, pp. 37–48.

2. Feyer, T., Kao, O., Schewe, K.-D., and Thalheim, B. Design of data-intensive web-based information services. In Proceedings of the 1st International Conference on Web Information Systems Engineering (WISE 2000), Q. Li, Z. M. Ozsuyoglu, R. Wagner, Y. Kambayashi, and Y. Zhang, Eds. IEEE Computer Society, 2000, pp. 462–467.

3. Feyer, T., Schewe, K.-D., and Thalheim, B. Conceptual modelling and development of information services. In Conceptual Modeling – ER'98, T. Ling and S. Ram, Eds., vol. 1507 of LNCS. Springer-Verlag, Berlin, 1998, pp. 7–20.

4. Kaschek, R., Matthews, C., Schewe, K.-D., and Wallace, C. Analyzing web information systems with the Abstraction Layer Model and SiteLang. In Proceedings of the Australasian Conference on Information Systems (ACIS 2003) (2003).

5. Kaschek, R., Schewe, K.-D., Thalheim, B., and Zhang, L. Integrating context in conceptual modeling for web information systems. In Web Services, E-Business, and the Semantic Web, Revised Selected Papers of Second International Workshop WES 2003 (2004), C. Bussler, D. Fensel, M. E. Orlowska, and J. Yang, Eds., vol. 3095 of LNCS, Springer-Verlag, pp. 77–88.

6. Kaschek, R., Schewe, K.-D., Wallace, C., and Matthews, C. Story boarding for web information systems. In Web Information Systems, D. Taniar and W. Rahayu, Eds. IDEA Group, 2004, pp. 1–33.

7. Ma, H., Schewe, K.-D., Thalheim, B., and Zhao, J. View integration and cooperation in databases, data warehouses and web information systems. Journal on Data Semantics (2005). submitted.

8. Schewe, K.-D. Querying web information systems. In Conceptual Modeling – ER 2001 H. S. Kunii, S. Jajodia, and A. Sølvberg, Eds., vol. 2224 of LNCS. Springer-Verlag, 2001, pp. 571–584.

9. Schewe, K.-D. The power of media types. In Web Information Systems – WISE 2004, X. Zhou, S. Su, M. Papazoglou, M. Orlowska, and K. Jeffery, Eds., vol. 3306 of LNCS. Springer, 2004, pp. 53–58.

10. Schewe, K.-D. Propositional reasoning about web information systems. submitted, 2005.

11. Schewe, K.-D., Kaschek, R., Wallace, C., and Matthews, C. Modelling web-based banking systems: Story boarding and user profiling. In Advanced Conceptual Modeling Techniques: ER 2002 Workshops (2003), vol. 2784 of LNCS, Springer-Verlag, pp. 427–439.

12. Schewe, K.-D., Kaschek, R., Wallace, C., and Matthews, C. Emphasizing the communication aspects for the successful development of electronic business systems. Information Systems and E-Business Management 3, 1 (2005), 71–100.

13. Schewe, K.-D., and Schewe, B. Integrating database and dialogue design. Knowledge and Information Systems 2, 1 (2000), 1–32.

14. Schewe, K.-D., and Thalheim, B. Conceptual modelling of internet sites. http://infosys.massey.ac.nz/~kdschewe/pub/slides/ER00tuti.ps with i = 0, . . . , 6, 2000. Tutorial Notes at the 19th International Conference on Conceptual Modelling (ER 2000).

15. Schewe, K.-D., and Thalheim, B. Modeling interaction and media objects. In Natural Language Processing and Information Systems: 5th International Conference on Applications of Natural Language to Information Systems, NLDB 2000, M. Bouzeghoub, Z. Kedad, and E. M´etais, Eds., vol. 1959 of LNCS. Springer-Verlag, Berlin, 2001, pp. 313–324.

16. Schewe, K.-D., and Thalheim, B. Reasoning about web information systems using story algebras. In Proceedings ADBIS 2004 (Budapest, Hungary, 2004).

17. Schewe, K.-D., and Thalheim, B. Structural media types in the development of dataintensive web information systems. In Web Information Systems, D. Taniar and W. Rahayu, Eds. IDEA Group, 2004, pp. 34–70.

18. Schewe, K.-D., and Thalheim, B. Conceptual modelling of web information systems. Data and Knowledge Engineering 54, 2 (2005), 147–188.

19. Schewe, K.-D., and Thalheim, B. Design and Development of Web Information Systems. Springer-Verlag, 2005. to appear.

20. Schewe, K.-D., Thalheim, B., Binemann-Zdanowicz, A., Kaschek, R., Kuss, T., and Tschiedel, B. A conceptual view of electronic learning systems. Education and Information Technologies 10, 1-2 (2005), 83–110.

Tutorial 5:
Schema and Data Translation

Paolo Atzeni

Università Roma Tre
atzeni@dia.uniroma3.it

1 Summary

Many application settings involve the need to exchange information between heterogeneous frameworks. In the database world, we often use different systems to handle data, following different models, and we therefore need to translate data and their description from one to another. The problem has been considered for decades in our field, but definitive solutions are not yet available. The problem is relevant at the schema level (for example, every designer works with a conceptual model, such as ER or UML, and then translates the conceptual schema into a logical model, usually relational), and at the data level, when we have databases, and we want to translate them into some other system, which may be similar (for example, with a slightly different version of the relational model) or completely different (for example, XML documents).

In current practice, translation problems are often tackled by means of ad-hoc solutions, for example by writing a program for each specific application, but this is clearly very heavy and hard to maintain.

A recent proposal by Bernstein argues for "model management," a high-level approach to problems of this kind, which require the management of descriptions of application artifacts. Indeed, the translation problem can be formulated in the form of a high level operator (called "ModelGen") that, given a source data model M1 (e.g., the ER model), a target data model M2 (e.g., SQL DDL or XML Schema), and a source schema S1 expressed in M1, generates a target schema S2 in M2.

An additional, more complex problem would also include the translation of instances: given also a source instance I1 for S1, we want to generate an equivalent instance I2 for the target schema S2.

Various approaches to ModelGen were proposed recently, including our own, based on a notion of metamodel.

The tutorial will discuss the major issues related to a generic translation features, including foundations (including the variety of models of interest and the "correctness" requirements for the target schema and instances), approaches and applications.

2 Outline

- Heterogeneous schemas and their translation: problems and issues
- Model management, a high-level approach

J. Akoka et al. (Eds.): ER Workshops 2005, LNCS 3770, pp. 468–469, 2005.
© Springer-Verlag Berlin Heidelberg 2005

- "ModelGen", an operator for schema translation
- A universe of models and a metamodel for describing them
- Information capacity and equivalence
- Applications: customization of automatically generated translations

References

1. Periklis Andritsos, Ronald Fagin, Ariel Fuxman, Laura M. Haas, Mauricio A. Hernández, C. T. Howard Ho, Anastasios Kementsietsidis, Renée J. Miller, Felix Naumann, Lucian Popa, Yannis Velegrakis, Charlotte Vilarem, Ling-Ling Yan: Schema Management. IEEE Data Eng. Bull. 25(3): 32-38 (2002)
2. P. Atzeni, P. Cappellari, and P. A. Bernstein. ModelGen: Model independent schema translation. In ICDE, Tokyo. IEEE Computer Society, 2005
3. P. Atzeni, R. Torlone: Management of Multiple Models in an Extensible Database Design Tool. EDBT 1996: 79-95.
4. P. A. Bernstein: Applying Model Management to Classical Meta Data Problems. Conference on Innovative Data Systems Research, Asilomar, CA, USA, January 5-8, 2003.
5. P. A. Bernstein, A. Y. Halevy, R. Pottinger: A Vision of Management of Complex Models. SIGMOD Record 29(4): 55-63 (2000) .
6. Richard Hull: Managing Semantic Heterogeneity in Databases: A Theoretical Perspective. PODS 1997, 51-61
7. R. Hull, R. King: Semantic Database Modeling: Survey, Applications, and Research Issues. ACM Computing Surveys 19(3): 201-260 (1987).
8. Renée J. Miller, Yannis E. Ioannidis, Raghu Ramakrishnan: Schema equivalence in heterogeneous systems: bridging theory and practice. Inf. Syst. 19(1): 3-31 (1994)

Tutorial 6:
Modeling and Simulation of Dynamic Engineering Design Processes

Vadim Ermolayev[1], Vladimir Gorodetski[2], Eyck Jentzsch[3],
and Wolf-Ekkehard Matzke[3]

[1] Zaporozhye National Univ., Zaporozhye, Ukraine
[2] Saint-Petersburg Institute for Informatics, RAS, Saint Petersburg, Russia
[3] Cadence Design Systems, GmbH, Feldkirchen, Germany

Design – a signature of human intelligence – was always a great challenge for researches in various disciplines. For example, observations of how humans act in design produced several fundamental ideas in AI and DAI – automated problem solving and reasoning [3]. In return, the researchers as the broad community attacked the problems of design domain by attempting to engineer systems and infrastructures that are capable of supporting humans in accomplishing tasks that require intelligence. Quite a big piece of this stake is of course the challenge of designing the concepts and the models of different aspects in design. Moreover, from data and knowledge engineering perspective the problems of Conceptual Modeling are design problems per se: the problem of designing intelligent artifacts [3], or exploring the design space of intelligence [1][4].

The complete process of design has not been fully automated yet in a satisfactory way. For example, agents or other "smart" software systems still do not design artifacts as humans do. Some attempts, however, have been undertaken. Some of these attempts have used agents (an engineering sub-area of DAI) to create intelligent software infrastructures for supporting engineering design processes performed by distributed teams and comprising contributions from various disciplines. The models of Engineering Design Processes produced in these developments are based on the paradigms of an Agent and an Agency.

The tutorial will survey these attempts in the period of the last 10-12 years structuring them alone the dimensions of complexity in Integrated Product Design as well as alone the time axis. We shall focus on how these dimensions of complexity affected the developed conceptual models. Some of the dimensions of complexity are: the boundaries between disciplines in multidisciplinary design, conflicting goals among the design team members, big chunks in design process, counter-intuitive behavior of the designers, etc. The time axis is divided into three topical periods: the "Antique" period, the "Middle Ages", and the "Renaissance". The Antique period is characterized by the substantial growth in interest in agent-based approaches to engineering design automation. The constellation of research projects undertaken at that time were raised by big, even romantic, expectations of a breakthrough, of a so called silver bullet in the field. We shall overview several of the most influential "Antique" projects: PACT+SHADE, ACDS, ABCDE, DIDE, SHARE, SiFA. These

J. Akoka et al. (Eds.): ER Workshops 2005, LNCS 3770, pp. 470–472, 2005.
© Springer-Verlag Berlin Heidelberg 2005

expectations, however, have not been fully backed up with the appropriately sound results. The tutorial analyses the reasons behind this. It is concluded that the main problem was the lack of the maturity in fundamental theories, basic frameworks, and underlying models, methodologies, and technologies. The tutorial then switches to the survey of the second historical period of "Middle Ages" which main focus was on the development of these basic theories, models, methodologies, and technologies like for example Dynamic Design Process Models (RAPPID project), Dynamic Distributed Planning and Coordination Mechanisms (ADN project). The developments of the "Middle Ages" formed what may be called a critical mass leading to the "Renaissance". Examples of several projects are given and analyzed in order to determine the realistic focus in engineering design automation activities which emerge in recent times. Recently launched projects and their accomplishments are overviewed in this part of the tutorial. Special attention is paid to the descriptions of the goals, the problems attacked, and the approaches to solutions. In the upcoming part of the tutorial the Productivity Simulation Initiative (PSI) of Cadence Design Systems, Inc. is presented in detail. This presentation is structured as follows:

- PSI Goals and objectives
- PSI Dynamic Engineering Design Process (DEDP) agent-based modeling framework: mechanisms and knowledge models
- PSI DEDP-MAS architecture, implementation methodology (Gaia), and MAS DK as the rapid prototyping toolkit
- Use cases, the testbed, and simulation experiments performed with PSI DEDP-MAS

PSI related part of the tutorial will demonstrate how agent-based models, principles, methodologies may be used for the intelligent support of dynamic, weakly defined engineering design processes in Semiconductor and Electronic Systems (SES) design domain providing for the increase in their productivity. Industrial opportunities of using multi-agent design process simulation tool will be outlined. The concluding part of the tutorial will present the general picture of the state of the art in agent-based engineering design automation as well as some future trends. The main question which will be proposed to the audience is: Are agent-based models of Engineering Design Processes really a kind of a silver bullet for engineering design automation? In the context of our PSI project this question may be reformulated as follows: Is there a chance to expect the order of magnitude increase in design productivity through employing agent-based models and simulation mechanisms? Some answers will be given based on the experience of the PSI project.

The **objectives** of the tutorial are:

- To survey the role of agent-based approaches in modeling dynamic, weakly defined processes in engineering design
- To outline the realistic focus, or the niche for agent-based approaches and solutions in managing engineering design
- To report how this focus has been addressed by the current accomplishments in PSI project of Cadence Design Systems
- To stimulate the audience to discuss if there might be a Silver Bullet in Engineering Design automation both in the broad sense and in some specific application areas like SES design

References

1. Davis, R.: What are intelligence? And why? AI Magazine 19(1), pp. 91-110.
2. Ginsberg, M.: Essentials of Artificial Intelligence. Morgan Kaufmann, San Francisco, CA. 1993.
3. Simon, H.: The Sciences of the Artificial. MIT Press, Cambridge, 1969.
4. Sloman, A.: Explorations in design space. In Cohn, A.G. (ed.): Proc. of the ECAI-94 Conf.; Wiley, New York, pp. 578-582.

Tutorial 7:
Modeling Enterprise Applications

Dirk Draheim

Institute of Computer Science,
Freie Universität Berlin Takustr. 9, 14195 Berlin, Germany
draheim@acm.org

1 Motivation

Enterprise applications are large and complex. Therefore, requirements elicitation and system analysis is especially important for the success of enterprise application development projects. Independent from the project organization, i.e., the used software process model, well-defined and easy-to-understand documents and work products are a cornerstone of successful communication of system documentation in each project team, both for the communication between the system analyst and the developer and the communication between the system analyst and the domain expert.

In practice, ad-hoc modeling techniques are often invented over and over again under the pressure of documentation needs. In our approach – form-oriented analysis - the conceptual underpinnings of form-based systems are externalized: from the viewpoint of strict submit/response style systems we explain techniques and technologies for the different stages of the software engineering life cycle.

2 Goals

In this tutorial participants will learn proven techniques for modeling form-based system in the semantically well-defined framework of form-oriented analysis. Concrete work products and activities will be presented by both practical advice and formal discussion. The tutorial also presents the model of a web shop as a comprehensive example.

3 Topics

- Requirement elicitation with page diagrams and form storyboards.
- Complete system modeling with formcharts and dialogue constraints.
- Layered data modeling.
- Model decomposition techniques.
- Case tools.
- Software process issues: a descriptive approach and work product orientation.

J. Akoka et al. (Eds.): ER Workshops 2005, LNCS 3770, pp. 473–474, 2005.
© Springer-Verlag Berlin Heidelberg 2005

- A comprehensive model: example web shop.
- Real-world case study: best practices in SAP project FUB/Campus.

4 Further Information

http://www.formcharts.org

Author Index

Lecture Notes in Computer Science

For information about Vols. 1–3664

please contact your bookseller or Springer

Vol. 3710: M. Barni, I. Cox, T. Kalker, H.J. Kim (Eds.), Digital Watermarking. XII, 485 pages. 2005.

Vol. 3709: P. van Beek (Ed.), Principles and Practice of Constraint Programming - CP 2005. XX, 887 pages. 2005.

Vol. 3708: J. Blanc-Talon, W. Philips, D. Popescu, P. Scheunders (Eds.), Advanced Concepts for Intelligent Vision Systems. XXII, 725 pages. 2005.

Vol. 3707: D.A. Peled, Y.-K. Tsay (Eds.), Automated Technology for Verification and Analysis. XII, 506 pages. 2005.

Vol. 3706: H. Fuks, S. Lukosch, A.C. Salgado (Eds.), Groupware: Design, Implementation, and Use. XII, 378 pages. 2005.

Vol. 3704: M. De Gregorio, V. Di Maio, M. Frucci, C. Musio (Eds.), Brain, Vision, and Artificial Intelligence. XV, 556 pages. 2005.

Vol. 3703: F. Fages, S. Soliman (Eds.), Principles and Practice of Semantic Web Reasoning. VIII, 163 pages. 2005.

Vol. 3702: B. Beckert (Ed.), Automated Reasoning with Analytic Tableaux and Related Methods. XIII, 343 pages. 2005. (Subseries LNAI).

Vol. 3701: M. Coppo, E. Lodi, G. M. Pinna (Eds.), Theoretical Computer Science. XI, 411 pages. 2005.

Vol. 3699: C.S. Calude, M.J. Dinneen, G. Păun, M. J. Pérez-Jiménez, G. Rozenberg (Eds.), Unconventional Computation. XI, 267 pages. 2005.

Vol. 3698: U. Furbach (Ed.), KI 2005: Advances in Artificial Intelligence. XIII, 409 pages. 2005. (Subseries LNAI).

Vol. 3697: W. Duch, J. Kacprzyk, E. Oja, S. Zadrożny (Eds.), Artificial Neural Networks: Formal Models and Their Applications – ICANN 2005, Part II. XXXII, 1045 pages. 2005.

Vol. 3696: W. Duch, J. Kacprzyk, E. Oja, S. Zadrożny (Eds.), Artificial Neural Networks: Biological Inspirations – ICANN 2005, Part I. XXXI, 703 pages. 2005.

Vol. 3695: M.R. Berthold, R. Glen, K. Diederichs, O. Kohlbacher, I. Fischer (Eds.), Computational Life Sciences. XI, 277 pages. 2005. (Subseries LNBI).

Vol. 3694: M. Malek, E. Nett, N. Suri (Eds.), Service Availability. VIII, 213 pages. 2005.

Vol. 3693: A.G. Cohn, D.M. Mark (Eds.), Spatial Information Theory. XII, 493 pages. 2005.

Vol. 3692: R. Casadio, G. Myers (Eds.), Algorithms in Bioinformatics. X, 436 pages. 2005. (Subseries LNBI).

Vol. 3691: A. Gagalowicz, W. Philips (Eds.), Computer Analysis of Images and Patterns. XIX, 865 pages. 2005.

Vol. 3690: M. Pěchouček, P. Petta, L.Z. Varga (Eds.), Multi-Agent Systems and Applications IV. XVII, 667 pages. 2005. (Subseries LNAI).

Vol. 3689: G.G. Lee, A. Yamada, H. Meng, S.H. Myaeng (Eds.), Information Retrieval Technology. XVII, 735 pages. 2005.

Vol. 3688: R. Winther, B.A. Gran, G. Dahll (Eds.), Computer Safety, Reliability, and Security. XI, 405 pages. 2005.

Vol. 3687: S. Singh, M. Singh, C. Apte, P. Perner (Eds.), Pattern Recognition and Image Analysis, Part II. XXV, 809 pages. 2005.

Vol. 3686: S. Singh, M. Singh, C. Apte, P. Perner (Eds.), Pattern Recognition and Data Mining, Part I. XXVI, 689 pages. 2005.

Vol. 3685: V. Gorodetsky, I. Kotenko, V. Skormin (Eds.), Computer Network Security. XIV, 480 pages. 2005.

Vol. 3684: R. Khosla, R.J. Howlett, L.C. Jain (Eds.), Knowledge-Based Intelligent Information and Engineering Systems, Part IV. LXXIX, 933 pages. 2005. (Subseries LNAI).

Vol. 3683: R. Khosla, R.J. Howlett, L.C. Jain (Eds.), Knowledge-Based Intelligent Information and Engineering Systems, Part III. LXXX, 1397 pages. 2005. (Subseries LNAI).

Vol. 3682: R. Khosla, R.J. Howlett, L.C. Jain (Eds.), Knowledge-Based Intelligent Information and Engineering Systems, Part II. LXXIX, 1371 pages. 2005. (Subseries LNAI).

Vol. 3681: R. Khosla, R.J. Howlett, L.C. Jain (Eds.), Knowledge-Based Intelligent Information and Engineering Systems, Part I. LXXX, 1319 pages. 2005. (Subseries LNAI).

Vol. 3680: C. Priami, A. Zelikovsky (Eds.), Transactions on Computational Systems Biology II. IX, 153 pages. 2005. (Subseries LNBI).

Vol. 3679: S.d.C. di Vimercati, P. Syverson, D. Gollmann (Eds.), Computer Security – ESORICS 2005. XI, 509 pages. 2005.

Vol. 3678: A. McLysaght, D.H. Huson (Eds.), Comparative Genomics. VIII, 167 pages. 2005. (Subseries LNBI).

Vol. 3677: J. Dittmann, S. Katzenbeisser, A. Uhl (Eds.), Communications and Multimedia Security. XIII, 360 pages. 2005.

Vol. 3676: R. Glück, M. Lowry (Eds.), Generative Programming and Component Engineering. XI, 448 pages. 2005.

Vol. 3675: Y. Luo (Ed.), Cooperative Design, Visualization, and Engineering. XI, 264 pages. 2005.

Vol. 3674: W. Jonker, M. Petković (Eds.), Secure Data Management. X, 241 pages. 2005.

Vol. 3673: S. Bandini, S. Manzoni (Eds.), AI*IA 2005: Advances in Artificial Intelligence. XIV, 614 pages. 2005. (Subseries LNAI).

Vol. 3672: C. Hankin, I. Siveroni (Eds.), Static Analysis. X, 369 pages. 2005.

Vol. 3671: S. Bressan, S. Ceri, E. Hunt, Z.G. Ives, Z. Bellahsène, M. Rys, R. Unland (Eds.), Database and XML Technologies. X, 239 pages. 2005.

Vol. 3670: M. Bravetti, L. Kloul, G. Zavattaro (Eds.), Formal Techniques for Computer Systems and Business Processes. XIII, 349 pages. 2005.

Vol. 3669: G.S. Brodal, S. Leonardi (Eds.), Algorithms – ESA 2005. XVIII, 901 pages. 2005.

Vol. 3668: M. Gabbrielli, G. Gupta (Eds.), Logic Programming. XIV, 454 pages. 2005.

Vol. 3666: B.D. Martino, D. Kranzlmüller, J. Dongarra (Eds.), Recent Advances in Parallel Virtual Machine and Message Passing Interface. XVII, 546 pages. 2005.

Vol. 3665: K. S. Candan, A. Celentano (Eds.), Advances in Multimedia Information Systems. X, 221 pages. 2005.